Nature's Pharmacopeia

NATURE'S PHARMACOPEIA

A World of Medicinal Plants

DAN CHOFFNES

COLUMBIA UNIVERSITY PRESS
New York

Columbia University Press
Publishers Since 1893
New York   Chichester, West Sussex
cup.columbia.edu
Copyright © 2016 Dan Choffnes
All rights reserved

Library of Congress Cataloging-in-Publication Data
Choffnes, Dan, author.
Nature's pharmacopeia : a world of medicinal plants / Dan Choffnes.
p. ; cm.
Includes bibliographical references and index.
ISBN 978-0-231-16660-7 (cloth : alk. paper)
ISBN 978-0-231-16661-4 (pbk. : alk. paper)
ISBN 978-0-231-54015-5 (e-book)
I. Title.
[DNLM: 1. Plants, Medicinal. 2. Ethnobotany—history. 3. Phytotherapy. QV 766]

RS164
615.3'21—dc23

2015020788

Columbia University Press books are printed on permanent
and durable acid-free paper.
Printed in the United States of America

c 10 9 8 7 6 5 4 3 2 1
p 10 9 8 7 6 5 4 3 2 1

COVER DESIGN: Milenda Nan Ok Lee
COVER IMAGE: @ Getty Images

References to Web sites (URLs) were accurate at the time of writing.
Neither the author nor Columbia University Press is responsible for URLs
that may have expired or changed since the manuscript was prepared.

Contents

Preface

Not far from the Madre de Dios River in eastern Peru, deep in a jungle rich with life, a gardener tends a plot of semiwild forest, where he propagates and protects several hundred types of medicinally useful plants. I approached his homestead on a dirt road rutted by the passing of motorcycles heavily laden with canvas sacks of Brazil nuts (*Bertholletia excelsa*) liberated from the dense wilderness and destined for grocery store shelves around the world. He met me at the mouth of a narrow path wearing calf-high rubber boots, and I followed him past all sorts of ferns, palms, and bromeliads, stepping over leaf-cutter ant trails, and toward an enduring education.

I was drawn to this place by the gardener's reputation. He was known throughout the region for his skill in collecting and maintaining living botanical specimens as well as for his ability to prepare remedies from what he harvests. As we walked through the forest, he pointed out plants useful for coughs, for skin problems, as insect repellents, and to treat joint pain. He also ran through recipes: take seven pieces of the stem from this plant, boil together in water with nine leaves from this other plant, and so on. This man's knowledge was rich.

"Maestro," I asked, "how did you learn all this about these plants?"

"The plants taught me," he responded.

What can I learn from the plants, I wondered, about their histories, their healthful and harmful properties, their chemistries, the ways that they intertwine themselves in our cultures? I wrote this book as an introduction to these topics, a point of departure for deeper study among those so inclined. These few stories about a handful of medicinal plants give just a glimpse into our long relationships with them.

As I learn from the plants, I am also grateful to my numerous teachers, the men and women who use medicinal herbs in traditional settings and have been generous to share some of what they know with me. I continue to learn so much from the work of countless researchers and scholars who have published their research, from among indigenous peoples, from the clinic, from the laboratory, and from archives, guiding me through the multiple facets and unending stories of medicinal plants.

The tales of these plants are woven into many aspects of human affairs. Readers will recognize threads of geopolitics, of social conflict, of the sacred, of the profane, and of an ongoing conversation about just what distinguishes a medicine from a drug of abuse. Many of these plants have psychoactive properties, their effects recognized early in human history and their biological mechanisms deeply explored.

I chose to write this book for a reader with little background in the topics considered. Those interested in pursuing the botany, pharmacology, and biochemistry of medicinal plants in greater depth are invited to seek out the cited works and other important contributions in these fields. Likewise, the historical passages can serve as impetus for a more serious study of primary and secondary sources on these subjects. Nor does this book substantially address clinical matters, for which there is a rich and fascinating literature. I regret that some readers may be disappointed by the limited space given to certain topics and by the decision to address broad themes at the expense of detailed analysis. While new data about medicinal plants and their properties have emerged daily from laboratories around the world, I generally avoided the temptation to present a comprehensive and up-to-the-minute account of new findings. As many important scientific studies have not been included in this book, interested readers will satisfy their curiosity in the pages of more expansive publications and in scientific journal articles. Finally, I will note that there are numerous widely consumed medicinal plants whose stories I could not contain in these chapters.

I am indebted to several anonymous reviewers whose suggestions have improved this manuscript. The organization and selection of themes benefited from the advice of many experts in the field, but in the end, I am responsible for any remaining lacunae and errors in the text. I would therefore appreciate hearing from readers to refine the content of the book.

Thank you to Amber Petersen, Jennifer Bickle, Christina Konecki, Josh Neukom, Nathalie Bolduc, and Julio Ramirez for their contributions to this manuscript at various stages, and to Michael Maher especially for his generous assistance with figures. For essential help, my appreciation to Mr. C.

I am grateful to Jonathan Marshall, Wenjie Sun, Paul Martino, Dan Miller, Penny Seymoure, and Jeff Roberg for such kind support over many years, to ASIANetwork for nurturing Asian Studies in the liberal arts, and to Judith Farquhar for much inspiration and assistance. I owe a lot also to Lila Vodkin, Don Briskin, and Jennifer Fletcher.

This project would have taken a much steeper path if not for the phenomenal skills of Kathy Myers, librarian extraordinaire. Patrick Fitzgerald, Kathryn Schell, and Bridget Flannery-McCoy at Columbia University Press skillfully guided this text from manuscript to published book. Thanks also to Milenda Lee, Irene Pavitt, and Robert Fellman.

To my partner, counselor, and collaborator on so many things, Winnie, my most profound appreciation. For all that is possible, Nicole and Alexander.

Note to the Reader

The information provided in this book is for educational, scientific, and cultural interest only and should not be construed as medical advice or as advocacy for the current use of medicinal plants under any circumstances outside of settings where their use is legally sanctioned by medical professionals. Neither the author nor the publisher assumes any responsibility for physical, psychological, legal or other consequences arising from use of these medicinal plants.

Nature's Pharmacopeia

Introduction

An herb merchant measuring medicinal plant powder in Bergama, Turkey.

Of all the diverse life-forms on the planet, plants have the unique ability among multicellular organisms to generate their own energy by capturing the energy of sunlight in chemical bonds. To carry out this feat, the green lineage has evolved complex biochemistries to sustain growth and reproduction fed only by air, water, earth, and light. Setting out roots to capture valuable water and minerals, unfurling leaves to claim a share of the sun, the plant form might seem well suited to a solitary existence. Yet the realities of life for plants in all the ecological niches in which they grow require them to compete for such limited resources with other organisms and to resist the assault of those species seeking to consume them. And while animals such as insects and mammals can walk, fly, bite, and chew their herbaceous prey, plants cannot flee their animal attackers. Rather, plants in fields and forests around the world have developed over millions of years highly sophisticated mechanisms to safeguard their bodies from herbivory.

Among the strategies plants use to deter attack is the channeling of their internal metabolisms to produce an enormous variety of chemical compounds that can interact specifically with animal physiology. Indeed, multiple families of plants generate compounds at great energetic cost that have little or no activity within their own cells, compounds that they accumulate or exude but are not critical for life. These molecules—including plant toxins—gave their producers an edge by allowing them to resist the rampages of insect pests just slightly better than their neighbors. In response to the incapacitating potential of plant poisons, the animal herbivores evolved resistance to these compounds, pressuring the plants to develop higher concentrations and greater diversity in their chemical weaponry. Thus this interspecies arms race has continued for millions of years, with trees, flowers, and grasses of all varieties producing a bewildering assortment of alkaloids, terpenoids, phenolics, and glycosides. These compounds can function as nerve poisons, steroid hormone mimics, heart toxins, and neurotransmitter-like molecules in animals from caterpillars to grazing livestock.

As humans formed societies and migrated into new areas, they learned of these poisonous plants and taught their children which leaves not to touch and which fruits never to taste. They also recognized that certain plants and plant parts were useful for flavor and aroma and, occasionally, for special medicinal or spiritual purposes. Shamans and priests developed a discipline in which plants were integral to the individual's state of health and the community's sense of purpose. At the appropriate dose, some plant poisons can generate profound physiological effects on humans by relieving pain, altering the mental state, and intensifying the senses. The wise practitioner held responsibility for distinguishing the desired plant from the deadly and the effective dose from the lethal.

Through the ages, peoples have harvested and cultivated plants that could serve to heal them, numb them, and stimulate them. Certain plant medicines, such as digitalin, a foxglove-derived heart drug, fell under the purview of doctors and pharmacists; others, such as cocaine and marihuana, drew strict regulation, and their possession was criminalized in modern times. Meanwhile, plant preparations in the form of coffee, tea, and chocolate have served as foundations of social and economic life for nations around the world for centuries.

The story of these plants and the compounds they produce is one that spans the globe and reaches back millennia. It encompasses some of the simplest life-saving power revealed to humankind and some of the most deadly chemicals ever uncovered. Wars of the pen and sword have been conducted to control these potent plants, and paintings and prose have been dedicated to extol them. The story of medicinal plants touches the best and worst moments of our civilization. Ultimately, it is a story of these plants' capacity to influence profoundly the human experience and to convince us to value them, cultivate them, and spread their seeds, and of our ability to learn from, profit from, and manipulate nature.

Chapter 1

Concepts of Ethnomedicine

An herbalist portrayed in Quechua folk art, Peru. (Paint on wood [twenty-first century])

Around 60,000 years ago, groups of humans began to venture out of their southern African center of origin and colonize new areas. Consummate explorers, some marched through eastern Africa and onward to Europe, others into Asia and beyond.[1] As they traveled, they encountered new plants and animals and perhaps new illnesses too. By nature curious, they undoubtedly tasted hundreds of leaves, roots, fruits, and seeds along the way. As they settled into lives in their new homelands, they developed a rich knowledge of which herbs were poisonous and at which times of the year. They learned which plants to gather for sustenance and eventually how to propagate them to support their growing communities. Existence was challenging for these early human explorers and colonizers. Fortunately, wherever they traveled, people discovered plants that fortified their bodies, healed their wounds, eased their pains, and affirmed their faith in the spirits that watched over them.

Communities maintained oral traditions, and in time some developed the ability to document their experiences with medicinal plants in art and writing. Archaeological evidence places the use of medicinal plants to as early as 5700 B.C.E. in Europe and approximately 4100 to 3500 B.C.E. in Asia.[2] Records describing the medicinal properties of plants date to at least 2500 B.C.E., when ancient medical-religious texts of India describe herbs as components of the "knowledge of life."[3] Around the same time, the Yellow Emperor in China is chronicled in legend as having documented an array of curative plants.[4] In Egypt around 1500 B.C.E., papyri record that garlic (*Allium sativum*) and juniper (*Juniperus* spp.) were used for their healing abilities.[5] This evidence in written form shadows the tradition linking herbal knowledge across generations in the development of a medicospiritual discipline. In North and South America, Australia, and Africa, people practiced medicine and passed on their expertise, and by the time the Europeans encountered these peoples, elaborate health beliefs and vast herbal resources existed.[6] Indeed, societies from Asia to Europe and the Americas likely independently developed their worldviews and ideas of wellness, philosophies in which plants were integral.

Before the systematic study of anatomy, notions of germs, and the advent of clinics, humans constructed detailed scenarios to explain the circumstances conducive to health and to remedy conditions of illness. In various forms of traditional medicine, people entrusted their physical, mental, and spiritual wellness to a framework of beliefs shared by members of the same community. Interestingly, some of these health-related ideas, while embraced by societies living far apart, share certain elements. For example, one such shared principle in traditional medicine is the belief that human health reflects a balance of forces or energies. When observing the world around them, early societies recognized that natural phenomena can frequently be described by terms in sets of opposites: light and dark, hot and cold, wet and dry, among others. A harmonious natural environment, these observers reasoned, was one in which neither heat nor cold is to an extreme, in which periods of dryness are followed by rain, and they expected a balance of such contrasting forces to promote life and vigor. The human body, being part of the natural world, also expresses such conditions. Thus when the body loses its balance, illness results, and balance can be restored through spiritual exercises, physical manipulation, and medicinal herbs. This equilibrium must occur in the individual as it does in the world and in the universe, in which the same forces occur and are usually at balance. The idea that health is a function of balance is among the most widespread of the traditional medical beliefs, evident in ancient China, India, the Mediterranean, and the Americas.[7] It also demonstrates that in many societies, medicine was inseparable from philosophy and religion.

As people settled in many different regions of the planet, they harvested native plants for medicinal purposes and cultivated those they brought from elsewhere or acquired in trade. The combination of locally sourced flora, particular landscapes and physical challenges, distinctive languages, cosmologies, and social structures together imparted unique characteristics to the world's many types of indigenous medicines. Rather than look at any region as uniform in terms of medical culture, it is worthwhile to consider the diversity in health-related beliefs and practices along several dimensions.

First, numerous ways of treating health can exist at the same time among a group of people—that is, medical plurality. For instance, different practitioners living in a single community may have vastly divergent approaches to addressing a patient's condition, and individuals may address medical concerns with a combination of professional assistance and self-care. Second, health-related ideas evolve over time, adapting to new illnesses, accommodating changing philosophies, and incorporating innovations. Therefore, a regional medicine as practiced now or in the past, though it may be dubbed "traditional," is not a static entity but rather dynamic. Third, cultural borrowing can lead to a synthesis between local medical knowledge and that appropriated from other people. While some forms of medicine have developed in isolated communities, many indigenous medical practices bear witness to years of commerce and exchange. Many of the world's major traditional medicines are complex amalgamations of beliefs and techniques, employing pharmaceuticals having originated in different places.

The following sections provide an overview of some of the traditional medical beliefs and practices of East Asia, South Asia, Africa, and the Americas and demonstrate the diverse ways that people have conceptualized health and the role of plant-based treatments in influencing it. The remainder of the discussion follows the European experience in medicine, where, as elsewhere, health was considered to be the product of a balanced physical, mental, and spiritual state. In recent centuries, an approach to medicine emerged in Europe in which the scientific testing of herbs offered new ways to gauge therapeutic activity while rejecting many previous ideas about disease causation. Now widespread, particularly in the industrialized world, this biomedical system of health care coexists with numerous traditional medical systems and countless informal and folk practices that also employ plants as medicines.

TRADITIONAL MEDICINE IN EAST ASIA

Many of the world's medical traditions developed concepts of health that viewed the person in the context of society, the local environment, and the universe as a whole. In these systems, the body is the beneficiary of natural energies (in the form of food and environment) and supernatural forces (such as spirit powers) that promote proper development. In China, people came to believe that the whole organism is healthy when it is in a state of balance and harmony with the world. In this system, health is considered a state of physical and mental well-being.[8]

Chinese traditional medicine[9] views the universe as permeated by the *qi* life force, which constantly flows through heaven, earth, and all living things. Since *qi* is present in the air, soil, food, and all parts of the environment, it can strongly influence human health. The properties of *qi* are believed to change according to the time of day and the seasons, and they can vary regionally as well. For example, *qi* has a warmer quality in the summer and a cooler quality in the winter, darker properties at night and lighter properties during the day. According to Chinese medicine, illness results when an individual is unable to adapt to the changing nature of *qi*.

These ever-fluctuating features of the universe are the foundation of Chinese medical thought. Chinese medicine recognizes that *qi* and all matter are endowed with two opposing qualities: *yin*, the dark and cool property, and *yang*, the light and warm property. In the body, as in the environment, neither quality should have complete reign. For example, when night falls, the sky becomes quite dark. But in the darkness, there is light in the coming dawn. The cycles of day and night, the four seasons, and the patterns of precipitation and drought are natural processes of a universe at balance. As *yin* properties increase, *yang* properties decrease, until the extreme, when *yang* properties appear again. Because human beings are part of the universe through which *qi* flows, the *yin* and *yang* qualities of the body, changing over time relative to each other, can affect the nature of its *qi*. As *qi* is the force for life, so too is it responsible for health and illness.

Chinese medicine views that the body processes *qi* to derive nutrition and protect itself from illness. Properly extracted from the universe, a type of *qi* known as orthopathic *qi* gives the body the means to resist illness. Meanwhile, the illness-causing heteropathic *qi* assaults the body from the exterior, putting two types of *qi* in opposition. The ability of orthopathic *qi* to resist heteropathic *qi* is considered a state of health. Any overabundance of heteropathic *qi* activity can lead to illness, as can an excess of orthopathic *qi*: the healthy state is a balance of these forces. Since *qi* is influenced by its *yin* and *yang* properties, illness is thought to emerge from changes in the environment (disrupting a balance by affecting heteropathic *qi*) and/or changes in the body (affecting orthopathic *qi*).

To promote the proper qualities of orthopathic *qi* (and thereby resist illness), practitioners of Chinese medicine are aware that *yin* and *yang* qualities in balance promote health. (This does not mean a balance of equal amounts. In Chinese medicine, the *yin–yang* relationship in a patient is dynamic and responsive to the state of illness and the environment.) To maintain health, they pay close attention to the emotional state, social activities, diet, and exercise regimen, all of which influence the type and movement of *qi* in their bodies (figure 1.1).

Belief in the role of *qi* in health influences lifestyle by encouraging balance in all activities: maintaining an even emotional keel, striving for social harmony, consuming cuisine with an appropriate representation of "warming" and "cooling" ingredients, and undertaking regular physical and mental pursuits. When illness strikes, however, doctors can identify patterns of colors (of face or tongue), temperatures, pulse profiles, and behaviors that indicate to them whether the patient suffers from an overabundance or deficiency of *yin* or *yang* qualities.

Medical interventions are developed to strengthen the patient's internal *qi* and improve its flow through the body by imparting to it the *yin*

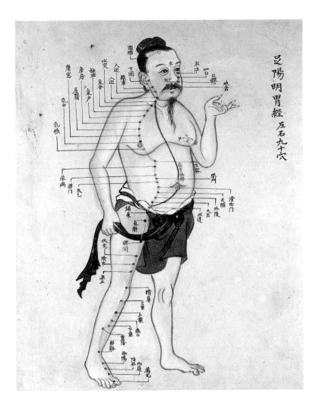

FIGURE 1.1 A chart of a meridian through which *qi* flows, according to Chinese traditional medicine. (Wellcome Library, London, L0012239)

or *yang* properties that would allow it to promote health and drive out illness.[10] Chinese pharmaceuticals, which are composed of plant material as well as some mineral and animal-based substances, are commonly given in mixtures of several ingredients, often prepared as soups or pills (figure 1.2). When choosing a treatment, doctors look to influence the balance of *yin* and *yang* activities. The Chinese herbal pharmacy is extensive, containing thousands of ingredients categorized by their warming or cooling properties and effects on the body's *qi*.[11] For example, the seed of milkvetch (*Astragalus complanatus*) is thought to support *yang*, and the stems and leaves of the dendrobium orchid (*Dendrobium* spp.) to strengthen *yin*.[12] Chinese medicine also values herbs that serve to reinforce orthopathic *qi* in general, such as ginseng (*Panax* spp.) root.[13] In summary, traditional East Asian medicine considers health to be a condition of balance, in which a person's body and mind are at

FIGURE 1.2 A pharmacist preparing a traditional Chinese herbal formula in Beijing, China. The wall behind the pharmacist is made up of hundreds of drawers containing dried plants.

harmony with universal forces. Medicinal plants, selected according to their *yin–yang* properties, are thought to reinforce the body's abilities to ward off illness.

TRADITIONAL MEDICINE IN SOUTH ASIA

Among the principal indigenous medicines of South Asia are *ayurveda*, a form of health care that originated in what is now northern India and Pakistan, and *siddha*, which is more widely practiced in the Tamil-speaking parts of southern India.[14] In *ayurvedic* medicine, human beings are considered to represent a microcosm consisting of the same energies and substances as the larger universe, and therefore health is fundamentally connected to the state of the macrocosm.[15] The basic matter of

the cosmos combines to create the internal forces that govern physiology and behavior, the *doshas*. The *doshas* regulate various bodily functions and are seated in different parts of the head, chest, and abdomen. When they are in balance, a person is healthy.[16] The state of equilibrium is influenced by, and can influence, the physical and mental constitution of the individual, so treatments frequently entail changes in religious or meditative behavior, emotional control, exercise, diet, and sensory experiences in the pursuit of balanced bioenergetic principles.[17]

Ayurvedic medicine recognizes that the flavors of food signal their elemental makeup and, therefore, their effect on the *doshas*. For example, sweet flavors increase one of the *doshas* while decreasing the other two, and pungent flavors raise two of the *doshas* while subduing the third. In total, *ayurvedic* medicine distinguishes six tastes that can be further classified by their hot-cold, oily-dry, heavy-light, and dull-sharp aspects.[18] Therapeutic approaches in Indian medicine often entail changes in diet and the use of herbal pharmaceuticals with flavor and other properties conducive to the equilibrium of the *doshas*. For instance, the three spices of black pepper (*Piper nigrum*) fruits, long pepper (*Piper longum*) fruits, and the underground stems of ginger (*Zingiber officinale*) are widely employed in South Asia for their culinary, and therefore medical, properties. In addition to taste, color is considered important in its effects on the *doshas*. Because sensory experience plays a role in health and illness, *ayurvedic* pharmaceuticals draw heavily from fragrant and colorful plant products, including saffron (*Crocus sativus*) and turmeric (*Curcuma longa*).

Drawing on some common sources with *ayurvedic* medicine and other influences, *siddha* is a prominent form of health care in the state of Tamil Nadu in modern-day southern India.[19] In this system of medicine, properties of the universe are at play in the human body in the form of matter (*shiva*) and energy (*shakti*), and they influence the health of the organs through their connections to the celestial zodiac.[20] In *siddha* medicine, health is a function of five elements that combine in various ways to form three bodily constituents, the *muppini*, whose equilibrium promotes health.[21] In addition to environmental, climatic, and hereditary factors, the diet and a vast array of pharmaceuticals can play a role in the balance of *muppini*. Among many mineral and metal-based treatments, *siddha* medicine also employs herbal medicines to purify inorganic drugs before use, cleanse the body internally and externally, and treat specific ailments and generally improve health.[22]

TRADITIONAL MEDICINE IN THE AMERICAS

In North America, the indigenous peoples developed approaches to health care that suited their diverse cultures and geographic settings. It would therefore be impossible to consider them as a bloc.

However, a few examples will illustrate that among some American Indian groups, illness and healing has centered on the role of the community and the spiritual world in the well-being of the individual. In these traditions, physical and mental health is seen to be a result of a good relationship with one's community, environment, and deities.[23]

Among the Iroquois (Haudenosaunee) of what is now the northeastern United States, for example, the health of the individual is inseparable from the state of the universe. Human illness is considered a response to harm that has occurred elsewhere in the social group or in nature.[24] The Iroquois generally do not view health in the material (by attributing wellness to the forces, elements, and opposing qualities comprising the universe). Rather, they see health encompassing abstract criteria ranging from physical and mental comfort to the maintenance of life and good luck. Therefore, Native American herbalists integrated medicine into the community-wide sense of fortune, harmony, and spiritual oneness. Plant drugs are chosen to treat the spiritual imbalances in individuals or their surroundings, which reduces the physical manifestations of those conditions.

Because the source of health and illness has often been seen as existing beyond the material world, communities relied on specialists who claimed to mediate and communicate with the intangible, spiritual component of the universe to effect healing. It is often the role of shamans to care for the ill, harnessing their special powers to identify and appease the source of illness. It is also clear that many tribes view health as a community matter, and so healing takes place in ceremonial settings, such as sweat lodges and talking circles, under the supervision of an individual thought to have the power (or skill) to guide the process.[25]

Among the Chippewa (Ojibwe, Anishinaabe) of north-central North America, promotion of health and long life has been thought of as a matter of the spirits as well. Shamans adept at diagnosing and treating illness, sometimes called medicine men, are esteemed in their communities for their ability to identify the appropriate treatment for an ailment, knowledge that they believe they receive in

FIGURE 1.3 Indigenous North American spiritual medicine. Among the Chippewa, shamans had special healing knowledge, including the use of medicinal plants: (*top*) a shaman preparing medicine; (*bottom*) a shaman treating a patient by drawing out illness. (Engravings from W. J. Hoffman, *The Mide'wiwin or "Grand Medicine Society" of the Ojibwa* [1891]; Project Gutenberg)

dreams.[26] In addition to remedies of physical ailments, Chippewa medicine men consider the preparation of good-luck charms as part of their healthcare duties. Therefore, they are known to provide patients with substances to attract mates or improve their hunting and fishing success (figure 1.3).[27]

The notion that an individual's mental, physical, moral, and spiritual health is interconnected with the well-being of the community is evident in the beliefs of the Lakota people of the North American Great Plains.[28] The Lakota consider that a person comprises "four constituent dimensions of self," including an individual soul or spirit, a divine spirit, a "vital breath" that sets the physical operations of the body in motion, and an intellectual-spiritual presence that guards against evil and helps overcome obstacles in life.[29] In these traditions, medicinal plants serve to not only alter the physiology but also contribute to the fortune and social well-being of an individual or a community.

Thousands of plants serve these roles in indigenous North American medicine, and they are often used for different purposes in different regions.[30] The big sagebrush (*Artemisia tridentata*) has been employed against respiratory illnesses such as the common cold, as a remedy for stomachache or headache, and as a spiritual cleanser.[31] In what is now the western United States, people turned to native yuccas (*Yucca* spp.) for medicine, and, like many medicinal plants of North America, they are thought to influence physical and spiritual health.

The Blackfoot, Cheyenne, and Lakota, for example, harvested the small soapweed (*Yucca glauca*) for use as a hair wash and baldness preventer, among other uses, and the Navajo (Diné) considered it a stimulant and contraceptive.[32] The Hopi prepared extracts of the related narrowleaf yucca (*Yucca angustissima*) in ceremonial purification rituals in addition to using it as a hair loss–preventing shampoo and laxative.[33]

It is difficult to determine the extent to which indigenous American medical practices were codified because most groups left no written records. It seems the role of healer or shaman was often a matter of bloodline and special knowledge.[34] Therefore, there are no ancient books describing practices as they occurred in the past and no way to determine the course of development of treatments, such as medicines shared between groups and how medicines may have diverged. Despite the enormous diversity of geographic settings and cultural histories, it is clear that indigenous Americans employed herbal medicines extensively for a variety of physical, mental, and spiritual conditions. Among many groups, plants served to promote health as a harmonious relationship between the person, his community, nature, and the spiritual universe.

The Aztec (Mexica) civilization, which flourished in present-day Mexico from the fourteenth to sixteenth centuries, left documents and descendents that shed light on its medical beliefs and practices.[35] According to indigenous mythology, the first humans were created from a divine mixture of material from the heavens and the earth, imbued with opposing properties that must remain in balance to sustain life. Disequilibria of these elements cause illness. The Aztecs viewed the earthly component as the Great Mother, instilled with feminine, cold, wet, and dark qualities. The Great Father was the heavenly contribution, a masculine, hot, dry, and light force.[36] At its core, the Aztec worldview framed a parallel between the organization of the human body and that of the universe, where each bodily function owed something to either the Great Father or the Great Mother.

Aztecs sought balance in their diet, moderation in their emotions and physical activity, and obedience in their relationship with authority as a way to procure the harmonious relationship among the opposing forces that formed their bodies.[37] The ultimate manifestation of harmony between a person, the community, the environment, and the cosmos was the *tonalli*, an energy-laden spirit from the gods that gave a person vigor.[38] Various factors, such as immoderate physical activity, were thought to diminish the *tonalli* and result in illness.

The Aztecs also believed that humans, being made of cosmic material, could influence the fate of heaven and earth. Therefore, they developed means to "give back" to their creators in the form of human sacrifice, which they believed supplied energy to the gods.[39] In addition to the belief that health derives from a balance of opposing properties in the body, the Aztecs held that spirits inhabited the forests, caves, rivers, and other features of the environment. These invisible beings, according to Aztec medicine, had a hunger for *tonalli*.[40]

Among the herbs that could protect the *tonalli* was *tlacopatli* (*Aristolochia anguicida*), "used to treat cold illnesses and to strengthen and revive people."[41] Children wore *tlacopatli* bead necklaces as a remedy, and its roots could be applied to the top of the head to restore *tonalli*.

Some illnesses were caused by, rather than cured by, plants, and the Aztecs harbored beliefs that affronts to plants or their spirits could invoke retribution. For example, people were forbidden to urinate on the *aquiztli* vine (*Paullinia fuscescens*), or else they would be afflicted with blisters all over their bodies.[42] (The same plant was used to treat blisters.) Someone who slept in the shadow of *aquiztli* would lose his or her hair.[43] Furthermore, some plants contained spirits that practitioners could take inside their bodies to derive other-worldly capabilities. For example, herbal drugs that induced visions were thought to function by transferring the god of the plant into the person ingesting it and by directing the user's *tonalli* toward the god's place of residence. Specialists sought

HEALTH-CARE PRACTITIONERS

Those who employed plants to treat illness and promote health bore many titles in various historical and geographic settings. In this book, the terms "doctor," "practitioner," and "physician" refer to a health-care provider in a general sense. The reality of health care is far more complex than it is possible to explore sufficiently here. Over time and in different parts of the world, health advice and pharmaceuticals have been dispensed by religious leaders, village herb gardeners, midwives, and university-educated specialists. The roles of such diverse medical providers were determined by cultural standards that changed over the centuries. In ancient Greece, for example, those who called themselves doctors worked alongside (and competed with) exorcists, bonesetters, root gatherers, priests, and others.[1] By the European Middle Ages and Renaissance, physicians exerted responsibility for a patient's health based on a knowledge of medical theory, and surgeons and barber-surgeons performed manual operations such as cutting, excising, bloodletting, and tooth pulling, the last profession also trimming hair. Meanwhile, druggists, apothecaries, and alchemists provided medicinal substances with or without physicians' prescriptions. At the same time, midwives and experienced female practitioners called wise women treated many patients, often with expertise in the areas of fertility and childbirth. Likewise, in Asia, Africa, and the New World, health practitioners include those with and without formal training, whether as doctors, drug preparers, shamans, tribal leaders, medicine men, priests, or scholars. Importantly, all these various specialists have knowledge of the use of plants to treat those under their care. Therefore, the study of medicinal plants can consider the ways that herbs are prepared and administered by people working at many levels and in numerous niches of a complex health-care environment.

1. Roy Porter, *The Greatest Benefit to Mankind: A Medical History of Humanity* (New York: Norton, 1997), 54.

such medicines to help them commune with deities and obtain secret knowledge to diagnose and treat patients in their care.[44] Therefore, among the Aztecs, plants (and their spirits) played important roles in both causing and curing illness.

TRADITIONAL MEDICINE IN AFRICA

The many forms of African medicine originated among peoples living throughout a vast continent, maintaining distinctive languages and social customs, practicing different religious traditions, and harvesting medicinal plants particular to their locations. To the West African Akan people of modern-day Ghana, illness results from a combination of social, environmental, and spiritual factors. Poor diet or overindulgence in certain foods can result in sickness, as can the forces of pathogen-laden wind, evil intentions, and the acts of evil people, spirits, and witchcraft.[45] Healers prepare a variety of plants as herbal teas, salves, baths, and so forth, and some have special abilities to communicate with the spirit manifestations of the creator residing in the ocean, rivers, mountains, and plants.[46] In addition, talismans play a role in helping practitioners select medicinal plants for their patients and improve the therapeutic outcome.[47] For instance, the root of a forest tree that goes by a local name signifying "executioner medicine" (*Mareya micrantha*), apparently because of its poisonous effects if prepared incorrectly, is employed in spiritual baths and enemas, and its leaves are used to treat stomach pains and constipation, among other conditions.[48]

Among some southern African groups, such as the Basotho of modern-day Lesotho, health is a product of a body and mind at ease with the environment and spiritual world. Illness can occur when a person is in a state of physical or social disequilibrium, such as having an overabundance of certain substances in the body or having a disturbed social relationship with certain living or deceased people.[49] The state of imbalance presents itself in a number of forms, such as having excess heat (in a figurative sense, rather than fever),

which can be treated with cooling drugs "such as ash, water plants, or aquatic animals."[50] Illness and misfortune can often be attributed to malevolent spirits and people with evil will, such as sorcerers, and therefore medical practitioners include herbalists, who are able to diagnose and treat many types of ailments, and diviners and seers, whose special powers to communicate with the spirit world render their diagnoses more accurate and their treatments (including herbal medicines) more effective.[51] The special capacity of medicinal plants to influence a person's luck is evident in the use of lemongrass (*Cymbopogon marginatus*) to procure love, as when a man adds it to his bathwater "while he calls out the name of the desired woman."[52] Other plants are less strongly associated with spiritual powers, such as *lengana* (the wormwood *Artemisia afra*), whose fragrant leaves are used as nasal plugs, in herbal teas, and for steaming in cases of colds, sore throat, and digestive problems, among others.[53]

To the ancient Egyptians, a body "in harmony with the cosmos . . . could serve as a receptacle for the vital forces that created the universe."[54] The Egyptians inhabited a universe composed of both physical matter and immaterial forces overseen by an assemblage of gods with varying responsibilities and powers. Accordingly, they sought to maintain health by regulating both physical aspects of their diet and behavior and by avoiding the ill will of deities who would harm them. Medical treatments commonly invoked the spirits to request relief from the various evils responsible for a person's ailment. Therefore, to the ancient Egyptians, medicine, religion, and magic were deeply intertwined.[55]

Numerous herbal medicines and plant-derived dietary ingredients have been documented in medical papyri and survive as archaeological specimens.[56] Among several hundred pharmaceuticals deciphered in ancient texts are many drugs that act in the bowels, such as the oil of castor seeds (*Ricinus communis*) and the fruits of the fig tree (*Ficus* spp.).[57] To the Egyptians, regularly cleansing the evil toxins that accumulated in food prevented the spread of their harmful properties to the rest of the body. Therefore, castor counts among drugs recorded in the Ebers medical papyrus (ca. 1550 B.C.E.) to "drive out suffering from the belly of a man."[58]

SHARED FEATURES OF MEDICAL SYSTEMS

There are common elements in some of these diverse medical traditions. In many parts of the world, health conceptions tend to link an individual's wellness to the state of the universe. The forces at play in the world also permeate the human body, contributing to vigor and longevity or, oppositely, illness. Good physical vitality and good fortune represent different aspects of health maintenance. Traditional philosophies stress the importance of balance in physical and mental activities, outlook on life, and choice of foods and spices. They often view psychological and physical health matters as deriving from common causes and sharing treatments. Plant-derived drugs tend to be classified within a scheme in which they serve to influence the vital force or restore equilibrium to the elements operating within a body, the effects of which reduce illness symptoms. In this sense, healers do not always seek to treat the ailment directly with their herbal preparations. Rather, they often treat the underlying imbalances causing the symptoms. In some cases, medicinal plants are used to communicate with and appease the deities or spirits responsible for illness.

The following sections will outline the evolution of the European lineage of medicine, from its origins in the ancient Mediterranean until the modern day.[59] It is clear that traditional European medicine has much in common with other world medical systems, including the classification of life forces or properties into opposing but complementary categories, the idea that physical and mental health are tied together, and the use of herbal substances to promote harmony within the individual and more broadly. During the past several centuries, a new, scientific tradition emerged in Europe that embraced a distinctive methodology for determining effective medical treatments and that now serves as the foundation for practice in much of the world.

TRADITIONAL MEDICINE OF THE MEDITERRANEAN

A group of Greek scholars including Hippocrates (ca. 450–370 B.C.E.) and his followers promoted the idea that the human being, as part of nature, "was subject to the same physical constraints as the rest of the ordered cosmos," and therefore illness could be understood and addressed according to observable phenomena at play in the environment.[60] Over many centuries, their theories were elaborated by doctors working throughout the Mediterranean region and served as the basis of an influential and widespread form of Greek and Roman medicine. To these thinkers, the universe revealed sets of opposing qualities—cold and hot, wet and dry, and so forth—and these principles joined in nature to create matter in the form of earth, air, fire, and water.[61] The relative abundance of these temperature and moisture properties changed over time (such as in the cycling of the seasons), but early physicians believed that the four elements that they produced were fundamentally at balance.[62] The human body generated from the elements a set of substances with properties corresponding to the principles of nature, which they called the humors.

In this system, blood (or sanguis) is hot and moist, phlegm is cold and moist, black bile (or melancholer) is cold and dry, and yellow bile (or choler) is hot and dry.[63]

> **The four humors of traditional European medicine**
>
> - Sanguis (blood)
> - Choler (yellow bile)
> - Phlegm
> - Melancholer (black bile)

These substances were present in the body from conception until death, influenced by diet, seasons, and geography, and affected a person's health and personality.

The four humors of blood, phlegm, black bile, and yellow bile were metaphorical, invisible substances, not actually fluids that could be isolated.[64] The blood humor is not the same as arterial blood, although when red blood emerges from a wound, the blood humor is instrumental. Similarly, phlegm is evident as a force behind watery secretions, black bile perhaps behind clotted blood and accretions in stool, and yellow bile in pus and vomit. People sought to achieve a relative balance among the humors, although individuals tended to be dominated by one of the four. Since the body generated humors from the elements present in the diet, and because the constitution drew influence from lifestyle choices, climate, and other factors, it was possible for imbalances to emerge that threatened the individual's health. Conditions associated with an excess of phlegm, for example, include lack of appetite and thirst, weak brain, miscarriages, dysentery, diarrhea, and chronic fevers.[65] Bronchitis and acute fevers would be rare among individuals with this excess.

The humors also played a role in shaping personality. The mental or psychological constitution of the individual—his or her temperament—could be classified according to the same principles. Someone with an excess of yellow bile was thought to be disposed to anger, particularly in the summer, and a person with too much phlegm might be passive and withdrawn, especially in the winter.[66] Evidence for the classical belief in humors abounds in the modern English language and corresponds to the four temperaments of the ancient Greeks and Romans. Someone with a sanguine disposition is cheerful and optimistic. Phlegmatic means calm and composed. Melancholy is a gloomy or depressed state of mind, and a bilious individual is irritable and cranky.

Although some humoral imbalances were corrected surgically or by lifestyle changes, people often regulated the balance of humors with diet, a practice that includes medicinal herbs.[67] The Greek writers who advanced the humoral system classified all foods and medicines by their fundamental qualities: hot, cold, dry, or moist. Remedies to illness were a matter of matching the personality and symptoms to a therapy with opposite properties.[68] In this view, an excess of black bile, associated with the cold and dry temperament and indicative of conditions such as constipation and depression, could be relieved with the consumption of hot herbs such as senna

(*Senna alexandrina*) and hellebore (*Helleborus* spp.).[69] These medicines were thought to function by inducing the body to eliminate feces and, with it, the offending humor.

It is therefore not surprising that many traditional European herbal medicines derive from treatments intended to regulate the humors, with a particular focus on controlling the secretions of the body and the digestive process. Moreover, since excess humors were thought to accumulate poisons, a goal of pharmacy was to remove them from the body. So to purge toxins of the phlegm, yellow bile, or black bile, people sought herbal agents to induce vomiting and defecation. (Even in healthy times, medical wisdom held that frequent vomiting and purgation promoted wellness.)[70] For example, scammony (*Convolvulus scammonia*) was recommended in early Greek texts as a purgative capable of effecting a rapid discharge of the bowels.[71] Scammony remained in use for this purpose in Europe and the United States into the twentieth century.[72]

Medical writers during this period often framed therapies in terms of the humoral system, recognizing that drugs have temperature and moisture properties that could influence the temperament and symptoms of the patient. The Greek physician Pedanius Dioscorides (ca. 40–90) broke with tradition by presenting pharmaceutical knowledge in terms of specific effects on the human body, without interpreting medicine in the theoretical context of the humors (figure 1.4).[73] His *De materia medica* (ca. 70 C.E.), a list of more than 500 herbs describing their names, appearance, and uses in the treatment of illness, was the first of its kind.[74] Such a thorough catalog of medicinal plants was previously unknown in the Mediterranean sphere, and it served as a key medical text for the next 1500 years throughout Europe and Asia Minor. The value of Dioscorides's work is in its descriptions of medicinal plants, which allowed others to gather the same herbs as he mentioned—an important step toward standardization in medical care—his instructions for preparing the medicines, and his inventory of the drugs' uses and risks.

Because most medicinal substances were harvested from plants, and because Dioscorides

FIGURE 1.4 Pedanius Dioscorides, pioneer herbalist. (Woodcut after André Thévet, *Dioscorides Arboriste* [1584]; National Library of Medicine, B07205)

Pharmacopeia

A list of medicines with their preparation and uses

produced the first book to treat drugs in this way, his influence was long lasting. A catalog of useful plant medicines became known as an herbal, and many important herbals have since been written. It is now commonplace to call all books listing medicinal plant properties *materia medica* in reference to the title of his work. Dioscorides's *materia medica* also emerged as the first pharmacopeia—a list of all accepted drugs, with their preparation and processing for medical use—in the Western world.[75] Since that time, national and international medical establishments have maintained written inventories of approved medicines. ("Approved" medicines are technically termed "official.") In modern times, approval is often noted on the label of a medicine. For example, "USP" indicates a drug listed in the United States Pharmacopeia.

THE HUMORAL SYSTEM: FIFTEEN CENTURIES OF MEDICAL DOCTRINE

The humoral understanding of health was first described by Greek physicians of the fifth and fourth centuries B.C.E. and elaborated in the writings of many others, including the celebrated physician Galen (129–ca. 216), living in the Roman Empire, on whose authority the practice of medicine in Europe depended for many centuries (figure 1.5).[76] Galen's theories on the nature of health satisfied the desire for a rational explanation of illness, and few

doctors of his era or of succeeding generations questioned the elegance of the system in which balanced elements ensured health. In the medical tradition that he championed, the role of diet, lifestyle, and environment in health was paramount. In the centuries after Galen, the humoral system that had taken hold

FIGURE 1.5 Galen, promoter of the humoral system of medicine. (Lithograph by Pierre Roche Vigneron [ca. 1865]; National Library of Medicine, B012561)

throughout Europe was codified into practice in Persia and then later throughout the Muslim world.[77] Even when the glory of Greek and Roman scholarship waned, Muslim doctors preserved Galen's medicine and reintroduced his writings to Europe hundreds of years later.[78] With the revival of classical learning in Europe during the Middle Ages and Renaissance, humoral medicine again became the subject of intensive study, and Galen's works were recopied and widely distributed. The humoral theories held sway in European medical education and practice until the eighteenth and nineteenth centuries, when they gradually began to subside. (It is worth noting that the Hippocratic-Galenic view of health was not the only form of medicine in existence in the classical Mediterranean region. Numerous schools competed for adherents during the era, a phenomenon that resulted in a large critical literature. In addition, many practitioners likely operated without an overarching theoretical framework or advocated religious or magical medicine. Ultimately, however, it was Galen's writings that were elevated to the position of primacy for many centuries.)

UNDERCURRENTS IN EUROPEAN MEDICAL PRACTICE: MYSTICISM AND THE DOCTRINE OF SIGNATURES

With the fall of the Roman Empire in the late fifth century, medical practice in Europe became an amalgam of ancient Greek and Roman tradition,

Catholic Church doctrine, and tribal folklore. The emphasis on natural forces and balance in lifestyle and dietary choices that Galen promoted became mixed with beliefs in the healing (or harmful) effects of spirit beings. For example, in the *Leech Book of Bald*, a tenth-century Anglo-Saxon herbal, treatments are described for illnesses caused by a "pagan charm" and "flying venom," a clear departure from the idea that imbalances of body substances were at the core of illness.[79]

In addition to their generally herbal remedies, medical treatises of medieval Europe included prayers and incantations as disease preventatives or treatments. The practice of medicine was closely linked to the Catholic Church: doctors were frequently monks or at least associated with monasteries, the sites of infirmaries. In addition to healing salves, herbal amulets, and surgery, people turned to holy relics such as the bones of saints for their restorative properties.[80]

European medical practice in the Middle Ages was often imbued with mysticism and drew on local folklore. Herbs such as wood betony (*Stachys officinalis*) and peony (*Paeonia officinalis*) saw frequent use for ailments of the skin, bowels, and other complaints. In the medicine of the era, these plants were thought to work in part by warding off evil spirits and through astrological influences whose activities were responsible for illness. For example, an early medieval English herbal from the eleventh century recommends wood betony for eye pain, earache, and constipation, and to protect "a person from dreadful nightmares and from terrifying visions and dreams."[81] Peony root is said to be good for the pain of sciatica when tied onto the body and to cure "lunacy."[82] Some plants were thought to be particularly useful against spirit-borne illnesses and creepy-crawly threats, such as the greater periwinkle (*Vinca major*), whose description in the same eleventh-century herbal claims its utility "for possession by demons, for snakes, wild animals, poisons, for any threats, envy, terror, so you will have grace, so you will be happy and comfortable."[83]

Many people also believed in the evil eye, a curse that penetrated their bodies and souls, bringing them misfortune and poor health.[84] Fear of the evil eye probably goes back millennia, and it has persisted alongside other forms of medical belief.[85] People produced charms from the leaves, stems, and flowers of medicinal plants to repel its malicious powers. Indeed, the same medieval English herbal suggests keeping mugwort (*Artemisia vulgaris*) in one's house, as "it turns away the evil eye."[86]

During the early medieval period, monasteries served as repositories of medical knowledge and centers of practice, where monks and nuns cultivated "physic gardens" of herbs and copied ancient Greek and Latin manuscripts.[87] When medical colleges became established in Europe, they primarily taught the principles of Hippocrates, Galen, and other classical physicians in the context that prayer and divine will were the ultimate source of health.[88] Complementing the church's role in medicine was the oral tradition of folk medicine and home remedies passed down generation to generation and dispensed by spice and herb merchants (apothecaries) and elders.

By the sixteenth century, the European medical practice bore evidence of a complicated mixture of influences, including Galen's humoral theory, Christian belief, and mystical folk traditions (figure 1.6).[89] Therefore, medicine encompassed beliefs in the fundamental forces of heat and moisture, whose balance allowed health, alongside divine providence, astrological forces, mystical energies, and curses. It was a complex discipline, and the bold Swiss alchemist-physician Paracelsus (Philippus Theophrastus Bombastus von Hohenheim, 1493–1541) sought to simplify it.

FIGURE 1.6 The four humors represented in a single person, with their astrological properties. (Woodcut by H. Steinmann [1574]; BIU Santé, 04059)

Paracelsus was a pioneering character in many ways (figure 1.7). He taught medicine in Switzerland, Germany, and Austria and broke with precedent by writing and lecturing in vernacular German rather than in Latin, as was customary.[90] Considered one of the earliest toxicologists, he was the first physician to record that the dose of a drug can render it either poisonous or therapeutic.[91] Most important, Paracelsus considered that the diversity of medical traditions, overrun by their various humors, spirits, incantations, and the like, obscured humans' ability to perceive the natural healing abilities of plants. The theory of Galen and the superstitions of some religious sects, he thought, stood in the way of matching a medicinal herb to the ailment it treats. Therefore, Paracelsus promoted a simple, folk classification scheme for medicinal plants. Known as the Doctrine of Signatures, this belief held that the Creator designed plants with clues for the ailments they treated (figure 1.8).[92] According to the Doctrine of Signatures, a plant growing yellow flowers is expected to cure jaundice, one with red sap would treat blood disorders, and one with fruit shaped like the brain, such as the walnut (*Juglans regia*), should be useful for improving mental abilities.

The long history of the Doctrine of Signatures in European medicine left a legacy in the common names of numerous plants that derive from their medical uses. For example, lungwort (*Pulmonaria officinalis*) grows leaves that bear a resemblance to diseased lungs with whitish spots and was used to treat lung infections such as bronchitis and tuberculosis.[93] Liverworts (Marchantiophyta) produce lobed leaf-like structures and were thought to be effective against disorders of

FIGURE 1.7 Paracelsus. (Woodcut by Tobias Stimmer [1587]; Wellcome Library, London, V0004456)

> **The Doctrine of Signatures**
>
> A plant's physical form offers clues about its medicinal uses.

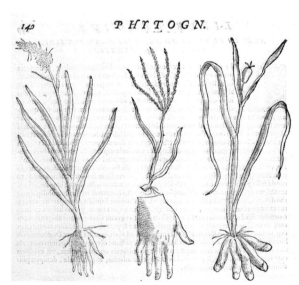

FIGURE 1.8 The Doctrine of Signatures: (*left*) lungwort leaves, thought to resemble lungs; (*right*) plants with roots thought to resemble hands. ([*right*] Illustrations from Giovanni Battista della Porta, *Phytognomica* [1588]; Wellcome Library, London, L0030485)

the liver, which is also a lobed organ. The flowers of birthwort (*Aristolochia clematitis*) are reminiscent of a woman's birth canal, and the plant was considered appropriate to treat obstetric concerns.[94] Mandrake (*Mandragora officinarum*), a plant whose root can look vaguely like a virile man, was taken as an aphrodisiac and a fertility drug.[95]

During the sixteenth and seventeenth centuries, medical authority continued to coalesce around major medical universities, where scholars lectured and published in Latin, while a number of rogue physicians published herbals in vernacular. In the English language, some of the most influential herbals of this period were by John Gerard (1597), John Parkinson (1640), and Nicholas Culpeper (1652).[96] The importance of plant medicines is underscored by the prominence of such herbals—catalogs of plants and their uses—as medical texts and field guides for apothecaries and herb gatherers. Because plants figured so prominently in human health, physician training included a thorough study of botany. It was also during this period that multiple new species were appearing in Europe for the first time, brought in from the east by traders and from the New World by conquistadors, missionaries, and natural-historian explorers. Many of these plants, too, had valuable medicinal properties. The description and classification of the diversity of plants on earth became a critical component of an advanced medical system.

THE LINNAEAN BIOLOGICAL CLASSIFICATION SCHEME

Although the herbals documented hundreds of medicinal plants and provided doctors and plant collectors a guide to their use, it was not always straightforward for a person to identify the plants described because the field lacked a uniform naming system. The organization of species into clear groups with universally acceptable names, rather than local names that varied regionally, became the life's work of the Swedish botanist-physician Carolus Linnaeus (Carl von Linné, 1707–1778). Among the general population, plants and animals bore local names that were not always geographically consistent, meaning two unrelated organisms could be described by the same name in different towns, and a single type of organism could go by many names. Before Linnaeus, a similar confusion could exist even within the trained botany community: physicians and herbalists sometimes used long Latin descriptions for plants that listed their attributes, but not in a universally standard format. As an example, take the description of the Persian buttercup in Gerard's *Herball, or Generall Historie of Plants*: *Ranunculus Asiaticus grumosa radice flore flavo vario* (Asian crow-foot with lumpy root and yellow striped flowers).[97] Gerard's seven-word identifier might just delineate a color variety that he happened to encounter but not one that applies to all individuals of the same biological type. In any case, the long label is cumbersome. Linnaeus developed a hierarchical system that organized all the specimens he studied by similarities and differences in structure and resulted in each unique type being assigned a single, simple Latin descriptor.

This advance, first published in *Systema Naturae* (1735) and elaborated in the monumental two-volume *Species Plantarum* (1753), gave all unique plant types two names: a genus name, a broad category encompassing many physically related varieties, and a species name, a precise name delineating just one type of organism.[98] This name, composed of a genus and species, is known as the Linnaean binomial, and it is the standard format by which scientists identify and distinguish organisms and varieties with precision.

Over the centuries, new knowledge has been applied to Linnaeus's classification scheme, and systematists (specialists in biological classification)

Binomial nomenclature

The scientific naming system that universally describes a single type of organism using two names, genus and species

The logic of Linnaeus

The genus name is a noun, and the species name is an adjective describing it. For example, the white oak is *Quercus alba* (oak white), black pepper is *Piper nigrum* (pepper black), and the opium poppy is *Papaver somniferum* (poppy sleep inducing).

refined the original hierarchies and species assignments. In particular, recent advances in molecular biology have allowed biologists to study the DNA of plants and ascertain evolutionary relationships not apparent to previous generations of researchers examining only the outward appearance of specimens. In some cases, molecular data have helped rewrite the classification of species by placing them among their closest genetic kin, despite differences in morphology.[99]

The binomial system pioneered by Linnaeus remains the only accepted way to identify an organism in the scholarly world. For example, the tomato plant can be properly described by its binomial *Solanum lycopersicum* (note the capitalized genus-name initial, lowercase species-name initial, and italics), which can be shortened to *S. lycopersicum* on subsequent references.[100] There are other members of the genus *Solanum*, relatives of the tomato such as *S. tuberosum*, the potato. Occasionally, distinctions within a species become apparent (sometimes called subspecies), necessitating a special notation, and the binomial is appended. For instance, *Origanum vulgare* refers to oregano, the aromatic perennial herb native to Europe and western Central Asia. *Origanum vulgare* ssp. *gracile* is a particular type of oregano from Central Asia with a distinctive leaf shape and flower color, and *Origanum vulgare* ssp. *hirtum* has been selected for its ease of cultivation and strong flavor.[101]

EMPIRICISM IN MEDICINE

By closely examining plant specimens and systematically describing their similarities and differences, Linnaeus demonstrated that careful observation, coupled with a willingness to break with long-standing intellectual customs, could lead to a new understanding of nature. Whereas much of European medicine sought to treat illness according to the prevailing conceptions

Empiricism

The notion that knowledge derives from systematic observation and experimentation

of the humors and employed plant drugs chosen by convention or "signature," the European Enlightenment period saw the growth of a movement to question and test medical ideas actively. This movement helped progress medical study from a discipline guided by theological and classical doctrine to one in which physicians refined old medicines and developed new ones by testing drugs and making careful observations of dosage, preparation, and patients' outcomes. To some doctors of this era, no longer were mysterious spirits and invisible substances the primary agents in human health. Rather, they began to see human disease as a manifestation of symptoms with reducible physical causes, and they realized that the constituents of plant medicines could serve as remedies.

During this era, some physicians challenged the spirit-, humor-, and life force–based conceptions of illness and came to regard human health as an essentially mechanical phenomenon that could be influenced by quantifiable medications and interventions, such as specific doses of experimentally tested medicines. An important example of this developing style of medicine is the work of William Withering (1741–1799), an English physician and botanist who pioneered a treatment for the age-old malady known as dropsy (figure 1.9).[102] The symptoms of dropsy include the accumulation of fluids in the extremities and lungs, shortness of breath, and a weak pulse. In today's terminology, these conditions indicate congestive heart failure. Prior to Withering, treatment of dropsy frequently entailed piercing the patient's extremities

FIGURE 1.9 William Withering. (Engraving by Henry Adlard [nineteenth century]; National Library of Medicine, 216167)

to release the excess fluid.[103] Withering suspected that dropsy could be treated differently and sought to establish a new medical intervention.

To develop a new treatment for this circulatory ailment, Withering turned to the experience of traditional herbal healers in England, who had been producing various plant concoctions for dropsy following folk practices dating back many centuries. He learned that among these treatments was the foxglove (*Digitalis purpurea* [figure 1.10]). "A lady from the western part of Yorkshire assures me, that the people in her country often cure themselves of dropsical complaints by drinking Foxglove tea," Withering wrote, and he set about to examine closely the plant's effects in patients.[104] For ten years (1775–1785), he performed extensive studies to establish the best season in which to harvest the medicine, the most effective part of the plant, the proper preparation of the material, and the effective dose in carefully measured quantities of powdered leaf extracts. As foxglove was known to be poisonous, he needed to establish the drug's therapeutic dose. At the correct dose, the plant's cardiac glycosides work to slow and strengthen the heartbeat, reduce fibrillation, and improve circulation to clear retained fluids. (Through the twentieth century, the foxglove-derived medicine digitalis remained an important treatment for congestive heart disease.)[105] Thus Withering's exacting measurements and reliance on physical, rather than spiritual, evidence for the source and treatment of disease set the stage for a new approach to medicine that took shape after the eighteenth century.

In a break with earlier beliefs, some in the medical community of the nineteenth century began to accept that physical entities, such as germs or chemical toxins, were the cause of disease, rather than harder-to-measure forces such as curses or excesses of yellow bile. The body was viewed in mechanical terms, and advances in anatomy and the understanding of cellular structures encouraged physicians to concentrate their attention on the minutely observable processes of movement, development, and biochemical transformation in human beings. Whereas previous generations of doctors prescribed treatments based on the recommendation of ancient Greek scholars, with diagnoses and dosage never formally tested, some in this era saw prudence in examining the patient for specific symptoms and the drugs for unique active components. Many began to recognize the difference between traditional conjecture and empirical evidence for the efficacy of natural drugs.

While successful herbal medicines have developed—and remain in practice—throughout the world drawing on ancient philosophies and traditional expertise, a great number of herbs and herb mixtures have not yet been tested in a systematic way.[106] Support for medicinal properties based on measured observation is central to the medical system that grew out of the European Enlightenment. In following practices elaborated during the nineteenth and twentieth centuries, biomedical researchers seek to test the effects on patients of carefully prepared and administered drugs across a broad spectrum of doses (including no drug at all), with data scrupulously recorded and shared with the wider medical community.

FIGURE 1.10 Foxglove flowers.

THE SCIENTIFIC METHOD

Based on the utility of empirical observation in gathering evidence about natural phenomena, biomedical research has evolved in recent centuries to develop a rigorous methodological platform from which to pose questions about the effectiveness of medicines and to generate answers to advance the discipline. This framework is known as the scientific method, and it serves to establish and support

Phases of the scientific method

1. Hypothesis formation
2. Experimentation and analysis
3. Critical review, publication, and retesting

claims of a medicine's value in treating disease. The scientific method consists of a multistep process.

• The first phase of this method is to develop, based on a great deal of prior research, a testable idea (known as a hypothesis) that can be addressed experimentally. For example, the notion that some chemical in foxglove is capable of reducing the symptoms of dropsy could be examined by giving dropsy patients known doses of the foxglove extracts and carefully observing the outcomes. Crucial to this step is the expectation that all measurements would be repeatable by others. That is, no "secret" plant extracts can be promoted, and all characterizations must be precise. Furthermore, all experimental outcomes must be based on natural observations and derive from physical explanations. In this sense, it is not appropriate to assign any biological activities to ghosts, spirits, life forces, hexes, or acts of God, because they cannot be measured or described with existing instruments.

• The second phase of the scientific method is the process of carefully designing and carrying out a controlled experiment. To do this, the researcher must be able to distinguish whether a drug is truly responsible for (that is, causes) a perceived effect. For example, consider another experiment to test the efficacy of purple coneflower (*Echinacea* spp.) extract, an herbal treatment available in many retail drugstores, on the severity of respiratory infections. In this fictional experiment, fifty children suffering common cold symptoms are given daily doses of purple coneflower extracts, perhaps in the form of capsules. On the second day, half the patients report feeling much better, and by the fourth day, nearly all the children feel fine, an assessment supported by objective measurements of their body temperatures and open nasal passages. It might appear that the herbal remedy is effective—after all, most of the children recover, and quickly, it seems. The patients may even report that this herbal treatment seems more potent than alternative treatments taken during previous illnesses. However, biomedical researchers might challenge the evidence for efficacy because of the lack of experimental controls. Without experimental controls, it is not possible to determine whether the patients would

have recovered, perhaps just as quickly, even without taking purple coneflower pills.

How can the researcher determine what the normal progression of the illness would be in the absence of any treatment? To establish this, each experiment can have a negative control treatment, in which a subset of the patients is given a false or mock medication intended to produce in itself no specific effect on the condition being studied. This treatment might consist of a preparation of medically inactive herbs or a sugar pill. The patients, however, are not informed which type of medication ("real" versus "mock") they receive. The treatment not containing active medicine is called the placebo. In this way, the investigator can assess whether the pharmaceutical under consideration has any therapeutic effect distinguishable from the placebo. Since some drugs produce a range of physiological effects in addition to the targeted condition (called side effects), clinical researchers can employ placebos that cause mild side effects, a way to disguise further the nature of the treatment to the patient. These are known as active placebos.

Use of a placebo group in a clinical study is a key element of modern biomedical research that allows researchers to document the physiological and psychological aspects of treatment in a single experiment. Interestingly, many patients respond symptomatically to the experience of being treated in a medical setting, regardless of taking an active medication. In studies of pain, depression, and numerous other ailments, on average 35 percent of patients experience an improvement in symptoms after taking an inert medicine.[107] This phenomenon, often called the placebo effect, can be traced not to the substance of the medication, which is thought to be biologically inactive, but to the treatment environment and the nature of the interaction between a patient and a caregiver.[108] That is, the patient feels listened to and that his condition is understood, the patient feels concern from others, and the patient feels a sense of control over his own condition. Thus the mental outlook and optimism of a patient can strongly affect the physical sense of health.

Because the placebo effect is such a powerful factor in the manifestations of illness, experiments should account for it. The best clinical studies test a medication of unknown efficacy against a placebo

THE CLINICAL TRIAL

The systematic, empirical approach to the study of nature employed during the eighteenth century in Europe contributed new methods toward determining the effectiveness of drugs. Researchers began to test carefully dosed medications given in similar ways to large numbers of patients, diligently recording the signs of disease before, during, and after treatment. Such observational activities were new to medicine and supplanted an earlier methodology of treating each patient individually following classical recipes and recording subjective outcomes, if any. A classic example of empirical medical research is the work of the English physician William Withering, who tested the effectiveness of precisely measured doses of foxglove extract on indications of the circulatory disorder dropsy by administering the drug to patients and recording their physiological responses. In this genre of observational study, a patient is given a drug, and then the outcome is monitored. Although the observational approach has noted merits, biomedical science recognizes flaws in its methodology.

Certainly, researchers want to know whether a drug has a physiological activity in the body. However, an observational study does not lead to clear cause–effect relationships. Is it the drug that causes the patients' symptoms to change or not? In such experiments, for instance, the patients receive a dose of what they believe to be medicine from a doctor and, in many cases, feel better at the end of the trial. This improvement in the patients' condition (objective and subjective criteria) is a perceived therapeutic effect. But can it be said that the patients got better as a result of taking the drug?

The perceived therapeutic effect, measured physiologically or as a subjective sense of wellness, may result from the *drug treatment* (that is, the components of the medicine itself) or from *other factors* that accompany such an observational study:[1]

Natural course of the disease. Many diseases, especially chronic ones, progress in periods of increased severity followed by remission (for example, cancers, arthritis, and depression), during which time the symptoms subside, even in the absence of medical treatment. A patient experiencing remission of a disease after taking medicine might erroneously attribute the lessening of symptoms to the curative power of the drug, even if the drug had no effect at all.

Concomitant treatments. Patients in observational studies sometimes engage in activities that may increase or decrease the severity of their symptoms in a way unrelated to the therapeutic effect of the drug administered. For example, they may change their exercise routine, alter their diet, or self-administer additional medications. In this situation, it is not possible to assert whether the particular medicine under study or the patient's other behaviors are responsible for any change in the disease.

Therapist–patient interactions. To many patients, the experience of being cared for in a clinical setting is itself therapeutic. The trusting relationship between medical practitioner and patient might account for an improvement in disease symptoms by improving the patient's mental state and sense of well-being. Therefore, the relationship between patient and provider might give rise to a general therapeutic effect in the absence of a specific (medicinal) therapeutic effect.

Hawthorne effect. This social-science phenomenon demonstrates that individuals who know they are under observation change their behavior. In short-term observational studies, patients assiduously follow their hygiene regimens and might perceive their health differently for no other reason than the awareness they are the focus of study.

among patients randomly assigned into the two categories. Neither the patients nor the caregivers know whether they are receiving or delivering the medication or the placebo, a design labeled "double-blind." Since the advent of scientific medical practice, countless double-blind, placebo-controlled studies have contributed to a contemporary pharmacopeia of natural and synthetic medications.

• The third phase of the scientific method is the publication and sharing of experimental findings with

Social desirability. Patients may seek to please their doctors by reporting their symptoms are less severe than actually experienced.

Biomedical research employs a particular methodology to test the efficacy of medicines in a way that accounts for potentially confounding factors of observational studies. By engaging in placebo-controlled, randomized, double-blind clinical trials, researchers seek to identify medicines with specific therapeutic effects and characterize their side effects in a robust manner.

Placebo-controlled. To ascertain whether a medicine under investigation produces a specific therapeutic effect, it is important to determine the effects of all other aspects of the test *in the absence* of the potential drug. This is known as a control, and it requires that all patients experience the same conditions of treatment except for the single element of the drug. Some individuals in such a test receive the drug, while some individuals receive a treatment that resembles the drug in all possible ways but that is thought to be physiologically inert. In clinical trials, for example, one group of patients takes a pill every week that consists of the herbal extract under examination, and the other group of patients takes an identically marked pill every week that contains simply starch. The mere administration of an inert medicine can have a therapeutic effect (known as placebo when a positive effect), and so this effect is accounted for as a placebo control. Among patients in a drug trial, any symptomatic improvement beyond the effect of placebo alone might be attributable to the effect of the medication.

Randomization. From a large population of patients, individuals are randomly assigned either to receive the medical treatment or to receive the placebo treatment. (In most cases, patients are advised that they have equal likelihood of participating in either test group but are never told to which group they belong.) The process of randomization avoids any selection bias in the medical treatment versus placebo treatment groups.

Double-blinding. Neither the patients nor the clinical staff are informed which individuals receive the medicine or placebo. In this way, none of the clinicians, through conscious or unconscious actions, behaviors, sympathies, or preferences, can influence the perceived therapeutic effect in either the medical treatment or placebo group disproportionately.

By clearly articulating an answerable clinical question, defining measurable, reproducible criteria for efficacy, and critically assessing the evidence, the clinical trial seeks to establish reliable causative links between a treatment regime and therapeutic usefulness. Such trials inform biomedical practice and establish the criteria by which pharmaceuticals in the United States and elsewhere are regulated. Much current research into determining the safety and efficacy of herbal remedies also employs these methods. Studies falling short of the standards of placebo control, randomization, and double-blinding call their findings into question.

Even the most carefully constructed trials face the challenge of critical assault, which is a desirable and positive aspect of the scientific process: researchers do not view their methods as perfect and constantly seek to improve them. Often many trials, enrolling a large number of patients and testing a drug across a wide range of doses, are necessary before scientific consensus can be reached on a drug's possible therapeutic effects and risks.

1. These categories are based on those of Edzard Ernst, "The Importance of Having a Robust Evidence Base—A Personal View," in *Homeopathic Practice*, ed. Steven B. Kayne (London: Pharmaceutical Press, 2008), 33–42.

the broader community. To facilitate this, numerous scholarly journals specialize in printing such medical studies, and their editors and reviewers vet the studies for adherence to scientific guidelines and clarity of conclusions drawn. Published experimental results are then subject to criticism and retesting, a process that strengthens accurate findings and disproves false findings or improperly conducted experiments. Consequently, medical science constantly renews and improves its understanding of human health.

The scientific method, consisting of a carefully composed hypothesis to test, a controlled experimental approach, and the publication of findings, marks an important methodological shift from the practice of traditional medicine in Europe and elsewhere. In recent decades, medical authorities in Europe, the United States, and other parts of the world have emphasized the role of this system, with its clinical trials and statistically determined outcomes, in improving health care in both industrialized and developing nations. Although originating from the European Enlightenment, the scientific medical discipline has spread to many countries, where doctors are trained in its techniques and drugs are studied in certain standardized ways. It is this form of health care, coupled with progress in biochemistry and the strength of recent surgical advances, that is practiced as biomedicine around the world.[109]

> **Criteria for high-quality clinical trials**
>
> • Placebo controls
> • Randomization of patients
> • Double-blinding

PLURALITY, EVOLUTION, AND SYNTHESIS: MEDICAL TRADITIONS IN PERSPECTIVE

The biomedical approach has joined numerous other traditions in offering its own understanding of health and therapy in a diverse and dynamic health-care ecology. In the modern-day United States, for example, patients can seek treatment in hospitals and doctors' offices, where they might be likely to receive pharmaceutical drugs prepared from natural or synthetic sources. Yet the same people may also visit a traditional herbal healer and purchase mixtures of plant products for their health concerns, grow their own herbs to make herbal teas, and follow their grandmothers' dietary advice. In today's China, medical universities train physicians to practice either biomedicine or Chinese traditional medicine, and patients can choose to seek treatment at clinics and hospitals specializing in either.[110] In addition, those same patients

might prepare cuisine at home according to traditional and religious principles, whether in terms of the heating and cooling properties of their foods or for their spiritual effects. They may also take medicines deriving from regional folk traditions. As in the past, many ideas about health and, consequently, many forms of herbal medicine coexist in the world's communities.

Medical beliefs and practices evolve, rendering any glimpse of a tradition as written in a text or observed among indigenous people merely a snapshot in time. Among the ancient Greeks, ideas about the humors were vigorously debated. Originally, the Hippocratic system accounted for only three fundamental substances, with black bile added later.[111] As these ideas were elaborated during Galen's time and afterward, differences in interpretation and application of shared concepts abounded. In the Muslim world, Galen's authority was reformulated by a series of influential writers over the course of several centuries, and forms of his teachings took hold in South Asia, Central Asia, and the Middle East, adapting to locally endemic illnesses and incorporating regionally abundant plants in therapy. In medieval and Renaissance Europe, Galen's notions of the humors variously took on astrological and religious inflections. As the voyages of discovery brought back never-before-seen plant specimens to the learned physician-botanists of Europe, these scholars accommodated the novel herbs in the preexisting scheme and assigned to them the qualities of hot, cold, dry, and moist that would enable them to be used alongside a growing number of plants both inherited from the ancients and adopted from folk practitioners. Likewise, biomedicine has gradually

> **Herbs in the context of medical treatment**
>
> An herbal medical treatment may yield outcomes, both positive and negative, resulting from the activity of the drug itself and from the setting of the treatment, the relationship between patient and caregiver, and the patient's expectations about the experience.
>
> Evidence-based medicine demands a rigorous, experimental demonstration of drug activity but cannot account well for other treatment effects.

THE SIGNIFICANCE OF EVIDENCE-BASED MEDICINE

A remarkable diversity of medicinal plant uses is evident in the written record of past cultures and among living practitioners who understand health according to their sometimes shared, sometimes divergent conceptions of the human being in nature. One goal of current scholarship is to learn about these people and their medicines in their own time and place. Another area of research seeks to apply the methodological standards of biomedical science to traditional medicines, to examine them for efficacy and risk. These approaches engage in experimental trials to obtain data that can be analyzed statistically, collect detailed case studies of patients in the clinic, and pursue biochemical assays in their quest for evidence of a treatment's value in medicine.[1]

To biomedical researchers, the statements "this tribe has used plant X to treat headaches for centuries" and "these healers prescribe mixture Y all the time, and there haven't been any problems with it" do not serve as medical evidence for the effectiveness of herbal remedies. They are anecdotes that document cultural practices but not data in support of the efficacy of any drug or treatment. The fact that a group of people exploits a plant or plant extract for a specific purpose does not demonstrate that it is effective for the intended or any other use. Nor does biomedical science view the medicine to be ineffective. Evidence-based medicine requires an experimental demonstration to support use of a novel drug. This foundation of methodological rigor has become institutionalized in the United States and elsewhere, for example, intrinsic to physician training and government regulatory decisions.

In recent decades, numerous plants drawn from traditional medical practices have been subjected to biochemical study and clinical trials to generate the sort of evidence expected by the biomedical establishment. Such experiments largely ignore some aspects of health so valued in many of the world's medical traditions, such as the relationship of a person to the spirit world, the role of a person in the community, and the customization of a treatment to suit an individual patient's condition.[2] Indeed, the scientific method requires researchers to set aside the supernatural, to standardize experimental treatments, and to consider data mostly in terms of statistical averages rather than individual outcomes. The tenets of experimental science notwithstanding, many biomedical practitioners have come to embrace some of the social, spiritual, and holistic aspects of therapy common in traditional health care.

The world's numerous medical traditions can treat those who believe in them, sometimes in ways that biomedical science has not yet fully addressed. On the one hand, some herbal drugs may have physiological activities thus far not demonstrated by clinical experiments. On the other hand, some medicine may work not because of the substance of the drug but through the social setting and expectations of those involved. In particular, academic investigators have only recently begun to study the role of the therapeutic relationship between patient and caregiver, one that can treat discomfort in the absence of drugs: "What is important is that doctors—healers of any sort or type—are convinced that their techniques are powerful and effective, and that there is undeniable evidence of this effectiveness. In some places, such proof comes from gods or spirits, in some places from personal experience, and in other places from the assertions of science."[3]

1. A defense of the scientific method and critique of complementary and alternative medicine is in R. Barker Bausell, *Snake Oil Science: The Truth About Complementary and Alternative Medicine* (New York: Oxford University Press, 2007).
2. A critique of evidence-based medicine is in Steve Hickey and Hilary Roberts, *Tarnished Gold: The Sickness of Evidence-Based Medicine* (CreateSpace, 2011).
3. Daniel E. Moerman, *Native American Ethnobotany* (Portland, Ore.: Timber Press, 1998), 43.

developed its robust methodology over the course of more than a century, discovering new treatments and discarding older ones, a process that will continue.

Humanity's many medical traditions have borne witness to significant cultural exchanges that produced new syntheses in various places and times (figure 1.11). For instance, South Asian *siddha* medicine is thought to be a fusion of an ancient southeastern Indian folk tradition with theories borrowed from *ayurveda*. Both *ayurveda* and *siddha* probably gained certain diagnostic tools, such

FIGURE 1.11 A Peruvian herbalist. This clinic treats illnesses such as cancer and anemia as well as the effects of witchcraft and evil wind, drawing on several indigenous herbal traditions.

as pulse diagnosis and uroscopy, from Greek medicine brought by Muslims.[112] In modern-day China, biomedical concepts of disease have been interpreted according to traditional medical tenets, such that a patient in a traditional medical clinic might be diagnosed in terms of both *yin–yang* imbalance and abnormal hormone levels and be offered an herbal prescription as treatment. After many years of setting aside the spiritual and psychological aspects of therapy, some biomedical practitioners are promoting the approach of "treating the whole person" and "mind–body medicine," the tenets of which clearly resonate with many traditional medical practices.

The medicinal plants of the world are employed in a tremendous variety of human settings. Some are harvested for use in extensive health-care systems grounded in elaborate theory, with specialized hospitals and pharmacies and a worldwide audience, such as those used in biomedicine and Chinese traditional medicine. Plants such as coffee (*Coffea* spp.) are turned into commodities, sold in markets large and small, and take on importance in locally accented concepts of health and sociability. Still other herbs are used by small communities, by people whose unique understanding of nature may never have been recorded, and whose medical traditions may reside only in the expertise of a dwindling number of practitioners. All these facets of the human experience with medicinal plants, whether found in cosmopolitan parts of the earth or in isolated villages, whether drawn from a large scholarly lineage or a folk custom practiced by few, demonstrate the capacity of medicinal plants to shape our culture.

In many parts of the world, people developed medical traditions that incorporated plants to promote health and treat illness. Among them are schemes where opposing forces or substances find balance in the human body and where plants can help restore disequilibria that develop. Other conceptions involve the intervention of spirit beings in the individual's health. Herbal remedies are employed in diverse therapeutic settings, some of which can be gleaned from ancient texts and others from the numerous peoples who have inherited a medicinal plant legacy. While medicine in Europe experienced many changes across the millennia, it had much in common with other world medical traditions until some of its practitioners developed an empirical and scientific approach during the eighteenth through twentieth centuries. As a result, biomedical researchers test drugs following a prescribed set of expectations in a controlled experiment. In this way, biomedicine has delineated a methodology that approaches disease as the product of physical dysfunction that can be remedied with specific pharmaceutical substances. In a world where numerous traditional medical practices coexist, the diversity of health-related beliefs and the mixture of practices provide a fascinating backdrop for medicinal plants in the modern day.

Chapter 2
The Regulation of Drugs

Mate de coca (coca tea) for sale at an Andean market in Humahuaca, Argentina. This product contains cocaine, which is restricted in the United States.

Wherever humans have settled, they have made plants important elements in their pursuit of physical, mental, spiritual, and community health. People have also recognized the value of toxic leaves, stems, and seeds and harnessed their poisons as weapons. To this day, medicinal plants serve mixed roles in society: they can rid patients of disease, and they can extinguish an enemy's life. They can be used to help a man fight an infection, or they can release his psyche to communion with a vast supernatural world. Human cultures variously define wellness at individual and group levels; see physical, mental, and spiritual health differently; and have assigned many roles to people specializing in diverse aspects of plant-based medicine. Because of the diversity of medicinal plants and the many roles these plants play across many cultural settings, people have struggled to incorporate them into the context of social norms, moral standards, and law. Which uses should be considered health promoting, and which are deemed harmful? What role should a government play in easing access to herbal remedies, and which activities should it restrict?

In different eras and regional contexts, communities have had differing approaches to these complex questions. Most societies assigned specialist roles to individuals with the knowledge to administer skillfully medicinal plant products. Over time, concepts of health were redefined, and governments and religious institutions began to limit the breadth of acceptable plant medicine and establish structures for the regulation of numerous drugs. How this came to be is a story that began more than 2000 years ago.

DRUGS AS POISONS, SPIRITUAL AIDS, AND AGENTS OF SOCIAL BONDING

Plant drugs, it was long known, could have healing or harmful properties, and the preparation of remedies and poisons often fell to specialists. The Greek and Roman scholars wrote extensive treatises on the preparation of medicinal plants in various forms of soups, ointments, and powders. Pedanius Dioscorides's inventory of *materia medica* was among the earliest of pharmacies, and those knowledgeable in its use served to remedy all manner of ailments.[1] However, skilled hands can also employ plants to fatal ends. While extracts of hemlock (*Conium maculatum*) were used medicinally by ancient Greeks and Romans, the practitioner must have exercised extreme caution, as this herb contains potent nerve toxins.[2] To this effect, hemlock was implemented in the dramatic execution of the Greek philosopher Socrates (469–399 B.C.E.).[3] Over the years, a class of learned individuals specialized not in the healing properties of plants but in their deadly effects. During medieval and Renaissance times, several families in Italy, such as the Borgia and Baglioni, either were themselves or employed experts in poisons. They developed ingenious methods of delivery and ensured their ascent on the social ladder over the bodies of their opponents. Apothecaries, whose knowledge of medicinal plants could be applied to the preparation of both therapies and poisons, served dual roles. ("Such mortal drugs I have," said the apothecary in Shakespeare's *Romeo and Juliet*.)[4] These potent preparations included mineral toxins, toad poisons, and extracts of plants such as monkshood (*Aconitum* spp., also known as wolfsbane and aconite) and nux vomica (*Strychnos nux-vomica*, also known as the strychnine tree).

Plant extracts have been used to heal and to harm. Through history, the use of medicinal plants has frequently occurred in conjunction with spiritual or religious appeals. This is because, in many parts of the world, health was characterized by inseparable spiritual and physical components. For example, a sixteenth-century drawing of an Aztec healing ritual pictures a patient in a steam bath, in the presence of medicinal plants, doctors, and an idol of the face of the goddess of medicine, Teteo Innan.[5] This imagery supports a well-documented medical-religious practice in ancient Mexico in which priests served to diagnose and treat illness.[6] The connection between medicine and faith is evident in Asia as well. In Tibetan herbalism, for instance, practitioners combined medicine with mantras to "energize" the herbs and promote their efficacy.[7] In modern-day Suriname, along the

Atlantic coast of South America, a survey of plant merchants revealed that 56 percent of medicinal plants sold at local markets were used for *winti*, Afro-Surinamese religious-spiritual health rituals.[8] Thus the medical uses of plants have mystical elements in many cultures.

In addition to roles in spiritual health, medicinal herbs have been used to promote comfortable social interactions. These special functions in community life are evident in numerous cultures around the world, and expertise in preparing these materials sometimes became the domain of a unique class of healer. Among some indigenous American groups, for example, tobacco (*Nicotiana* spp.) was important in social-bonding rituals and in such settings was dispensed by elders and spiritual leaders.[9] To this day, rites of passage in many regions are marked by the consumption of medicinal plants. Some medicinal plants play a more ubiquitous role in the social life of people: for example, the leaves of the tea plant (*Camellia sinensis*) or coffee seeds (*Coffea* spp.) serve to produce beverages around which people form friendships and business relationships. Thus medicinal plants can have roles in social health.[10]

DIVERSE PROFESSIONS SPECIALIZED IN SUPPLYING MEDICINAL PLANTS

During the Roman era, while some useful herbs fell under the auspices of physicians, people also incorporated medicinal plants in their everyday cuisine and visited spice merchants to obtain the balance of flavors and temperatures deemed healthy according to their medical beliefs. Similarly, in Asia, the maintenance of health through choice of food was a day-to-day function of the meal preparer. In China, specialized herb shops became established, where patients seeking treatment could turn in their physician's prescription for carefully measured packets of herbs to brew into teas and soups at home. These specialist herb shops still exist in Chinese communities around the world and serve as trusted pharmacies. Today, pharmacies in China often have separate sections, one selling either herbal, animal,

and mineral drugs for use in traditional medicine and the other selling prepared and synthetic biomedical drugs and antibiotics. Since medicine and cuisine have been so closely linked for so long, and in many parts of the world, food merchants and spice and herb sellers have played an important and continuing role in providing medicinal plants and perpetuating knowledge.

In Europe, the roles of plant medicines became gradually defined by the various specialists who tailored their use to heal, to poison, for cuisine, and for religious and other ends (figures 2.1 and 2.2). By the Middle Ages, a distinction developed among spiritual-medical practitioners that resulted in herbalism falling into the hands of a diverse group of specialists. Midwives and

FIGURE 2.1 Pierre Quthe, an apothecary. This portrait shows a medicine provider with his most important tool: a book of herbal medicines. [Painting by François Clouet [1562]; BIU Santé, 66442x13)

village herb gatherers provided folk medicinal materials to a great many individuals during this period, even as monasteries remained the repository of classical and religious knowledge, maintaining Greek- and Roman-inspired herb gardens for this practice. Pepperers and spicers were merchants in herbs for culinary and medicinal uses. Apothecaries and alchemists developed a trade in mixing herbs, metals, and animal parts into solid and liquid concoctions, frequently for medicinal goals.[11] Meanwhile, the production of poisons by sorcerers became a lucrative trade, dependent on great insight into drug action—yet separate from the medical-chemical professions. Finally, expertise in the use of certain types of mind-altering herbs fell to the so-called witches, who developed potent cocktails for their spiritual rituals.[12] Later, during the era of the Enlightenment, these divisions became

FIGURE 2.2 Medicinal herbs are prepared in many different settings and have been largely unregulated until recent times: (*left*) a woman mixing an herbal recipe at home for an ill relative; (*right*) patients visiting a shop operated by a monk to obtain eye medicine. ([*left*] Illuminated manuscript, Bruges [late fifteenth century]; British Library Royal 15 D, i. fol. 18; [*right*] illuminated manuscript, Austria or southern Germany [ca. 1675]; Wellcome Library, London, L0042099)

further entrenched, since empirical medicine had no accommodations for spiritual or mystical healing. However, these specialist titles certainly should not imply exclusive roles. Apothecaries sold remedies and poisons. Folk herbalists had knowledge to diagnose and treat patients and probably to induce mind-altered states.

As a result of the European experience, many people viewed the broad array of medicinal plant practices as composing two categories: on the one hand, the use of herbs to treat bodily ailments and, on the other, all other uses. In the latter category are many social and religious herbal practices now considered outside the realm of medicine in Europe, the United States, and elsewhere. In the European context, the ancient doctors and herbalists of the classical era gave rise to medieval apothecaries and modern physicians and pharmacists, administering treatments in response to primarily physical disorders.

Among many non-European cultures, the separation of physical, mental, and spiritual health is not recognized. In such societies, the role of treating illness, protecting the spirit, and expanding the mind frequently falls to the same priest, shaman, sage, or medicine man.

THE AMERICAN HERBAL MEDICAL TRADITIONS

It is a mixture of many traditions that characterized the American people of the eighteenth and nineteenth centuries: a blend of European, indigenous American, Asian, and African beliefs, an amalgam that enriches and complicates the modern-day American sense of health. Importantly, many of the divergent health beliefs of America's peoples developed in the absence of formal government regulation. Indeed, the practices of herb collecting, medicine preparation, and spiritual healing evolved without state edicts or legislation controlling how the health trades could be practiced.[13]

Among the many schools of medical practice that coexisted in the nineteenth-century United States were those based on theories drawn from Europe (including humoral notions of health) and a growing cadre of physicians who embraced the biochemical and anatomical advances of early biomedicine. America's pioneering culture and expansive, still largely wild natural environment gave rise also to domestic herbal practices that gained large numbers of adherents. One such school was founded by Samuel Thomson (1769–1843), a self-taught herbalist who rejected many of the techniques, such as bloodletting and mineral purges, still practiced at that time by formally trained European doctors.[14] Instead, Thompson proposed that herbs alone could restore balance to an ill body, and his firsthand experience tasting North American plants convinced him that local herbs were better suited to treat local ailments than imported plants. He especially valued Indian tobacco (*Lobelia inflata*) for its capacity to induce a therapeutic bout of vomiting.[15] As an author, businessman (he sold proprietary herbal pills), and promoter of an American herbal medical practice, his influence persisted through much of the nineteenth century among his followers and the general public.

After Thomson, another school of American herbal medicine developed on the notion that plant-derived treatments were safer and more effective than many of the harsh methods of European medicine. Its practitioners called themselves the eclectics, on the belief that useful herbs could be found in a variety of settings, whether from North American Indian expertise or from the traditional medicines of other continents. (The eclectics most vigorously advocated the use of American medicinal herbs.) Early advocates, some of them former Thomsonians, sought also to formalize the American herbal medical tradition by establishing colleges for physician training, in which students learned physiology and chemistry in a curriculum emphasizing the use of herbs. Eclectic medical colleges trained physicians in the United States from the 1820s into the early part of the twentieth century.[16] Among their legacy were numerous pharmaceutical companies grounded in the eclectic tradition, whose technicians prepared and sold a wide variety of herbal medicines and extracts in pharmacies and by catalog. By the late nineteenth century, Americans obtained medicines from a diverse assortment of medical practitioners and herb sellers, in addition to those they grew themselves.

At the turn of the twentieth century, the U.S. government—and those of most nations—had very few regulations on the manufacture and consumption of drugs. Medications of all varieties could essentially be harvested or produced by anyone, sold without a formal prescription, and marketed by specialist shops or general salesmen. The use of medicinal plants in food and drink or as drugs was considered to be an individual's decision.[17] The use of alcohol, in contrast, was much criticized during the nineteenth century, condemned by women's groups, on the one hand, which warned of the dangers of inebriated and violent husbands, and by industrialists, on the other, who sought to sober up and increase the productivity of the workforce. Yet, particularly in the United States, individuals and interest groups had little power to effect change, as conservative forces maintained that the cultivation and harvesting of plants and the formulation of medicines was not governed by the Constitution and that any restriction in these activities would represent an affront to personal liberties.[18]

DRUG REGULATION IN THE UNITED STATES

By the late nineteenth century, public concern about medicinal plants mounted in two areas. First, people became sensitive to the possible risks of widely distributed home-remedy elixirs and other such popular medicines. Through the nineteenth century, numerous physicians, pharmacists, and entrepreneurs claiming medical authority produced and marketed concoctions advertised to treat all manner of ailments. Such medicines often contained medicinal plant substances, but their contents were unlabeled, as manufacturers were not required to disclose this information. Furthermore, producers could make any type of claim they wished on labels or in advertising, promising to cure the most debilitating conditions and to be free of any harmful substances, without the expectation

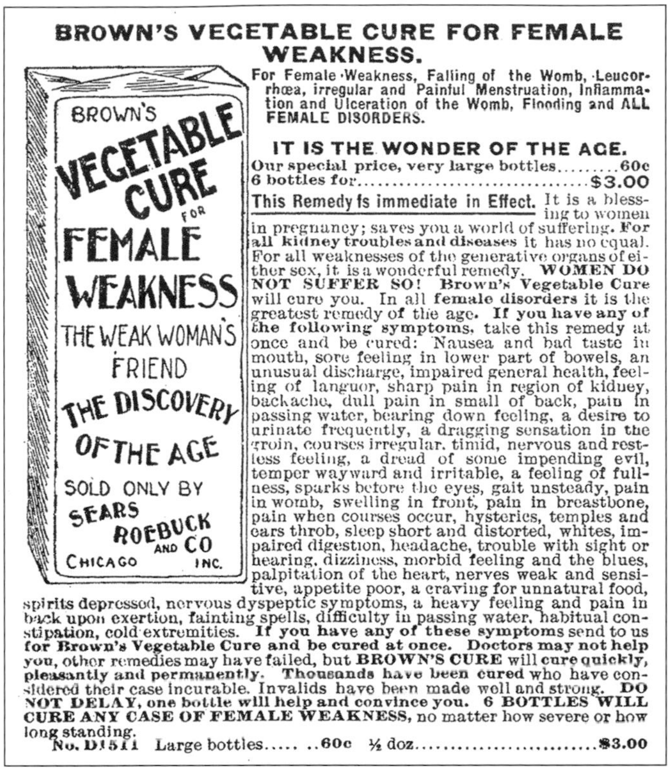

BROWN'S VEGETABLE CURE FOR FEMALE WEAKNESS.
BROWN'S
VEGETABLE CURE
FOR
FEMALE WEAKNESS
THE WEAK WOMAN'S FRIEND
THE DISCOVERY OF THE AGE
SOLD ONLY BY
SEARS ROEBUCK AND CO
CHICAGO INC.
For Female Weakness, Falling of the Womb, Leucorrhœa, irregular and Painful Menstruation, Inflammation and Ulceration of the Womb, Flooding and ALL FEMALE DISORDERS.
IT IS THE WONDER OF THE AGE.
Our special price, very large bottles........60c
6 bottles for.................................$3.00
This Remedy is immediate in Effect. It is a blessing to women in pregnancy; saves you a world of suffering. For all kidney troubles and diseases it has no equal. For all weaknesses of the generative organs of either sex, it is a wonderful remedy. WOMEN DO NOT SUFFER SO! Brown's Vegetable Cure will cure you. In all female disorders it is the greatest remedy of the age. If you have any of the following symptoms, take this remedy at once and be cured: Nausea and bad taste in mouth, sore feeling in lower part of bowels, an unusual discharge, impaired general health, feeling of languor, sharp pain in region of kidney, backache, dull pain in small of back, pain in passing water, bearing down feeling, a desire to urinate frequently, a dragging sensation in the groin, courses irregular, timid, nervous and restless feeling, a dread of some impending evil, temper wayward and irritable, a feeling of fullness, sparks before the eyes, gait unsteady, pain in womb, swelling in front, pain in breastbone, pain when courses occur, hysterics, temples and ears throb, sleep short and distorted, whites, impaired digestion, headache, trouble with sight or hearing, dizziness, morbid feeling and the blues, palpitation of the heart, nerves weak and sensitive, appetite poor, a craving for unnatural food, spirits depressed, nervous dyspeptic symptoms, a heavy feeling and pain in back upon exertion, fainting spells, difficulty in passing water, habitual constipation, cold extremities. If you have any of these symptoms send to us for Brown's Vegetable Cure and be cured at once. Doctors may not help you, other remedies may have failed, but BROWN'S CURE will cure quickly, pleasantly and permanently. Thousands have been cured who have considered their case incurable. Invalids have been made well and strong. DO NOT DELAY, one bottle will help and convince you. 6 BOTTLES WILL CURE ANY CASE OF FEMALE WEAKNESS, no matter how severe or how long standing.
No. D1511 Large bottles.......60c ½ doz.........................$3.00

FIGURE 2.3 An advertisement for Brown's Vegetable Cure for Female Weakness, a patent medicine, from Sears, Roebuck and Company, 1897. The contents of these purported cures were known to only the manufacturers until the Pure Food and Drug Act (1906) required labeling.

that these assertions would be tested for truthfulness (figure 2.3). These nostrums went by the name of patent medicine.

During the second half of the nineteenth century, patent medicines were increasingly under scrutiny, especially for their unlabeled contents. Some of them, it was found, had very few medicinal ingredients but high levels of alcohol; others packed potentially dangerous doses of drugs such as cocaine (from coca [*Erythroxylum coca*]) and morphine (from the opium poppy [*Papaver somniferum*]). The public and lawmakers sought to control this burgeoning industry of cure-all patent medicines (figure 2.4).[19] Beginning with the Pure Food and Drug Act of 1906, which required the food and medicine industries to label the contents of their products, the government's role in protecting public health took root.[20]

A second factor that influenced the course of medicinal plant regulation in the United States was the sensitivity of some Americans to the changing face of their society. As immigration advanced in the nineteenth century and domestic racial minorities pursued their lives alongside whites, the largely white media and government institutions targeted the (real and imagined) drug-taking customs of certain groups and developed laws to restrict particular types of medicinal plant use. For example, many were troubled by the Chinese custom of opium smoking (which grew stylish among Americans of all races) and the supposed nefarious connection between drug use and crime in immigrant communities.[21] Racial motives in the regulation of medicinal plants are evident in the public discourse and in legislation surrounding poppy, coca, and marihuana (*Cannabis sativa*), among others. In short, from an era in which medical regulation was essentially absent, widespread concern over the potentially dangerous contents of unlabeled patent medicines, coupled with a desire to target certain practices among America's ethnic groups, led to the development of a framework in which drug control became conscionable.

While the Pure Food and Drug Act required manufacturers to label their products, it did not restrict the producers' choice of ingredients. (Manufacturers facing scrutiny of post-1906 labeled contents often chose to eliminate dangerous ingredients voluntarily.) Policies advanced during the twentieth century sought to restrict the production, distribution, or consumption of particular drugs based on the government's growing responsibility to protect Americans from the potential dangers of medicines. Using legislative powers and executive privileges, the government had enormous latitude in

Collier's
THE NATIONAL WEEKLY
THE PATENT MEDICINE TRUST
PALATABLE POISON FOR THE POOR
DEATH'S LABORATORY

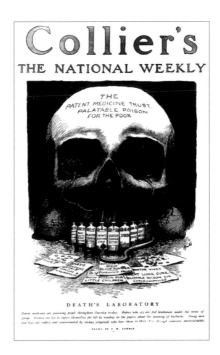

FIGURE 2.4 "Death's Laboratory": a magazine cover critical of unlabeled, dangerous patent medicines. (Illustration by E. W. Kemble [1906])

THE FOOD AND DRUG ADMINISTRATION

Before the twentieth century, the production and sale of drugs was left to the free market, and the consumption of medicines was largely considered a matter of personal choice in the United States. The regulatory landscape changed in 1906 when the government passed the Pure Food and Drug Act, which required the labeling and enforced the testing of food, drinks, and medicine in interstate commerce. Products became subject to rigorous government testing for strength and purity, and from 1912 on, the law prohibited the sale of medicines under fraudulent claims of effectiveness. Originally monitored by the Department of Agriculture, such products came under the management of the Food and Drug Administration (FDA) in 1930.

In 1938, the Food, Drug, and Cosmetic Act (FD&C)[1] required drug manufacturers to demonstrate to the FDA that a drug was safe (by relatively loose standards) before it could be sold.[2] An amendment in 1951 established a system whereby certain drugs became labeled for sale by prescription only. Despite these regulations, most drug companies performed perfunctory drug trials that fell short of scientific credibility; in short, many medicines had not been demonstrated *safe* and *effective* by properly controlled study. To resolve this deficiency, Congress passed amendments to the FD&C in 1962 establishing rules by which drug manufacturers tested their products. The FDA served to approve or disapprove the marketing of these drugs based on a number of scientific criteria demonstrating their safety and efficacy.[3]

The Fair Packaging and Labeling Act of 1966 gave the FDA purview over the honest and informative labeling of products. In 1970, the FDA required that paper inserts accompany medicines to inform patients of the risks and benefits of medications, and in 1972 the FDA began to evaluate the claims of safety and effectiveness of over-the-counter medications.

In the modern era, drug companies seeking to market new drugs must apply to the FDA after performing substantial animal testing and propose their methodology for conducting human trials. Only after approval of this *investigational new drug application* can pharmaceutical firms commence human clinical trials.[4]

Along with oversight of prescription and over-the-counter drugs, the FDA regulates the labeling and sale of foods and dietary supplements, many of which have traditionally been regarded as medicinal. A 1958 amendment to the FD&C established that a large number of food additives and food colors (following certain guidelines) could be regulated under a framework of Generally Recognized as Safe (GRAS). GRAS products required a "reasonable certainty" of safety as demonstrated by "recognized experts," the burden of proof held by the product's manufacturer.[5]

Grandfathered into the GRAS standard were products in common use in food before 1958, provided an expert consensus existed that they were safe for human consumption. Thus most herbal preparations that were not explicitly sold as drugs came under regulation as food additives and therefore were required to meet the GRAS standard. While the GRAS standard provided a mechanism whereby manufacturers demonstrated the safety of their herbal products, the FDA found itself examining the health-related claims on the labels of such products and preventing their sale as unlawful, unapproved drugs. While such actions ensured (in the FDA's view) public protection from potentially unsafe dietary additives and unsupported drug claims, the public's perspective was of poor access to nutritional health supplements in a climate of burdensome regulation.

The response of Congress was to enact the Dietary Supplement Health and Education Act (DSHEA) of 1994,[6] which loosened the FDA's regulation of such supplements in several ways. First, DSHEA explicitly defined dietary supplements and ingredients as foods, preventing the FDA from treating them as drugs or GRAS additives. Second, DSHEA allowed manufacturers to assume the safety of their dietary ingredients, placing the burden to demonstrate lack of safety on the FDA. Third, dietary supplements are subject to a "reasonable expectation" of safety, in contrast to the "reasonable certainty" of safety for GRAS, a substantially less emphatic standard. Any dietary supplements in common use before October 15, 1994, are not considered new and are therefore exempt from any review process under DSHEA. Thus herbal supplements commonly used during the twentieth century can be marketed as dietary supplements without demonstrating safety, as long as they are labeled as dietary supplements, provide recommended dosage on the label, and do not make declarations of effectiveness to treat or prevent illness (or other drug claims).

An illustration of how DSHEA changed the regulatory environment for herbal products is in the marketing of the sweet-tasting compound stevioside from the South American sweetleaf (*Stevia rebaudiana*) plant (figure B.1).[7]

FIGURE B.1 Sweetleaf plant, source of a nonsugar sweetener.

The plant has a long history of use by indigenous South Americans, particularly the Guaraní of Brazil, Paraguay, and Argentina. Under the GRAS standard, the burden of proof for the safety of food additives is on the manufacturers. For stevioside, the FDA warned that so little toxicology data had been accrued in scientific studies that producers had failed to demonstrate its safety adequately. Thus the FDA considered stevioside a food additive incompliant with the required "reasonable certainty" for safety. As a dietary supplement, however, manufacturers were mandated to notify the FDA only of stevioside's impending release onto the market, and the FDA presented no objection based on the "reasonable expectation" of safety in the statute. Therefore, marketing of stevioside was banned under food-additive standards but permitted under dietary-supplement standards, where it currently enjoys commercial success as a diabetic-friendly alternative sweetener with about 200 times the sweetness of table sugar (sucrose).

FDA oversight of herbal products sold as dietary supplements allows three types of claims: health claims, structure-function claims, and nutrient-content claims. Health claims (such as "Scientific evidence suggests that X may reduce the risk of Y") must be vetted by the FDA and supported with scientific authority. They rarely appear on dietary-supplement labels.

Structure-function claims ("Promotes mental acuity") are allowable only if they make a general health statement and do not purport to treat a disease. Indeed, dietary-supplement labels must also clarify: "This statement has not been evaluated by the Food and Drug Administration. This product is not intended to diagnose, treat, cure, or prevent any disease." Manufacturers are not permitted to make a disease claim ("Protects against heart disease"), as this is seen to mislead the consumer into viewing a dietary supplement as a drug.

Nutrient content claims ("High in vitamin C") are acceptable for many supplements for which the FDA has established recommended daily intake values; for others, a relative claim ("20 percent omega-3 fatty acids") can appear.

Most herbal-supplement manufacturers assert structure-function claims on their products' labels, and while they do not explicitly purport to treat disease, some members of the medical community are increasingly concerned by the off-label use of dietary supplements against specific illnesses.[8] Yet some practitioners and the public feel empowered to be able to access plants and plant

FIGURE B.2 Dietary supplements for sale at a pharmacy in the United States.

extracts for health-related purposes, drawn by stories of traditional uses and a growing body of investigational science (figure B.2). In the current regulatory environment, the FDA certifies neither the *safety* nor the *effectiveness* of herbal remedies when sold as dietary supplements.[9]

1. Food, Drug, and Cosmetic Act of 1938, P.L. 75–717, 52 Stat. 1040, effective June 25, 1938.

2. Philip J. Hilts, *Protecting America's Health: The FDA, Business, and One Hundred Years of Regulation* (New York: Knopf, 2003), 93.

3. Hilts, *Protecting America's Health*, 160–161.

4. Hilts, *Protecting America's Health*, 229–230.

5. George Burdock, "Dietary Supplements and Lessons to Be Learned from GRAS," *Regulatory Toxicology and Pharmacology* 31 (2000): 69.

6. Dietary Supplement Health and Education Act of 1994, P.L. 103–417, 108 Stat. 4325, effective October 25, 1994.

7. Burdock, "Dietary Supplements," 68–76.

8. Institute of Medicine, *Complementary and Alternative Medicine in the United States* (Washington, D.C.: National Academies Press, 2005).

9. The FDA is responsible for monitoring adverse reactions to dietary supplements after they are released in the market and reviews health-related claims and labeling "as its resources permit" (Food and Drug Administration, *FDA 101: Dietary Supplements*, August 4, 2008, http://www.fda.gov/ForConsumers/ConsumerUpdates/ucm050803.htm). The dietary supplement industry largely self-regulates with regard to the safety of its products. The medical community is also expected to report adverse events related to dietary-supplement use to the FDA.

SCHEDULES OF CONTROLLED SUBSTANCES
(CONTROLLED SUBSTANCES ACT, 1970)

Schedule I

(A) The drug or other substance has a high potential for abuse.

(B) The drug or other substance has no currently accepted medical use in treatment in the United States.

(C) There is a lack of accepted safety for use of the drug or other substance under medical supervision.

Schedule II

(A) The drug or other substance has a high potential for abuse.

(B) The drug or other substance has a currently accepted medical use in treatment in the United States or a currently accepted medical use with severe restrictions.

(C) Abuse of the drug or other substances may lead to severe psychological or physical dependence.

Schedule III

(A) The drug or other substance has a potential for abuse less than the drugs or other substances in schedules I and II.

(B) The drug or other substance has a currently accepted medical use in treatment in the United States.

(C) Abuse of the drug or other substance may lead to moderate or low physical dependence or high psychological dependence.

Schedule IV

(A) The drug or other substance has a low potential for abuse relative to the drugs or other substances in schedule III.

(B) The drug or other substance has a currently accepted medical use in treatment in the United States.

(C) Abuse of the drug or other substance may lead to limited physical dependence or psychological dependence relative to the drugs or other substances in schedule III.

Schedule V

(A) The drug or other substance has a low potential for abuse relative to the drugs or other substances in schedule IV.

(B) The drug or other substance has a currently accepted medical use in treatment in the United States.

(C) Abuse of the drug or other substance may lead to limited physical dependence or psychological dependence relative to the drugs or other substances in schedule IV.

Source: 21 USC Sec. 812. The section of the Comprehensive Drug Abuse Prevention and Control Act of 1970 dealing with control and enforcement is known as the Controlled Substances Act.

curbing a range of practices across the spectrum of American life, from traditional spiritual rituals, to destructive narcotic abuse, to the consumption of home-remedy elixirs.

In 1909, the federal government banned the importation of opium for smoking under the Smoking Opium Exclusion Act, and the Harrison Narcotics Tax Act of 1914 regulated all forms of opium and cocaine.[22] A number of further restrictions, such as controls on marihuana, hallucinogens, and synthetic drugs, accumulated in a patchwork fashion during the next half-century until comprehensive reform in 1970.[23] In that year, Congress passed the Comprehensive Drug Abuse Prevention and Control

Act,[24] which, with amendments, presently governs American drug policy. Under one of its sections, the Controlled Substances Act, the federal government generates listings of chemicals and chemical analogs that have potential for abuse and establishes who can legally access them.

According to this system, schedule I drugs, such as heroin and marihuana, are considered under federal law to have no accepted medical use and are banned. Schedule II drugs, such as cocaine and morphine, have some restricted medical use and are allowed by a physician's prescription under strict regulations. Schedule III drugs and schedule IV drugs include those that are regularly prescribed by physicians. Finally, schedule V drugs have a relatively low abuse potential and are available without a prescription. More recent state laws, such as those in California, Colorado, Washington, and other jurisdictions allowing prescribed marihuana or eliminating some penalties for recreational use, directly challenge federal drug policy by preventing local prosecutions for marihuana possession under certain circumstances. (Growers, sellers, and users remain subject to federal prosecution.) The consequences of this mixed legal status are playing out in the courts.[25]

Recent government policies on drug control have generally maintained strict punishments for the production, sale, possession, and use of scheduled substances. During the Ronald Reagan presidency (1981–1989), legislation mandating minimum sentences for drug offenses took effect, requiring judges to adhere to stringent prison terms for those convicted of such crimes, regardless of individual circumstances. This framework increased average prison time for drug-related convicts nearly threefold. It also took particular aim at the use of cocaine, imposing a five- to forty-year mandatory sentence for possession of 500 grams of powder cocaine or five grams of crack cocaine. Since black, urban users were thought more likely to possess cocaine in its crack form and white users were more closely associated with the powder form, this hundred-fold disparity in the threshold for mandatory prison sentences was long criticized for harsh treatment of America's black cocaine users.[26] In 2007, the Supreme Court ruled the sentencing inequality unconstitutional.[27] This case demonstrates the subtle ways that racial notions of drug use can influence legislation.

In many parts of the world, diverse specialists have come to harvest and administer medicinal plants for spiritual, culinary, therapeutic, social, and sometimes sinister purposes. In the United States, an atmosphere of unregulated access to medicinal plants by the public gave way to one of increasing state involvement. The result of 100 years of American drug policy is an evolving framework of prohibitions, constructed first of pieced-together controls for various particular aims—some of them steeped in contemporary political culture—and later by more comprehensive policies. Yet even this unified legislative arsenal has been confronted by state and local laws and legal challenges that render this field ever changing. Interesting questions abound in the choice of drugs scheduled (for example, marihuana but not tobacco) and the effectiveness and fairness of domestic drug policies. At their core, the policies seek to protect individuals and society from the risk of drugs of abuse, drugs that can have profound effects on the human body.

Major laws regulating medicinal plants in the United States

1906 Pure Food and Drug Act
1909 Smoking Opium Exclusion Act
1914 Harrison Narcotics Tax Act
1937 Marihuana Tax Act
1938 Food, Drug, and Cosmetic Act
1970 Controlled Substances Act
1994 Dietary Supplement Health and Education Act

Notable plant-derived controlled substances

Schedule I	Marihuana, peyote, heroin
Schedule II	Opium, codeine, morphine, cocaine
Schedule III	Weak preparations of opium and opiates
Schedules IV and V	Mild opiates

Chapter 3

The Actions of Medicinal Plants

The characteristic stems of jointfir. Plants of the genus *Ephedra* produce the powerful stimulant drug ephedrine.

With the advent of scientific methodology during the twentieth century, it became possible to determine with greater certainty the causal relationships between medical treatment and perceived physiological effects. A new understanding of anatomy and biochemistry gave investigators the tools to assess how pharmaceuticals act in the human body. While traditional medical systems described the role of herbs in regulating the humors, directing *qi*, and so forth, the biomedical system seeks to establish how molecules and cells interact and influence health. The specific ways that medicinal plant constituents affect the body at the biochemical level—their mechanisms of action— are of great interest, as they form the basis of many modern-day therapies, suggest new avenues for pharmaceutical research, and elucidate how the body carries out its complex functions.

For many age-old herbal remedies, science has not yet established just how these treatments may work. While the scientific approach questions all claims of efficacy grounded solely in folk wisdom and philosophical dictates, it also provides a means to test whether therapeutic actions exist and to describe how they exert themselves physiologically. Applying the methods of science to traditional medicine can be challenging. To begin, the outcomes of therapy in traditional medicine and biomedicine can be difficult to align. For example, people of the indigenous North American Houma tribe in modern-day Louisiana employed a liquid made from the roots and bark of the coastal plain willow (*Salix caroliniana*) as a "blood medicine," taken "for 'feebleness' due to thin blood."[1] It is unlikely that the Houma and biomedical practitioners have the same idea of how "thin blood" can be assessed, so any attempts to subject the coastal plain willow to laboratory testing would probably redefine "thin blood" as some measure of cell count or blood chemistry. Testing traditional herbs in a biomedical setting usually requires the investigator to reframe the health-related outcomes from the conceptual milieu of their historical, indigenous uses and expectations into objective laboratory assays involving cell counts, chemical analyses, heart rate, blood pressure, and the like. The process of redefining traditional and historical medical ideas in biomedical terms is fraught with potential biases and mistranslations, especially if researchers lack sufficient understanding of the cultural setting from which the idea comes.

Biomedical researchers working with traditional medicinal plants must also define the source, dose, and route of administration of the herbs they wish to test, and they must apply the material uniformly across many test subjects while delivering inactive treatments (placebo) to another set of subjects. This sort of experimental approach lends itself to strong inferences and is widely regarded as crucial to determining therapeutic efficacy. However, it cannot account for the way medicines are prepared in many traditional settings, where doses are formulated according to an individual patient's symptoms and modified through the course of treatment. Furthermore, when researchers in different locations and at different times subject medicinal plants to testing, they frequently choose dosing schemes and delivery methods that do not match those of fellow workers, which results in a diverse set of experimental outcomes that are difficult to compare. Nevertheless, more than a century of laboratory research and decades of clinical study have demonstrated that many herbs long employed in medicine are efficacious by the standards of experimental science. For those herbs whose medicinal properties are purported, speculated, or held on faith but not demonstrated scientifically, future work will certainly attempt to resolve methodological differences between studies and ultimately establish whether they are effective in ways that can be measured and, if so, by what mechanism.

At the same time, it is possible that some plants valued in traditional medicine might never demonstrate therapeutic effects in well-designed experiments (figure 3.1). Undoubtedly, a great number of herbal remedies must derive a large part of their ascribed activities from the strength of the patient's belief in the medicine.[2]

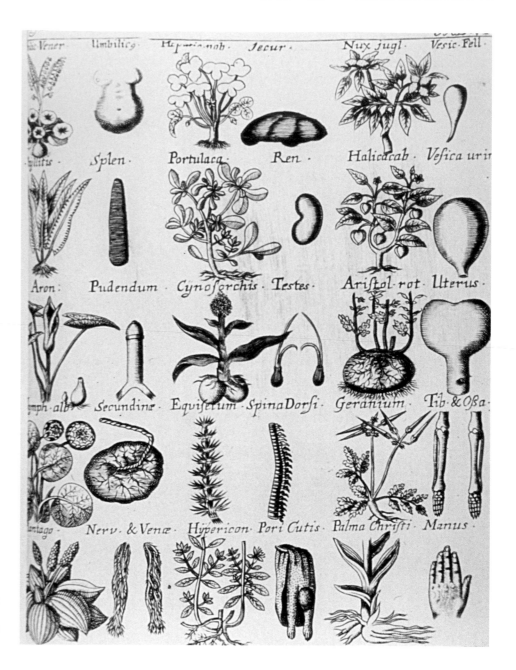

FIGURE 3.1 Some schools of traditional European medicine applied the Doctrine of Signatures to relate particular plants to the body parts for which they were thought effective. (Illustration from Michael Bernhard Valentini, *Medicina Nov-Antiqua* [1713]; National Library of Medicine, A030218)

ACTIONS OF MEDICINAL PLANTS AND THEIR DERIVATIVES

Once a medicinal plant's physiological effects are demonstrated, researchers attempt to identify the chemical component responsible for causing them; this component is termed the active principle. Ultimately, the goal of such work is to determine at the biochemical level how the active principle exerts its effects on the system.

This approach to isolating pharmacological agents and establishing their mechanism of action is possible because of the chemical basis of life. All living things—from bacteria, to plants, to people—are composed of tiny atoms bonded together in particular ways to form molecules.

Active principle

A chemical responsible for the biological activity of a medicinal preparation

While atoms of the element carbon are the most abundant components of biological molecules, atoms of oxygen, nitrogen, hydrogen, sulfur, and other elements also contribute, forming compounds—that is, molecules incorporating more than one type of atom (figure 3.2). Compounds can vary tremendously in size and complexity, and the ways that the atoms form together give them their unique properties in living systems (figure 3.3). Importantly, when atoms of carbon and those of other elements link together in compounds, they take on particular shapes that can interact with other molecules in a biological setting.

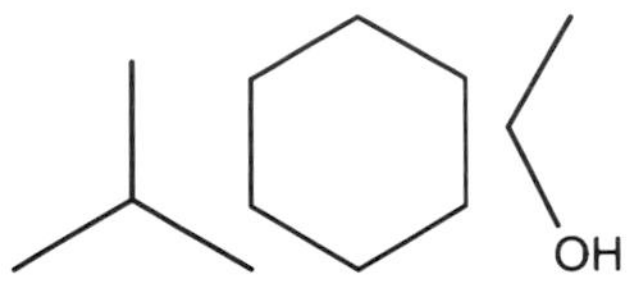

Life depends on the specific interactions of molecules. For example, very large molecules called enzymes have three-dimensional shapes that match closely the shapes of other molecules on which they act. Because of their structural affinities, enzymes can bind to their target molecules and do their work, which often involves the transformation of the target molecule in some way; for instance, its shape may be changed, or it may have atoms added or removed. Enzymes in human tissues process molecules taken up in food into other molecules of a different form, molecules capable of delivering cellular energy, for instance. Enzymes can build up smaller molecules into larger and more complex forms or break down large molecules into simple ones. The activity of the body's enzymes is called metabolism.

Another example of a specific molecular interaction takes place in biological signaling. Microscopic chemical messages travel between cells and tissues in living systems, and these signals can influence development, growth, and behavior. For example, hormones are complex compounds that can circulate in the bloodstream and are perceived in cells having the appropriate receptors, which are relatively large molecules that reside either inside a cell or embedded in its outer membrane. Because hormones have a specific structure (spatial arrangement of atoms), they match up with the three-dimensional shape of their corresponding receptors. Some biochemists think of this relationship as analogous to pieces of a jigsaw puzzle that line up together precisely, or the exact fit of a key in a lock. The specific interaction (binding) of hormone to receptor can cause an array of effects inside the cell, including a change in expression of genes, movement, physiological changes, and so on.

Plants produce a large set of chemical compounds that support their own metabolism, growth, and development, compounds that enable them to convert sunlight into biological energy, reproduce, and serve their important ecological roles. Some plant molecules can also affect human health, and it is the work of pharmacologists to discover their mechanisms of action. Through their active principles, medicinal plants can exert a diverse array of effects on the body and psyche. Some plant-derived chemicals work by directly altering human metabolism or by mimicking hormones. Others specifically prevent the growth of harmful bacteria, fungi, or viruses in human tissues. Still others act on the nerve cells with a whole host of downstream effects. A theme common to the active principles of medicinal plants is that many of them function by binding to human receptors, initiating or blocking signals in the body, in a sense "tricking" the human body's physiology to respond by circumventing or overpowering the normal, human-derived

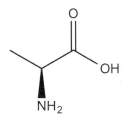

FIGURE 3.2 A structural formula (chemical diagram) conveys the spatial arrangement of atoms in a molecule. Atoms of carbon are represented as the points at the end of lines. The lines, usually either single or double, stand for chemical bonds. Other atoms are indicated by abbreviations: oxygen (O), hydrogen (H), and nitrogen (N): (*left*) molecule with four carbon atoms; (*center*) molecule with six carbon atoms in a ring; (*right*) compound with two carbon atoms and an alcohol (-OH) portion: ethanol (or fruit/grain alcohol).

FIGURE 3.3 Molecular bonds occupy three-dimensional space, symbolized by triangular lines joining atoms. The two molecules pictured here are distinct because their -NH$_2$ groups are attached at different angles.

CATEGORIES OF PLANT MEDICINAL COMPOUNDS

Plants produce an enormous diversity of chemical products. Some of them serve central roles in plant growth and development, such as enabling the photosynthetic processes that capture the sunlight's energy to fuel metabolism and building the microscopic structures that sense the environment around them, to allow their roots to penetrate deeply into the soil below and their shoots to reach upward toward the light. Some molecules help support and protect plants, such as those that form the strong fibers of their cell walls and the waxes that coat their leaves. Others act as signals to communicate with nearby plants, bacteria, and fungi. In an age-old back-and-forth with the animal kingdom, evolution has also shaped the form of countless plant molecules that affect the fauna around them. Some compounds serve as attractants, such as the fragrance and color products of flowers that appeal to insect pollinators, or as repellents, such as the bitter-tasting chemicals that herbivores abhor. Some are poisons, honed to inconvenience an offending creature by paralyzing it, stopping its heart, disorienting it, or causing some other misfortune. Collectively, a single plant can produce many thousands of different chemicals, a profile that differs among roots, leaves, flowers, fruits, seeds, and so forth, and that depends on age, growing condition, and other factors. Some of these compounds may affect human health, and botanists, chemists, and physicians are working to understand their functions both in the plant and in the human body. Medicinal plant products are diverse in composition, size, and biological roles and can be classified into three groups.[1]

Alkaloids

About 12,000 alkaloids have been inventoried in plants, many of which are of medical importance (figure B.1). They usually contain one or more atoms of nitrogen (N) as part of a ring structure. Alkaloids can accumulate in various parts of the plant, and some are synthesized in response to injury.

Phenolics

Most of the approximately 8000 phenolic compounds derive from biosynthetic pathways using phenol as a base unit (figure B.2). This class of compounds contains among the largest molecules in plants, including products with many repeating units of simpler structures. (When there are many phenol units in a molecule, it is called polyphenolic.) The phenolics include a diverse group of molecules with the flavonoid core structure, among them anthocyanins, proanthocyanidins, and isoflavones. Other types of phenolic compound are the coumarins and stilbenes. Many phenolic and polyphenolic molecules oxidize (often observed as turning brown) when exposed to air. Also, many are tannins—chemicals with the property of binding to protein.

Terpenoids

There are over 25,000 terpenoids in plants, comprising a heterogeneous group of molecules that share a basic building block called isopentane (figure B.3). By joining together isopentane units in various configurations, adding or deleting chemical adornments, plants can generate these compounds, many of which may be found in the essential oil.

Other Types of Compounds

In addition to the alkaloids, phenolics, and terpenoids, medically active plant compounds can take many other forms. For example, there are numerous carbohydrates with health-related properties, including complex molecules forming gels, starches, and fibers, as well as the sugars in plants. There are also protein-based plant products that have effects on physiology, and many of the building blocks of proteins, amino acids, also play roles. Certainly, the numerous vitamins that plants provide are essential to human health and are of great interest to those studying herbal medicine and nutrition.

1. Rodney Croteau, Toni M. Kutchan, and Norman G. Lewis, "Natural Products (Secondary Metabolites)," in *Biochemistry and Molecular Biology of Plants*, ed. Bob B. Buchanan, Wilhelm Gruissem, and Russell L. Jones (Rockville, Md.: American Society of Plant Biologists, 2000), 1250–1318.

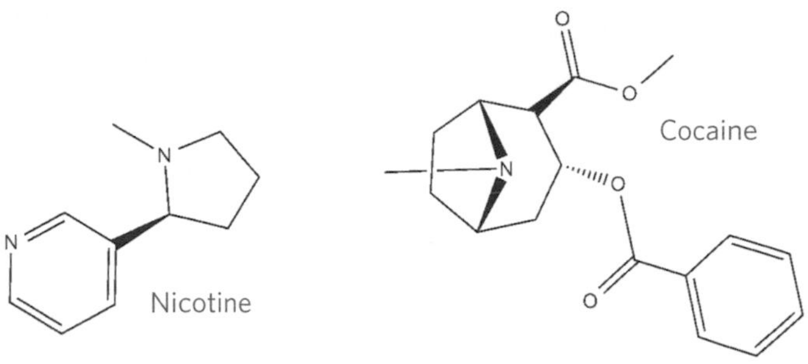

FIGURE B.1 Alkaloids contain atoms of nitrogen, often in a ring structure. Examples include nicotine, cocaine, and caffeine.

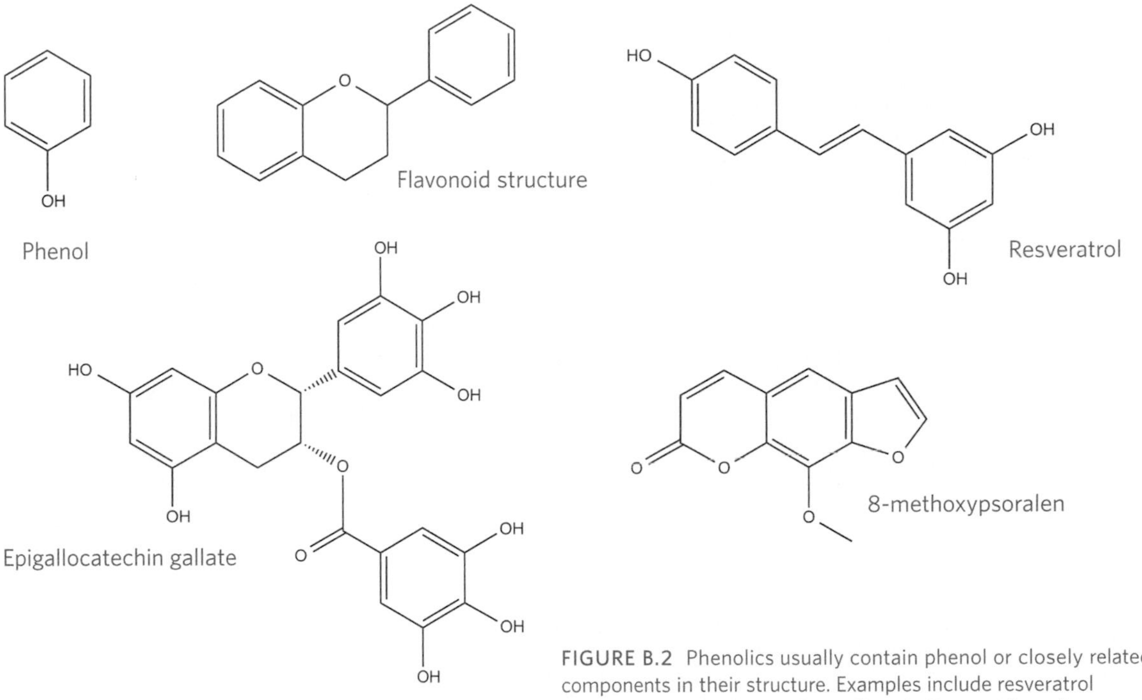

FIGURE B.2 Phenolics usually contain phenol or closely related components in their structure. Examples include resveratrol (a stilbene), epigallocatechin gallate (a flavonoid), and 8-methoxypsoralen (a coumarin).

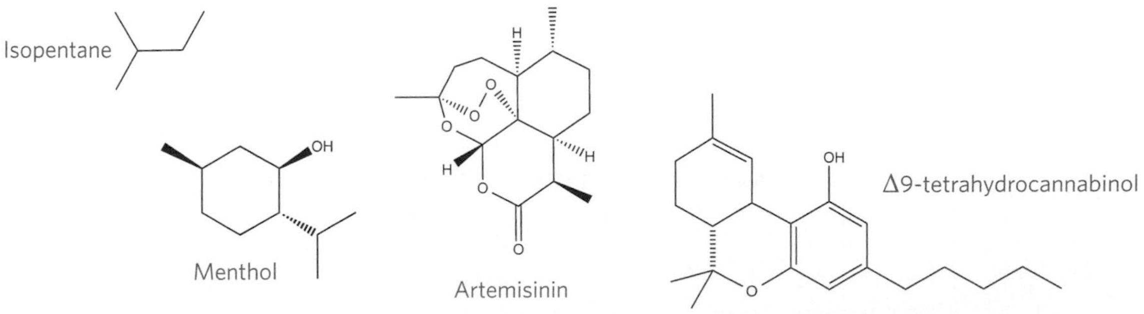

FIGURE B.3 Terpenoids are produced from units of isopentane. Examples include menthol (a volatile oil), artemisinin, and Δ9-tetrahydrocannabinol, whose "tail" classifies it as a terpenoid.

PLANT CHEMISTRY IN THREE DIMENSIONS

The particular bonding arrangement of the atoms in a molecule determines how the chemical occupies three-dimensional space and, ultimately, how it interacts with structures that it encounters. The bonding of carbon atoms, nitrogen atoms, hydrogen atoms, and so forth is usually represented as schematic drawings including lines and letters. Much can be communicated in such structural formulas, and chemists are generally accustomed to this form of depiction. While such two-dimensional schematics have proved to be a useful shorthand, many of those working in the field of medicinal plant research also seek to visualize the important three-dimensional aspect of a molecule's form (figures B.4 and B.5).

FIGURE B.4 Chemists have devised several methods to represent the arrangement of atoms in molecules. Taking the molecule caffeine as an example, its structure can be depicted in letters and lines (A), as colored balls joined by sticks (B), or in a computer-generated model of how its atoms occupy three-dimensional space (C).

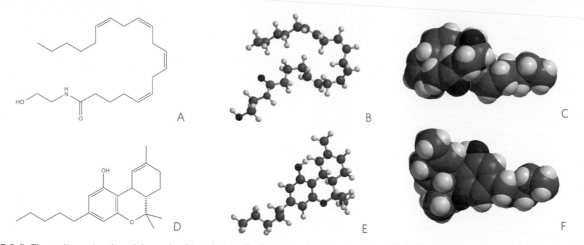

FIGURE B.5 Three-dimensional models can lend insight into the function of active principles. While the human molecule anandamide (A, B, and C) and the cannabis-derived active principle Δ9-tetrahydrocannabinol (D, E, and F) are thought to act similarly in the brain, very little resemblance is apparent in their structures until they are modeled in three dimensions (C and F). The similar positions of the oxygen atoms (in red) and the carbon–hydrogen tails (protruding to the right) imply a common mechanism of action in the body.

Using computational methods, chemists can produce models of how physiologically active molecules likely occupy space in their microscopic cellular environments. Such informed predictions have given scientists special insight into the mechanism of action of countless plant-derived chemicals, as the shape of a molecule can suggest how it might interact with specific contours of the human biological machinery. Many physiologically active molecules exert their effects by binding to sensitive parts of human proteins. As a result of these interactions, some botanical chemicals can block enzymes or mimic the fit of a natural signaling molecule into a particular surface of a receptor, initiating a suite of responses.

In some cases, a three-dimensional appreciation of a chemical's structure can be key to understanding its biological function. For example, scientists studying the effects of marihuana (*Cannabis sativa*) attributed many of its mind-altering properties to the interaction of the compound Δ9-tetrahydrocannabinol (THC) with a human receptor in the brain that had evolved to detect native brain-signaling molecules such as anandamide. Yet THC and anandamide bear little resemblance to each other in two-dimensional structural formulas. When visualized in three dimensions, however, it becomes clear that the molecules occupy space similarly, each protruding analogous bulges. This observation implies that the molecules may act very much alike, binding to the same cellular receptors by virtue of a shared portion of their three-dimensional structures.

signal molecules. Still, many herbs appear to exert effects through their major chemical constituents, although by means thus far unknown. For those plants whose active principles have been identified but whose mechanisms of action have not yet been established, discoveries await. The following sections will explore some of the body's systems in which plants and their constituents have been found useful.

DIGESTIVE SYSTEM

The digestive system consists of the body's apparatus to bite, chew, digest, absorb, and eliminate material ingested for sustenance. It is essentially a long muscular tube starting at the mouth, where food is taken in, and ending at the anus, through which waste passes. Various structures along this path perform critical roles in extracting nutrition from the diet. After swallowing, material enters the stomach, a highly acidic compartment, where enzymes and muscular activity substantially break down complex foods. The bolus of semidegraded organic materials enters the small intestine, where continued enzymatic activity further breaks down biomolecules, and nutrients pass into the bloodstream via the semipermeable intestine wall. The large intestine absorbs much of the remaining water from the contents, resulting in a thick paste of undigestible waste products that accumulate at the terminal end of the digestive canal: the colon and rectum. From there, the feces pass as stool.[3] In cases of exposure to toxins or illness, muscular contractions of the stomach can lead to vomiting (emesis). Other disorders of digestion include overly loose or watery stool (diarrhea) and overly hard or difficult-to-pass stool (constipation).

In many parts of the world, people developed treatments that helped them expel the contents of their stomach and bowels, whether to draw up perceived unhealthy matter in vomit, rid themselves of their waste when constipated, or eliminate unwholesome internal substances as part of a medical ritual. For example, a well-documented aim of traditional medicine of the European pedigree is to regulate the balance of bodily humors by periodically vomiting and cleansing the bowels.[4] Apothecaries produced herbal concoctions to ensure prompt emesis. In some cases, doctors recommended enemas to remove perceived toxins or overabundances from the body. Also in widespread practice was the use of laxative agents, which gently cause the feces to pass, and purgative agents, which do so violently. Within the framework of the humoral system, physicians treated a large number of ailments in this manner, the release of stool serving to allow the evils within to escape.

The ancient Egyptians were probably the first to document an arsenal of laxatives in the Ebers papyrus of around 1550 B.C.E. Among the most important is one still in use for this purpose: they chewed the oil from castor (*Ricinus communis*) seeds and consumed it with beer to procure a bowel movement.[5] Ricinoleic acid, a component of castor oil, acts in the large intestine to reduce water resorption and triggers rhythmic muscle contractions (peristalsis) by binding to receptors on the intestinal lining (figure 3.4).[6] Among the ancient Greeks and Romans, aloe served as a laxative, one of a large number of its medicinal uses. Aloe's role in bowel motility is ascribed to the compound aloin, which accumulates in the plant's yellow latex (but not in the leaf's juice or gel). When eaten and passed through the digestive tract, aloin is activated by bacteria in the large intestine to form the chemical aloe-emodin. Aloe-emodin reduces the uptake of water by the colon and stimulates

FIGURE 3.4 Castor oil has been used as a laxative for thousands of years. This humorous, early-twentieth-century postcard plays on its well-known physiological effects. (National Library of Medicine, A28757)

muscle contraction. Aloe latex was used for this purpose until modern times, although since 2002, the Food and Drug Administration has banned aloe as an ingredient in over-the-counter laxatives, safety data lacking.[7]

Whether for ritual purgation, as in traditional medicine, or to treat constipation, as frequently practiced in the modern day, plants produce a variety of compounds that strongly stimulate bowel motility. Europeans first learned of the purgative use of Alexandrian senna (*S. alexandrina*, also known as tinnevelly senna) from the Arabs during the ninth or tenth century.[8] Plants of the genus *Senna* (and related *Cassia*) are widely distributed, and indigenous medical systems around the world have recognized the potent laxative nature of their leaves, stems, seeds, and roots. The young leaves of Alexandrian senna from North Africa are a primary source of the purified senna drug, sold commercially as Ex-Lax and Sennekot. Senna's active compounds are sennoside A and B, which are acted on by bacteria in the colon to produce stimulatory agents.[9]

Buckthorns, such as the glossy buckthorn (*Rhamnus frangula*) of Europe, North Africa, and northern Asia and the North American cascara buckthorn (*R. purshiana*), also produce potent purgatives documented in traditional herbal medical practice.[10] The American buckthorn bark was apparently more effective than that of the Old World species, and it was commercialized during the nineteenth century by American pharmaceutical firms and sold worldwide.[11] Buckthorn stands in the northwestern United States and in Canada became threatened by overharvesting, and plantations were established beginning in the 1920s to maintain a supply of the drug.[12] Much like other plant-based laxatives and purgatives, buckthorn's frangulin A and B molecules are converted to the active emodin compound by intestinal bacteria. Along with aloe, the FDA has banned cascara buckthorn from over-the-counter laxative use.[13] Aloe, buckthorn, senna, and other traditional laxatives such as Chinese rhubarb (*Rheum* spp.) act in the colon through similar mechanisms, as their active principles bear a close structural resemblance.[14]

Diarrhea can result from a variety of causative agents, including gastrointestinal infection, food poisoning, structural abnormalities, or metabolic defects. In cases of diarrhea caused by microbial infection (dysentery), herbal compounds that break up microorganism colonies may help alleviate symptoms and shorten the duration of the disease. In modern medicine, it is common to use natural microbial-derived or synthetic antibiotics to treat dysentery. In traditional health systems, people have used the astringent plant resins known medically as kino, often prepared from the stem exudates of trees.[15] Kino contains complex plant chemicals called tannins that bind to biological molecules such as proteins and act as general antimicrobials and drying agents.

> **Tannins**
>
> Plant-derived polyphenolic compounds that can bind to proteins and coat cell surfaces. They tend to be bitter and astringent.

(Tannins derive their name from their long use in the tanning process, which preserves and toughens animal skin to produce leather.) The bark of white oak (*Quercus alba*), among other oaks, contains about 10 percent tannin. Indigenous Americans used oak bark extensively against diarrhea. As European settlers moved across the continent, they too employed oak bark to alleviate dysentery.[16]

In addition to the tannins, which are widely distributed among both woody and herbaceous plants, the peoples of Europe, Asia, and northern Africa historically used extracts of the poppy (*Papaver somniferum*) to treat diarrhea. The poppy compounds morphine and codeine slow the rate of intestinal peristalsis by binding to receptors on the nerves controlling the muscles that line the intestine. This allows more time for the intestine to absorb liquid from the stool. Because of the risk of addiction, it is now considered preferable to use the synthetic compound loperamide (sold as Imodium, and others), which acts similarly to the poppy's active principles but without effects on the central nervous system at usual doses.[17]

In modern-day Western medicine, the induction of vomiting is generally restricted to emergency poison control or in cases of aversion therapy. In traditional medicine, however, emesis is sometimes utilized in medicinal cleansing rituals and to regulate the internal humors and other substances related to health. A great number of plants serve as emetics, and most are also highly poisonous. Therefore, the people who prepared them for therapeutic and spiritual uses must have developed great expertise in selecting the correct dose of such medicines. One of the best known emetic plants is the Brazilian native ipecacuanha (*Carapichea ipecacuanha*), whose roots can be boiled into a syrup known as ipecac (figure 3.5).[18] The active principles emetine and cephaeline both irritate the lining of the stomach and stimulate the chemoreceptor trigger zone of the brain, which leads to vomiting. It was also discovered that emetine is toxic to the amoebic microorganisms responsible for some forms of dysentery. Along with these properties, ipecac can have unpleasant effects on a person's nervous system and can alter the heartbeat, and for that reason its use has been discouraged since the 1990s.[19]

FIGURE 3.5 An eighteenth-century apothecary jar that once contained ipecacuanha root, a potent emetic. (Smithsonian Institution, National Museum of American History, 1991.0664)

CARDIOVASCULAR SYSTEM

One of the primary functions of the circulatory system is to maintain the flow of blood throughout the body in order to collect oxygen at the lungs and deliver it into the tissues. The blood circles in a closed loop, returning to the lungs to give up the waste product carbon dioxide formed by the body's cells and release it via the lungs to the atmosphere. The blood also serves to distribute the nutrients gathered from digestion and as a conduit for the defensive immune system's cells to reach sites of foreign invasion wherever they occur. Propelling the constant movement of blood is the muscular heart, composed of chambers separated by valves that prevent fluid from moving against the direction of flow. The blood is nearly always contained in vessels: the arteries, which transport blood away from the heart under pressure; ultrathin capillaries, which permeate the tissues and allow rapid nutrient exchange; and the veins, which allow blood to return to the heart.[20] The cellular components of blood—the oxygen- and carbon dioxide–carrying red blood cells, the defensive white blood cells, and others—usually stay within the vessels. The liquid fraction of blood, plasma, can leak out of the thinner vessels as lymph and thus bathes the tissues in fluid. When tissue is injured, whole blood can escape from its vessels. It usually forms a clot, a dense network of blood cells and fibers that prevent further blood loss and allow the tissue slowly to heal.[21]

While the heart and blood have long been considered important in the world's traditional and folk medicines, various historical and indigenous interpretations of the role of the heart, blood, and circulation are generally different from the biomedical understanding. In Chinese traditional medicine, for example, the Heart is considered to be *yin* in nature, the center of thought and emotion. (Heart is capitalized to distinguish the traditional Chinese term from the Western anatomical structure.) In classical European medicine, blood is one of the four critical humors. According to beliefs that held sway for centuries in Europe and beyond, illness could be remedied by bleeding a patient and allowing the body to regain its appropriate balance of humoral qualities. Despite an awareness of the heart's structure and the existence of blood, Western physicians did not link anatomy with function until the seventeenth century, when the role of the heart in the circulation of blood through vessels was recognized.

PLANT CHEMISTRY IN THE ECOSYSTEM

While humans have been keen to exploit plants for their medicinal properties, many of the active chemicals that exert such potent effects in people evolved long before the advent of civilization. Fated to a stationary life under the sun, surrounded by all variety of creatures, plants at an early stage in their history developed the capacity to produce chemical compounds that contributed to survival in their diverse environments. Some of the chemicals now valued in medicine probably evolved to shield plants exposed to potentially harsh surroundings. Among certain plant lineages, chemicals came about that profoundly affected the behavior of creatures of all types, to the benefit of the plants. As active participants in complex ecosystems, plants employed their chemical capabilities in a way that helped them persist and thrive.[1]

Although plants need the sun to propel the life-sustaining process of photosynthesis, which converts carbon dioxide in the air into useful, energy-rich sugar molecules, the sun's intense rays can actually damage plant tissues under some conditions. Many plants produce protective compounds that help guard their cells against such injury. For example, certain phenolics, such as the flavonoid kaempferol, are thought to protect against the ravages of ultraviolet radiation.[2] As the energy transformations inherent to plant life can produce potentially dangerous chemicals called free radicals that tend to destroy delicate cellular structures, it is fitting that plant tissues generate a wide range of antioxidant compounds with the property of quenching free radicals.[3] It is possible that the antioxidant properties of plant-derived chemicals might have similar protective effects in humans, which is why numerous studies have investigated their potential benefits against diseases in which free radicals are implicated.[4]

Plants also produce chemicals that can alter the behavior of, or even physically harm, animals. In general, plants are thought to accumulate toxic or repellent chemicals as a form of defense against herbivory.[5] Certain alkaloid and polyphenolic compounds, for example, are thought to taste bitter to insects and mammals and to deter the creatures from making a meal of the plants that synthesize them.[6] Meanwhile, some of the strongly scented and flavored terpenoid compounds, such as camphor, discourage herbivory by many animals.[7] Beyond deterrence, a great number of plant-produced chemicals are toxic to insects or mammals, some especially affecting their nervous systems.[8] For example, the pyrethrin terpenoid compounds produced by the Dalmatian chrysanthemum (*Tanacetum cinerariifolium*) are neurotoxic to insects but not to mammals or birds, an observation that has led to the development of naturally derived pesticides and synthetic analogs that are safe to use on food and around many noninsect animals.[9] Strikingly, many alkaloid compounds exert especially potent effects on animals that ingest them. For instance, nicotine (in tobacco, *Nicotiana* spp.) can paralyze insects, probably by interfering with nervous system signaling.[10] Cocaine (in coca, *Erythroxylum* spp.) and morphine (in poppy, *Papaver* spp.) are likewise thought to be toxic to the nervous systems of hungry pests. It is therefore not surprising that the synthesis of many defensive compounds is induced by damage to the plant tissue, such as might occur during insect or mammal feeding.[11] Furthermore, plant compounds such as digitalis (in foxglove) can interfere with circulatory function in grazing animals, and hormone-mimicking chemicals can interrupt normal insect development.[12]

Differences of mechanical understanding aside, traditional medicine identified a number of plants to treat a range of circulatory ailments. One such illness was a widespread condition known as hydrops or dropsy. Sufferers of dropsy accumulate fluid (a condition called edema) first in their extremities and finally throughout their bodies, puffing them like balloons and terminating in death (figure 3.6). Likely unaware that the condition was caused by a heart too weak to propel the return circulation of blood plasma (congestive heart failure), surgeons drained the swollen tissues by knife, which treated the symptoms but not the source of the illness.[22] Folk medicine, though, had an herbal treatment,

Since plants that accumulate more defensive poisons in their tissues tend to be better protected against the attacks of herbivores than plants with a lower concentration of these compounds, evolutionary forces would select plant lineages with increasing levels and potency of such chemicals.[13] At the same time, animals whose diet includes these types of plants develop the physiological capacity to neutralize or sequester the potentially dangerous products, a phenomenon that can lead to further escalation of toxin levels in plants. This biochemical tit-for-tat between plant and herbivore species is a type of evolutionary arms race.[14]

In addition to producing toxic compounds that deter herbivory, plants can synthesize chemicals that attract beneficial animals. Numerous scent compounds and floral color patterns are associated with cues for pollination, and nectar acts as a reward for the service of the animals that carry pollen from one flower to another, assisting in perpetuating the plant species. Interestingly, some members of the *Citrus* and *Coffea* genera secrete a low level of the alkaloid caffeine in their nectar, which, rather than being toxic to insects, serves a useful role in pollination.[15] The trace of caffeine seems to help insects remember where they obtained the nectar and encourages them to revisit the plant multiple times and spread its pollen more widely.

Among the diverse array of plant chemicals that protect against environmental threats and advance the species' survival in the ecosystem are compounds that humans have found also to serve as medicines. By learning of their properties, manipulating the dose, and recording health-related outcomes, practitioners and researchers have transformed such naturally occurring chemicals into pharmaceuticals, a fascinating and unprecedented event in an ancient evolutionary story.

1. The evolution and ecological roles of numerous plant compounds are explored in David O. Kennedy, *Plants and the Human Brain* (Oxford: Oxford University Press, 2014).

2. Kaempferol protects against ultraviolet-B radiation. See Rodney Croteau, Toni M. Kutchan, and Norman G. Lewis, "Natural Products (Secondary Metabolites)," in *Biochemistry and Molecular Biology of Plants*, ed. Bob B. Buchanan, Wilhelm Gruissem, and Russell L. Jones (Rockville, Md.: American Society of Plant Biologists, 2000), 1303–1334.

3. Elizabeth A. Bray, Julia Bailey-Serres, and Elizabeth Weretilnik, "Responses to Abiotic Stresses," in *Biochemistry and Molecular Biology of Plants*, ed. Buchanan, Gruissem, and Jones, 1189–1191. Plant antioxidant compounds include chemicals such as ascorbate (vitamin C), tocopherol (vitamin E), and certain terpenoids and polyphenols.

4. A sampling of such studies is reviewed in, among others, Daniele Del Rio et al., "Dietary (Poly)phenolics in Human Health: Structures, Bioavailability, and Evidence of Protective Effects Against Chronic Diseases," *Antioxidants and Redox Signaling* 18 (2013): 1818–1892; and Christine M. Kaefer and John A. Milner, "The Role of Herbs and Spices in Cancer Prevention," *Journal of Nutritional Biochemistry* 19 (2008): 347–361.

5. Peter Scott, *Physiology and Behaviour of Plants* (Chichester: Wiley, 2008), 243–251.

6. Croteau, Kutchan, and Lewis, "Natural Products," 1274, 1303.

7. Jonathan Gershenzon and Rodney Croteau, "Terpenoids," in *Herbivores: Their Interactions with Secondary Plant Metabolites*, 2nd ed., ed. Gerald A. Rosenthal and May R. Berenbaum (San Diego: Academic Press, 1991), 1:165–219.

8. David O. Kennedy and Emma L. Wightman, "Herbal Extracts and Phytochemicals: Plant Secondary Metabolites and the Enhancement of Human Brain Function," *Advances in Nutrition* 2 (2011): 32–50.

9. Gershenzon and Croteau, "Terpenoids"; Walter Lewis and Memory P. F. Elvin-Lewis, *Medical Botany: Plants Affecting Human Health* (Hoboken, N.J.: Wiley, 2003), 597.

10. Thomas Hartmann, "Alkaloids," in *Herbivores*, ed. Rosenthal and Berenbaum, 1:112–113; Anke Steppuhn et al., "Nicotine's Defensive Function in Nature," *PLoS Biology* 2 (2004): e217.

11. Croteau, Kutchan, and Lewis, "Natural Products," 1272–1274.

12. Stephen B. Malcolm, "Cardenolide-Mediated Interactions Between Plants and Herbivores"; and M. Deane Bowers, "Iridioid Glycosides," both in *Herbivores*, ed. Rosenthal and Berenbaum, 1:262–263, 275–278; 312–314.

13. Paul Feeny, "The Evolution of Chemical Ecology: Contributions from the Study of Herbivorous Insects," in *Herbivores: Their Interactions with Secondary Plant Metabolites*, 2nd ed., ed. Gerald A. Rosenthal and May R. Berenbaum (San Diego: Academic Press, 1992), 2:1–44.

14. Judith X. Becerra, Koji Noge, and D. Lawrence Venable, "Macroevolutionary Chemical Escalation in an Ancient Plant-Herbivore Arms Race," *Proceedings of the National Academy of Sciences USA* 106 (2009): 18062–18066. For a nuanced review of plant-herbivore interactions in evolution, see Douglas J. Futuyma and Anurag A. Agrawal, "Macroevolution and the Biological Diversity of Plants and Herbivores," *Proceedings of the National Academy of Sciences USA* 106 (2009): 18054–18061.

15. G. A. Wright et al., "Caffeine in Floral Nectar Enhances a Pollinator's Memory of Reward," *Science* 339 (2013): 1202–1204.

in preparations of the European native purple foxglove (*Digitalis purpurea*) plant (figure 3.7). It is not known who first harvested foxglove for this purpose or where in Europe it occurred, but it was likely employed as a remedy for circulatory problems for many hundreds of years before being recorded in medical texts. The herbalist John Gerard's treatise of 1597 recommends foxglove "boiled in water or wine, and drunken" to "cut and consume the thicke toughnesse of grosse and slimie flegme and naughtie humours."[23] It is hard to tell whether this description is a reference to the symptoms of dropsy or of other conditions producing watery mucus. In any case, an application to dropsy was

FIGURE 3.6 A woman suffering from dropsy, with swollen belly and limbs. (Lithograph from Jean-Louis-Marie Alibert, *Nosologie naturelle* [1817]; National Library of Medicine, A012332)

creases the amount of fluid removed via the kidneys (that is, it acts as a diuretic), reduces edema, and helps the heart overcome structural weaknesses.

The active principles of foxglove are a group of related molecules called cardiac glycosides, which block the channels regulating the electrochemical state of heart muscle cells.[26] One effect of this activity is the generation of increased pressure in the heart's pumping ability. The principal cardiac glycosides are digoxin (figure 3.8) from the foxglove *D. lanata*, and the *D. purpurea/D. lanata* compounds digitoxin, lanatoside C, acetyldigitoxin, and deslanoside. Although digoxin remains widely prescribed in the United States and both digoxin and digitoxin are used abroad, in recent years new agents have been developed to treat congestive heart failure.[27] Administration of digitalis must be undertaken carefully, since toxicity occurs just beyond the effective dose, resulting in irregular heartbeat.[28]

While the pumping action of the heart and muscular walls of the blood vessels ensure that the blood courses through its circulatory system under pressure, an excess of blood pressure (hypertension) or an insufficiency (hypotension) is considered a disease state. Extracts of the Indian snakeroot (*Rauvolfia serpentina*, also called serpentine wood and snakewood) and African poison devil's pepper (*R. vomitoria*) contain the active principle reserpine, an important hypertension reducer employed during the twentieth century (see figures 3.8 and 3.9).[29] Reserpine interferes with the transmission of nerve signals from the brain to the muscles lining the blood vessels, allowing them to relax and dilate.[30] This action, combined with its depressive action in the vasomotor center of the brain, reduces blood pressure, which is accompanied by pupil constriction and a lowering of body temperature.[31] Other side effects include depression, difficulty in concentration, and other psychological changes. Its use has declined since the late twentieth century because of its unpleasant side effects and with the development of more effective synthetic drugs.[32]

not universally recognized: like most texts of the era, another English herbal of 1666 recommends foxglove to cleanse wounds, heal sores, and as a purgative.[24]

The English physician William Withering was the first to test doses of foxglove systematically on patients with dropsy, the results of which he published in 1785.[25] Through this important work, Withering described the first cardiotonic agent, a drug that specifically strengthens the pumping action of the heart muscle, making it useful against congestive heart failure. By increasing blood flow, foxglove in-

FIGURE 3.7 Foxglove. (Woodcut from Rembert Dodoens, *Histoire des plantes* [1557]; Wellcome Library, London, L0021171)

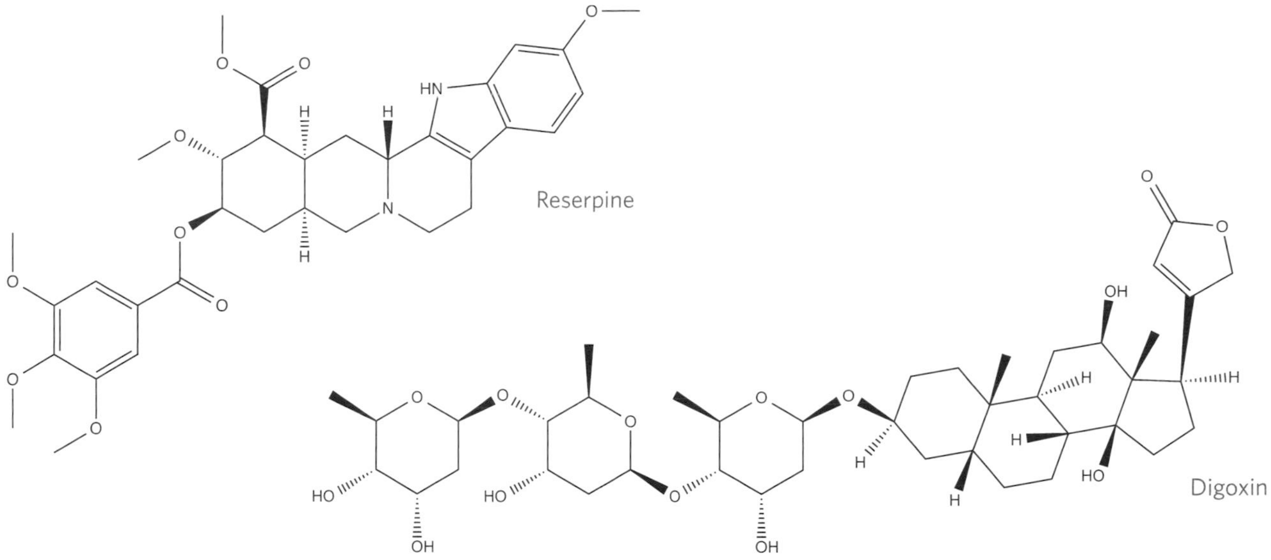

Reserpine

Digoxin

FIGURE 3.8 Plant-derived active principles that affect cardiovascular health: reserpine, from Indian snakeroot, among others; digoxin, from foxglove.

Vasopressors (drugs that can increase blood pressure) of plant origin include compounds from the genus *Ephedra*, such as ephedrine and pseudoephedrine, which increase the heart rate and constrict the blood vessels (figure 3.10). These compounds have numerous physiological effects, including the alteration of respiratory function.[33]

FIGURE 3.9 *Rauvolfia* leaves and fruits. Members of this genus produce the active principle reserpine, which can lower blood pressure.

RESPIRATORY SYSTEM

Humans, like many animals, breathe to draw in fresh oxygenated air, which is necessary for cellular activities, and exhale to remove the metabolic waste gas carbon dioxide. The action of breathing is under both voluntary and involuntary control through the action of the diaphragm muscle, situated at the lower edge of the ribcage. Air enters the mouth or nostrils and passes through a cartilage-reinforced tube (trachea) that branches into two trunks (bronchi) in the chest cavity. The branches split several times over into a treelike series of increasingly smaller tubes (bronchioles) ending in membrane-thin air sacs (individual units called alveoli) where gas exchange takes place. The network of branched tubes and air sacs constitute the two lungs. The bronchioles have muscular walls that can constrict or expand the air passage.[34] The nasal passages and bronchi also secrete mucus, which traps dust and potentially infectious particles before they can enter the deepest parts of the lungs.[35] Diseases

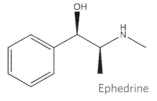

Ephedrine

Pseudoephedrine

FIGURE 3.10 Jointfir compounds: ephedrine and pseudoephedrine.

Inflammation

Localized response to tissue damage or irritants, usually characterized by swelling and pain

of the respiratory tract include overly constricted bronchial tree (asthma), inflamed bronchial tree (bronchitis), damage related to smoking or other inhalation hazards (for example, emphysema), and respiratory infections.

A number of plants in traditional medicine serve to treat chest congestion, although few have been examined for efficacy in clinical studies, nor have mechanisms of action been determined. For example, indigenous people of the Pacific coast of North America such as the Chumash steeped the leaves of the yerba santa (*Eriodictyon californicum*) shrub to treat chest pain and other respiratory concerns.[36] In South Asia, practitioners of traditional and folk medicine use roots and leaves of the Malabar nut tree (*Justicia adhatoda*) to treat various lung conditions.[37] However, the medicinal properties of these plants have not yet been thoroughly studied experimentally.

In the case of excess mucus and respiratory irritation, some patients seek relief from the impulse to cough. Antitussives serve this purpose by suppressing the brain's cough signals or by soothing the throat. Opium poppy–based drugs are used for the former, and a variety of cough syrups, powders, and teas are used for the latter. People of numerous eastern North American Indian tribes prepared hot-water infusions of cherry (*Prunus serotina* [black cherry] and *P. virginiana* [chokecherry]) bark as a cough medicine.[38] The Cherokee taught early Appalachian settlers to chew the stem or brew a root tea of yellowroot (*Xanthorhiza simplicissima*) as a treatment for sore throat.[39] In European traditional and folk medicine, a syrup of horehound (*Marrubium vulgare*) was valued against cough, sore throat, and asthma.[40] The efficacy and mechanisms of action of these herbs have not yet been systematically tested.

Medicines that relax the smooth muscles lining the bronchioles, thereby allowing them to pass more air (bronchodilators), treat chronic and acute asthma and other conditions of poor respiration. A plant with a long history of use against these symptoms is the traditional Chinese medicinal herb *ma huang* (the jointfir *Ephedra sinica*), first mentioned in a medical text 2000 years ago.[41] According to Chinese medicine, *ma huang* "disseminates and facilitates the Lung *qi*, calms wheezing, and stops coughing."[42] (Lung is capitalized to distinguish the traditional Chinese physiological element from the Western anatomical structure.) Related species of ephedra have similar properties, although species vary in their bioactive chemistries.[43] In northern India and Pakistan, traditional medicine values the dried stems of Gerard jointfir (*E. gerardiana*) against asthma.[44] The jointfirs contain chemicals that act as central nervous system stimulants by altering nerve cell communication and circulating hormones, tricking the body into a higher state of alertness and more rapid energy metabolism. The active principles ephedrine and pseudoephedrine are responsible for bronchodilation and increase blood pressure, with side effects of restlessness and insomnia (see figure 3.10).[45] Because of these strong central effects, doctors and patients must take care in their use: the toxic dose is approximately 30 to 45 grams of plant material (a regular dose is in the range of 2 to 9 grams), which is equivalent to about 15 to 30 milligrams of the active principles.[46] (Clinical preparations of ephedra consist of carefully measured doses of ephedrine.) Between the 1930s and the 1970s, ephedrine, delivered as a vapor by inhalation or swallowed as a tablet, was the leading biomedical treatment for asthma. It was made obsolete during the 1970s and 1980s by synthetic molecules with a more specific set of therapeutic actions and fewer side effects, but ephedrine and pseudoephedrine remain useful as nasal decongestants.[47]

Another class of bronchodilator drugs is based on the structure of the theophylline molecule from the tea (*Camellia sinensis*) plant. In cases of asthma and chronic obstructive pulmonary disease, inhaled theophylline can open the airways, probably by modulating specific signaling pathways inside the cells lining the bronchioles.[48] Theophylline also acts as an anti-inflammatory agent via a separate mechanism, which can further ameliorate such breathing conditions. While theophylline is employed less frequently in the treatment of the airways in recent decades in favor of newer agents, it remains in wide use as part of combination therapy.[49]

Other distresses of the respiratory system originate as microbial infections, such as influenza, tuberculosis, the common cold, many forms of bronchitis, and viral and bacterial diseases that result in lung inflammation and the accumulation of liquid and pus (pneumonia). For cold and flu symptoms—such as nasal discharge, sore throat, low-grade fever, and chills—a variety of plants containing volatile oils seem to reduce the severity of illness. For example, the Ojibwa of the northern Great Lakes region heated the needles of balsam fir (*Abies balsamea*) over coals in sweat baths and inhaled the aromatic fumes to treat colds.[50] A common folk remedy in much of the world involves herbal teas or lotions including mint (*Mentha* spp.), whose oil, menthol, serves as a soothing agent that gives the feeling of easier breathing in cases of respiratory discomfort (figure 3.11). The menthol molecule binds specifically to receptors in the nasal passage that signal the sensation of cool.[51]

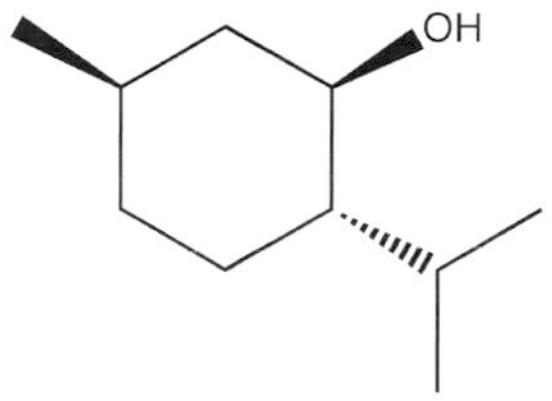

FIGURE 3.11 Menthol, a volatile oil that produces a cooling sensation.

The potential antibiotic and antiviral properties of certain traditional herbals have been tested in the laboratory; however, the clinical trials conducted to date have not consistently demonstrated efficacy in patients. Efforts are ongoing to identify herbal compounds that boost immune system health, reduce the severity of the cold nuisance, and treat or prevent the flu.

URINARY SYSTEM

The kidneys, a pair of organs located to either side of the spinal column in the lower back, regulate blood volume and chemical balance as well as excrete metabolic waste products and toxins. The excreted chemicals and fluid filtered from the blood (urine) travel through thin collecting tubes (ureters) to the bladder, where the urine is stored. During urination, urine travels through another tube, the urethra, and exits the body.[52] As the urinary system is critical to the proper hydration state of the body and the removal of unhealthy substances, it has long been the target of medical treatment.

In medieval Europe, for example, examination of the urine was considered a key diagnostic aid to physicians, who believed that the health of a patient could be assessed by wisely interpreting its color and consistency.[53] Therefore, it is not surprising that many plant-based medicines were selected for their effect—direct or indirect—on the urine. Chinese medicine employs numerous herbal treatments affecting the urinary system, such as *tong cao* (rice-paper plant [*Tetrapanax papyrifer*]), which is said to allay "urinary difficulty and dark urine due to damp-warm disorders," according to the indigenous framework of health.[54]

One of the longest-standing uses of medicinal plants for the urinary system is to increase the quantity of urine by exerting a diuretic effect.[55] Raising the urine volume can help ease the symptoms of high blood pressure and edema (such as associated with congestive heart disease) by reducing blood volume.[56] Although dozens of diuretic herbs are catalogued by traditional medicine, their efficacy has generally not been tested in clinical trials, and their mechanisms of action are largely unknown.[57] Certainly, some may function as stimulants that increase the rate of blood flow, and thus filtration, through the kidneys. Others may act in more specific ways, but these remain to be scientifically resolved.

For example, traditional European, Middle Eastern, and East Asian medicine employed dandelion (*Taraxacum* spp.) for a wide variety of therapeutic uses, including the improvement of urine flow.[58] (The French, Italian, and English vernacular names for the plant—*pissenlit*, *piscialetto*, and piss-a-bed—tend to reinforce this folk idea.) However, very little experimental evidence for dandelion's diuretic properties has been gathered.[59] In contrast, a potent family of diuretics contains the xanthine chemicals produced by the coffee (*Coffea arabica*), tea, and cacao (*Theobroma cacao*) plants, whose activities have been well characterized.[60]

Bacterial infections of the urinary tract usually begin at the urethra and ascend toward the bladder. Extracts of cranberry (*Vaccinium macrocarpon*), blueberry (*V. corymbosum*), bilberry (*V. myrtillus*), and lingonberry (*V. vitis-idaea*) fruits are considered to have therapeutic benefit, although experimental evidence is mixed.[61] The mechanism of action by which *Vaccinium* products either prevent or treat infections remains unclear. It is speculated that polyphenolic compounds in these fruits interfere with the adherence of pathogenic bacteria to the walls of the urethra and bladder.[62] Bacteria then are washed out in the urine.

REPRODUCTIVE SYSTEM

The male reproductive system consists of two testes, which produce the reproductive sperm cells, and a suite of accessory glands that manufacture the seminal fluid. During intercourse, the penis engorges with blood, becomes rigid, and releases semen into the female vagina. Only a small number of sperm cells reach an "egg cell" (ovum), one of which may penetrate into the cell, resulting in conception (fertilization). The male reproductive system is regulated by the nervous system and by male sex hormones, including testosterone.[63]

The prostate is an organ located under the urinary bladder and surrounding the urethra in the lower abdomen. It serves a role in the production of seminal fluid and can become enlarged as men age (the noncancerous condition benign prostatic hyperplasia), which sometimes leads to painful, intermittent urination.[64] During the nineteenth century, American herbal practitioners harvested the fruit of the North American saw palmetto (*Serenoa repens*), a shrub native to the southeastern United States (figure 3.12). They used the fruit dried, in a tea, and in other forms to treat a variety of male reproductive system ailments, including the characteristic urination symptoms of an enlarged prostate.[65] In modern times, manufacturers have produced capsules containing fat-soluble fruit extract and whole fruit that are thought to contain a potent array of compounds capable of treating benign prostatic hyperplasia. However, a recent review of clinical evidence concluded that under the conditions tested in clinical trials, saw palmetto extract is not more effective than placebo

FIGURE 3.12 Saw palmetto: (*left*) plant; (*right*) dried berries. Evidence for its effect on male reproductive health is mixed.

for the treatment of urinary symptoms associated with benign prostatic hyperplasia.[66] Although saw palmetto does not appear to improve symptoms in men with enlarged prostate, a mechanism of action has been speculated. While the male sex hormone dihydrotestosterone promotes prostate enlargement, various fat-soluble compounds in the plant are thought to interfere with an enzyme that produces dihydrotestosterone from its precursor, testosterone.[67]

The central and southern African stinkwood (*Prunus africana*, also called African cherry and pygeum) is widely used in the medical practices of indigenous people for a variety of health concerns, including as a remedy for witchcraft, treatment for stomachache and intestinal parasites, and for male and female sexual health.[68] Since the mid-twentieth century, stinkwood bark and bark extracts have been used outside Africa to treat benign prostatic hyperplasia, particularly in Europe. Clinical evidence is gathering that the herb modestly reduces the severity of urinary symptoms in men with enlarged prostate.[69] However, the mechanism of action is not yet clear. Perhaps it acts by blocking the prostate's response to dihydrotestosterone, possibly together with other effects on inflammation and cell proliferation.[70]

The female reproductive system is composed of a pair of ovaries that alternate in the release of an ovum monthly, in response to regular hormonal fluctuations, and the apparatus to nurture a fertilized ovum through the development and birth of a child. The egg, once released, descends one of the narrow uterine (fallopian) tubes into the womb (uterus), where, if fertilized by a sperm, it implants into the uterine wall and rapidly develops. If not fertilized, the ovum and the uterine lining are shed (menstruation) through the vagina. At term, the fetus is delivered by muscular contractions of the uterus. The monthly hormonal changes that govern the female reproductive cycle are attributable to the activities of the ovaries and reproductive control regions in the brain.[71]

Since the earliest known times, women have sought pharmaceutical means to improve their fertility, on the one hand, and prevent conception or terminate a pregnancy, on the other. For example, the traditional Chinese medicine *ai ye* (leaf of the mugwort *Artemisia argyi*) is an ancient treatment for irregular menstruation and infertility.[72] Meanwhile, the author of the Egyptian Ebers papyrus (ca. 1550 b.c.e.) prescribed the following recipe to induce abortion: "dates, onions, and the fruit-of-the-Acanthus" (*Phoenix* spp. palm fruit, *Allium cepa* bulb, *Acanthus* spp.), crushed with honey and applied to the genitals.[73] For many centuries in Europe, pennyroyal (*Mentha pulegium*) has been considered effective to bring about menstruation in women whose periods are delayed, perhaps even those delayed by pregnancy.[74] As the seventeenth-century English herbalist John Parkinson relayed, a pennyroyal tea "provoketh womens monthly courses [and] expelleth the dead child and afterbirth."[75] Other medicinal plants have been used to improve the outcomes of pregnancy and expedite labor. For example, in Central Africa, Cameroonian midwives prepare the fresh leaves of *Vernonia guineensis* (also known as *Baccharoides guineensis*) to ease delivery, among many other uses of the plant.[76] These are just a small number of the hundreds of fertility-related traditional medicinals identified over thousands of years around the world. What remains unknown is their safety and efficacy. Furthermore, their mechanisms of action—if truly active as ascribed—have not yet been demonstrated. Therefore, the following examples consider medicines for which some evidence exists.

The chaste tree (*Vitex agnus-castus*), native to the Mediterranean region, has long been regarded as an important medicinal plant for the female reproductive system, employed in ancient times to treat discomfort of the uterus and promote menstruation (figure 3.13).[77] Furthermore, it also has documented use in preventing male and female sexual desire,

FIGURE 3.13 Chaste tree flowers and leaves.

from which it derives its English name.[78] In the late sixteenth century, Gerard wrote that chaste tree leaves are "a singular medicine and remedie for such as would willingly live chaste," preventing "all desire to the flesh."[79] The chaste tree's dry fruits were once called "monk's pepper," in reference to their role in helping medieval monks maintain their vows of celibacy.[80] In modern times, chaste tree fruits or chemical extracts have been used to treat a range of menstrual concerns, such as irregular menstruation, pain, breast tenderness, and symptoms associated with menopause.[81] Recently, some of the present-day uses have been subjected to clinical testing. Although studies vary in size and quality of design, treatment with chaste tree fruit extract has largely been demonstrated effective against symptoms of premenstrual syndrome, premenstrual dysphoric disorder, and other measures of discomfort.[82] Current models suggest that chaste tree compounds bind to specific receptors for signaling molecules involved in pain and stress, thereby alleviating the anxiety and discomfort associated with the condition.[83] Chaste tree's roles in suppressing sexual desire and regulating menstruation remain to be validated in the laboratory and clinic.

Given the long-standing human interest in sex for procreation and pleasure, it is not surprising that some of the oldest medicines claim to improve sexual ability or desire. Aphrodisiacs comprise treatments to increase libido or enhance performance. Some of the herbs associated with this use, such as poppy, seem to function not as specific agents to enhance the sex drive but by reducing inhibitions and lowering tactile sensitivity. Other treatments, such as plants containing strong stimulants, including cocaine, from coca (*Erythroxylum coca*), and caffeine, from a variety of plant sources, may have some general effect on sexuality by improving alertness and increasing blood flow throughout the body. Many of these substances produce pleasurable feelings that might be enhanced by sex.

In some cases, there is gathering evidence for the roles of certain traditional herbal medicines as aphrodisiacs. West African folk medicine employs the bark of the yohimbe (*Pausinystalia johimbe*) tree as an aphrodisiac.[84] Studies in animals indicate that yohimbe extracts increase sexual activity, although human research has been troubled with poor experimental design.[85] With mixed results, recent trials have shown some possible clinical efficacy in treating impotence (erectile dysfunction).[86] Yohimbe's active compound, the alkaloid yohimbine, alters the nervous system's regulation of blood flow in the body, probably improving erection to some degree through secondary effects.[87]

The early traditional Chinese texts recognized value in the aerial portions of *yin yang huo* (*Epimedium* spp.), which often goes by the loose English translation "horny goat weed" (figure 3.14). About 2000 years ago, the medical texts declared that this herb "governs impotence, infertility, pain in the penis, facilitates urination, [and] augments the power of *qi*."[88] Numerous Chinese *materia medica* prescribe its use for male sexual dysfunction, and even its name dates to the fifth century in Sichuan province, when writers noted that livestock that ate this plant copulated frequently.[89] While studies involving rats demonstrate

FIGURE 3.14 Horny goat weed, native to Asia and speculated to improve male sexual performance.

the effectiveness of *Epimedium* extracts in causing erections, well-designed human trials have not yet been conducted.[90] Some of the therapeutic value, if it is found to exist, might come from the dozens of compounds that structurally mimic the animal sex hormones estrogen and testosterone. By binding to receptors for these hormones in the body, it is possible that *Epimedium* constituents may alter sexual development and response. However, the precise mechanisms have not been determined.[91] In addition to hormone-like molecules, the plant produces the phenolic compound icariin, which in laboratory studies appears to inhibit the constriction of blood vessels in the penis, facilitating erection.[92] With centuries of plant-related knowledge documented in ancient texts and perpetuated in the medicines of many cultures, there will no doubt be ongoing interest in examining the clinical effectiveness and mechanisms of action of traditional aphrodisiacs.

MUSCULOSKELETAL SYSTEM

The skeleton is a living structure composed of cells and their calcium-rich matrix, making bones stiff and supportive. They are the base of attachment for the muscle fibers that allow the body to move. The muscles are supplied with blood to provide the sugars and oxygen that fuel their contraction and relaxation. The three types of muscle fibers—skeletal, smooth, and cardiac—are structurally and functionally distinct. Skeletal muscle is attached to the bones and generally under conscious control by nerve fibers originating in the brain. Smooth muscle, which lines the gastrointestinal tract and other organs, and cardiac muscle, the tissue of the heart, are under unconscious control. The nerve fibers signal muscle contraction by releasing the neurotransmitter-signaling molecule acetylcholine at the nerve–muscle junction, which triggers an electrical change in the muscle fiber and ultimately causes it to tighten or contract.[93]

The muscles normally respond to the neural signals that induce their movement, but interfering pharmacological agents can block this and paralyze the body. Several South American tribes recognized this phenomenon long ago and developed toxic plant mixtures with which to tip their hunting spears or darts. When stalking animal prey, people such as the Achuar, Huambisa, and Aguaruna in Amazonian Ecuador and Peru have used poison-tipped weapons to immobilize and eventually kill small animals by means of respiratory failure (figure 3.15). The poison, called curare, is usually prepared as a mixture of many plant extracts.[94] The most commonly used plants are of the genera *Strychnos*, *Curarea*, and the curare vine (*Chondrodendron tomentosum*). The poison applied to blow darts and used to such deadly effect includes the active principle of *C. tomentosum*, the alkaloid compound tubocurarine.[95] Once injected into the blood and distributed to muscle tissues, molecules of this chemical bind to the skeletal muscle fiber's nicotinic acetylcholine receptors, blocking the transmission of signals from the brain. The muscles affected by curare are unable to contract, resulting in paralysis that spreads slowly from the site of

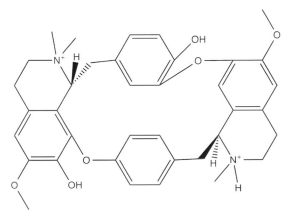

FIGURE 3.15 Curare: (*left*) a group of hunters, photographed in the late nineteenth century in the Brazilian Amazon, carrying weapons tipped with curare poison; (*right*) tubocurarine, an active principle of curare poison. ([*left*] Library of Congress, Prints and Photographs Division, LC-USZ62-83657).

injection to the entire body. As the curare toxins are poisonous only at the muscle–nerve junction, game felled in this manner is safe to consume.

During the twentieth century, the application of tubocurarine to medicine allowed tremendous advances in surgery. Prior to the advent of tubocurarine, surgeons used high doses of agents such as ether and chloroform to induce a deep anesthesia in their patients and prevent reflexive muscle twitching that might hinder the operation. However, such strong anesthesia carried a significant risk of death. By administering controlled doses of tubocurarine to their patients, surgeons beginning in the 1940s were able to procure a safer "balanced anesthesia," one where the dose of anesthetic is low and the patient's muscles are relaxed, prevented from spasms, by the neuromuscular block.[96] In the 1980s, pharmaceutical firms introduced the synthetic compounds atracurium and vecuronium, which are widely used today and serve the same function as the natural tubocurarine.[97]

Damage or strain in the body's tissues is transmitted back to the brain through a series of molecular signaling events subjectively perceived as discomfort or pain. Painful, chronic disorders of the bones, ligaments, tendons, and muscle (rheumatism), including inflammation and pain in the joints (arthritis), are among the most common diseases to strike an aging population.[98] Undoubtedly, people long ago suffered from these ailments and found plants to ease their discomfort. In Chinese traditional medicine, the symptoms of arthritis are attributed to the cold and damp properties of nature, causing obstruction of the *qi* channels and undernourishment of the joints.[99] Numerous herbs are employed to counter the cold and moist aspects of the patient and to improve the movement of *qi*.

In South Asia, the shrublike tree guggul (*Commiphora wightii*) produces a gum resin that has long been used in medicine to treat a wide variety of concerns, including the pain and swelling characteristic of rheumatism.[100] While human trials have not yet demonstrated the efficacy of guggul for these conditions, laboratory research lends support to the notion that constituents of this age-old

remedy might act at the molecular level to reduce inflammation.[101]

One of the more useful medicines to treat pain is a substance that itself, paradoxically, causes pain. The heat of the American native chili pepper (*Capsicum annuum*) is attributable to the compound capsaicin, which binds and activates pain receptors in the skin (figure 3.16).[102] It has a particularly harsh effect on the mucous membranes of the mouth, nose, and eyes, which is why people chopping peppers for cooking must be careful not to touch sensitive areas without having thoroughly washed their hands. When used in a topical patch on the skin above a muscle or joint, capsaicin binds to the tissue pain receptors and dampens pain signals sent to the brain, thus reducing the sensation of discomfort.[103] Medicinal preparations contain approximately 0.075 percent capsaicin, sometimes in concert with other pain- and inflammation-reducing drugs.[104]

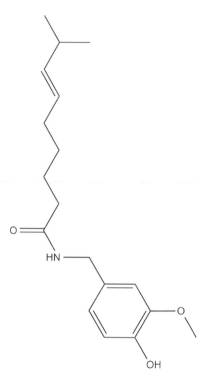

FIGURE 3.16 Capsaicin, a compound from chili pepper that activates pain receptors.

IMMUNE SYSTEM AND INFECTIOUS DISEASE

The body's natural defenses against infections include the skin and mucous membranes, which act as a physical barrier to potentially harmful microbes; various secretions of the body, such as mucus and tears, which can trap and disable infective agents; and the immune system, an active cellular system that attacks infections inside the body. The immune system functions by allowing specialized cells to circulate throughout the body, identify invading microbes (such as bacteria, fungi, or viruses) or chemicals (such as proteins or particles), and selectively destroy the foreign material.[105] If the immune system prevents illness by resisting the harmful agent before it can exert effects or spread in the body, the response is called immunity. When the system is unable to control the growth or spread of the microbe, the result is infectious disease. When the system mounts an inappropriately strong response to a foreign substance, the effect is allergy. Cases of allergy can range from mild swelling and localized irritation of tissues to severe swelling and loss of blood pressure, a dangerous, incapacitating condition called anaphylaxis.[106]

The immune system derives from specialized cells whose role is to recognize, via cell-surface proteins, a wide variety of human and nonhuman substances. They can then identify foreign materials in the body and eliminate them. During early development, the body produces millions of lymphocytes (one class of specialized immune cells) that have cell-surface proteins of a wide variety of shapes. The cell-surface proteins act as sensors for a nearly unlimited number of molecules. On the one hand, the body uses these cell-surface proteins to learn which molecules are "self," belonging to the body that produced them. That way, the body's immune system does not mount a protective reaction against its own tissues. (When this safeguard fails, a self-destructive autoimmune disease results.) On the other hand, the immune system can recognize a tremendous variety of potentially pathogenic agents and target them specifically for destruction.

The sensor proteins capable of identifying foreign substances, called antibodies, also circulate freely in the bloodstream. When the body experiences the invasion of a microbe, for example, specialized antibodies bind to parts of the microbe and serve as flags to beckon phagocytes, the cells that engulf and remove the foreign object. Pharmaceutical agents can enhance or inhibit the ability of these systems to function, altering the body's responses to self- or non-self-structures in the body.

Through most of history, people had no microbial and immunological explanation for infectious disease. Instead, traditional medicine attributed such illnesses to the natural properties of the environment, temperament, and diverse supernatural origins. Regardless of the cause, people facing infections sought treatments, many of them plant-based. For example, medieval Europeans considered leprosy (now called Hansen's disease, known to be caused by the bacterial pathogen *Mycobacterium leprae*) to result from moral decay or wickedness.[107] While prayer and virtuous living were certainly part of the prescription, herbalists such as Gerard offered plants including dodder (*Cuscuta* spp.) and black hellebore (the author suggests both black hellebore [*Helleborus niger*] and false black hellebore [*Veratrum nigrum*]).[108] It is now possible to investigate such herbs in the laboratory and clinic.

North American Plains Indians employed a native herbaceous plant, the purple coneflower (*Echinacea purpurea*, and also *E. angustifolia* and *E. pallida*), for a wide variety of purposes (figure 3.17).[109] They chewed the roots

FIGURE 3.17 Purple coneflower, which may improve immune system function.

Infectious disease agents (pathogens)

Bacteria	Simple single-celled organisms
Fungi	Yeasts and multicellular organisms related to mushrooms
Protists	Single-celled organisms of complex structure
Viruses	Nonliving particles with genetic instructions to infect living cells and replicate

Since all these pathogens are very small, they are considered microorganisms, or microbes. Although viruses are not technically organisms, they are still loosely grouped with other pathogens.

to treat toothache and sore gums and applied the leaves and juice externally for burns and snakebite. The Sioux were known to have used the plant to reduce the sores of syphilis, caused by a bacterium. The Crow tribe of Montana and eastern Wyoming harvested the purple coneflower to treat the common cold, a viral infection.[110] It was white settlers who picked up on the native use of the plant against infection, and by the late nineteenth century, purple coneflower was a mainstream herbal remedy.[111] During the mid-twentieth century, purple coneflower extracts for injection were sold in Europe under the name Echinacin, supposedly effective against infections and cancer.[112] Today, purple coneflower is widely available in the United States as a dietary supplement in various forms of liquid extract and capsules, suggested to boost resistance to the common cold and other respiratory infections.

Experiments in the laboratory have established a wide spectrum of possible activity. Using various types of extracts and test conditions, researchers have found that constituents of purple coneflower may activate phagocytes, suppress the inflammatory response to infection, and kill bacteria and viruses directly.[113] However, active principles have not yet been identified. Numerous human studies have attempted to test whether purple coneflower extracts improve immune system function or reduce the intensity or duration of infections such as the common cold. Most suffer from problematic experimental design, including nonstandardized extracts or dosing schemes and reliance on small numbers of patients. An analysis of several purple coneflower experiments showed enormous diversity in the type of coneflower extract used, the presence of concomitant supplements, the outcome measures, the duration of the experiment, and the size of the experimental groups.[114] The support for purple coneflower's activity in boosting the immune system and fighting respiratory disease remains anecdotal and the scientific evidence mixed. The medical consensus is that purple coneflower extracts are neither helpful nor harmful to patients wishing to reduce the incidence and duration of infection.

An herb long employed in European and Asian medicine, licorice (*Glycyrrhiza* spp.) has been used to treat wounds, diabetes, cough, stomachache, digestive ailments, and sexual concerns, among other health matters (figure 3.18).[115] Traditional Chinese medicine uses the root for a wide variety of symptoms, viewing it as a moderator of many other herbs when formulated together as a mixture. As the most commonly used plant in the Chinese *materia medica*, it is not surprising that it would be applied for coughs and infection among so many

FIGURE 3.18 Licorice flowers. Licorice has a long history in Asian and European medicine.

other ailments.[116] Some laboratory studies support the notion that the sweet-tasting terpenoid compound glycyrrhizin and its derivatives from licorice can reduce the severity of viral infections, including those targeting the respiratory tract, skin, and liver. The mechanism of action appears in these studies to be a combination of reduction of the virus's ability to bind to and infect cells as well as the stimulation of the body's immune defenses.[117] Despite a long history of use and some bioactivity observable in the laboratory, therapy by licorice root extract has not yet been thoroughly demonstrated through clinical trials.

It is evident that many plants have traditional uses as broadly described immunomodulators that somehow boost the body's defenses against infectious agents. Some proponents of such herbs might not be concerned with the mechanisms by which they improve natural resistance, but scientific evidence will be required before these plants and their constituents gain wider use through biomedical health-care channels. There have not been any herbal compounds yet identified that can specifically improve the ability of phagocytes to attack invading microbes or the propensity of antibodies to flag their targets for destruction.

However, the antibody-producing cells and phagocytes themselves are responsive to the overall level of physical or emotional stress, which can suppress the immune system. Perhaps it is through the modulation of stress that some herbal immune stimulants function. Whether through physical or psychological means (that is, direct cellular effect or placebo effect), plant medicines may reduce the anxiety associated with illness and thereby allow an improved immune response.[118] Herbs reported to promote the body's ability to cope with stress are termed adaptogens.[119]

In contrast to plants without strong clinical support for immune system function, there are many herbal components with antimicrobial and antiparasitic functions. While most modern pharmaceutical antibiotics (drugs primarily targeting bacteria) are fungal or bacterial in origin, some plants produce compounds that may be useful to combat such infections. However, the use of plants as specific antibiotic agents remains limited. Numerous traditional uses of plants include broad anti-inflammatory, antioxidant, and analgesic functions, but the technical challenges to identifying single antimicrobial agents in complex botanical chemical mixtures have meant that few plant-derived "penicillins" have emerged. Moreover, many of the antimicrobial plant compounds consist of volatile oils and agents that are generally (rather than specifically) toxic to cells, such as those of billy-goat weed (*Ageratum conyzoides*, also known as tropical whiteweed) or the noxious antibacterial sulfur-containing compounds of onion and garlic (*Allium cepa* and *A. sativum*).[120]

Although malaria was once thought to originate from the unwholesome influence of a moist, foul environment (*mala aria* [bad air]), it is now known that the disease is microbial in origin. Malaria, which remains a significant threat in the tropics, is caused by a mosquito-borne *Plasmodium* parasite that infects the human victim's liver and blood, replicating and causing often-fatal fever and anemia. Yet long before the advent of biomedicine, traditional medicine identified a host of active agents that combat this ailment. The most renowned of these is the Peruvian fever tree (*Cinchona* spp.),

native to tropical South America. This plant produces the alkaloid quinine in its bark and is effective against the parasites responsible for malaria. The medicinal use of fever tree bark extract became known to Europeans following their sixteenth-century conquest of South America and soon served as an important pharmaceutical shield to missionaries and explorers wishing to protect themselves from malaria as well as to those facing its characteristic intermittent fevers in Europe. The active principle quinine was isolated in the early nineteenth century, and commercial production rapidly destroyed the native fever tree stands. Breeding programs were established to increase the quinine yield from 4 percent to more than 13 percent of bark, which maintained a steady plantation-derived supply of quinine into the early twentieth century.[121] Chemists synthesized quinine-like compounds, such as chloroquine and quinacrine, beginning in the 1930s, which allowed for a greater supply of drug to treat the millions living under malarial peril (figure 3.19). As much of the *Plasmodium* parasite threatening the tropics has become resistant to chloroquine, the original quinine has again been advanced for malaria treatment.[122]

Fever tree bark influenced European history in a significant way by serving the military and economic expansion of France, Britain, Holland, and Spain during the seventeenth through twentieth centuries. As these powers brought tropical territories in the Americas, Africa, and Asia under their control, they did so with the protective medicine quinine coursing through their veins. It is no wonder that the nineteenth-century British military forces added quinine to their tonic water during the conquest of India—drinking it with gin to create a medicinal cocktail. The quinine molecule and modern synthetics based on its structure are toxic to the parasite reproducing in human red blood cells. While quinine does not protect a person from becoming infected by the parasite, the drug does halt the organism's replication and maturation, limiting the development of symptoms and spread of infection. Quinine and its analogs have some degree of toxicity in the human body, experienced as nausea, thought disturbances, and

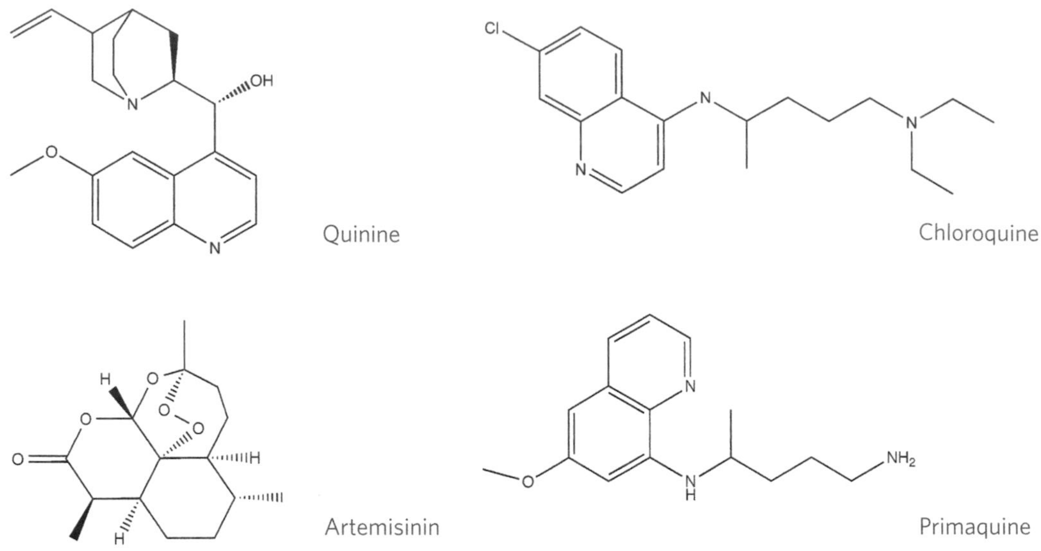

FIGURE 3.19 Antimalarial active principles and synthetic derivatives: the natural alkaloid quinine, from trees of the genus *Cinchona*; the synthetic chloroquine; the synthetic primaquine; the natural terpenoid artemisinin, from sweet wormwood.

other effects. However, the side effects are generally considered mild and balanced by the benefit of malaria protection.

Asia has its traditional antimalarial as well: the sweet wormwood (*Artemisia annua*), native to temperate Asia and described in Chinese herbals more than two millennia ago.[123] In Chinese medicine, the herb is known as *qinghao* and is used against the intermittent fevers associated with malaria. Its terpenoid active principle, artemisinin, is particularly valued in areas where the *Plasmodium* parasites have become resistant to quinine-based medications (see figure 3.19). Artemisinin is toxic to *Plasmodium* maturing in the red blood cells and highly effective in resolving a patient's symptoms and reducing the transmission potential of an infection.[124] Although resistance is also developing against artemisinin and its derivatives, it remains largely effective when used as part of combination therapy.[125]

The world's traditional medicines

include countless herbs thought to treat illnesses and promote good health. Although people have assigned varying roles to the body's systems and attributed ailments to diverse causes, the collective experiences of so many have yielded a wealth of plants potentially useful in biomedicine. Armed with the methods of clinical trials, researchers can determine whether a treatment causes its purported effects. Employing the laboratory techniques of chemistry and biology, scientists can decipher how plants' active principles interact at the molecular level with targets in the body. While many plant-derived compounds have entered the Western biomedical pharmacopeia, a great number of plants' uses remain to be subjected to testing. The results of such efforts yield a better understanding of historical and traditional plant uses as well as biochemical agents that can be harnessed against some of the most challenging health concerns that humans face. Importantly, by elucidating the molecular mechanisms of action of herbal agents, investigators are able to decipher much about the workings of the human body. These efforts are perhaps most advanced in the study of the brain and nervous system, where plant-derived chemicals enabled progress in unraveling the molecular and anatomical basis of sensation, perception, and behavior.

Chapter 4

The Actions of Medicinal Plants on the Nervous System

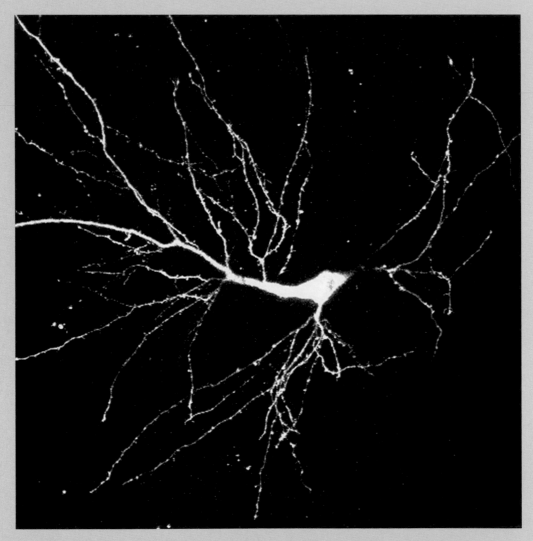

The dendrites of neurons, here shown magnified and stained with a fluorescent dye, carry nerve impulses that control movement and generate consciousness. The signals they propagate can be influenced by plant-derived chemicals. (Micrograph by Amber Petersen and Nashaat Gerges)

The nervous system—responsible for human consciousness, cognition, sensation, and movement—is composed of a complex network of cells that together gather information from all parts of the body, integrate diverse sources of input, and then signal the appropriate responses. The system enables us to perceive our complex environment, learn, think, and have emotions.

The central nervous system consists of the brain and spinal cord. Its role is to collect and integrate sensory information, coordinate muscle movement, store memories, process language and imagery, and generate feelings. The peripheral nervous system, a nerve network that penetrates all reaches of the body, is largely responsible for gathering the sensations that an individual experiences and initiating muscle contractions according to instructions from the central nervous system. The peripheral nervous system receives input, for example, by touch or taste, and mediates responses, such as by causing muscles to move or glands to release hormones. The active principles of some herbal medicines can affect the function of the nervous system, and those that alter the behavior of the brain are called psychoactives. The basic unit of the nervous system is the nerve cell (neuron), and many medicinal compounds exert their effects by specifically interacting with these fundamental cells and the structures they form.

Neuron

A nerve cell, the basic unit of the nervous system

Psychoactive drug

A drug that produces an effect on the central nervous system, especially on the mind

CENTRAL NERVOUS SYSTEM: STRUCTURE AND FUNCTION

The brain can be considered in four regions: the brain stem, the overlying thalamus-hypothalamus structures, the cerebellum, and the cerebrum (figure 4.1). The brain stem carries sensory and motor messages between the brain and the spinal cord, acting as a conduit for information between the central nervous system and the remainder of the body. It also plays a role in generating and modulating messages. For example, the portion of the brain stem closest to the spinal cord (medulla) regulates the vital functions of heart rate, digestion, breathing, blood pressure, and coughing. Slightly above that portion is the pons, which shares information related to movement between the cerebrum and the cerebellum. The midbrain, located above the pons and near the center of the brain, processes sensory information from the eyes and ears and helps regulate movement of the body.[1] Importantly, the pons and midbrain also contain neurons involved in aspects of arousal, sleep, and emotion. Drugs that act on cells in this region, such as some central stimulants, can make a person vigilant and sleepless.

Above the brain stem, the thalamus serves to distribute incoming sensory information for processing in the cerebrum, making it partly responsible for sensory awareness and attention. It also coordinates signals related to movement originating in various parts of the brain for transmission to the rest of the body. The hypothalamus governs several core biological functions, including temperature control, thirst and hunger, and aspects of sexual function and emotion. It also releases hormones into the bloodstream.[2] Numerous plant-derived chemicals can affect these processes.

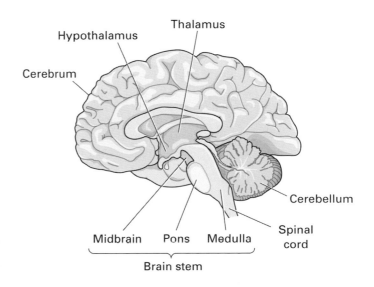

FIGURE 4.1 The regions of the brain.

REGIONS AND PARTS OF THE BRAIN: FUNCTIONS OF SENSORY PERCEPTION, MOTOR CONTROL, EMOTION, AND COGNITION

Region	Part	Function
Brain stem		Regulates vital functions such as breathing and heart rate; transmits and processes information between spinal cord and brain; collects sensory information and provides motor control for the head
	Medulla	Controls breathing, blood pressure, heart rate, swallowing, and respiration; contains the vomiting center, responsive to toxins in the bloodstream
	Pons	Influences transition between sleep and wakefulness and between stages of sleep; controls breathing rates and pattern; houses neurons that govern arousal, vigilance, and attention; conveys information about movement between cerebral hemispheres and cerebrum
Thalamus and hypothalamus		Processes information between brain regions; regulates unconscious body functions and controls release of some hormones
	Thalamus	Processes and distributes sensory information between brain stem and cerebral cortex
	Hypothalamus	Regulates body temperature and salt–water balance; modulates hunger, thirst, and sexual behaviors
Cerebellum		Coordinates body movements in force and range; contributes to learning and memory
Cerebrum		Plans and executes movement; stores memory; enables language, reasoning, and creativity
	Cerebral cortex	Divided into frontal, parietal, occipital, and temporal lobes; processes sensory information; site of complex reasoning, planning, personality, speech, reading, and conscious body movement
	Basal ganglia	Regulate muscle movement
	Amygdala	Coordinates emotional responses; generates feelings of pleasure, punishment, sexual arousal, rage, and fear
	Hippocampus	Establishes new long-term memories
	Limbic system	Includes the cingulate cortex, amygdala, hippocampus, and other structures; integrates emotional responses and reward; regulates motivated behavior and learning

Sources: Walter Lewis and Memory P. F. Elvin-Lewis, *Medical Botany: Plants Affecting Human Health* (Hoboken, N.J.: Wiley, 2003), 632; Gerard J. Tortora and Bryan H. Derrickson, *Principles of Anatomy and Physiology*, 11th ed. (Hoboken, N.J.: Wiley, 2009), 496–522; Eric R. Kandel et al., *Principles of Neural Science*, 5th ed. (New York: McGraw-Hill, 2013), 8–11.

The cerebellum is in the back of the head and connects to the brain stem. Its role is to gather sensory input regarding motion and balance and then to provide the signals to ensure smooth and coordinated muscle movements. It also plays a role in learning, particularly of motor skills. The effects of alcohol intoxication on the cerebellum are evident as staggering and slurred speech.

Occupying most of the brain's volume is the cerebrum, comprising two hemispheres with wrinkled surfaces. This structure is responsible for sensory awareness, personality, memory storage, planning of action, and voluntary movements. Much of the cerebrum's activities take place in the upper few millimeters of its surface (cerebral cortex), where neurons form numerous connections with one another and with other parts of the central nervous system, including cells in the brain stem, thalamus and hypothalamus, and cerebellum.[3] Deep within the cerebrum are several structures (basal ganglia, amygdala, and hippocampus) involved in motor control, learning and memory storage, and motivation.[4] Because the brain's structures are intricately connected by bundles of long neurons (nerve fibers) that cross between regions, sensation, perception, and movement can be coordinated and the intriguing functions of learning, emotion, and thought can emerge.

The sensory information that comes in from the body's extremities is channeled through the spinal cord, as are signals that go out to the muscles and other tissues, causing actions to occur. The relationship between the spinal cord and the nerve network that permeates the tissues (peripheral nervous system) is a complex one, and one that medicines can alter to great effect.

PERIPHERAL NERVOUS SYSTEM: STRUCTURE AND FUNCTION

The peripheral nervous system carries messages around the body through two nerve networks: the somatic nervous system, which controls the skeletal muscles and senses the external environment, and the autonomic nervous system, which regulates the activity of the internal organs. The somatic nervous system provides the ability to feel the heat of a cup of coffee, control the movement of hand muscles while writing, and sense vibration when a bus passes nearby. The autonomic system carries the signals that, without conscious effort, direct the heart to beat, the intestine to pass along food, and the glands to release their hormones at the appropriate times. This division of labor in the nervous system essentially distinguishes whether the movement is voluntary and the sensation conscious or whether the action is not a deliberate decision but a basic, unconscious activity of the human body. Via the central and peripheral nervous systems, nerve fibers control the function of nearly all the body systems.[5]

For example, the flow of blood throughout the body is regulated by the peripheral nervous system

in many ways: the autonomic nervous system can increase or decrease the heart rate, depending on cues from the body and from the central nervous system; it can cause the smooth muscle lining the major blood vessels to dilate or constrict; and it can even control the ability of blood to reach the smallest of vessels in the skin, causing someone to blush. Likewise, the peripheral nervous system plays key roles in the activity of the digestive system, the reproductive system, respiration, and so forth. Indeed, a great many of the body's functions are connected to the state of the nervous system, which may be why so many medicinal plants act there. These psychoactive drugs and peripheral nervous system modulators would not function if not for their effects on the nervous system's basic units, the nerve cells—neurons.

Functions of the peripheral nervous system	
Somatic	Conscious sensation and muscle movement
Autonomic	Unconscious control of the organs

NEURONS

Neurons, like all living cells, are self-contained units of biological activity that can synthesize myriad chemical compounds, receive input from their microenvironment, and send out signals of many types. Neurons can be quite small, on the order of one-tenth to one-twentieth of a millimeter in length, although most neurons tend to be long and thin, with multiple extensions.[6] Some neurons, such as those in the spinal cord, can be tens of centimeters long but a fraction of a millimeter in diameter.

Neurons are specialized to receive, integrate, and send signals, functions that are reflected in their structure (figure 4.2). The neural cell body (soma) contains the nucleus, a repository of genetic information, and much of the machinery to sustain biological activities. The neuron gathers information, whether by sensing the environment or by communicating with other neurons, through highly branched cellular extensions called dendrites.

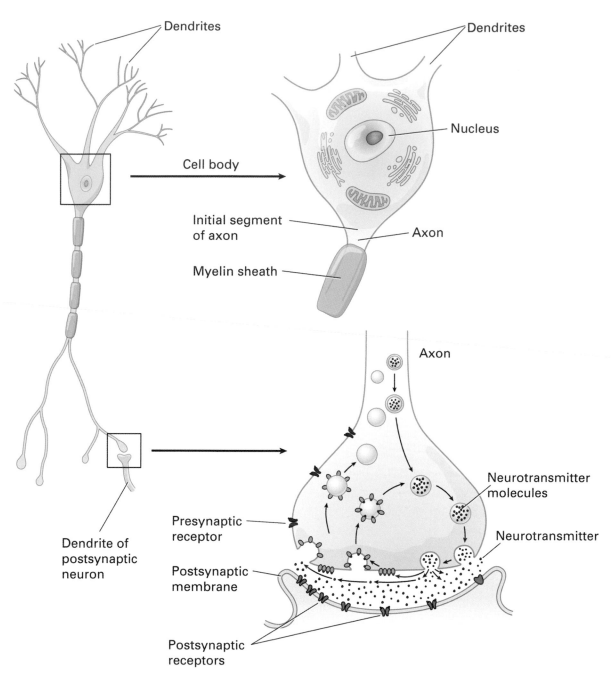

Dendrites

Dendrites

Nucleus

Cell body

Initial segment
of axon

Axon

Myelin sheath

Axon

Neurotransmitter
molecules

Presynaptic
receptor

Neurotransmitter

Dendrite of
postsynaptic
neuron

Postsynaptic
membrane

Postsynaptic
receptors

FIGURE 4.2 Neuron structure and function. Chemical messages are emitted in the form of neurotransmitter signals from the end of an axon. Transmitters are released into the synapse between cells and perceived by specialized receptors on the surface (of the dendrite, for example) of a neighboring neuron. Following their release into the synapse, structures on the presynaptic membrane take up and recycle excess neurotransmitters. (Adapted from Robert M. Julien, Claire Advokat, and Joseph E. Comaty, *A Primer of Drug Action*, 12th ed. [New York: Worth, 2011], fig. 3.7)

On receipt of a signal, the neuron processes the stimulus and sends a new message to neighboring neurons along the length of an extended thin tube, the axon. The axon propagates its message electrically, which is why axons are typically coated with a sheath of the jelly-like substance myelin. Myelin serves as an insulating material to preserve the integrity of the axon's electrical signal. The end of the axon is branched, and the tip of each branch is a presynaptic terminal, a point where the neuron communicates with the soma or dendrites of neighboring cells by converting electrical impulses into chemical messages.

Parts of a neuron	
Cell body	Contains the genetic material and most of the cell's volume; also called the soma
Axon	Long and thin cellular extension that transmits signals electrically; axon terminals communicate chemically with neighboring cells
Dendrite	Branched extension that receives sensory input
Synapse	Point of communication between neurons

The terminus of an axon is typically separated from the dendrites and soma of neighboring cells by narrow gaps (synapses) across which neurons signal using molecules called neurotransmitters. Communication takes place when an axon end, the presynaptic terminal, sends neurotransmitters across the synapse to the receiving cell, at the postsynaptic terminal. Some neurons, such as those in the central nervous system, generate hundreds of thousands of synapses, a condition that reflects the importance of intercellular communication in nervous system function.[7] Since the brain contains 100 billion or more neurons, each making an average of several thousand synaptic connections, the sensitive and specific neuronal information sharing that produces human sensation, movement, and higher-order functions owes itself to enormously complex intercellular interactions.[8] Psychoactive compounds can interfere with cell-to-cell signaling at this level.

While communication along an axon and between certain specialized neurons is electrical, propagated by minute changes in voltage that can travel down the axon at speeds in the neighborhood of 100 meters per second, most intercellular signaling at synapses is mediated chemically.[9] The presynaptic terminal maintains a reserve of neurotransmitters, sequestered inside the cell, that when released into the synaptic cleft between neighboring neurons, pass across to the postsynaptic cell's membrane and are sensed by specialized neurotransmitter receptors. In some cases, the binding of a neurotransmitter to its receptor is excitatory, causing an impulse to propagate; in others, the message is inhibitory, causing the target cell to reduce its signaling activity. The diversity of known neurotransmitters reflects the complex nature of the nervous system.

At the microscopic level, the scale at which neurons operate, chemical compounds have three-dimensional structures determined by the nature of the bonding between their atoms. Most biological molecules are composed of atoms of carbon, oxygen, nitrogen, hydrogen, and a few other elements connected together to form units of distinct shapes and with unique functional properties. All the building up or breaking down of various molecules in and near cells is carried out by enzymes, specialized minuscule structures whose task is to assemble or disassemble very specific chemical configurations while leaving all others untouched. It is important that cells be able to detect with precision a wide variety of chemical compounds, for so many of these are critically important to their development and ability to communicate. To recognize the chemical signals around and within them, cells produce molecular sensors known as receptors, which are capable of receiving chemical input and informing the cell of the nature of the signal.

NEUROTRANSMITTERS

Human neurons produce a diversity of neurotransmitters that enable communication within the nervous system and between the nervous system and various tissues. These molecules fall into a handful of categories based on chemical structure, each

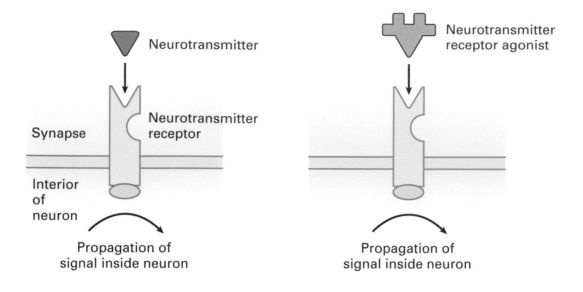

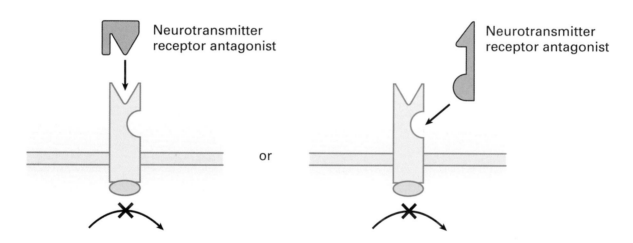

FIGURE 4.3 Neurotransmitter agonists and antagonists. Neurotransmitters bind to specific receptors, propagating a message inside the target neuron. Neurotransmitter receptor agonists can bind to neurotransmitter receptors and provoke a similar signal inside the neuron. Neurotransmitter receptor antagonists bind to neurotransmitter receptors and block their ability to propagate a signal.

perceived by particular receptors that are present in different subsets of cells and initiating different processes. The role of a neurotransmitter receptor is to recognize the presence of a specific neurotransmitter chemical in the fluid outside the neuron and then to signal that chemical's presence to the interior of the neuron.

In some cases, however, molecules whose chemical structures resemble neurotransmitters can also bind to receptors and give rise to biological effects (figure 4.3). Since the receptors have an affinity for the shape of the neurotransmitter to which they normally bind, they can sometimes be activated by a closely related chemical structure (agonist) or blocked by a chemical that binds but fails to activate the receptor (antagonist). Neurotransmitter action is a function of multiple activities occurring at the synapse: signaling can be increased by releasing more neurotransmitter into the synaptic cleft or by inhibiting its degradation or reuptake; conversely, signaling can be reduced by hastening the compound's destruction or departure from the synapse. Indeed, these and other complex phenomena serve to modulate the level of signal between neurons and are frequently engaged in the presence of psychoactive medicinal compounds.

Numerous plant products, including many potent psychoactive chemicals, function by activating or inhibiting the body's natural (endogenous) signaling mechanisms. Herbal drugs can increase the heart rate by fooling the nervous system into a stimulated state. They can also cause numbness by preventing sensory neurons from communicating at all. Some of these chemicals function systemically; others work locally. What the plant psychoactive chemicals have in common is their direct action on the human nervous system and their propensity to perturb, hijack, or block the normal processes of cellular communication.

Neurotransmitter receptor agonist

A chemical that binds and activates a receptor in place of a neurotransmitter

Neurotransmitter receptor antagonist

A chemical that binds and blocks signaling by a receptor

Amino Acid Neurotransmitters: Glutamate, GABA, and Glycine

The amino acid neurotransmitters are active throughout the central and peripheral nervous system and serve to increase or decrease the activities of many types of neurons. As a result, they have diverse functions in the human systems. Glutamate is the principal excitatory neurotransmitter in the brain and spinal cord, and it mediates signals induced by numerous other neurotransmitters (figure 4.4).[10] Among its many roles in such functions as muscle control and sensory perception, glutamate signaling is also important in forming new neural connections, such as during the process of learning.[11]

Gamma (γ)-aminobutyric acid (GABA), for its part, serves as an inhibitory molecule, dampening the signals generated by other neurotransmitters (see figure 4.4).[12] Among GABA's main effects are the promotion of fluid muscle movements and a sense of calm in the mind. Some natural chemical products can interfere with normal signaling via the GABA receptors, by both activating and blocking GABA responses. For example, the fly agaric (*Amanita muscaria*) mushroom yields muscimol, a GABA agonist that inhibits a broad range of neurotransmitter signaling (see figure 4.4). Traditional shamanic practices of eastern Siberia include the consumption of fly agaric to induce stupor and altered sensory perception.[13] The fish-berry plant (*Anamirta cocculus*), a shrub traditionally used in India as a fishing poison and medicine, produces the highly potent central nervous system stimulant picrotoxinin, a GABA antagonist (see figure 4.4).[14]

Similarly to GABA, the neurotransmitter glycine has roles in maintaining a low level of nerve cell stimulation, primarily in the brain stem and spinal cord (see figure 4.4). In addition to its activity as a GABA antagonist, picrotoxinin blocks signaling via certain types of glycine receptors.[15] The South Asian strychnine tree (*Strychnos nux-vomica*) has long been employed in indigenous medicine as a tonic and stimulant, and its seeds were traded widely in antiquity.[16] Classic writers also warned of its risks. For instance, the English herbalist John Gerard (1545–1611?) described "the Vomiting Nut" as "not to be given inwardly" because "the dangers are great."[17] The seeds contain the glycine antagonist strychnine, which prevents glycine's normally calming effects (see figure 4.4). Strychnine causes agitation at a low

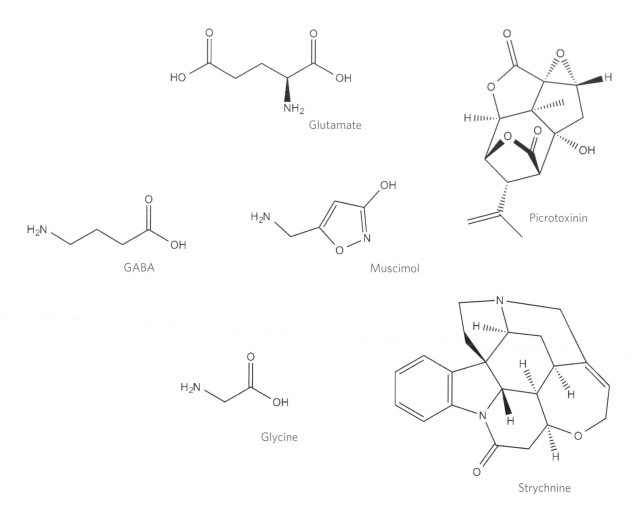

FIGURE 4.4 Amino acid neurotransmitters: glutamate; γ-aminobutyric acid (GABA); the GABA receptor agonist muscimol, from the fly agaric mushroom; the GABA receptor antagonist picrotoxinin, from the fish-berry plant; glycine; the glycine receptor antagonist strychnine, from the strychnine tree.

dose, convulsions at a moderate dose, and death through paralysis at a higher dose.

Acetylcholine

The neurotransmitter acetylcholine is involved in diverse processes. In the central nervous system, nerve cells transmitting acetylcholine (called cholinergic neurons) are found in the cerebrum and midbrain, with projections that communicate with neurons throughout the cerebral cortex, cerebellum, thalamus, and other parts of the brain. In these regions, acetylcholine serves a role in learning and memory, among other functions (figure 4.5). In the periphery, acetylcholine induces muscle movement. Acetylcholine is released from the presynaptic terminals of motor neurons, binds to receptors in muscle cells, and causes them to contract.[18] (In much of the body, such as in the smooth and skeletal muscles, acetylcholine initiates muscle contraction. In the heart muscle, however, acetylcholine reduces the heart rate.)[19]

Some plant compounds can interfere with normal acetylcholine signaling. Among the most widely known is the alkaloid nicotine (see figure 4.5), which accumulates in the leaves of the New

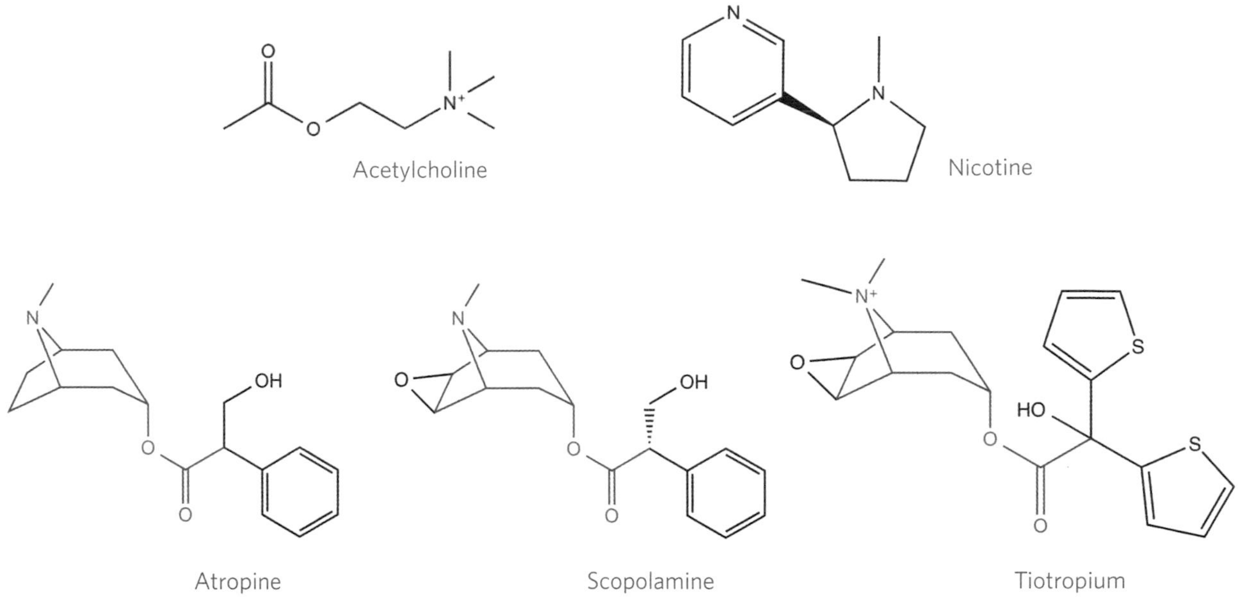

FIGURE 4.5 The neurotransmitter acetylcholine; the acetylcholine receptor agonist nicotine, from tobacco; the acetylcholine receptor antagonists atropine and scopolamine, from the nightshade plants; the synthetic respiratory drug tiotropium. Structural similarities are shown in purple.

World tobacco plant (*Nicotiana tabacum* and other species). Indigenous peoples of the Americas inhaled tobacco smoke, chewed its leaves, and took leaf powder into their noses for a range of spiritual-medical purposes.[20] Beginning in the sixteenth century, tobacco use spread to Europe, Africa, and Asia, ultimately becoming one of the most ubiquitous customs on earth. Nicotine's effects are widespread in the body. In the periphery, the compound generally binds as an agonist to acetylcholine receptors in the nerves that control the skeletal muscles, smooth muscles, heart, and glands, although its actions are complex and depend on dose. Its broad physiological effects include increased tone of the muscles lining the gastrointestinal tract and a rise in heart rate and blood pressure, in part because of its stimulation of the adrenal gland. Nicotine also binds and activates acetylcholine receptors in the central nervous system, resulting in increased alertness and mild analgesia.[21]

In contrast to the agonist effect of nicotine on acetylcholine receptors, the alkaloid atropine, produced by the Eurasian deadly nightshade (*Atropa belladonna*), antagonizes acetylcholine receptors (see figures 4.5 and 4.6). This plant also goes by the name belladonna, a reference to its historic use as a cosmetic drug in Italy, dropped into the eyes to dilate ladies' pupils (*bella donna* [beautiful woman]). Its effects vary greatly, depending on dose: it slows the heart rate at a low concentration and increases the heart rate at slightly higher levels. It also blocks the activity of sweat and salivary glands, slows peristalsis of the intestines, and at very high doses can disrupt coordinated signaling in the brain, resulting in hallucinations. Overdose (greater than 10 milligrams) can be lethal.[22] The nightshades have long been employed in the traditional medicine of Europe. For example, the English herbalist John Parkinson (1567–1650) recommended the family of nightshade plants as cooling herbs to treat inflammation, headache, shingles, dropsy, and other complaints but warned that *A. belladonna* "[is] held more dangerous than any of the other," noting the "lamentable" deaths of children who ate the plant's black berries

FIGURE 4.6 Deadly nightshade, a source of the acetylcholine receptor antagonist atropine.

or drank a broth in which its leaves had soaked.[23] When carefully prepared and administered, atropine remains a useful agent in medicine in the modern day. The chemical is used to dilate the pupil during eye examinations, as it relaxes the muscles of the iris.[24] Atropine also blocks the constriction of smooth muscle in the lungs and can be given to ease breathing in patients with asthma and chronic obstructive pulmonary disease. In addition to atropine, the nightshade plants produce a closely related acetylcholine receptor agonist, scopolamine, which shares a number of physiological effects (see figure 4.5). Pharmacologists have developed a semisynthetic and longer-lasting drug based on the chemical structure of atropine and scopolamine, tiotropium (marketed as Spiriva), which is now considered the preferred bronchodilator for patients with chronic obstructive pulmonary disease (see figure 4.5).[25]

Monoamines: Norepinephrine, Dopamine, and Serotonin

Although the monoamine neurotransmitters are grouped together based on their structural similarity, their roles in human physiology are diverse. These chemicals broadly participate in the processing of sensory information, decision making, emotion, and many aspects of human intelligence and creativity. Produced by a relatively small number of neurons located in the central nervous system, monoamine neurotransmitters exert wide-ranging effects because of their cells' many projections throughout the brain and (sometimes) spinal cord and their numerous synapses, which modulate signaling at a fine level.[26] A great number of psychoactive plants exert their effects through

signaling via these neurotransmitters' receptors or by blocking their natural functions.

Norepinephrine neurons reside in the brain stem, and their axons project through many regions of the cerebrum, cerebellum, brain stem, and spinal cord. Together with other neurotransmitters, norepinephrine regulates alertness and focus, plays a role in the wake–sleep cycle, and is responsible in part for feelings of fear and anxiety. It is also thought to play a role in basic instinctual animal behaviors, including the search for food and water.[27] The closely related compound epinephrine (also known as adrenaline) acts systemically and increases the heart rate, constricts peripheral blood vessels, opens air passages, and dampens sensations of pain. Such effects are particularly important when facing a threat and are frequently felt when one is startled.[28] Both norepinephrine and epinephrine bind to the same class of neurotransmitter receptors, called the adrenergic receptors.

A number of plant-derived chemicals interact with the adrenergic system. For example, the jointfir *ma huang* (*Ephedra sinica*), employed in Chinese medicine for millennia, produces the potent stimulant alkaloid ephedrine (figures 4.7 and 4.8). Ephedrine is an agonist at adrenergic receptors in the central and peripheral nervous system, which causes an increase in heart rate and blood pressure, in addition to bronchodilation and other effects. The drug also enhances the release of norepinephrine from neurons in the periphery.[29] Since it stimulates the heart rate and gives the sense of increased energy, ephedrine (in the form of pills containing jointfir herb) was marketed in the United States during the late twentieth and early twenty-first centuries as a performance-enhancing and weight-loss supplement. Most ephedrine-containing supplements were ultimately banned by the Food and Drug Administration in 2004 because of deaths associated with the use of *Ephedra*-containing products.[30]

The bark of the yohimbe tree (*Pausinystalia johimbe*), thought to be an aphrodisiac, contains the alkaloid yohimbine, which exerts its activities by antagonizing a subset of adrenergic receptors.

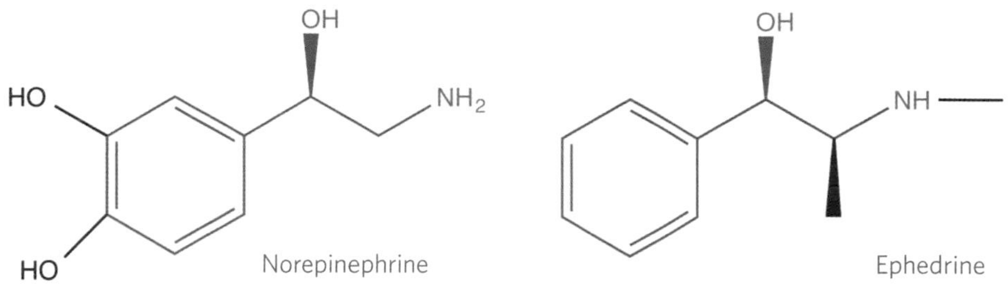

OH

HO

HO   NH₂
                        Norepinephrine

OH

                                                NH ——

                                                    Ephedrine

FIGURE 4.7 The neurotransmitter norepinephrine; the adrenergic receptor agonist ephedrine, from the jointfir. Structural similarities are shown in purple.

FIGURE 4.8 Jointfir, source of the adrenergic receptor agonist ephedrine.

Interestingly, blocking one type of adrenergic receptor induces the release of norepinephrine systemically, which increases the heart rate and blood pressure.[31] Perhaps it is a combination of these physiological effects and an indirect influence on mood that gives the yohimbe herb its reputation for improving male sexual function.[32]

The molecule reserpine from the Indian snakeroot (*Rauvolfia serpentina*) and other plants was widely used in the twentieth century to treat high blood pressure. Reserpine binds to the storage structures inside neurons that contain norepinephrine, dopamine, and serotonin, ultimately resulting in their depletion. Reduced signaling by norepinephrine leads to a decrease in blood pressure.[33]

The neurotransmitter dopamine is structurally related to epinephrine and norepinephrine, but it serves distinct roles in many aspects of cognition and behavior (figure 4.9). Most dopamine-producing neurons reside in the midbrain and project into the cerebrum, where they play an important role in learning and memory, control

of movement, and motivation.[34] Dopamine's contribution to voluntary movement is evident in the symptoms of Parkinson's disease, which results from the loss of a subset of dopaminergic neurons and impairs the patient's muscle control.[35] A role for dopamine in influencing mood is revealed by the effects of reserpine, which, in addition to its function in lowering blood pressure, reduces dopamine signaling and can induce psychological depression.[36]

Conversely, an increase in dopamine signaling can lift a person's mood. A striking example of this effect is the use of cocaine, the active principle of the South American coca shrub (*Erythroxylum coca*). The alkaloid blocks the reuptake of dopamine, norepinephrine, and serotonin at presynaptic nerve terminals and therefore artificially increases the concentration of these neurotransmitters in synapses.[37] Together, these neurotransmitters produce an intense stimulation of the central nervous system, resulting in increased heart rate and blood pressure along with elevated alertness, self-confidence, and lightened mood.[38] Among the neurotransmitters, dopamine is most directly responsible for producing a sense of well-being (euphoria), a feeling associated with the use of a number of psychoactive drugs.

Serotonin (sometimes referred to by its alternative name, 5-hydroxytryptamine) is produced by neurons in the brain stem that project into the spinal cord, medulla, cerebrum, and cerebellum (figure 4.10). Serotonergic neurons are thought to make so many connections that essentially every

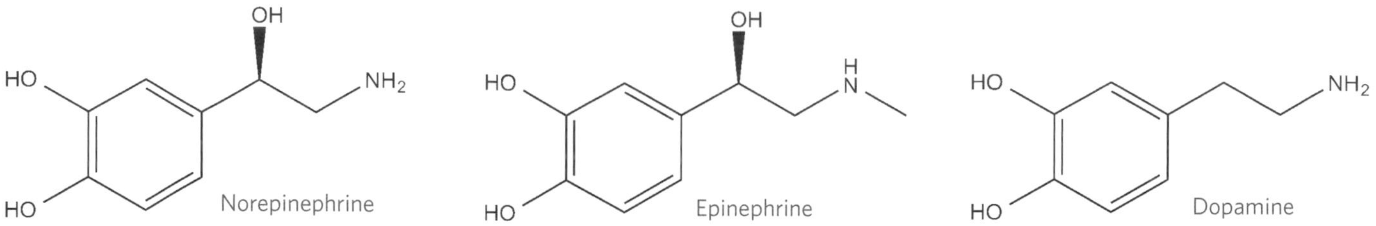

FIGURE 4.9 The neurotransmitters norepinephrine, epinephrine, and dopamine share a structural resemblance but differ in physiological properties.

neuron in the brain forms a synapse with one.[39] Serotonin has an important and widespread role in human behavior and higher-order functions, as it governs attention, mood, aspects of emotion, and the integration of sensory information.[40] While neurobiology researchers continue to dissect serotonin's multiple functions, it is clear from the effects of serotonin-like drugs that this neurotransmitter is crucial in processing stimuli and creating an individual's private reality.

Numerous plant compounds have been found to bind to serotonin receptors and induce a range of abnormal sensory experiences. For example, the ayahuasca beverage, composed of several plant ingredients including the ayahuasca vine (*Banisteriopsis caapi*) and chacruna (*Psychotria viridis*), is employed by indigenous peoples in tropical South America in spiritual healing practices. The prepared drink contains dimethyltryptamine, which closely resembles serotonin and likely acts as an agonist at its receptor (see figure 4.10).[41] Among its many effects on perception, ayahuasca induces visual anomalies consisting of complex geometric patterns. Likewise, a number of fungal active principles affect the serotonin system and produce changes in mood and sensory alterations, such as the compounds psilocin and psilocybin from *Psilocybe* and other genera of mushrooms (see figure 4.10).[42]

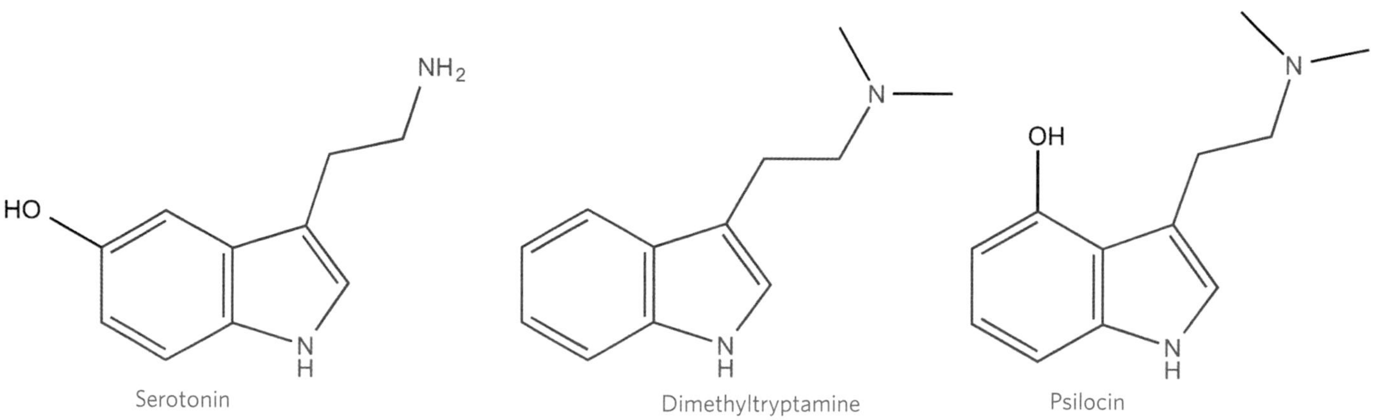

FIGURE 4.10 The neurotransmitter serotonin; the serotonin receptor agonists dimethyltryptamine, from the ayahuasca beverage, and psilocin, from *Psilocybe* and other genera of fungi. Structural similarities are shown in purple.

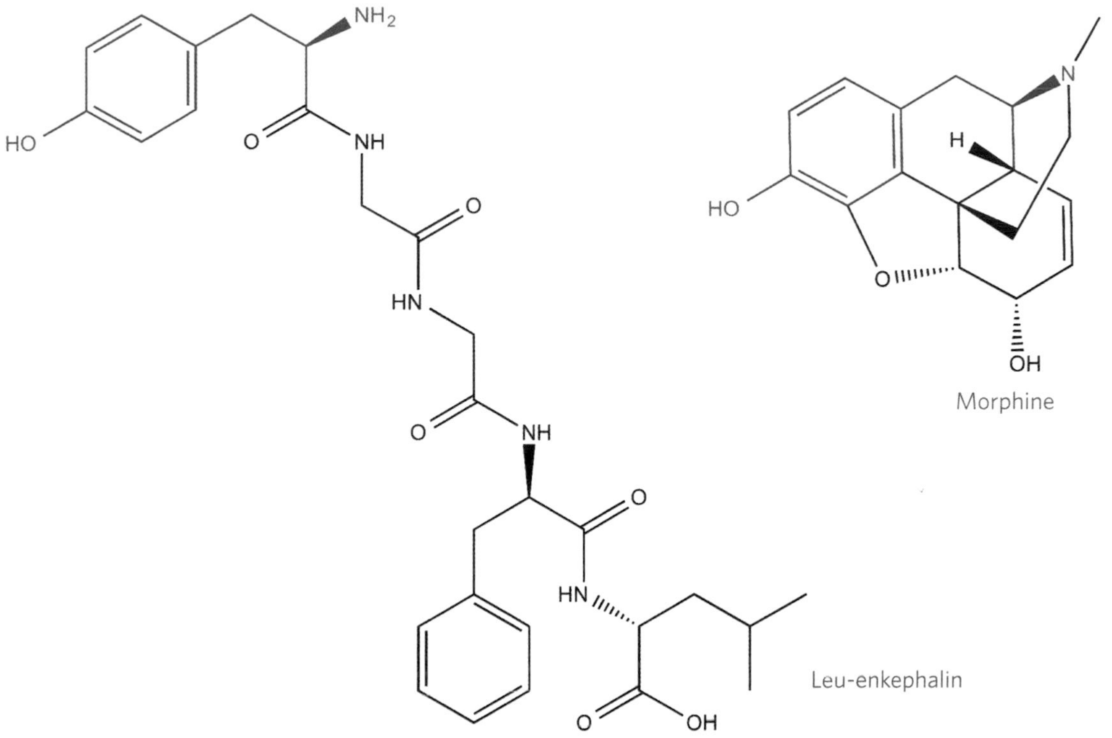

FIGURE 4.11 The endogenous neuropeptide leu-enkephalin; the opioid receptor agonist morphine, from poppy. Structural similarities are shown in purple.

Endorphins

The neuropeptide (small protein) endorphins (figure 4.11) and related neurotransmitters bind to opioid receptors, which are distributed widely in the brain and spinal cord as well as in the neurons that regulate the intestinal smooth muscle and elsewhere in the periphery.[43] Among their numerous roles, these chemicals inhibit the perception of pain and are produced at higher levels during stressful experiences, such as childbirth, and at physically demanding times, such as while exercising. Opioid agonists can counteract the transmission of pain sensation by blocking signals at the spinal cord or dampen pain perception in the medulla or midbrain.[44] In other circumstances, opioid agonists can depress cardiac and respiratory activity, induce vomiting, reduce activity of the gastrointestinal tract, suppress the cough reflex, and interfere with motor control.[45] The opium poppy (*Papaver somniferum*) has a long history of use for its pain-reducing, sleep-inducing, cough-suppressing activities. These properties are attributable to alkaloids, including morphine, that bind to endorphin-sensitive opioid receptors as agonists and produce analgesia, central nervous system depression, and other effects (see figure 4.11).

Endocannabinoids

Endocannabinoid neurotransmitters such as anandamide play a widespread role in modulating mood, hunger, body temperature, coordination, memory, and the perception of pain (figure 4.12). Their receptors, the cannabinoid receptors, are abundant in the central nervous system, and in many settings they function by down-regulating excitatory signals, such as by suppressing the release of glutamate from presynaptic nerve terminals. Cannabinoid receptors are found at greatest density

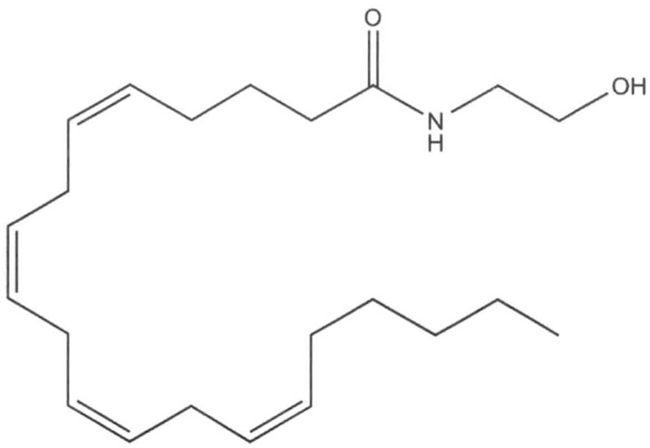

Anandamide

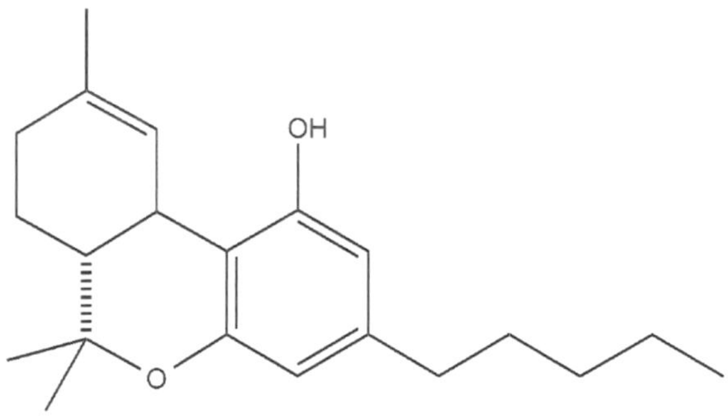

Δ9-tetrahydrocannabinol

FIGURE 4.12 The neurotransmitter anandamide; the cannabinoid receptor agonist Δ9-tetrahydrocannabinol, from hemp.

in the cerebral cortex, basal ganglia, hippocampus, and spinal cord, where they influence thought, memory, and movement.[46]

Traditional medical and religious practices across Europe and Asia valued hemp (*Cannabis sativa*) for its effects on a person's sensory experiences, now understood to be caused by its active principles, including the cannabinoid receptor agonist Δ9-tetrahydrocannabinol (see figure 4.12). This terpenoid compound produces symptoms of altered sensory perception, anxiety or calm (depending on dose and mental state), increased appetite, and analgesia.[47]

COMPLEXITY OF PSYCHOACTIVE DRUG ACTION

Since neurons of the central nervous system are intricately interconnected, making synapses with many other cells and projecting widely across regions of the brain, the multiple neurotransmitter pathways allow cells to integrate diverse sources of input and modulate signals appropriately. For example, orchestrating a simple voluntary movement such as picking up a pen requires the participation of many parts of the brain: those involved in object recognition, planning and execution of movement, and the entire motor-sensory pathway into the periphery. Ultimately, a great number of neurons are involved

in such an activity, and they are spread throughout the cerebral cortex, basal ganglia, cerebellum, thalamus, spinal cord, and elsewhere, signaling through many types of neurotransmitters.[48] The integration of so many brain regions and communication pathways as the basis of human experience (for example, consciousness) remains an important problem of neuroscience research.

Because of the interconnectedness of neural signals and the "cross-talk" between neurotransmitter functions, psychoactive drugs can produce a range of physical and mental changes attributable to primary actions (such as agonism and antagonism of neurotransmitter receptors) and to secondary interactions through additional neurotransmitter signaling pathways. For example, the peyote cactus (*Lophophora williamsii*), which grows in Mexico and the southern United States, has a long history as an aid to American Indian religious and medical practice, producing excitement and a sense of communion with the spirit world. Its active principle, mescaline, binds to a type of serotonin receptor found mostly in the frontal cerebral cortex, which is likely responsible for the altered sensory perceptions it induces. As a secondary effect, it appears to modulate the levels of glutamate and dopamine.[49] Meanwhile, a sense of exhilaration and central stimulation is mediated by acetylcholine and other neurotransmitters.

While acting through diverse neurotransmitter pathways, many psychoactive drugs share the characteristic of being pleasurable. The feeling of satisfaction is closely linked to brain structures in the limbic system that integrate learning, memory, emotion, and behavior (figure 4.13). From an evolutionary perspective, it makes sense that the animal brain would develop circuitry to associate food, sex, and goal-oriented behavior with pleasure. By connecting the sense of reward with pathways of memory, animals can learn to anticipate and seek out objects and activities that will advance their chances for survival. The major components of the limbic system communicate via dopaminergic transmission and draw input from cholinergic, opioidergic, and many other neurons.[50] By modulating dopamine signaling, numerous plant-derived active principles engage the brain's responses to pleasurable stimuli.

For example, cocaine inhibits the reuptake of norepinephrine and dopamine from synapses, resulting in an intense central stimulation (mostly via norepinephrine) and feelings of euphoria caused by increased dopaminergic signaling in the limbic system. Moreover, convergent connections in the nervous system result in a great number of neurotransmitters acting in part through the dopamine pathway. The drug scopolamine, produced by jimsonweed (*Datura stramonium*, also called thorn apple) and its relatives, is an acetylcholine receptor antagonist that gives rise to diverse physiological effects, ranging from dry mouth and blurred vision to increased heart rate. It disrupts acetylcholine signaling in the brain, inducing sensory illusions, amnesia, and cloudy thoughts.[51] It also produces euphoria by secondary effects that activate dopamine signaling.

Some active principles of the poppy, such as morphine and codeine, induce dopamine release in the limbic system.[52] Nicotine, cannabinoids, alcohol, and many stimulant drugs, despite the vast differences in their physiological effects and mechanisms of action, all also provoke a sense of euphoria. It is clear that for those compounds that activate the brain's reward center, dopamine is a

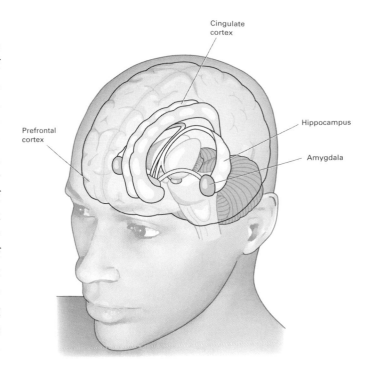

FIGURE 4.13 The limbic system plays an important role in emotion, motivation, and memory. Its interconnected structures signal largely via the neurotransmitter dopamine and communicate extensively with other neurotransmitter pathways. (Some structures are not shown.) (Adapted from John P. J. Pinel, *Biopsychology*, 9th ed. [Boston: Pearson, 2014], fig. 3.27)

key player in the pleasure of drug taking and in the development of addiction.

TOLERANCE, DEPENDENCE, AND ADDICTION

Many medicinal plants produce active principles that interfere with the normal biological processes of the nervous system, whether by modulating the intracellular processes of neurons, blocking synaptic communication, or substituting for the body's own neurotransmitters and initiating abnormal signals. As a result, the effects of psychoactive drugs can be rather striking for the initiate. On repeated and long-term exposure, however, the body compensates for the drug-induced state of neural function by altering the activity of neurotransmitter receptors and neurotransmitter levels, among other changes. This condition, called tolerance, amounts to a reduced effect of the drug at a constant level of administration. Because of tolerance, some drug users ratchet their consumption upward over time simply to obtain a similar level of effect.

Since the body adjusts to increased drug levels over time, the discontinuation of a drug can unmask the compensated state, with its natural neurotransmitter systems unable to perform their full range of predrug functions. For people accustomed to drug taking, this can produce unpleasant symptoms. For example, if opiate drugs such as morphine suppress pain, cause the skin to flush, and induce euphoria, their cessation can reveal the modified physiological state, resulting in pain, chills, and dysphoria. The symptoms of withdrawal, which differ depending on the drug, dose, and individual, are hallmarks of drug dependence. For some people, the experience of withdrawal, while uncomfortable, eventually passes, and the normal physiological state returns over time.

Addiction to a drug is a pathological drive to take it despite the negative consequences. It can exist in the absence of tolerance and dependence and can persist or reappear after withdrawal symptoms have subsided, called relapse. Addiction researchers, therefore, believe that the condition is closely related to the neural processes of reward and associative learning.[53]

CLASSIFICATION OF PSYCHOACTIVE DRUGS

By acting on neural pathways governing cognition and sensation, herbal products can profoundly affect a person's alertness and sensitivity to stimuli. Many plant-derived substances can excite the nervous system and give the sense of elevated mental sharpness and focus. Such compounds are known as stimulants. Conversely, drugs that dull the mind and produce the sensation of numbness are called narcotics. (It is worth bearing in mind that among government agencies and in the press, the term "narcotic" is frequently used to refer to any mind-altering drug, without distinction to its physiological effects.) Drugs that produce a feeling of well-being are euphoriants.

The central nervous system processes sensory input in the cerebrum to develop an awareness of

EFFECTS OF PSYCHOACTIVE DRUGS

Euphoriants	Create a sense of well-being and pleasure
Stimulants	Increase mental alertness and (usually) elevate physiological activity
Narcotics	Dull the senses and (usually) reduce physiological activity, induce sleepiness
Phantasticants	Alter senses such that sounds, sights, smells, and the like seem different than they actually are
Hallucinogens	Provoke sensation of stimuli that are not actually present
Psychedelics	Cause a sense of liberation from one's mind
Entheogens	Lead to a spiritual or religious experience

A number of these outcomes can occur simultaneously, depending on the drug and the user's subjective experience.

the world. However, plant-derived chemicals that resemble neurotransmitters and act at their receptors can alter its ability to generate an accurate interpretation of reality. For instance, a number of herbal active principles create visual distortions or the impression of light, sound, or other inputs that do not exist. Such chemicals form special classes of psychoactive compounds. Phantasticants are substances that alter the sensory input in such a way that sounds, images, smells, and other stimuli seem different than they actually are. They can make colors, for example, appear more intense than under normal circumstances. In contrast, hallucinogens cause the mind to experience sensations and perceive stimuli that do not exist, such as geometric patterns in the sky and dialogue with animals. Another class of psychoactives, the psychedelics, produce feelings of liberation from one's mind coupled with false sensory imagery. Finally, the entheogens produce experiences of spiritual or religious awakening, sometimes with visions. By altering in various ways the ability of the mind to process its complicated inputs, such chemicals can produce poignant illusions.[54]

PSYCHOACTIVE PLANT COMPOUNDS IN CONTEMPORARY NEUROSCIENCE

For most of human history, making sense of the mind was the domain of philosophers and theologians. With the advent of cell theory and the rise of biological chemistry, scientists employed new tools to pry apart neurons and investigate the mysteries of their communication. Yet the fragile axons and ramified dendrites did not lend themselves to surgical manipulation, and teasing apart the complexity of chemical transmission was beyond the technical capabilities of the nineteenth- and early-twentieth-century laboratory. The specific, potent actions of plant-derived compounds led investigators to search for receptors, deduce the biochemical basis of neurotransmission, and establish how synapses relay neural messages. By studying the effects of herbal active principles, neurobiologists determined how neurotransmitters function at their receptors.[55] Indeed, many neurotransmitters were discovered only because a plant-derived chemical produced effects that had not been explained.

For example, a century and a half of chemical analysis from the early nineteenth century through the 1980s had identified hemp's active principles (including $\Delta 9$-tetrahydrocannabinol) and elucidated their chemical structures. However, their mechanism of action could only be speculated on. Ultimately, investigators used plant cannabinoids as a sort of "bait" to see what they bound in brain tissue, and in 1988 they came up with the cannabinoid receptor. Surely the cannabinoid receptor did not exist solely to detect hemp consumption, researchers contended, and they set out to identify the neurotransmitter compound produced in the brain that normally signaled via the cannabinoid receptor. By the mid-1990s, several endocannabinoid (for endogenous cannabinoid) neurotransmitters had been discovered, and their functions could be described at the molecular level.[56]

Similarly, the identification of opioid receptors (sensitive to plant-derived opiate active principles) led investigators to seek molecules in the nervous system that acted similarly to the poppy compounds, ultimately resulting in the discovery of the endorphins and related neuropeptides.[57] The action of acetylcholine in the central nervous system and periphery would be poorly understood if not for the experimental utility of nicotine, atropine, and other agents. Armed with a chemical toolkit of natural psychoactive compounds, neuroscience researchers teased apart the structure and function of the nervous system, a legacy that yielded a detailed and improving understanding of the biological nature of the mind.[58]

The nervous system's basic units,

the neurons, communicate with one another using chemical signals that ultimately govern the body's ability to sense its surroundings, move its muscles, and feel emotions. The nervous system accomplishes these tasks by coordinating the activities of the central and peripheral divisions. The action of plant-derived drugs outside the brain accounts for many physiological effects.

Within the central nervous system, psychoactive compounds in medicinal plants can have a tremendous influence on a person's perception and emotional state. Plants have evolved to produce dozens of chemicals that bind to neurotransmitter receptors, acting as agonists to promote signaling or as antagonists to block it. Through their many connections, neurons integrate diverse messages and generate an individual tableau that psychoactive compounds can distort. Many psychoactive drugs stimulate the reward center of the brain, the limbic system, providing a sensation of contentment and gratification that is highly alluring but that can lead to abuse. In addition to their long use in medicine and spiritual pursuits, plants that produce compounds acting on the nervous system have provided researchers with new tools to explore the cellular basis of sensation, perception, and the human experience.

Chapter 5

Poppy
Papaver somniferum

A poppy flower

The poppy is an herbaceous annual plant that produces upright flowering stems up to 1.5 meters in height and leaves with toothed edges that grow close to the ground.[1] Each flowering stem is tipped by a single flower that bears (usually) four large, thin petals. Poppy flowers are white, red, or purple. The flowers give rise to spherical seed pods known as capsules. Each poppy capsule matures to contain hundreds to thousands of small black, gray, or white seeds. These nutritious seeds have been used in European and western Asian cuisine since antiquity, pressed for oil and baked into bread and pastries.[2] Domesticated poppy, recognized by its larger capsule and tendency to retain mature seeds for harvest, was propagated from a wild ancestor, probably in the European western Mediterranean region.[3] The plant tolerates a range of temperate and tropical climates (at elevation in the latter) and now grows on all inhabited continents.

ANCIENT USE IN THE MEDITERRANEAN REGION

Archaeological evidence supports the notion that the poppy was widespread in Europe and the Near East in ancient times. Among the earliest finds are poppy seeds at a 7700-year-old Neolithic settlement in northern Italy and dried poppy capsules associated with woven baskets at a burial cave in southern Spain dated to 2200 B.C.E. A number of Swiss, German, Polish, Czech, and other sites throughout northern and central Europe have turned up remains of poppy dating back 4000 or more years. Such botanical relics indicate that the poppy was collected, probably for use as food, across a vast expanse of Europe. Artifacts also demonstrate the poppy's significance to ancient peoples. For example, Mycenaean Greek (1550–1100 B.C.E.) jewelry items from modern-day Greece and Turkey depict what appear to be poppy capsules.[4] Artifacts such as poppy capsule–shaped juglets hint that poppy may have been prepared as a drug as early as the sixteenth century B.C.E. (figure 5.1).

While it is unknown when the poppy's medicinal properties were discovered, by the time of the Egyptian Ebers papyrus of 1550 B.C.E., the herb was documented as a pharmaceutical substance of great strength.[5] For example, the plant figures in a prescription for a "remedy to stop the crying of a child." This recipe requires "pods-of-the-poppy-plant" mixed with "fly-dirt-which-is-on-the-wall," eaten for four days. According to the text, "it acts at once!"[6]

The Greeks and Romans left extensive records detailing the uses of the poppy in cuisine and pharmacy.[7] The early Greek botanist Theophrastus (371–287 B.C.E.) mentioned the plant in *Enquiry into Plants*, and the influential herbalist Pedanius Dioscorides (ca. 40–90) noted that "its seed is baked into bread to use in a health-inducing diet." He also recommended the boiled leaves and capsule to help one sleep and ground seeds consumed in wine to treat diarrhea and gynecological infections. Furthermore, he explained that the *opos* (juice) expressed from a cut plant was a particularly potent medicine. "It is analgesic, soporific [sleep-inducing], helpful for digestion, and it comes to the aid of coughs and abdominal conditions," Dioscorides wrote.[8] Galen (129–ca. 216) agreed that the "seed of the cultivated poppy is useful as a seasoning spread on bread, just like sesame seed." As a plant with a cooling property, according to Galen, "it is soporific and, if taken to excess, causes lethargy."[9] In these and many other passages, classical physicians documented some key pharmacological properties of what modern-day medicinal chemists recognize to be an important class of plant active principles: the opiates. The opiates, including morphine and

FIGURE 5.1 An earthenware vessel from Cyprus believed to have contained opium, ca. 1600–1400 B.C.E. The shape resembles an upside-down poppy capsule, and the groove opposite the handle might represent a cut made for opium harvesting. (Science Museum, London. Wellcome Images, L0058861)

codeine, are poppy-derived alkaloid compounds grouped together because of their shared numbing effects. In his detailed treatment of the poppy, Dioscorides noted the pain-relieving, sleep-inducing, cough-suppressing, and antidiarrheal effects of the herb. The only major medicinal effect not described is the feeling of pleasure it gives to those who consume it.

The method of preparation affects its pharmacological properties. The aerial portion of the plant (leaves and stems) is known as the poppy herb, and it accumulates relatively little opiate content. Dioscorides recommended poppy herb lozenges for coughs and abdominal conditions.[10] The poppy seeds contain edible oils and are rich in protein, making them a nutritious element of cuisine.[11] They have only a minuscule level of opiates.[12] Ancient poppy growers must have noticed that the capsule was more medicinal than the rest of the plant and that its thick, milky latex (the fluid exuded from wounds) was more potent still (figure 5.2). Dioscorides described a method to obtain this substance, called opium, in a form still practiced today.[13] "It is necessary to scratch all around the capsule with a knife in a way as not to pierce through its inner part, and to make superficially straight cuts at the side of the capsule," he explained, "then wipe up the tear that flows with the finger into a spoon."[14] By allowing the latex to flow and harden, then carefully collecting it after sev-

Latex

An opaque, milky fluid that exudes from certain plants when cut

FIGURE 5.3 A man in Turkish dress preparing opium. (Woodcut from Angelo Sala, *Opiologia* [1618]; Wellcome Library, London, M0010469)

eral hours, cultivators accumulated significant quantities of a potent, sticky, solid opium product (figure 5.3). To obtain the medicinal effect desired, people ate this thickened poppy latex alone or mixed it with foods or drinks. While the ability of opiates to soothe discomfort earned the poppy its prominent role in the classical *materia medica*, the dangers of opium and its concentrated active principles reminded physicians then (as in the modern day) that the line between medical benefit and toxic effect can be simply a matter of dose. Dioscorides was well aware of this risk, as he warns of opium, "but when too much of it is drunk, it plunges into a coma and it is deadly."[15]

Such lethal effects intrigued those whose intents were to harm rather than heal. Poppy proved itself a subtle agent of death. Pliny the Elder (23–79), writing in Rome, relayed the story of an elderly man with an incurable illness who took his own life with an overdose of opium, his "malady having rendered existence quite intolerable to him."[16] Opium served in assassination as well. Mixed with the victim's food, opium might be undetected until the target fell into a deep, permanent sleep. It is thought that the Emperor Claudius's wife, Agrippina, in 55 C.E. put opium in the wine of her fourteen-year-old stepson, Britannicus, to procure the throne for her own son, Nero.[17]

The Greek and Roman period saw widespread use of the poppy for food, for its medicinal properties, and for its symbolism in artifacts and mythology. For example, the Greek goddess Demeter (Roman Ceres), representing fertility and agriculture, is often depicted holding stems of poppies along with stalks of grain. Worshippers are said to have made offerings of poppy to her in hopes of a bountiful harvest.[18] The Greek Hypnos (Roman Somnus), the god of sleep, is frequently portrayed

FIGURE 5.2 Latex exuding from incised poppy capsules. (Photograph by Toni Kutchan, Donald Danforth Plant Science Center; from Marion Weid et al., "The Roles of Latex and the Vascular Bundle in Morphine Biosynthesis in the Opium Poppy, *Papaver somniferum.*" *Proceedings of the National Academy of Sciences USA* 101 [2004]: 13957–13962. © 2004 National Academy of Sciences USA)

holding a bunch of poppies, highlighting the plant's potency as a sleep inducer.[19] The brother of Hypnos, tellingly, is Thanatos (Roman Mors), the god of death.

OPIUM AS WORLD MEDICINE

Apparently, the Romans did not participate in long-range trade in poppy, and after the fall of their empire, very little record exists of the poppy's role as a major medicinal or economic plant.[20] However, among the Muslims, whose influence began to spread during the eighth through tenth centuries, commerce in opium was an important component of their cultural domain. Poppy was cultivated in Persia, Anatolia, and elsewhere in the Islamic world and traded from Moorish Spain through North Africa, the Middle East, and Asia. The Islamic physicians (many of whom, such as Abu Ali al-Husayn ibn Abd Allah ibn Sina [Avicenna, 980–1037], were widely respected in Europe and Asia for centuries) incorporated the poppy into their medical textbooks as painkillers and sleep inducers, much as the Greeks and Romans had. The Arab and Persian caravans and trading vessels brought the poppy to India and Southeast and East Asia, where cultivation yielded opium as a valuable medical commodity.[21]

While knowledge of opium had not totally disappeared during the European Middle Ages, its prominence certainly faded. The European Crusaders of the eleventh through thirteenth centuries also learned more of the poppy's medicinal values from the Muslims against whom they fought.[22] As the Venetian merchant mariners developed an extensive and profitable Mediterranean trade in the fifteenth century, they sought sources of opium for importation. At first, the Arabs supplied Venice with opium and spices, goods that the Venetians traded throughout Europe. Eventually, the Portuguese eclipsed Venice by sailing around Africa and purchasing opium directly from dealers in India. It was during this time that opium regained an important role in the European medicine chest.[23]

Keenly aware of its capacity to alleviate their patients' suffering, many physicians came to view opium as an essential weapon in their therapeutic arsenal. In time, they developed new formulations of opium. In its original state, opium is the gummy, brownish dried exudate of the poppy capsule, and various men came up with inventive admixtures and concoctions to offer their patients. The radical Swiss doctor Paracelsus (1493–1541) is credited with having invented a type of opium tablet that he named laudanum, from the Latin word *laudare* (to praise).[24] By the early seventeenth century, laudanum was reformulated as an alcoholic tincture (opium dissolved in alcohol) containing various medicinal plant extracts, sometimes mixed with honey and spices for ease of ingestion.[25] Over time, laudanum was standardized to contain simply opium dissolved in alcohol.[26] The unquestioned effectiveness of opium against pain and its increasingly popular, simple-to-take formulations led many to praise its usefulness—the patriarch of English medicine Thomas Sydenham (1624–1689) said that "medicine would be a cripple without it"—but led some to recognize a growing dependence on its soothing yet barbed hold.[27] The combination of physical tolerance and opium's activation of pleasure pathways in the brain gave rise to many long-term (probably addictive) relationships between patients and the poppy.

To the British physician George Young (1692–1757), opium's overuse posed significant risks to the public, not from acute toxicity but from the consequences of chronic abuse. "Opium is a poison by which great numbers are daily destroyed; not, indeed, by such doses as kill suddenly, for that happens very seldom," he wrote, "but by its being given unseasonably in such diseases and to such constitutions for which it is not proper."[28] Another eighteenth-century writer noted that patients taking opium often require increased doses over time, advising physicians to consider "the constitution and habit of the patient" when prescribing the drug.[29]

While opium and its derivatives have had striking effects on all the societies in which they were used and abused, the story of opium in China is particularly poignant.

OPIUM IN CHINA

The poppy probably reached China via Arab traders over the Silk Road by the eighth century, and it saw use for centuries as a medicinal herb.[30] During the Tang dynasty (618–907), it was cultivated in the southwestern Chinese province of Sichuan, and descriptions of the poppy flowers appear in poetry of the era.[31] By the time of the Song dynasty (960–1279), Chinese scholars recommended the seeds, capsules, and herb of the poppy as powerful medicines. For example, a book of pharmacy from 973 declared that "its seeds have healing powers," especially suggested to counteract the toxic effects of mercury-containing drugs.[32] Later, the government official and poet Su Shi (1037–1101) praised the convenience of poppy in medicinal herbal teas useful against sunstroke, writing, "even a child can prepare the *yingsu* [poppy] soup."[33] Other texts recommended the poppy capsule to treat dysentery and pain.[34]

By the fourteenth century, the poppy had become integrated into medical practice and was considered useful against a broad range of ailments. However, its dangers had also become apparent. In a passage reminiscent of the warning expressed by the Greek herbalist Dioscorides, writing many centuries earlier, the physician Zhu Zhenheng (1282–1353) explained: "The poppy capsule is used extensively for cough at the present time in the case of those who are weak and consumptive. It is employed to take away the cough. It is used also for diarrhea and dysentery accompanied with local inflammation. Though its effects are quick, great care must be taken in using it because it kills like a knife."[35]

It is not clear when the Chinese began using the poppy in its dried-latex opium form, but during the Yuan dynasty (1271–1368) and afterward, opium was undoubtedly used as a tribute good and reached China from nearby nations.[36] Produced in China and imported from vassal states in Southeast Asia, opium was increasingly available throughout the country during the fourteenth and subsequent centuries.

In addition to its use in treating pain, cough, and intestinal problems, opium transitioned to a role in spiritual sexual health. By the early fifteenth century, opium was being offered by herbalists as both *yao* (medicine) and *chunyao* (spring medicine, or aphrodisiac).[37] In the *Grand Materia Medica* of 1596, the landmark medical book by Li Shizhen (1518–1593), opium was listed to treat diarrhea and to "help control the essence of men; ordinary people use it for the art of sex."[38] In the medical interpretation of the era, opium was useful to control the male sexual response, as retention of ejaculate was thought to regenerate the spirit.[39] It is not surprising, then, that even the Chinese emperors indulged frequently in such medicine. In fact, the Ming emperor who ruled as Wanli from 1573 to 1619 was long suspected to have been addicted to opium.[40] In a report published in 1997, the Chinese Ministry of Public Security confirmed that the emperor's bones contained high levels of the opium compound morphine.[41]

Until the sixteenth century, the Chinese took opium much as the Europeans did: by eating the opium paste or mixing it with water or alcohol to drink. After contact increased with Europeans, especially in coastal cities, the Chinese learned to smoke, and they applied that knowledge to the consumption of opium.

Sometime during the second half of the sixteenth century, tobacco (*Nicotiana* spp.) entered China through trade with the Spanish or Portuguese.[42] By the 1620s and 1630s, Chinese writers noted that smoking tobacco extended through all strata of society, and many individuals in the government and military were frequent users.[43] The Chinese acquired pipes and soon innovated all varieties of smoking instruments, including water pipes, pipes with long stems, and short pipes for tobacco consumption. While opium was typically prepared in soups and as pills to be swallowed until the seventeenth century, during the seventeenth and eighteenth centuries smoking opium—through the Chinese-conceived long pipes—became commonplace. Among the lower classes, bamboo pipes substituted for the more elegant devices employed by the nobles.[44]

Opium prepared for smoking is specially processed to promote its rapid vaporization under the heat of a lamp. First, the raw opium is boiled and strained to remove traces of poppy herb. In the process, the opiates become concentrated in a dark

FIGURE 5.4 Workers rolled Indian opium into balls and dried them on large wooden racks before export. (Lithograph after Walter S. Sherwill, *The Stacking Room, Opium Factory at Patna, India* [ca. 1850]; Cornell University Library, RMC_2005_0215)

paste. Then, a small bead of smoking opium is teased apart with small needles and worked, under heat, into a ball that fills the minute cavity of a porcelain or an earthenware bowl at the end of the long pipe. By placing the bowl directly over an oil flame, the smoker more easily vaporizes and inhales the opium.[45] By the eighteenth century, opium smoking was widespread through China and among its Asian trading partners.

Much of the opium smoked in China was cultivated in India, a territory gradually entering British trade domination during the late eighteenth century. With the assistance of the British government, the British East India Company obtained a monopoly on the Indian opium trade (figure 5.4). With that, the East India Company was able to sell its Indian opium in exchange for silver to merchants, ostensibly independent agents who shipped the product to stations off the coast of China for distribution. At the same time, the East India Company used the silver to purchase Chinese tea for delivery to Great Britain.[46] By the nineteenth century, the East India Company employed thousands in the processing and packaging of Indian opium and held monthly opium auctions in the cities of Calcutta and Bombay to supply largely Chinese wholesalers. The Chinese elite continued to promote the health-restoring virtues of opium use, and this philosophy permeated the urban centers, from the seafaring south to the Qing (1644–1912) court in Beijing. And while the Qianlong emperor (r. 1736–1799) did issue an edict to ban opium, little could be done to enforce it in a culture of booming trade and consumption.[47]

FIGURE 5.5 Opium smoking as a Chinese cultural practice: (*left*) two men smoking opium; (*right*) upper-class (Manchu) woman smoking opium, ca. 1900. ([*left*] From George Morrison, *An Australian in China* [1895]; [*right*] Library of Congress, Prints and Photographs Division, LC-USZ62-25834)

The British, for their part, tried to exert diplomatic pressure to open up China's vast market to free trade. In 1793, and again in 1817, representatives of the Crown visited Beijing demanding an embassy, access to the Chinese market, and a concession of land for foreign merchants. Both visits were graciously accommodated, but the Qing emperors refused the requests. Official policy notwithstanding, the smuggling trade of opium from India to China grew from 4244 chests in 1820, to 18,956 in 1831, to 40,200 in 1839.[48] The opium-smoking habit in China originated as an imperial medicinal-recreational sex aid and spread through cities among government officials and merchants, from the urban port areas of the east and south to areas north, west, and inland (figure 5.5). Members of the lower classes emulated their elite contemporaries in this trendy practice, and, coupled with prostitution, smoking seduced millions.

Numerous historical, economic, and social factors converged in the Chinese government's ill-fated effort to exert control over a rapidly degenerating opium and foreign trade situation.[49] The Daoguang emperor (r. 1821–1850) sent agents to enforce the prohibition, seizing and destroying a large quantity of British-owned opium. A number of skirmishes ensued, culminating in the arrival of a British expeditionary force in June 1840 that was victorious over the Chinese opposition. The Treaty of Nanjing (1842) ended this First Opium War (known as a "War of Free Trade" in Britain). In China's defeat, Britain received an indemnity, access to five ports for trade, and Hong Kong. Although the treaty obligations did not legalize the opium trade, after the first war, Qing officials only sporadically enforced its ban.[50]

Under constant pressure from Western diplomatic forces, China acceded to demands from the French and Americans for access to trade. In October 1856, Chinese officials boarded a Hong Kong–registered vessel in southern China in search of smugglers and pirates, an affront to the British colonial authorities.[51] The British retaliated by attacking Guangzhou and were joined by the French after a missionary was assaulted. This Second Opium War resulted in

The Opium Wars

FIRST OPIUM WAR (1839–1842)

Outcome	China ceded Hong Kong to British; opened five treaty ports for international trade; indemnity.

SECOND OPIUM WAR (1856–1860)

Outcome	China opened ten more treaty ports; allowed unimpeded access to European missionaries; indemnity.

strikes on major cities of the south and on Beijing's imperial palaces. In 1860, China ratified the Treaty of Tianjin, agreeing to open ten more ports of free trade, grant unimpeded access to the country by Western missionaries, legalize opium importation, and pay a substantial indemnity to Britain and France.[52]

The Opium Wars brought about an increased trade in opium, both imported and of domestic production, since even the large volume of Indian opium was insufficient to meet domestic demand, and Chinese farmers saw the poppy as a lucrative cash crop.[53] The ports of free trade—including British Hong Kong and the concessions to Western powers in Shanghai—provided one avenue for the drug to reach markets, and Chinese merchants effectively met the inland demand for opium.

The treaty port of Shanghai, with its booming trade and foreign population, became the center of a drug-recreation culture that soon spread worldwide. During the mid-nineteenth century, the opium shop transitioned from one where the average person could purchase opium for consumption at home to a veritable "flower-smoke house" (opium den), where opium and sex could be cheaply and easily purchased.[54] In the second half of the nineteenth century, China—battered by two asymmetrical wars, impoverished by unbalanced foreign trade, and increasingly dependent on narcotics—saw itself drawn into a period defined in great part by a rampant opium lifestyle.[55]

It is difficult to estimate the effect of opium on Chinese society during this period. Statistics of opium use are available only in the subjective diaries of witnesses, so there is no way to be certain how extensive opium consumption really was. A British government official estimated that 15 to 20 percent of the adult population in Xiamen, along the southeastern Chinese coast, smoked opium in 1870.[56] Other estimates for the era ran as low as one in 166 to as high as nine in ten opium users.[57] Among Westerners, some of the strongest critics of opium were missionaries. In 1900, one English writer lamented, "I hold that the opium vice is the most colossal in its pernicious effects that the world has ever known."[58]

OPIUM IN NINETEENTH- AND TWENTIETH-CENTURY WESTERN MEDICINE

As opium entered mainstream Chinese culture, it also became further established in the pharmacopeias of Europe and the United States. As a painkiller of unequaled power, with the singular ability to calm persistent coughs, ease the belly, and provide rest, opium and its alcoholic formulation, laudanum, had been extolled since the Renaissance by doctor and patient alike (figure 5.6). Yet opium's extensive use led some to require ever-increasing doses to meet their addiction. There was little effort, though, during much of the time that opium reigned as the frontline medicine for the challenges of a difficult existence, to identify the particular chemical compounds responsible for its effects.

As the study of pharmacy grew more empirical during the late eighteenth century, scientists began to investigate the chemical makeup of medicines.[59] In isolating chemicals from medicinal plants and testing their effects (on animals and people), these pioneer researchers sought the specific chemicals that caused physiological effects in humans. In these particular chemical compounds, isolated from all the other chemicals present in plants—the active principles—researchers hoped to discover the material basis of the actions of medicinal plants.

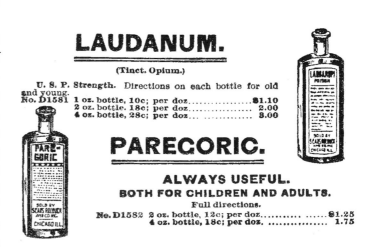

FIGURE 5.6 Opium in alcohol (laudanum) was an important medicine until the early twentieth century. Paregoric is a weaker, camphorated opium solution, here sold alongside laudanum by mail order from Sears, Roebuck and Company, 1897.

The process of identifying and characterizing the pharmacologically active chemicals in plant medicines is one of the key steps in understanding their physiological roles. Through such studies, medicine has gained insight into the function of the nervous system and attained safer dosing regimens. The isolation of active principles is also an important step in purifying and distributing such drugs for therapeutic (and illicit) use.

The first to tease out the active principle from the poppy—the first plant active principle to be so studied—was the German organic chemist Friedrich Sertürner (1783–1841), who accomplished the task in 1803.[60] Many more alkaloids were later determined from a wide range of medicinal plants by chemists working in Europe and elsewhere. Sertürner named the chemical he extracted from poppy morphine, after Morpheus, the Greek god of dreams (figure 5.7).[61]

Following the chemical isolation of morphine, European and American firms began to produce large quantities of it to meet the medical demand.[62] In the form commonly prepared during the nineteenth century, morphine consisted of a powder that could be added in precise amounts to alcohol or water for ingestion by patients seeking help sleeping, relief from pain, or alleviation of myriad other concerns.[63] With the invention of a practical hypodermic syringe in the 1850s, a quicker route of administration of morphine was available, useful for intense pain such as that faced during surgery.[64] Doctors or their patients dissolved morphine powder in water and injected it directly into the tissue or bloodstream, which resulted in faster pain relief than offered by the oral route.[65] Thus morphine ascended as a key anesthetic in Western medicine. During the Civil War (1861–1865), both opium and morphine administered by mouth and probably (to a lesser extent) morphine administered by injection were used to treat the pain of war wounds, diarrhea associated with dysentery, malarial fevers, and other battlefield afflictions.[66] While both orally administered opium and injected morphine had the capacity to produce dependence in its users, the direct route of administration delivered the active principle more quickly to the brain, producing a more intense feeling of reward and therefore a higher risk of addiction.

During the second half of the nineteenth century, opium and morphine were components of numerous patent medicines, drugs advertised to cure any number of ailments but whose ingredients remained secret. As a result, countless women fed their children morphine-laced syrups to calm their teething pains, many men used such products to help them sleep at night, and all took these medicines to ease the pain of a difficult life. Indeed, the morphine in these patent medicines made them enjoyable to take and deviously addictive.[67] As an American physician commented in 1871, "the popular nostrums of the day, the cholera-drops, the pain-killers, the lung-troches and other pectorals, draped as they are in a flaunty incognito . . . owe whatever inherent virtue (if any) they possess to the omnipresent leaven, opium. The stereotyped cautionary phrase, *Caveat Emptor*, should be the statutory appendix to every one of their trade-marks."[68] It is not known how many middle-class, small-town families became dependent on patent-medicine morphine during the nineteenth century, but the number is no doubt very large.[69]

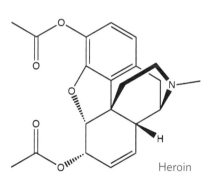

FIGURE 5.7 Opium-derived drugs: the principal opiate morphine; the opiate codeine; the semisynthetic opioid heroin.

In morphine, physicians discovered a new, sure way to alleviate their patients' troubles, and armed with a hypodermic syringe, many could at last offer something beyond the violent purges and blood-letting that persisted in some medical circles in nineteenth-century North America and Europe.[70] Morphine soothed those suffering from coughing diseases and helped agitated patients find calm and rest. However, morphine was a more effective painkiller than it was a cough suppressant, and it was highly addictive at therapeutic doses. There-fore, chemists sought to produce or isolate opiate compounds that treated pain or cough without the addictive properties.

Codeine, a compound in poppy latex less abun-dant than morphine, is composed of the morphine chemical structure with an additional small carbon-containing methyl group (see figure 5.7). This drug is a useful painkiller and cough suppressant, both slower acting and less addictive than morphine.[71] In the modern day, it often is prescribed for post-operative pain, for example, in combination with the anti-inflammatory drug acetaminophen.

The industrial era of the late nineteenth and early twentieth centuries saw a population explosion in crowded city districts characterized by cramped quarters and poor sanitation, fertile grounds for the spread of the bacterial lung disease tuberculosis. Morphine elixirs were somewhat useful to combat this scourge by suppressing the intense coughing that spread the bacteria in aerosolized sputum, yet the risk of drug dependence led chemists to develop new molecules with more potent painkilling and cough-suppressing effects and less potential for addiction. Chemists in London in the 1870s and researchers at Bayer Laboratories in Germany in the 1890s added two simple acetyl groups (just a few more carbon, hydrogen, and oxygen atoms) to the structure of morphine. The molecule they produced, in chemical terms diacetylmorphine, turned out to be a much more potent analgesic than morphine, and to the German pharmaceutical firm a *heroisch* ("mighty" or "powerful") medicine.[72] Bayer mar-keted the drug as heroin (see figure 5.7).

A white powder, heroin can be mixed with water and injected, formulated into pills for oral consumption, snorted into the nose, or vaporized and smoked. Heroin's rapid action is attributable to its acetyl adornments, which render the molecule more fat soluble and therefore more readily taken up into the brain from the bloodstream. Effects of a dose of heroin, particularly when injected, are felt intensely within seconds. Contrary to the English and German chemists' predictions, their modified morphine was far more addictive than any of the other opiates. Heroin was briefly marketed as a more effective alternative to morphine for pain relief and cough suppression in the late 1890s and early 1900s but soon fell out of favor in mainstream medical prac-tice as its human toll became clear (figure 5.8).[73] Yet its relative simplicity of manufacture and unequalled euphoriant effects have assured its persistence as a recreational drug into the twenty-first century.

OPIUM CULTURE
IN THE UNITED STATES

Americans had access to the opium-based alco-holic medicine laudanum from the early years of the colonies, but its use for medicine and recre-ation was probably limited by cost and availability.[74] The American experience with opium was shaped strongly by a wave of Chinese immigration to the United States and by homegrown demand for pat-ent medicines and morphine that mounted during the mid-nineteenth century. During the 1850s and later, thousands of Chinese men left their ances-tral villages, mostly in the south of China, to seek employment in American mines and railroads. Oth-ers found their way to cities along the West Coast and further inland, eventually settling in cities such as San Francisco, Seattle, Chicago, and New York. The men were usually poor and rarely brought their families, but some did bring with them the habit of opium smoking.[75] Chinese entrepreneurs, some of whom had originally immigrated for the purpose of prospecting for gold or working railroad jobs, estab-lished opium dens in the Chinese quarters of Ameri-can cities and offered their clientele an experience inspired by the decadent establishments of their homeland.[76] Attracting Chinese, Americans, and visitors alike, opium dens developed a reputation

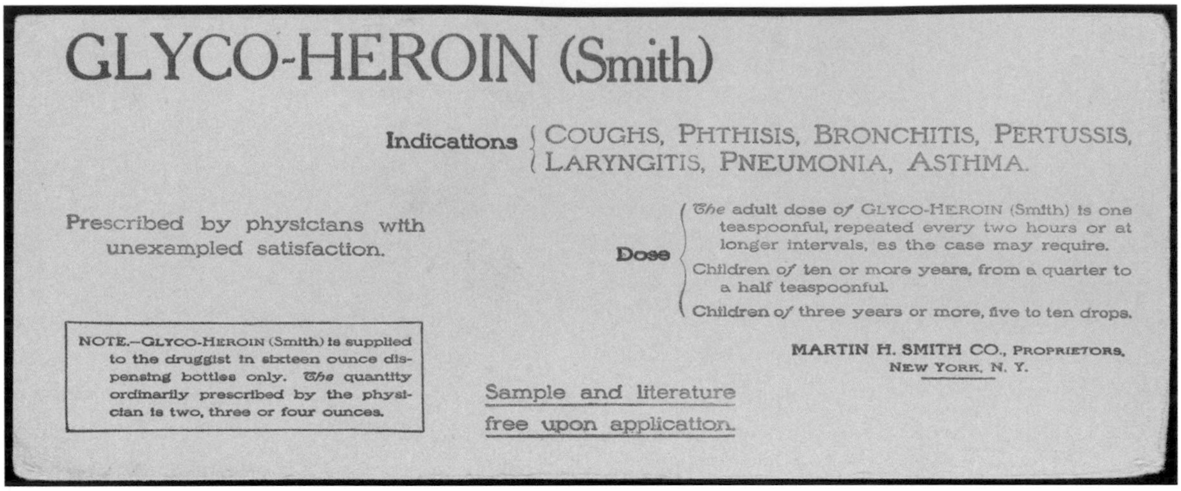

FIGURE 5.8 A trade-card advertisement for heroin, early twentieth century. (Wellcome Library, London, L0064712)

of exoticism that, along with the opium and other pleasures they offered, fueled their business during the late nineteenth century.[77]

A typical opium den consisted of rooms with beds or mats that allowed smokers to lie on their side while smoking. The clients smoked through long pipes with a small bowl at one end where the opium paste was heated. They inhaled the smoke through the length of the tube, which was packed with grass, hair, wood shavings, or other material to reduce the harshness of the smoke. After a long session, smokers made use of the padded floor or bed to experience a prolonged euphoria and sleep.

Over the course of the last few decades of the nineteenth century, the opium den's mythologized association with the underworld and its renown as a place where the races and sexes commingled earned it a place in the crosshairs of anti-opium crusaders (figure 5.9). Bans on smoking opium (that is, the form of opium specially prepared for the pipe) and opium dens were instituted in various localities during the 1870s, 1880s, and later, with some small effect.[78] Ultimately, the first federal legislation in U.S. history targeting any drug was written specifically to restrict the form of opium preferred by the Chinese and went into effect in 1909.[79] While the drug's physical effects

on the body and economic toll loomed large in the effort to restrict it, there was undoubtedly also a moral and racial motivation behind the antiopium movement. As Hamilton Wright (1867–1915?), a U.S. delegate to the International Opium Commission wrote, "One of the most unfortunate phases of the habit of opium smoking in this country is the large number of women who have become involved and were living as common-law wives of or cohabiting with Chinese in the Chinatowns of our various cities."[80]

MECHANISM OF ACTION

Opium's principal psychoactive compound, morphine, comprises approximately 10 percent of the raw poppy latex; codeine makes up about 0.5 percent.[81] Absorption of morphine and other opiates is slow and incomplete through the gastrointestinal tract. The effect of morphine is more rapid and pronounced when injected or smoked. Morphine can also be administered rectally.[82] Because of the semisynthetic opioid heroin's capacity to cross from the blood into the brain quickly, it produces the most immediate effects.

These differences in route of administration account, in part, for the varying levels of addictiveness of poppy compounds. When ingested, opium

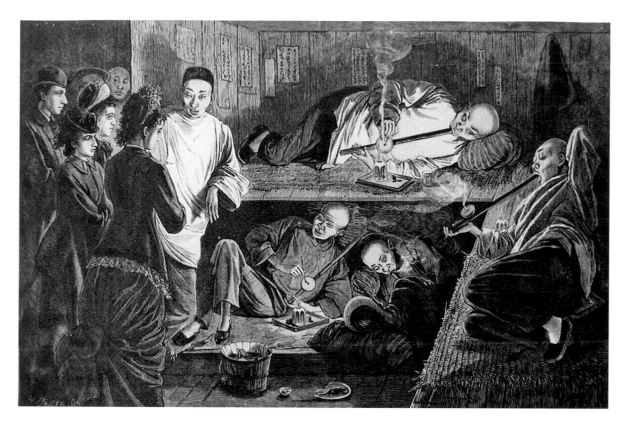

FIGURE 5.9 Many white American men feared the consequences of their wives, sisters, and daughters frequenting opium dens in urban Chinatowns. (Caricature from *Frank Leslie's Illustrated Newspaper,* April 24, 1878)

or opium-containing laudanum produces mild and sustained euphoria. However, when smoked or injected, the opium (or morphine) euphoria begins more quickly and reaches a higher level, and it is thus more likely to lead to physical dependence. The euphoria of injected heroin is more intense still.

PHARMACOLOGICAL EFFECTS OF THE OPIATES

Analgesia
Shallow breathing
Relaxation and sleep
Reduced intestinal motility
Decrease in body temperature
Constriction of the pupils
Peripheral vasodilation
Cough suppression
Euphoria

Morphine is administered in the modern-day clinical setting as an analgesic for severe or chronic pain, and its side effects include sleepiness, constipation, nausea, flushing of the skin, itching, and respiratory depression.[83] Morphine blocks the transmission of pain sensations from the periphery to the brain, which is why anesthesiologists sometimes inject it directly to the surface of the spinal cord (epidural) to reduce the pain of labor and delivery.[84] This treatment allows women to remain alert and maintain normal vital signs while experiencing less pain. The locally acting morphine does not reach the brain and therefore does not produce systemic effects.

Morphine, codeine, and other opiates exert their analgesic effects by binding as agonists to opioid receptors in the brain and at the interface between sensory neurons and the spinal cord. They reduce the sensation of pain and increase the tolerance for pain. Synthetic chemicals that have similar effects (called opioids) are widely prescribed. They also come with a significant risk of addiction and are subject to overprescription and diversion into the illicit drug market.[85] Some of them are based closely on the structure of morphine, such as hydrocodone (sold as Vicodin) and oxycodone (sold in a slow-release tablet as OxyContin). Oxycodone is also combined with the analgesic anti-inflammatory drugs acetaminophen (sold as Percocet) and aspirin (sold as Percodan).[86] Other synthetic molecules have structures that differ from morphine's but that nonetheless act as agonists at opioid receptors. Examples are meperidine (sold as Demerol), methadone, and fentanyl.[87]

Morphine and its chemical cousins also induce drowsiness and a state of tranquility, which leads to a shallow sleep. Opiates and opioids produce a feeling of euphoria through opioid neuronal connections to the dopamine network in the brain's limbic system. Opioid receptors located on presynaptic terminals of neurons respond to agonist binding by inhibiting the release of GABA, a neurotransmitter among whose functions is the suppression of dopamine signaling.[88] The decrease of GABA in the synaptic cleft in turn increases dopamine's presence there, which results in a sense of extreme pleasure and reward. Long-term use gives rise to tolerance, which requires increased doses to maintain the initial pleasurable state and can accompany addiction. Withdrawal symptoms (including dysphoria, pain, diarrhea) also become increasingly severe after continued, high-dose use.[89]

As antidiarrheals, opiates and opioids activate opioid receptors at the interface of the peripheral nervous system with the smooth muscle of the intestine, where they slow the rate of intestinal muscle contraction (peristalsis) and decrease intestinal secretions, resulting in a greater level of fluid reabsorption.[90] Synthetic drugs such as loperamide (sold as Imodium) and diphenoxylate (sold as Lomotil) are used to treat gastrointestinal symptoms.[91] Diphenoxylate generally does not produce central effects except at very high dosage, and loperamide's chemical structure prevents it from reaching the central nervous system in any significant concentration. Therefore, it has no painkilling or euphoriant properties.

From its origins in the Mediterranean region to worldwide distribution as a clinically important source of medicine, the poppy has long entwined itself in human history. It is one of the earliest documented medicinal herbs, having helped the Romans face pain and the Chinese combat diarrhea. Its active principle, morphine, was the first effective painkiller in a new era of biochemical medicine, but its use—and abuse—also led countless individuals, captured by the drug's euphoriant effects, to addiction and ruin. Europeans waged wars in China that boosted its commercial success, as it meanwhile spread through the West as laudanum and emerged in many cities at the end of a pipe. Even today, the poppy remains a lucrative cash crop, yielding many tons of the potent, simple, and seductive morphine derivative heroin for the illicit trade. Since ancient times, the poppy has exercised its gentle force over humanity.

Chapter 6

Coca
Erythroxylum coca

An Andean man with a coca offering in Cuzco, Peru. Among the Inca and their descendents, coca is a sacred plant.

Coca grows as a perennial shrub with simple, oval, hairless dark green leaves. It reaches a height of 2 to 3 meters in cultivation and can grow taller if left unpruned. Small whitish five-petaled flowers appear on short stems arising from the leaf nodes and produce olive-shaped, pea-size fruits that turn red at maturity (figure 6.1). The leaves are harvested for medicinal use. The genus *Erythroxylum* contains hundreds of species, only a handful of which are used medicinally. *E. coca*, the most widely cultivated type, exists in two geographically distinct varieties. *E. coca* var. *coca* grows in the humid tropical mountain forests in the eastern Andes region of South America, currently occupied by Ecuador, Peru, and Bolivia. *E. coca* var. *ipadu* grows in the lowland Amazon basin. The closely related *E. novogranatense* is cultivated in drier forest and desert areas in Venezuela, Colombia, Ecuador, and Peru.[1]

FIGURE 6.1 Coca: (*top*) plant; (*bottom*) fruit.

ANCIENT USE IN THE ANDEAN REGION

Ancient people recognized the stimulating properties of coca, and the practice of chewing its leaves is at least 2000 years old (figure 6.2).[2] Coca produces a feeling of increased energy and suppresses thirst and hunger, properties that must have been useful in societies established around mountain life. Even in the modern era, individuals facing strenuous mountain hikes chew coca leaves to sustain their endurance and cope with changes of altitude.

Archaeological relics and folklore indicate that coca had deep spiritual significance for people in northern South America. Artifacts from Peru including human figures and decorated vessels dating to at least 500 C.E. depict what are thought to be priests chewing coca, a connection between the plant and religious practice. Among ancient peoples and their descendents, coca was thought to be a physical manifestation of the divine.[3]

Evidence of coca use is present in Andean mummies predating by many centuries the rise of the dominant Inca Empire. During the time of the early Inca civilization in roughly the thirteenth century, agriculturalists began growing coca in plantations and systematically harvesting the leaves for use in trade. The Incas probably used coca as a sort of tribute to the hierarchy of nobles and as a reward to workers and travelers, sustaining their difficult labors. Much of the Inca Empire extended high into the Andes and deep into gold and silver mines, and coca served to reenergize tired muscles and enhance endurance.

Traditionally, coca has been consumed by chewing the fresh

FIGURE 6.2 Coca chewing: (*top*) Andean peoples produced statuettes depicting coca chewing more than 1000 years ago; note the bulging cheeks in this Jama-Coaque clay figurine, ca. fifth century; (*bottom*) a *coquero* of the South American high plains. ([*top*] Photograph © Justin Kerr K7886, www.mayavase.com)

Chapter 7

Peyote

Lophophora williamsii

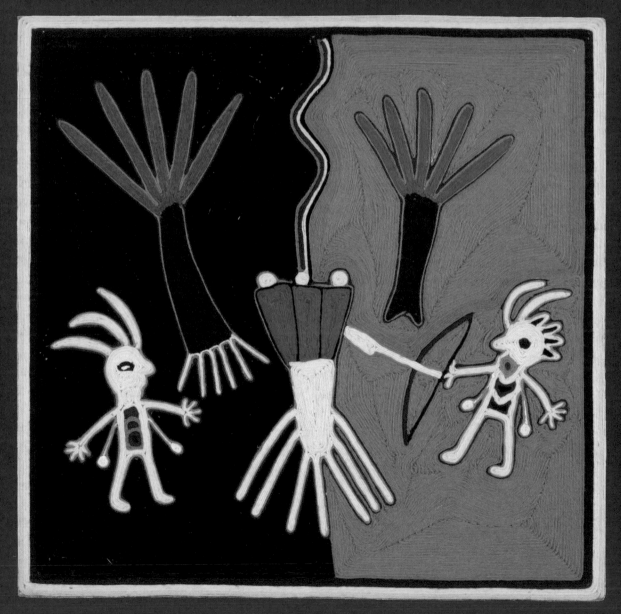

A hunt for peyote in Wirikuta. (Yarn painting by the shaman Ramón Medina Silva [twentieth century]; Fowler Museum of Cultural History, UCLA. Purchase courtesy of the Ford Foundation)

The perennial thornless peyote cactus grows low to the ground and produces flat-topped, round shoots 4 to 10 centimeters in diameter, and only a few centimeters from the surface of the soil (figure 7.1). Its root is thick and deep, shaped like a carrot or turnip. The shoot portion of the plant, in the form of a disc of ribbed nodes consisting of the plant's vegetative bud and leaves, can be harvested by cutting at the base, yielding the medicinal peyote "button." The shoot can regrow if cut in this way and is more likely to produce multiple new branch shoots in response. Each rib is topped by a tuft of whitish hairs (trichomes). Peyote produces a small number of prominent pinkish-white flowers (occasionally yellow) at the center of the stem that give rise to thin red fruits no longer than 2 centimeters.[1] Peyote grows wild in far southern Texas and much of northern Mexico, generally at low elevation in desert scrub ecosystems.[2] It can also be cultivated, although it grows slowly. Its genus consists of two species, the less common type occurring rarely in one part of Mexico.[3] In literature and common speech, peyote is sometimes mistakenly called mescal, a name that refers instead to both a North American psychoactive legume (*Sophora secundiflora*) and the distilled fermented sap of the Mexican maguey plant (*Agave americana*).[4]

FIGURE 7.1 Peyote: (*top*) cactus; (*bottom*) flower. (U.S. Department of Agriculture, Forest Service)

ORIGINS AND ANCIENT USE IN NORTH AMERICA

Although there are no written records and few archaeological clues to the antiquity of peyote use in pre-Columbian times, early Spanish writings and more recent ethnographic accounts indicate that it probably served a role in indigenous North American medical-spiritual practices for many centuries before the arrival of Europeans. The small handful of archaeological investigations that have come across intact peyote demonstrate that peyote use probably extends back more than 5000 years.[5] For example, an analysis of a string of buttons associated with a burial site in northern Mexico revealed it to be about 1000 years old.[6] Peyote buttons that probably came from an archaeological site near the Rio Grande in southern Texas have been subjected to chemical analysis, which gave an age of around 5700 years.[7] In both ancient samples, investigators were able to detect peyote alkaloids, lending evidence to the notion that people harvested peyote for its psychoactive properties. The principal alkaloid, mescaline, is concentrated in the crown of the cactus and is likely responsible for many of the mystical and medicinal powers with which the plant has been associated.

As the Spanish took account of their freshly subjugated New World colonies during the sixteenth century, they recorded descriptions of peyote use among the indigenous people. According to one such report by the botanist-physician Francisco Hernández (1514–1587), *peyotl* (from the Aztecs' Nahuatl language) "is a medium size root, bringing forth no branches nor leaves above ground, but with some sort of wool adhering to it." Explaining what he learned from the local people, he wrote:

It seems to be of a sweet taste and of moderate heat. If ground and applied it is said to cure pains of the limbs; this wonderful thing is said about this root (provided one gives credence to a thing which is most popular among them), that by eating it they can foresee and predict anything; for instance, whether enemies are going to attack them the following day? whether they will continue to be in favorable circumstances? who has stolen household goods or something else? and other things of this sort, which the Chichimeca [seminomadic peoples of northern Mexico] try to know by means of this kind of medication. When they want to find out whether the root is hidden in the ground, and where it is growing, or whether it will be harmful, they learn by eating another one.[8]

In this brief account, peyote comes across as an herb with healing abilities and visionary powers used in various divination practices. Around the same time, the missionary Bernardino de Sahagún (1499–1590) described peyote's mind-altering, mood-modifying, and hunger- and thirst-suppressing capabilities. "Those who eat it or drink it see visions, horrible or laughable," the priest wrote. "This intoxication lasts two or three days, and then it goes away. It is like a food to the Chichimeca, which supports them and gives them courage to fight, and to have neither fear nor thirst nor hunger, and they say that it keeps them from all harm."[9]

These early descriptions of peyote outlined a range of uses among indigenous people in therapy, prognostication, and bolstering the resolve, and whether or not such accounts accurately described the local peoples' beliefs and practices, they certainly must have raised concerns among the Spanish, many of whom were in the business of saving indigenous souls. In its missionary zeal, the Inquisition prohibited peyote use in 1620, declaring it "opposed to the purity and integrity of our Holy Catholic Faith." According to the Holy Office, the supposed visions experienced by peyotists could not possibly be caused by peyote itself, "nor can any [herb] cause the mental images, fantasies and hallucinations on which the above stated divinations are based. In these latter are plainly perceived the suggestion and intervention of the Devil."[10] Prohibited by the Catholic Church, the so-called *raíz diabólica* was excluded from the spiritual and medical practices of those indigenous people living under Spanish supervision.[11]

Among some Mexican indigenous groups that escaped conversion by missionaries and led lives apart from Spanish administration, peyote use persisted into the modern era, offering ethnographers the opportunity to observe customs that may preserve aspects of a pre-Columbian tradition. For example, for the Tarahumara and Huichol of northwestern and west-central Mexico, peyote plays an important role in cosmology and health beliefs.[12] In the late nineteenth century, a Danish anthropologist observed Tarahumara practices closely resembling some of peyote's uses as recorded by the Spanish three centuries earlier: "[Peyote is] applied externally for snake-bites, burns, wounds, and rheumatism; for these purposes it is chewed, or merely moistened in the mouth, and applied to the afflicted part." To the Tarahumara, peyote's remarkable medicinal strength served as both treatment of poor health and prophylactic: "Not only does it cure disease, causing it to run off, but it also so strengthens the body that it can resist illness, and is therefore much used in warding off sickness." The power of peyote is "to give health and long life and to purify body and soul." Among this group, peyote is thought to protect the good fortune of those who consume it, provide luck, and keep enemies at bay. Peyote has a virtuous spirit, according to the Tarahumara, and sits next to Father Sun. Therefore, it must be handled gently and saluted when encountered.[13]

PEYOTE AMONG THE HUICHOL

A rich account of the role of peyote among indigenous Mexicans appears in the work of the anthropologist Barbara Myerhoff (1935–1985) with the Huichol.[14] Having spent many months over a period of several years living with them, she was in a good position to observe their practices, learn about their beliefs, and participate in their ritual activities. As Myerhoff relayed, to the Huichol, three living, godlike beings—maize, peyote, and the deer—form a sacred union on which life depends. "They are one, they are unity, they are ourselves," the Huichol say.[15] These three entities represent the totality of Huichol history and social life, connecting the people spiritually to their homeland and way of life, and they are thus interdependent aspects of a revered whole. The deer symbolizes the ancient way among the Huichol, which their tradition relays as a life of hunting in a faraway ancestral land. The maize represents the agricultural lives

FIGURE 7.2 The Huichol shaman (*mara'akame*) Juan Hernández González from San Sebastian, collecting peyote during his annual pilgrimage to the sacred land of Wirikuta. (Photograph by Heriberto Rodriguez)

they now lead, cultivating grain, dependent on unpredictable rainfall, in which men, women, and children toil tending the fields. Peyote mediates the Huichol's past and present realities, allows them to perform religious rites, and guides their individual spiritual experiences.[16]

According to oral accounts, the Huichol eat peyote to relieve pain; they apply it, ground into a paste, on wounds.[17] They take it for endurance, courage, and energy. As a Huichol *mara'akame* (spiritual leader) explained, "One eats it like medicine or for whatever purpose one wants to eat it. If one feels weak, if one feels tired, if one feels ill, if one needs strength, then one eats it."[18] To the Huichol, peyote can "read one's thoughts" and cast judgment on a person's moral righteousness. A person who has been evil, who has not lived harmoniously in society, is punished. The path to redemption requires a confessional visit to the *mara'akame*, who can

cleanse the transgressor's soul and put him on good terms with peyote.[19] Thus peyote has a profound role in the Huichol sense of physical, spiritual, and community health.

A significant manifestation of the Huichol worldview is the annual peyote hunt, in which, under the guidance of the *mara'akame*, participants make a long pilgrimage to their ancestral sacred land, called Wirikuta (figure 7.2).[20] When there, the *mara'akame* follows a set of deer tracks to their origin, the peyote-deer, which he symbolically slays with his bow and arrow before digging the crown, slicing it into pieces, and distributing portions to all participants. In Wirikuta, people revisit the paradise of their creation, become deities, and attain a state of biological, social, and spiritual oneness. As Myerhoff describes it, "when the peyote is eaten in the Sacred land of *Wirikuta*, distinctions are overcome between plant and animal, between man and animals, and between the natural world, the human world, and the supernatural world."[21]

Participants consume peyote frequently during their ritual hunt, and in small amounts it produces exhilaration and a sense of well-being, no doubt intensified by the religious significance of the event. In the evenings, the pilgrims consume larger amounts of peyote, after which "people see beautiful lights, lovely vivid shooting colors, little animals and funny creatures." The visions of the *mara'akame* are especially significant because he or she alone can communicate with the protector deity Grandfather Fire to learn the lessons of the other worlds to share with the other Huichol.[22] On the hunt, the participants harvest enough peyote for the year, including some to sell to tribes that do not travel to collect it. While fraught with danger—Huichol who are not spiritually prepared for the journey risk losing their souls in Wirikuta—under the guidance of an experienced *mara'akame* and with the aid of the peyote, deer, and maize, the pilgrimage serves a most important role in health. In the words of a *mara'akame*, "We find our life over there."[23]

THE NATIVE AMERICAN CHURCH

While peyote's natural range extends into modern-day Texas, there is little evidence that it was in use among the indigenous tribes of today's United States and Canada when Europeans first arrived in the New World.[24] As American Manifest Destiny became more aggressive during the nineteenth century and increasing numbers of Native people were constrained to reservations, they found themselves uprooted from ancestral lands, their spiritual beliefs persecuted, and their social order threatened. By the late nineteenth century, the "civilizing" activities of the U.S. government and Christian missionaries had made headway in replacing old Indian notions of faith and nature with their own. In such an environment, the tribes set aside traditional alliances and antipathies and instead began to band together in common cause, sharing a newfound identity as indigenous people whose ways were distinct from those of the whites.

Previously hostile groups began to send emissaries to one another's lands, and intermarriages also resulted in an increased exchange of ideas, goods, and people between tribes. A form of peyote religion was practiced among the Carrizo of northeastern Mexico as early as the eighteenth century, characterized by all-night ceremonies and ritual drumming and singing in a circle. During the 1870s, a Lipan Apache (from the borderlands crossing Texas and Mexico) brought peyote to the Kiowa Apache living in Indian Territory (now Oklahoma) and introduced it in a spiritual-medical ritual.[25] By about 1880, peyote had spread to other nearby groups, such as the Kiowa, and by the mid-1880s to the Comanche.[26] Among the Plains Indians, peyote probably suited the needs of shamans, who might have found it an effective supplement to their traditional healing and prognosticating rites. It also served in the individualistic vision quests of American Indian spirituality, allowing people to communicate more readily with the supernatural.[27] The ceremonial consumption of peyote developed into a religion that quickly spread from community to community, replete with a set of roles for participants during worship, ritual paraphernalia, and, eventually, corporate organization (figure 7.3).

FIGURE 7.3 A peyote ceremony among Kiowa, 1892. Such gatherings are held to cure diseases, celebrate important life events, and perform other forms of worship. (Photograph by James Mooney; Smithsonian Institution, National Anthropological Archives, NAA INV 06275300)

In the late nineteenth and early twentieth centuries, the peyote religion extended to tribes in the American Southwest, across the Great Plains and into the Midwest, and across the Canadian border to indigenous groups living there.[28] The peyote religion spread in part at the hands of charismatic prophets, to whom the plant revealed its spiritual virtues.[29] Ultimately, the peyote religion diversified into numerous forms, when individual leaders chose to modify, adapt, and invent new ceremonial flourishes to the rites they conducted, and congregations harmonized peyote worship with their ancestral beliefs and those they learned from white Americans.

American Indian adherents believed peyote to be imbued with the Great Spirit, a manifestation of a supreme deity or god, and that this entity has unique curative power.[30] Indeed, the spiritual and health-related aspects of peyote cannot be easily separated. For example, a scholar studying the Menominee of Wisconsin wrote that "the advantages of Peyote are two-fold: it is a 'medicine for the soul' . . . and it is a catholicon, a universal remedy which cures all diseases."[31] Among the Kiowa, peyote was said to treat "tooth-ache, hemorrhages, head-ache, consumption, fever, breast pains, skin disease, hiccough, rheumatism, childbirth, diabetes, colds and pulmonary diseases in general."[32] A Prairie Potawatomi woman showed a visitor a small vial of ground peyote and explained that she used it "just like aspirin."[33] While peyote served as medicine in daily life, ground up and applied externally, taken by mouth as a dried button, or brewed into an herbal tea, it was thought to be particularly effective when incorporated into a worship service.

Peyote ceremonies take numerous forms, at times scheduled regularly on Saturday evenings, at times on Christian and secular holidays, and at times in response to a particular community need, such as the curing of an illness. Although variations abound, a typical setting for such a rite is in a tipi specially raised for the purpose. Inside the structure, a ridge of soil in the shape of a crescent serves as a type of altar, on it inscribed a line representing the "peyote road," the correct way to live. In the center of the crescent is placed a large peyote button, dubbed the Peyote Chief, through which prayers to the supernatural world are channeled. (Many people keep their own Peyote Chiefs outside of such ceremonies, and they are thought to have the power "to ward off evil and bring good luck.")[34] Participants are expected to follow certain ceremonial protocols, alternately singing and praying, taking peyote buttons, and engaging the spirit world through an overnight service.

In a typical peyote ceremony, participants might consume between four and twelve dried buttons, moistening them in their mouths, chewing, and swallowing.[35] A detailed firsthand anthropological account describes the taste as "bitter," like "dried pieces of orange peel."[36] About an hour after taking four peyote buttons, sensory alterations ensue, including modified visual and auditory perception and a feeling of bodily transcendence "with no distinction between internal and external aspects of experience." Some people become nauseated and vomit after ingesting peyote, an effect that American Indians can explain in medical terms. "After eating Peyote," a Menominee man explained, "a sick person usually vomits, and the sickness may be vomited up along with the Peyote, this cleansing the body." After vomiting, the patient "should eat more Peyote in order to gain strengthening power."[37]

In some communities, the peyote ritual took on some of the symbolism and theology of Christianity. In a fusion of Christian and indigenous religious practices, ceremonial leaders frequently place the Bible next to the Peyote Chief and invoke Jesus as well as animal spirits in their prayers.[38] While missionaries and the U.S. government were generally hostile to the peyote religion during the early part of the twentieth century, adherents from many tribes argued that the ceremony was in fact a form of Christian worship. They pointed to certain shared theology and symbols as evidence: the "peyote road" reflects fundamental Christian values of brotherly love and temperance, the tipi poles represent Jesus and his disciples, and ceremony leaders make the sign of the cross, for example (figure 7.4).[39]

To legitimize their beliefs and protect them against harassment from American authorities, communities of peyote faithful registered themselves formally as religious organizations. In 1915, a group of Omaha from Nebraska formed the Omaha Indian Peyote Society. A few years later, in 1921, they organized the Peyote Church of

Christ, whose charter declared: "We recognize all people who worship God and follow Christ as members of the one true church."[40] The most influential of the peyote organizations was the Native American Church, chartered in 1918 by representatives of American Indian tribes from Oklahoma (figure 7.5).[41]

By the 1940s, the Native American Church had established branch chapters in various states, and it became the Native American Church of the United States in 1944. In 1954, the Native American Church of Canada was formed.[42] Membership in the Native American Church stood at 13,300 in 1922, 225,000 in 1960, and 300,000 in the early 2000s.[43] Peyote and its active principle, mescaline, remain listed as controlled substances under schedule I of the Controlled Substances Act of 1970, which allowed an exception for the religious use of peyote by American Indians. However, states differed in their enactment and enforcement of peyote-related laws. The American Indian Religious Freedom Act of 1978, which asserted the rights of indigenous Americans to exercise their faiths, was amended in 1994 to strengthen protections for the sacramental use of peyote.[44]

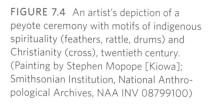

FIGURE 7.4 An artist's depiction of a peyote ceremony with motifs of indigenous spirituality (feathers, rattle, drums) and Christianity (cross), twentieth century. (Painting by Stephen Mopope [Kiowa]; Smithsonian Institution, National Anthropological Archives, NAA INV 08799100)

FIGURE 7.5 A ceremonial peyote plate used in rites of the Native American Church. (U.S. Department of Agriculture, Forest Service)

ISOLATION OF PEYOTE'S ACTIVE PRINCIPLE

White Americans and Europeans generally took little interest in peyote as a potential pharmaceutical, and, not distributed beyond its native range until the late nineteenth century, the cactus remained of botanical, rather than medical, relevance. During the Civil War (1861–1865), a story goes, members of a Confederate volunteer force known as the Texas Rangers were taken prisoner. With no recourse to alcohol, the soldiers inebriated themselves on what they called "white mule": peyote buttons soaked in water.[45] Despite their experiment born of necessity, the veterans apparently had no inclination to enter the "white mule" business, and peyote remained largely uninvestigated.

By the 1890s, some chemists in the United States and Germany had obtained samples of peyote and began describing its properties. For example, the Parke-Davis drug company of Detroit conducted an analysis of a water extract of peyote and found it to be "an intensely poisonous substance," lethal when injected into frogs, pigeons, and rabbits. Despite the convulsions and rapid death witnessed in such animals, the company pressed ahead with a trial of peyote extract in a human patient, albeit at a much lower dose. From a one-day test, the chemist conducting the study gathered that "uncombined and alone I believe it to be the best concentrated cardiac tonic we possess."[46] Although Parke-Davis scientists made no note of peyote's effects on the mind, *The Dispensatory of the United States of America* described its "psychical symptoms" as "not only overestimation of time, sense of dual existence, and delirium, but also pronounced visual hallucinations with undulatory motion of light . . . and a regular kaleidoscopic play of colors."[47]

These psychoactive properties sparked the interest of German pharmacologists, who were among the first to experiment with peyote and prepare purified extracts.[48] Ultimately, it was the German chemist Arthur Heffter (1859–1925) who first isolated the active principle of peyote, mescaline, in 1896 (figure 7.6). Mescaline is responsible for peyote-induced hallucinations and accumulates to between

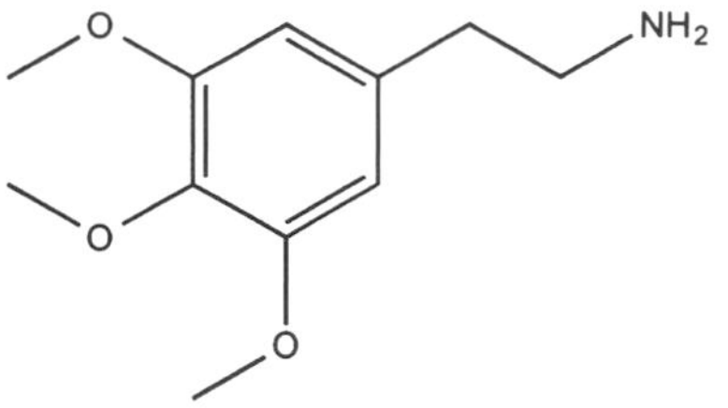

FIGURE 7.6 Mescaline.

approximately 1 and 4 percent by weight in peyote buttons.[49] Mescaline is also present in the South American San Pedro cactus (*Echinopsis pachanoi* [figure 7.7]) and other cacti. Although chemically extracted and tested for physiological activity, the translucent white crystals of mescaline roused little pharmaceutical interest. As the *Dispensatory* concluded in 1907, "the value of mescal buttons as a remedial agent is doubtful."[50] Curiosity returned in the 1950s, however, when writers began experimenting with the drug as a tool to understand human consciousness.

OPENING THE DOORS OF PERCEPTION

The English-born author Aldous Huxley (1894–1963) was fascinated by the power of certain types of intense religious experience to transcend the bounds of the ordinary and draw the worshipper into a more expansive, knowing reality, a phenomenon he saw most dramatically expressed by mystics. Artistic genius, the true meaning of things: these were aspects of an insight inaccessible to most people. Hoping to glimpse the world beyond his everyday senses, Huxley took some mescaline in pill form and wrote the short book *The Doors of Perception* (1954) about his experience. Huxley argued that mescaline revealed the "naked existence" of the world around, unfiltered by the normal barriers constraining the senses of all but the few whose minds allow them to take in more. "The other world to which mescalin admitted me was not the world of visions," he wrote. "It existed out there, in what I could see with my eyes open. The great change was in the realm of objective fact.

FIGURE 7.7 San Pedro cactus, native to Andean South America and for sale at an herb market in Peru. It contains mescaline and is employed in divination and healing rituals by indigenous people.

NUTMEG

The essential oil of the seeds of the Southeast Asian nutmeg tree (*Myristica fragrans*) contains, among other components, a significant concentration of the potentially psychoactive chemical myristicin, structurally similar to mescaline. The seeds are ground for use as a flavoring and used medicinally (figure B.1). In small amounts (0.3–1 gram), the powdered seeds are employed in traditional Southeast Asian medicine to improve digestion, treat coughs, and as a sedative, among other uses. In slightly higher doses, headaches result. At a dose above 5 grams (for example, one or two teaspoons mixed into water as an herbal tea), nutmeg induces severe vomiting, tremors, heart palpitations, and, occasionally, hallucinations.[1] Despite anecdotal reports that nutmeg can be consumed as a recreational mood-altering, vision-inducing agent, myristicin has not been demonstrated to be responsible for these effects. If nutmeg causes changes in perception, it occurs only very close to the toxic dose and may be a result of the activities of multiple components in nutmeg or its essential oil.

FIGURE B.1 Nutmeg.

1. Ben-Erik Van Wyk and Michael Wink, *Medicinal Plants of the World* (Portland, Ore.: Timber Press, 2004), 210; Christian Rätsch, *The Encyclopedia of Psychoactive Plants: Ethnopharmacology and Its Applications* (Rochester, Vt.: Park Street Press, 2005), 371–375. See also the interesting discussion of nutmeg in Michael J. Balick and Paul Alan Cox, *Plants, People, and Culture: The Science of Ethnobotany* (New York: Scientific American, 1996), 132–141.

PHARMACOLOGICAL EFFECTS OF MESCALINE

Visual and auditory hallucinations
Sleeplessness
Altered sense of time
Tremors
Increase in blood pressure
Increase in heart rate
Euphoria

What had happened to my subjective universe was relatively unimportant."[51]

Around the same time, the poet Allen Ginsberg (1926–1997) experimented with the inspirational qualities of peyote, taking it to accent his strolls through the 1950s urban landscapes of San Francisco and New York City.[52] The transformed world around him gave rise to descriptions both beautiful and bizarre. For example, Ginsberg's encounter with the Sir Francis Drake Hotel in San Francisco, the "Drake Monster," as his mescaline-influenced mind saw it, evoked prose rich in color and meaning, as in a letter about the experience written to his friend the writer Jack Kerouac (1922–1969). "We wandered on peyote all downtown," Ginsberg recalled in 1955. "Saw Moloch Moloch smoking building in red glare downtown . . . with robot upstairs eyes and skullface, in smoke, again."[53] Ginsberg's revolutionary poem "Howl" (1956) took the art in a new direction, both criticized and lauded for its gritty language and evocative imagery. By breaking through boundaries of theme and form, Ginsberg helped shape an influential generation of Beat writers whose contributions to American literature rank among the most important.

MECHANISM OF ACTION

Mescaline is water soluble and can be taken by mouth in fresh or dried peyote, steeped in a peyote herbal tea, consumed orally as purified mescaline, inhaled by smoking, or injected into the bloodstream. A typical dose of peyote eaten in a religious context is perhaps four to twelve buttons, although some people take very few and others many more, following their experience and intended effects. Because of the inherent variability of mescaline concentration and the size of the buttons, the actual dose received is rather unpredictable. In general, a dose of about 350 milligrams of mescaline might approximate an average peyote session.[54]

When ingested, the active principle is taken up rapidly by the stomach and small intestine. Mescaline bears a striking structural resemblance to the neurotransmitter norepinephrine, but, interestingly, its hallucinogenic properties arise from its partial agonist activity at serotonin receptors, resulting in a marked increase in activity in the serotonin neuron–rich frontal region of the cerebral cortex.[55] While much of mescaline's activity remains to be explained, its structural aspects and neurotransmitter receptor binding pattern probably account for the active principle's diverse effects: it causes an increase in blood pressure, heart rate, pupil dilation, and anxiety within an hour, followed by an eight- to ten-hour period of euphoria, psychological insight, breakdown of spiritual-natural barriers, and intense hallucination of colors, geometric designs, animals, and acoustic phenomena.[56] Mescaline itself probably does not induce nausea, but this effect and the associated urge to vomit may be caused by other chemicals in peyote.

While very little clinical research has investigated the effects of mescaline on human subjects, its chemical analogs have proved to be useful probes in various laboratory studies of the serotonin receptor's involvement in sensation and perception. It is speculated that mescaline may activate some of the same pathways affected in the human psychiatric conditions of psychosis and schizophrenia.[57] Future work will undoubtedly reveal much about the delicate circuitry of the brain by examining how mescaline brings about such a profound alteration of the senses.

The peyote cactus probably has been a part of American Indian religious practice for millennia. Indigenous peoples consume the plant in the form of peyote buttons to treat all manner of illnesses and to maintain contact with a spiritual world. During the twentieth century, writers experimented with the hallucinogenic effects of the active principle, mescaline, and produced an influential genre of mold-breaking work. Today, peyote continues to serve a spiritual community that communes with god through a sacred cactus, and the neurological pathways it has helped uncover may soon shed new light on the brain's sensory apparatus. Through peyote religion, pharmacologically inspired art, and the neurobiological laboratory, mescaline is helping people learn what it is to be human.

Chapter 8

Wormwood

Artemisia absinthium

Édouard Manet, *The Absinthe Drinker* (1859). (Ny Carlsberg Glyptotek, Copenhagen; photograph by Ole Haupt)

Absinthe wormwood grows as a perennial herbaceous plant up to 1 meter in height, has highly dissected silver-green compound leaves, and produces many small yellow flower heads on its branches (figure 8.1). It grows wild in Europe, North Africa, and Asia, and it has now been introduced in North and South America.[1] *Artemisia absinthium* is one of many members of the genus *Artemisia* that grow worldwide, throughout the tropical and temperate zones. These related plant species include African wormwood (*A. afra*, eastern Africa), sweet sagewort or annual wormwood (*A. annua*, eastern Europe and Asia), mugwort (*A. vulgaris*, Europe, Asia, and North Africa), and gray sagewort (*A. ludoviciana*, North America). There are over 100 species and subspecies under *Artemisia*, making taxonomy a challenge in this widespread group of plants. The common names for many of these species reflect their long history of use and diverse geographic-cultural environments. A single botanic species can bear a dozen or more vernacular names (in English alone) based on variations of wormwood, mugwort, sagewort, and so on. Interestingly, dozens of wormwood species are associated with traditional medical use in those areas in which they are found.

FIGURE 8.1 Absinthe wormwood.

ANCIENT USE IN MEDICINE

The use of wormwood likely predates the earliest recorded medicine of the Mediterranean, where it is documented as an ingredient in numerous pharmaceutical preparations in the early Egyptian medical work known as the Ebers papyrus (ca. 1550 B.C.E.). According to the text, ancient physicians employed wormwood in remedies for constipation and abdominal obstruction, intestinal worms, and menstrual pain.[2] More than fifteen centuries later, Pliny the Elder (23–79) described in detail the known properties of wormwood during his time, recognizing it first as a rather familiar and versatile herb, particularly among his countrymen: "As to its general utility, a plant so commonly found and applied to such numerous uses, people are universally agreed; but with the Romans more particularly it has been always held in the highest esteem."[3] He detailed the many ways that it was prepared, including a wine made by soaking the stems and leaves with grape must, called *absinthites*; a decoction produced by soaking or boiling the shoots in water; and mixed in vinegar or honey.[4] As for remedies, Pliny counted forty-eight, including to strengthen the stomach, treat nausea, eliminate intestinal worms, prevent seasickness, reduce flatulence, improve eyesight, heal bruises, and cure scorpion bites. Pliny also recommended that wormwood be kept with clothes to prevent them from being damaged by pests, as an oil or a fumigant to repel insects, and mixed with ink to prevent manuscripts from being nibbled by mice. In contrast to his scholar brethren of a much earlier time in Egypt, Pliny warned that "it must never be administered in fevers."[5]

An herb in widespread use in the Mediterranean, wormwood figures prominently in the works of many influential ancient physician authors, including Hippocrates (ca. 450–370 B.C.E.) and Galen (129–ca. 216).[6] The pioneering herbalist Pedanius Dioscorides (ca. 40–90) also wrote extensively on wormwood's properties. Calling it "astringent and warming," the prolific Greek physician noted *apsinthion*'s diuretic property and recommended it for earaches, to regulate menstruation, as an antidote to hemlock (*Conium maculatum*) poisoning, and for sore throat, in addition to many of the therapeutic and prophylactic uses offered by Pliny.[7] While recommending for medical use the steeped shoots and the vapor produced by boiling the liquid, both Pliny and Dioscorides warned against using wormwood juice as an internal medicine. "We disapprove of using it in drinks," Dioscorides wrote, "because it is bad for the stomach and gives headaches."[8]

During the Middle Ages, wormwood was used for a variety of concerns, from the veterinary to the hygienic. For example, in France during the thirteenth century, it was given as an oil called *absince* to dogs suffering from flatulence. In another veterinary application, the fifteenth-century *Saint Albans Book of Hawking* recommended administering wormwood juice to kill mites infesting a hawk.[9] Of course, people took the herb as well, as a treatment for assorted stomach and lower digestive issues.[10] The twelfth-century German medical writer and mystic Hildegard of Bingen (1098–1179) named it the "principal remedy for all ailments" and considered it useful to warm the stomach, purge the bowels, and improve digestion.[11] Its English name derives from an important use in Britain as a treatment for internal parasites. In an herbal published in 1597, the physician-botanist John Gerard (1545–1611?) stated, "Wormewood voideth away the wormes of the guts" while also declaring it useful to strengthen the stomach and stimulate the heart.[12]

The plant was employed in medieval Europe, as Greek and Roman writers had recorded many centuries earlier, to repel vermin. A sixteenth-century book from England provided the following poetic housekeeping advice: "Where chamber is sweeped, and wormwood is strown, no flea for his life dare abide to be knowne."[13] Stuffed under pillows, hung from the rafters of a home, or burned to fumigate it, people also used the herb to fend off the plague.[14] From digestive concerns to epidemic shield, wormwood saw use for so many medical conditions that it is of little surprise that the seventeenth-century courtesan Madame de Coulanges, after taking wormwood for a stomach illness, declared, "My little wormwood is the remedy for all ills."[15]

ABSINTHE

Wormwood drinks during pre-Renaissance times were bitter beverages often made by soaking the leafy parts of the plant in wine for many days. The resulting harsh medicines were typically sold by apothecaries for therapeutic use or made privately, the wormwood plant being widely available in the wild. The bitterness of wormwood probably gave it its Greek name, as *apsinthion* means "undrinkable."[16] Its flavor alone rendered it useful as an aid to the weaning of children, as captured in a few

lines from William Shakespeare's (1564–1616) *Romeo and Juliet*, in which Juliet's nurse recalls applying wormwood to her breast many years earlier to wean the girl of her milk:

> When it did taste the wormwood on the nipple
> Of my dug and felt it bitter, pretty fool
> To see it tetchy and fall out with the dug![17]

Although wormwood is bitter, some consumers must have found it agreeable, as both medicine and refreshment, as evidenced by an expansion of the plant into the arena of recreation. The use of wormwood for both mental and physical health inspired a seventeenth-century French aphorism by Martin: "Wormwood calms the nerves, and is also good for worms."[18] One place into which wormwood entered as a beverage of social enjoyment is the tavern. In an era when freshwater was unsafe to drink, mildly alcoholic fermented grain ales, flavored and preserved with plant extracts, served as liquid sustenance. Wormwood ale, known as purl, was widely consumed in seventeenth- and eighteenth-century Britain, eventually giving way by the nineteenth century to bitter ales and beers flavored with plants such as hops (*Humulus lupulus*).[19] In addition to purl, Britons enjoyed purl-royal, a wormwood wine that also fell out of favor in more recent times.[20] The advance of wormwood beverages into the social life of Europe during the Renaissance and early modern era presaged the use of absinthe as a recreational drink in the nineteenth century.

In addition to the growing acceptance of wormwood-infused beverages as social drinks beginning before the seventeenth century, the development of distillation technology allowed apothecaries and merchants to offer wormwood in a novel and more potent form.[21] Distillation concentrates the alcohol and volatile oils from a water-based solution because their lower boiling point allows them to vaporize before water. The cooled vapor, partitioned into a separate vessel, is richer in alcohol than the original solution, with purified and concentrated flavors and medicinal compounds. According to one of the early proponents of medicinal distillation, Hieronymus Brunschwig (1450?–1512?),

FIGURE 8.2 Wormwood. (Woodcut from Hieronymus Brunschwig, *Vertuose Boke of Distylacion* [early sixteenth century]; Peter H. Raven Library, Missouri Botanical Garden, St. Louis)

distilled wormwood had many therapeutic uses (he enumerated thirty-five), such as to treat malarial fevers, moisten the mouth, and heal sores (figure 8.2).[22] During the sixteenth, seventeenth, and eighteenth centuries, distilleries produced concentrated absinthe liqueurs and potent elixirs of wormwood steeped in wine and mixed with spirit alcohol, adding an inebriating punch to the medication.

The birth of modern absinthe was a stroke of advertising rather than invention. By the late eighteenth century, residents of the western Swiss Alpine countryside had for generations made wormwood elixirs of various formulas. One particularly successful manufacturer, a woman by the name of Henriette Henriod, sold a distilled alcoholic extract of wormwood that found its way into the hands of a Frenchman, Major Daniel-Henri Dubied, who used it frequently to treat fevers and prevent indigestion.[23] In 1797, Dubied purchased the formula from Henriod, and with his son-in-law Henri-Louis Pernod set about producing his own absinthe in the French border town of Pontarlier in 1805. Dubied and Pernod aggressively marketed their liqueur in France as a tonic of unrivaled virtue and established a brand with enduring appeal.

The wormwood liqueurs of the day were of diverse recipes, always with wormwood and alcohol to a strength of at least 65 proof, but otherwise variable in ingredients.[24] The recipe for Henriod's concoction has been lost to history, but likely included an assortment of traditional herbs such as anise (*Pimpinella anisum*), hyssop (*Hyssopus officinalis*), dittany (probably *Dictamnus albus*), sweet flag (*Acorus calamus*), melissa (*Melissa officinalis*, also known as lemon balm and sweet balm), coriander (*Coriandrum sativum*), veronica (*Veronica officinalis*, also known as speedwell), chamomile (probably *Matricaria chamomilla*), parsley (*Petroselinum crispum*), and spinach (*Spinacia oleracea*).[25] The distilled beverage resulting from this herbal mélange was bitter but richly flavored, strongly alcoholic but with a spicy and intriguing aroma. In addition to its unique taste and smell, absinthe was colored green by the chlorophyll in the leaves used to create it. Furthermore, the high alcohol content gave rise to one of its most captivating features. In a high-alcohol solution, the oils distilled from the herbal ingredients in absinthe easily dissolve, rendering a clear green liqueur. In common practice, drinkers add cold water to the absinthe before drinking, causing the oils to fall out of solution. Minute droplets become suspended in the green liquid, resulting in a milky greenish-white beverage that many find entrancing (figure 8.3). Among its devotees, such distinctive flavor, aroma,

FIGURE 8.3 Henri Privat-Livemont, advertising poster for Absinthe Robette, 1896.

and color earned absinthe the nickname *la fée verte* (the green fairy).[26]

Pernod and other absinthe makers successfully marketed their product as a health tonic during the early nineteenth century, gaining a strong following particularly in France. During the French conquest of Algeria in the 1840s, absinthe was issued in soldiers' field rations to fight off fevers, and when mixed with water, it was thought to kill germs.[27] Later, the French military employed absinthe against the illnesses encountered throughout their growing colonial domain, from Africa to Indochina.[28] Some French doctors of the nineteenth century suggested the drink to treat gout and dropsy and even to stimulate intellectual activities.[29] The Pernod factory continued producing absinthe, from 400 liters a day in the early nineteenth century, to 20,000 liters a day around 1850 and 125,000 liters a day by 1896.[30] Rising to the challenge of supplying a worldwide demand for absinthe, hundreds of companies in France and Switzerland began producing the beverage using a variety of names and processes.

In general, the fashionable nineteenth-century recipe included wormwood (both *la grande absinthe* [*A. absinthium*] and *la petite absinthe* [probably *A. pontica*, also called Roman wormwood]); anise, star anise (*Illicium verum*) or fennel (*Foeniculum vulgare*) for a prominent aniseed flavor; and hyssop for color, among other possible herbal additives. These flavorings were macerated with wine alcohol and then distilled, or occasionally added after distillation to add particular colors and aromas to the absinthe.[31] Among the less reputable firms, grain or beet alcohol—or worse, wood alcohol—might have served as the spirit basis of the drink.[32] The beet alcohol (principally ethanol) was seen as a cheaper and less elegant form of alcohol than that from wine (also ethanol); wood alcohol (methanol) was not only cheap but also utterly dangerous, as it is highly toxic and can cause blindness and respiratory failure. Chemical analyses on vintage bottles of unopened absinthe to determine the original formulations of the nineteenth century found that the best-known brands, which probably also sold the more expensive bottles and produced in the largest

volume, contained ethanol of high quality and methanol in a range similar to today's standard.[33] There is some evidence that smaller producers may have added toxins such as copper salts (to intensify the green color) and antimony trichlorate or zinc sulfate (to augment the desired cloudiness of the prepared beverage).[34] Whether absinthe—low grade or high—posed significant health problems because of its inherent ingredients or its adulterants was a contentious issue during the late nineteenth century and has attracted much scholarship in recent years, to no unambiguous conclusion.[35] During the twentieth century, many absinthe manufacturers reformulated their drinks without wormwood to comply with new anti-wormwood laws. The beverages maintain much of the herbal flavor and character. For example, Pernod's pastis, a twentieth-century wormwood-free anise liqueur, is flavored with star anise, coriander, and mint (*Mentha* spp.).[36]

The "Green Muse"

The second half of the nineteenth century in France gave rise to a revolutionary movement in art and literature, nucleated in Paris and reverberating throughout bohemian social circles in Europe and the United States. While European public art took on neoclassical proportions and glorified the modern empires that its nations were building, the Paris counterculture rejected its techniques. Rather than represent idealized subjects of ancient and patriotic virtue, a generation of writers, painters, and thinkers preferred to explore the gritty reality of nineteenth-century life, cutting across lines of gender, class, and convention. Among the earliest such artists was the poet and essayist Charles Baudelaire (1821–1867), who defied the standards of his day by passing time with his black girlfriend and publishing works of erotic poetry, critiques of religion, and depictions of drug use, an oeuvre that ultimately resulted in censorship and fines for offenses against public morality.[37] Indeed, the artistic countermovement that Baudelaire represented came to be known as la Décadence for its decadent, self-indulgent culture. The French poets Paul Verlaine

FIGURE 8.4 Edgar Degas, *Dans un café* (1876). (© RMN-Grand Palais/Art Resource, N.Y.)

and alcohol of all sorts to frequent excess. It is not surprising that so few of the most influential men of the era survived past their forties. Baudelaire's own work attests to his experiences with alcohol and other drugs; in 1860 he published a book of poetry on themes of altered reality under the title *Les paradis artificiels* (*Artificial Paradises*).

Absinthe figures strongly in the artwork of the era, as it was a prominent part of these artists' milieu—at times subject, verb, and object in the creative process (figure 8.4). Painters documented their grisly world of hazy cafés and corner bars, where the working classes mingled with merchants and intellectuals, women exerted their independence from traditional roles by entering the social scene without men, and the milky emerald of the wormwood liqueur inspired all: the "green muse."[38] Painters and poets alike devoted much attention to the ritual of absinthe preparation. In the nineteenth-century tradition, the bitter, green spirit was often served in a particular type of glass that allowed the drinker to measure precisely the desired ratio of absinthe to cold water. The *absintheur* slowly poured the water into the glass through a sugar cube placed on a flat, perforated spoon spanning the glass's rim, and into the absinthe below, an action that both sweetened the drink and caused it to cloud into a characteristic creamy jade. Much of the imagery of the era includes such artifacts of absinthe preparation, which evoke a process that transforms the physical nature of the drink from clear to opaque and perhaps transforms as well the creative essence of those who performed the ritual.

Critics of the era and commentators since have speculated that absinthe's unique psychoactive properties contributed to the inspired hedonism of la Décadence and gave rise to a whole host of societal ills.[39] As the use of absinthe and other distilled liquors permeated the bourgeois and toiling classes, absinthe became associated with poverty and ill health (figure 8.5). Whether absinthe's chemical attributes or the social environment that accompanied its use gave rise to an artistic revolution in nineteenth-century Europe—and, indeed, whether it may have prevented an even greater achievement of literary and visual talent—remains

(1844–1896) and Arthur Rimbaud (1854–1891) and short-story writer Guy de Maupassant (1850–1893) are among its major contributors, along with Oscar Wilde (1854–1900) working in Britain and France, as well as others. Some impressionist and postimpressionist painters, such as Édouard Manet (1832–1883) and Vincent van Gogh (1853–1890), also contributed to the movement.

A common element in the work of these artists is a sense that barriers should be crossed in art—barriers between the corporeal and the spiritual and between propriety and indecency. Drawing inspiration from nature as well as nightclubs, communities of artists centered on literary salons, gatherings where ideas and absinthe flowed freely. Of course, many of those engaged in professional decadence lived the part, taking opium, *hashish*,

FIGURE 8.5 Louis Emile Benassit, *L'absinthe!* (1862). In this lithograph, absinthe issues from a fountain of death. (National Gallery of Art, Washington, D.C., 2010.75.2)

a topic of debate in scholarly circles. As for Baudelaire, the forerunner of a movement that broke through entrenched norms and created an environment ripe for a flourishing of expression in the nineteenth and twentieth centuries, he left his comrades some advice in the title of one of his last poems, published posthumously in 1869: "Get Drunk."[40]

The End of Absinthe

The psychoactive and toxic effects of absinthe and its associated lifestyle alarmed many Europeans, who sought to limit its dangers. In time, these activists joined forces with a preexisting temperance movement, in effect taking aim at an entire genre of alcoholic beverages by maligning one of its most hazardous examples. Unlike the temperance movements in the United States and Britain, which targeted all drinks containing alcohol, the

French temperance movement vilified only the distilled spirits, such as absinthe, whiskey, and cognac, while considering beverages such as beer and wine to be quite wholesome.[41] The French physician Valentin Magnan (1835–1916) was one of absinthe's most lettered detractors. After performing a series of experiments, he concluded that the essence of the wormwood plant produced muscular shocks, vertigo, hallucinations, and epileptic attacks in animals, effects that he asserted were fundamentally different from, and faster acting than, those of alcohol alone.[42] He examined a few case studies of hospitalized alcoholics and extended his findings to the human animal, warning of absinthe consumption's dire consequences.

The essential oil of wormwood, first studied chemically during the 1840s after careful distillation of the wormwood plant's leaves, is principally composed of thujone, a terpene compound also found in arborvitae (*Thuja* spp.), sage (*Salvia* spp.), and a number of other plants.[43] Magnan and others concluded that thujone, one of many constituents of absinthe, was responsible for delirium and hallucination among its drinkers. They did not consider the possibility that habitual absinthe drinkers—or alcoholics in general—might have an underlying psychiatric disorder, nor did they consider that commercial absinthe was formulated with several herbal extracts in addition to wormwood. In cheap varieties, bootleggers and counterfeiters doctored their drinks with copper compounds to make them greener and used industrial alcohols of dubious quality. (Despite Magnan's apparent shortcuts of logic, he was an accomplished and influential psychiatrist and one of the first to detail the effects of—and possible connections between—drug use and epilepsy.)

Some of the other traditional ingredients in absinthe may have medicinal effects of their own. For instance, a study conducted in 1889 by two French chemists labeled hyssop oil a "dangerous . . . convulsant" along with anise and fennel, and the essential oils of melissa and mint were considered sedative and "stupefying" substances.[44] These claims, based on self-experimentation using high doses of isolated oils, nevertheless recognized the

FIGURE 8.6 "Another Imported Fashion": an engraving depicting the evils of absinthe and alcoholism. (From *Harper's Weekly*, September 15, 1883; National Library of Medicine, 139741)

complexity of absinthe and the challenge to identifying any single active principle in the wormwood drink. Furthermore, it is difficult to gauge the toxicity of black-market additives and alcohol, although they likely did much more damage than did absinthe's herbal extracts.[45] These inconsistencies notwithstanding, European and American leaders saw wormwood-containing absinthe as a dangerous poison and engaged themselves to ban it (figure 8.6).

One by one, nations of the West enacted legislation to restrict absinthe production or sales (figure 8.7). Belgium outlawed absinthe in 1905, followed by Holland and Switzerland in 1910 and the United States in 1912.[46] France, the nation consuming the largest quantity of absinthe (33.9 million liters annually at its peak at the outset of World War I), was among the last to criminalize it in 1915.[47] Absinthe remained legal in Spain and a number of other countries, but the absinthe culture faded, and with production in Switzerland and France so limited, the economics of the business changed. Producers in Europe reformulated to produce the anise liqueur pastis and vermouth

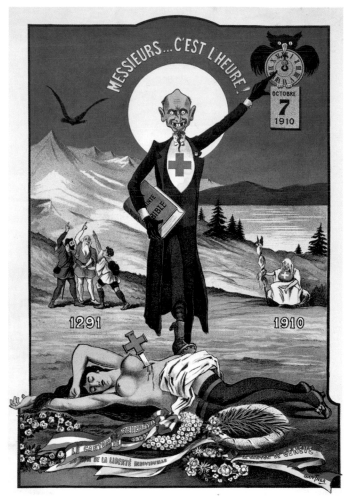

FIGURE 8.7 "La fin de la 'fée verte'": a satirical poster marking the end of the absinthe era in French Switzerland (the text reads: "Gentlemen, it's time!"). (Lithograph after A.-H. Gantner [1910]; Wellcome Library, London, L0030543)

(an ancient fortified wine traditionally made bitter with wormwood) without the absinthe wormwood plant. (The restrictions were imposed on the absinthe drink but never the wormwood plant: the plant and its seeds are widely available in horticulture.)

According to U.S. law, food and drink must be thujone free, interpreted by the government as under 10 parts per million (ppm).[48] Until recently, suppliers ventured to sell only wormwood-free products such as pastis. Since 2001, some companies have developed absinthe drinks using varieties of wormwood naturally low in thujone, and in small amounts, and thus have re-created much of the flavor and aroma of the nineteenth-century beverage

SWEET WORMWOOD (*ARTEMISIA ANNUA*)

Sweet wormwood, also known in English as Chinese wormwood, annual wormwood, and sweet sagewort, and in Chinese as *qinghao*, is an annual herbaceous plant native to temperate Asia. It grows to about 2 meters and produces highly dissected yellow-green leaves and numerous small yellow flowers (figure B.1).[1] The use of Chinese wormwood is documented over 2000 years ago, when it was listed in a medical manuscript of the second century B.C.E. as part of a treatment for hemorrhoids.[2] In later texts, the plant is recommended to heal wounds, stop pain, dab on bee stings, control joint pain, reduce fevers, grow hair on the head, and eat as a vegetable.[3] The herb was deemed cold in nature, according to the framework of Chinese medicine, and the fresh shoots and seeds were employed to treat heat-related illnesses, "pernicious *qi* and demonic poison."[4] By the time of the influential Chinese physician Li Shizhen (1518–1593), the role of *qinghao* had been thoroughly established in the prevailing medical system.

FIGURE B.1 A field of sweet wormwood. (Photograph by Jorge Ferreira/ Wikimedia)

In Li's *Grand Materia Medica* of 1596, the author summarizes the uses of the plant by previous generations of herbalists and then enumerates eighteen medicinal formulas using sweet wormwood to treat a wide range of symptoms, from diarrhea to bruises to the effects of "depletion and overexertion."[5] Among the descriptions of illnesses are some that resemble what biomedicine would call malaria. Briefly, malaria is caused by a mosquito-borne parasitic protozoan (single-celled microorganism) that enters the bloodstream after a bite, then multiplies first in the human liver and later in the red blood cells, ultimately rupturing them after a few days. The cycle of red blood cell infection and bursting produces alternating periods of fevers and chills in the patient. Li recommended *qinghao* to treat "intermittent heat and coldness due to the intermittent fever illness."[6] Importantly, Li and some previous authorities instructed physicians to prepare the herb by squeezing the fresh leaves and stems to extract the juice rather than making an herbal tea in boiling water. This observation indicates that the medically active component is not water soluble, a notion that corresponds well with the biochemistry of sweet wormwood's active principle.[7]

In the mid-twentieth century, China faced a growing population and a resurgence of malarial infections after the rise of parasites resistant to existing antimalarial treatments, a situation that threatened the nation's military effectiveness in subtropical Asia and the interests of its allies. In a stroke of remarkable authority, China's leader, Mao Zedong (1893–1976), ordered more than 500 scientists in about sixty locations to search secretly for new antimalarials both from synthetic chemicals and among China's traditional medicines, an endeavor code-named Project 523.[8] One of the most promising chemicals to emerge from the search was artemisinin, a compound of the terpenoid family that is highly concentrated in the glandular trichomes (leaf hairs) of sweet wormwood and extractable in organic solvents but not water (figure B.2).

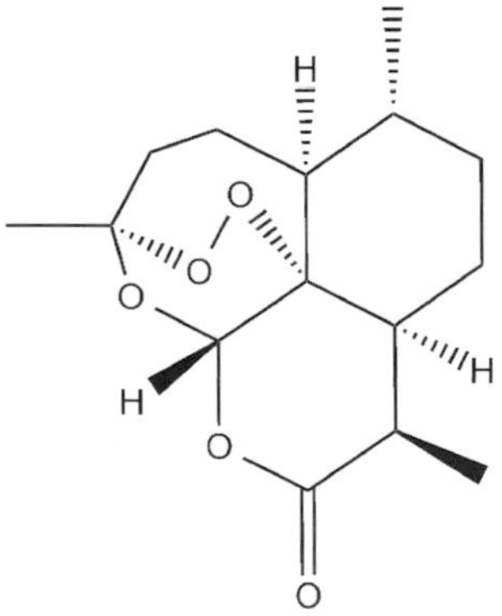

While originally commissioned to aid China's military, Project 523 ultimately yielded a product that was of great utility among civilians in malaria-prone areas, in Asia and elsewhere. Purified artemisinin was found effective to treat malarial infections in laboratory mice and rats, acting in the infected red blood cells to increase the level of oxidation and damage the parasite.[9] (Because artemisinin targets the parasite in the red blood cell, it can treat but not prevent malarial infection.)

FIGURE B.2 Artemisinin.

In high-yielding cultivated varieties, artemisinin accumulates to approximately 1 percent in leaves. Artemisinin cannot yet be synthesized from basic chemical building blocks and must instead be prepared from biological material.[10] Therefore, vast amounts of sweet wormwood are grown in China and Southeast Asia to meet the demand, particularly in Asia and sub-Saharan Africa. As effective antimalarial drugs are in short supply, the World Health Organization recommends the use of artemisinin as part of a combination therapy, paired with another antimalarial drug. This strategy, it is hoped, will delay the spread of artemisinin-resistant parasites and extend the drug's useful lifetime.[11] Because such parasites were already detected in Southeast Asia late in the first decade of the twenty-first century, artemisinin resistance is being met by renewed efforts to semisynthetically modify the artemisinin molecule and evade the parasite's defenses. Furthermore, there is concern that counterfeit or diluted-strength artemisinin or artemisinin combination-therapy pills in the developing world may risk further resistance and increased malaria deaths.[12]

The story of sweet wormwood demonstrates how ancient practices can provide clues for the biochemical study of active principles from medicinal plants. The classical authorities, by explaining the preparation of qinghao's juice, rather than as a boiled herbal tea, had deduced some of the features that led twentieth-century technicians to purify artemisinin from the oil-rich hairs of the shoots. By transforming ancient knowledge into biomedical therapeutics, the Chinese reshaped a centuries-old relationship with a medicinal plant to address a long-standing problem of public health, saving countless lives.

1. A detailed study of qinghao in ancient Chinese medical literature is Elisabeth Hsu, "Qing hao 青蒿 (Herba Artemisiae annuae) in the Chinese Materia Medica," in Plants, Health, and Healing: On the Interface of Ethnobotany and Medical Anthropology, ed. Elisabeth Hsu and Stephen Harris (New York: Berghahn, 2010), 83–130. The name qinghao 青蒿 means "blue-green hao herb." Hsu outlines a lexicographic curiosity in that the botanical description of qinghao in an influential sixteenth-century medical text corresponds more closely to the species Artemisia apiacea than A. annua, yet the latter is widely accepted today as qinghao. A. annua, whose leaves tend toward yellow-green, is more strongly medicinal than A. apiacea, whose leaves are blue-green.
2. Hsu, "Qing hao," 88.
3. Hsu, "Qing hao," 90–109.
4. These last two conditions are quoted from the Tang-era scholar Da Ming's Rihuaizu Bencao (Materia Medica of Master Sun Rays), in Hsu, "Qing hao," 96.
5. Hsu, "Qing hao," 107.
6. Hsu, "Qing hao," 107.
7. Hsu, "Qing hao," 116; C. W. Wright et al., "Ancient Chinese Methods Are Remarkably Effective for the Preparation of Artemisinin-Rich Extracts of Qing Hao with Potent Antimalarial Activity," Molecules 15 (2010): 804–812.
8. Louis Miller and Xinzhuan Su, "Artemisinin: Discovery from the Chinese Herbal Garden," Cell 146 (2011): 855–858.
9. Joseph M. Vinetz et al., "Chemotherapy of Malaria," in Goodman and Gilman's The Pharmacological Basis of Therapeutics, 12th ed., ed. Laurence L. Brunton, Bruce A. Chabner, and Björn C. Knollmann (New York: McGraw-Hill, 2011), 1395; Miller and Su, "Artemisinin." The precise mechanism of artemisinin toxicity has not yet been established. See Paul M. O'Neill, Victoria E. Barton, and Stephen A. Ward, "The Molecular Mechanism of Action of Artemisinin—The Debate Continues," Molecules 15 (2010): 1705–1721.
10. Geoffrey D. Brown, "The Biosynthesis of Artemisinin (Qinghaosu) and the Phytochemistry of Artemisia annua L. (Qinghao)," Molecules 15 (2010): 7603–7698.
11. Despite the World Health Organization's recommendation of combination therapy, many malaria patients take only artemisinin, which acts so quickly (within hours) that, feeling better, they discontinue malaria treatment, leaving a small number of artemisinin-resistant parasites in their blood, which multiply and infect other victims—a situation lamented by Miller and Su, "Artemisinin"; and R. M. Fairhurst et al., "Artemisinin-Resistant Malaria: Research Challenges, Opportunities, and Public Health Implications," American Journal of Tropical Medicine and Hygiene 87 (2012): 231–241.
12. Gaurvika M. L. Nayyar et al., "Poor-Quality Antimalarial Drugs in Southeast Asia and Sub-Saharan Africa," Lancet Infectious Disease 12 (2012): 488–496.

while remaining under the government thujone threshold.[49] By the end of the first decade of the twenty-first century, several wormwood liqueurs were approved for sale in the United States, including those of both domestic and foreign manufacture, provided that the name and label do not imply any mind-altering property.[50] While thujone levels remain low in today's products, it is interesting to note that recent tests of vintage absinthe bottled in the late nineteenth and early twentieth centuries yielded thujone concentrations in the 20 to 50 ppm range.[51] While certainly this figure is above the 10 ppm threshold of U.S. regulators, the notion that fin-de-siècle absinthe contained extremely high levels of thujone cannot be supported.[52] Nor has thujone unequivocally been demonstrated as the main psychoactive chemical in absinthe.

MECHANISM OF ACTION

Thujone exists in α and β forms (figure 8.8), differing in their stereochemistry (the three-dimensional arrangement of their atoms). The psychoactive α-thujone is a GABA receptor antagonist, binding

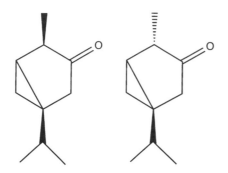

FIGURE 8.8 α-thujone (*left*) and β-thujone (*right*) differ in their stereochemistry.

cocculus). Picrotoxinin is probably responsible for some of the fish-berry plant's medicinal properties: it is used in traditional South Asian medicine as a stimulant and to reduce the nausea of travel sickness.[56]

The GABAergic system is also implicated in mood and anxiety, which might help explain some of the changes in feelings experienced by absinthe drinkers, in concert with the role of alcohol.[57] However, animal studies have not yet explained how or whether absinthe alters sensory perception in such a profound way that it might inspire the work of artists and draw so many others to its peculiar effects.

Since ancient times, people have valued

wormwood as a digestive aid and vermifuge, among other uses. During the nineteenth century, Swiss and French manufacturers produced ample quantities of the wormwood-based alcoholic drink absinthe, a libation that helped fuel a generation of avant-garde artists. The vivid imagery they created by pen and by brush is steeped in their absinthe experience. Although many Western governments restricted the wormwood content in absinthe and similar beverages during the twentieth century under suspicion of the chemical thujone's toxicity, modern days have seen a resurgence in the popularity of the absinthe drink, now produced (relatively) thujone free. *La fée verte*—the green fairy—has returned to inspire another generation.

to and preventing signaling by this neurotransmitter.[53] GABA normally is responsible for reducing the level of signaling through a number of different neurotransmitter systems, thereby lowering the level of stimulation. By blocking the GABA receptor, thujone acts as a stimulant and can be a strong convulsant in animals, lethal at an injected dose around 45 milligrams per kilogram of body weight.[54] Since absinthe drinkers take thujone at a very low oral dose, they would not reach this level, even among the most ardent of habitués.[55] However, subtle, biologically relevant effects are detectable in laboratory assays within the range that might occur by drinking absinthe. As for the role of β-thujone, if any, in the absinthe experience, it has yet to be elucidated.

Interestingly, tests of the wormwood compound produce a range of effects similar to those of picrotoxinin from the fish-berry plant (*Anamirta*

Chapter 9

Hemp
Cannabis sativa

The window of a medical marihuana clinic in Los Angeles, California. (Photograph by Neeta Lind/Flickr)

Hemp grows as an annual and can reach a height of 5 meters.[1] It is highly branched and produces leaves with multiple long, thin, serrate (jagged-edged) leaflets (figure 9.1). Uncommon among plants, hemp is dioecious, having male individuals producing only pollen-bearing flowers and female individuals producing only egg-bearing flowers. Fertilization takes place on the female plants, yielding seeds that have culinary uses.[2] Hemp stems produce a versatile fiber, and the leaves, leaf buds, and flower buds are most commonly harvested for their medicinal properties.

FIGURE 9.1 Hemp leaves.

Human selection and geographic isolation have distinguished *C. sativa* ssp. *sativa*, used for fiber, from the more strongly medicinal *C. sativa* ssp. *indica*.[3] It grows both as a weed and in cultivation throughout the temperate and subtropical zones. In recent decades, medicinal hemp varieties have been subject to intensive breeding efforts to select characteristics allowing the plant to thrive in tropical areas and in indoor settings. Cannabis, hemp, and (particularly since the twentieth century) marihuana are all common English names for this plant and sometimes also refer to its parts prepared for fiber (hemp) and medicine (marihuana).[4]

ORIGINS AND ROLE IN ANCIENT EURASIAN RITUAL AND MEDICINE

The hemp plant probably originated in Central Asia and may have been under human cultivation as early as 20,000 years ago.[5] Stone Age peoples recognized numerous useful properties in the plant, from its nutritious seeds, to the versatile fibers that can be coaxed from its stem, and perhaps to its effects on the senses.[6] Cannabis seeds have been found in archaeological sites across Europe and Asia dating as early as 5000 to 10,000 years ago, and

farmers in China processed hemp stalks into ropes, cords, and fabrics at least about 5000 years ago.[7] By about 1000 B.C.E., hemp was widely employed as a food and source of fiber for rope, cloth, and paper, from western Europe, through Central Asia, to South and East Asia (figure 9.2).

In addition to these uses, ancient people also noticed that the leaves, flowers, and especially buds of hemp altered the senses when ingested. Much has been written about the relationship between ancient societies and the many psychoactive plants used for ceremonial and religious communion,[8] and it is likely that hemp was among those employed in shamanic rituals in Central Asia by nomadic people thousands of years before the present.[9] Long ago, people recognized that the leaves and flowers, particularly of the female plants, generated the strongest mind-altering experience. They learned to prepare and eat the harvested material or inhale its smoke to achieve a numbing, exhilarated state that they might have interpreted as otherworldly. These effects are attributable to the oily resin produced on the glandular surfaces of leaves and flowers, especially concentrated in leaf and flower buds.

FIGURE 9.2 A Hmong woman teases hemp fibers into thread in Sa Pa, Vietnam.

In time, hemp consumption was taken up throughout Europe and Asia, and evidence of the use of hemp for spiritual-medical purposes comes from several sources in diverse geographic settings. As early as 1500 to 1000 B.C.E. in what is now India, a book containing healing charms called the *Atharva Veda* described hemp (*bhang*) as a "sacred grass" that "may release us from distress," indicating its perceived role in physical and spiritual health.[10] Later, the Greek historian Herodotus (484–425 B.C.E.) related a custom among the Scythians, nomadic tribes that inhabited eastern Europe and the Crimea, in which at funerals they burned the seeds or flowering tops of cannabis on red-hot stones and inhaled the smoke as part of a cleansing ritual. In their lamentations, according to the chronicler, "the Scythians howl in their joy at the vapor-bath."[11] The ancient Persians (perhaps as early as the seventh century B.C.E.) also likely used cannabis to achieve states of shamanic ecstasy in religious ceremonies, and among the Germanic tribes, hemp served as a sacred ritual aphrodisiac.[12] Archaeological evidence supports the notion that Central Asians used cannabis for spiritual-medical purposes 2400 to 2700 years ago, as grave sites of nobles and shamans have been found containing hemp flowering tops and seeds placed in a fashion suggesting ritual importance.[13]

In ancient Egypt, cannabis is documented in stone carvings and medical papyri as useful for its fiber (to prepare bandages) and for its pharmaceutical properties. In ancient texts dating to a period about 3500 years ago, cannabis was recommended to treat the eyes and aid childbirth.[14] Among the Assyrians of seventh-century B.C.E. Mesopotamia, hemp was described as "the drug which takes away the mind," and it was used as an ointment for bruises and swellings, eaten or boiled in water, and drunk to alleviate depression, cure impotence, and protect against witchcraft hexes. Its smoke was said to relieve "poison of all limbs," perhaps a reference to arthritis.[15]

Although hemp was widely used for textiles and food in China and elsewhere in East Asia dating back millennia, it is difficult to ascertain when its psychoactive properties were first noted. Interestingly, the Chinese character for hemp, 麻, can mean "fiber," such as that obtained from hemp and flax; "pockmarked" or "spotty," like the exterior pattern of some hemp seeds; "sesame," a parallel to the small size of its oil-rich seeds; and "numbing" or "tingling," in reference to the altered sensory experiences that the plant can induce.[16] A single Chinese character thus evokes the ancient roles of hemp as fiber, foodstuff, and psychoactive material. Chinese medicine records the use of hemp at least 2000 years ago, and medical texts recommend hemp boiled in wine as a numbing agent before abdominal surgery, to treat diarrhea, and as a remedy for rheumatism

(figure 9.3).[17] In contemporary Chinese medicine, hemp seed is also valued as a laxative.[18] However, an ancient Chinese text warns that cannabis seeds taken in excess can cause a person to begin "seeing devils."[19]

In the Greco-Roman world, as elsewhere, hemp was known for its fiber, oil, and diverse ascribed medicinal properties, although generally writers of the era who addressed hemp warned of its ill effects on health.[20] The Roman encyclopedist Pliny the Elder (23–79) described the plant in his *Natural History*, writing about its usefulness in rope making and in human and veterinary medicine. He suggested the "juice of this seed" to eliminate worms and other parasites from the ears, "though at the cost of producing head-ache." He also recommended hemp infused in water to treat diarrhea in livestock.[21] Pedanius Dioscorides, a Greek herbalist working in the first century C.E., wrote that the "juice extracted from it when green and instilled is appropriate for earaches."[22] These authors and the influential physician Galen (129–ca. 216) warned of hemp seed's dangers when eaten. As a strongly warming, drying herb, they cautioned, hemp could render men impotent.[23] Furthermore, Galen said the seeds are "hard to digest, disagree with the stomach, cause headaches and contain bad juices."[24] It seems that he was also aware of cannabis's

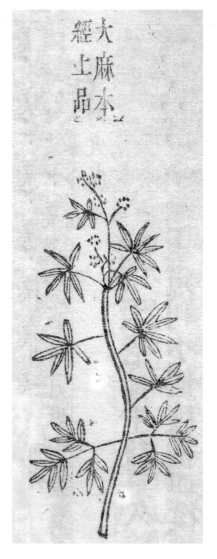

FIGURE 9.3 Hemp in a Chinese herbal. (Woodcut from Li Zongzi, *Origins of the Materia Medica* [1612])

influence on the mind, as he wrote: "[Hemp seeds] are particularly heating and so affect the head, when just a few too many have been eaten, by sending up to the head a hot and medicinal vapor."[25] (Since hemp seeds themselves do not contain a psychoactive resin, it is possible that the "seeds" employed in Chinese and Mediterranean medicine included some surrounding leaves.)

CANNABIS CULTURE IN SOUTH ASIA

In South Asia, hemp assumed an important role in indigenous spiritual life, employed to transcendent, purifying, and therapeutic ends.[26] Many centuries ago, people recognized that the psychoactive properties of cannabis are stronger in female plants than in male and are concentrated in the resin that exudes from specialized leaf hairs covering the leaves, stem, and flower buds. The resin glands being denser at the growing tip of the plant, ancient cultivators learned to harvest hemp and manipulate it in particular ways for ritual consumption.

In the Hindi language, *bhang* refers to the hemp plant and a variety of medicinal products made from it. *Bhang* consists of the dried leaves and flowering tops of cannabis plants (male and female, or selectively female), which can be eaten directly or prepared into small balls of chopped, flavored leaves. Alternatively, *bhang* can be added to dishes such as curries and dumplings; mixed with sugar, black pepper, nuts, and other additives as a confection; and blended with water and yogurt or milk as a beverage.[27] Long associated with Hindu religious practice, *bhang* was thought to ward off evil spirits and bring good luck. As early as 1000 years ago and continuing through more recent centuries, *bhang* and other cannabis preparations were employed as a form of worship, an avenue to communion with the divine.[28] Cannabis consumption in this spiritually engaged way often took place in social settings, with ritualized behaviors established for sharing the product (figure 9.4).[29]

A more potent form of cannabis is *ganja*, consisting of the resin-rich female flowers, buds, and young leaves plucked before producing seed (figure 9.5). Some cultivators recognized they could increase the yield of *ganja* by limiting the amount of airborne pollen, and therefore the fertilization of female flowers, learning to remove male plants from hemp fields as early as possible. In this way, the female plants produce more numerous, denser resin-rich flowers.[30] (This technique is now known as *sinsemilla*, from the Spanish for "seedless.") *Ganja* is traditionally eaten and,

FIGURE 9.5 A resin-rich female flower bud. Glandular hairs coat the surface of the leaves. (Photograph by eggrole/Flickr)

after the development of smoking pipes, probably in the sixteenth century, also smoked. The practices of smoking and pipe making likely reached Asia during the early or mid-sixteenth century, borne by European merchants carrying tobacco.[31]

The most strongly medicinal cannabis preparation is composed of the resin-rich glands, scraped or rubbed from the flowering tops of hemp plants and compressed into a thick paste called *charas*.[32] (In much of the world, the resinous paste is instead called *hashish*, Arabic for "grass.")[33] This material can be eaten or smoked.

Those who developed the techniques to prepare potent cannabis extracts in ancient times gleaned insights into biochemistry that can now be explained in modern terms. The most strongly psychoactive chemical in cannabis resin is Δ9-tetrahydrocannabinol (THC), a fat-soluble compound produced when fresh or dried hemp is heated (figure 9.6). Interestingly,

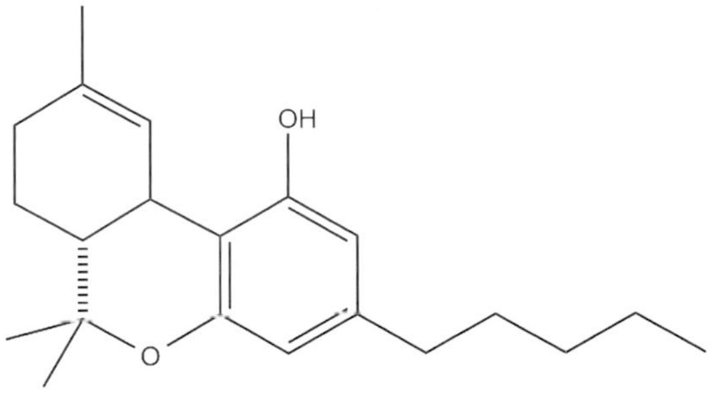

FIGURE 9.6 Δ9-tetrahydrocannabinol (THC).

FIGURE 9.4 Social-spiritual cannabis consumption in India. One woman in the hut prepares *bhang* while another smokes a water pipe. The gathering includes a musician, holy men, and soldiers. The men in the foreground have consumed *bhang*. (Painting, Pahari school [eighteenth century]; British Museum, 1940.0713-0.49)

newly harvested leaves contain THC in an inactive THC carboxylic acid form, and traditional preparation techniques—such as drying in the hot sun, boiling, cooking, or heating on charcoal—convert the inactive structure to the active THC.[34] Early South Asian manufacturers also recognized that the psychoactive agents in cannabis were not well carried in water alone, and they therefore used fatty and oily ingredients in their *bhang* preparations, such as dairy. THC's low solubility in water also renders cannabis poorly psychoactive as an herbal tea.[35]

In addition to the chemical transformation of inactive to active THC that occurs in hemp exposed to heat, the route of consumption greatly affects its physiological actions in the human body. When cannabis resin vapor is inhaled, it passes rapidly into the bloodstream and exerts an array of effects on the system, including altered mood and sensation. If cannabis products are eaten, they are slowly absorbed by the stomach and small intestine, and THC is carried to the liver, where metabolic enzymes convert it to 11-hydroxy-THC, a more potent and longer-lasting psychoactive chemical.[36] Therefore, a large quantity of *bhang*, eaten as an oil-rich confection or mixed into a milky drink in the traditional South Asian way, might provoke rather powerful effects.

REDISCOVERY OF MEDICINAL HEMP IN EUROPE

While the social, ritual, spiritual bond of cannabis remained strong in India and parts of Central Asia during much of history, in Europe, hemp cultivation was primarily directed at producing fiber for rope, cloth, and paper. Through the Middle Ages, the Renaissance, and the centuries that followed, medicinal hemp was little documented in scholarly literature, except largely for variations on the themes of Galen, Dioscorides, and Pliny. As exceptions, there are a few surviving manuscripts from England in the eleventh century and Italy in the thirteenth that recommend hemp leaves in an ointment to treat breast swelling or tenderness.[37] Certainly, hemp must have also been employed in folk medicine; there are stories of medieval Germans

THE LANGUAGE OF HEMP

Cannabis	The plant *Cannabis sativa* and its products
Hemp	The plant *C. sativa*, particularly when used for food or fiber; the fiber itself
Marihuana	The plant *C. sativa*, particularly when used for psychoactive purposes; its products
Bhang	The leaves and flowers of cannabis; drinks, confections, and snacks produced from them
Ganja	The female flowers, buds, and young leaves of cannabis
Charas	The resinous exudate of cannabis, also known as *hashish*
Sinsemilla	Seedless, high-potency marihuana produced from isolated female plants
THC	Δ9-tetrahydrocannabinol, the primary active principle of cannabis

using sprigs of hemp as charms to ease childbirth and of eastern Europeans inhaling the vapor of toasted hemp seeds to assuage toothache.[38] Generally, however, a systematic knowledge of cannabis preparation and use for psychoactive ends was not well documented.

It is possible that cannabis played a role in the countercultural religious-medical practices of witchcraft and sorcery, employed, probably in concert with other psychoactive herbs, for its ability to alter the senses and evoke feelings of spiritual release and flight. In Catholic western and central Europe, people suspected of pursuing such practices would have been considered morally corrupt and persecuted by the Inquisition. In 1484, Pope Innocent VIII issued an edict that in effect declared the use of cannabis as a tool of the satanic mass, further suppressing medicinal, spiritual, and recreational hemp from the European consciousness.[39] Perhaps the force of Catholic Church doctrine dissuaded experimentation with the physiological properties of hemp and its record in writing. It is also possible that the varieties of cannabis selected by generations of farmers for making textiles in Europe were poorly psychoactive and might not have yielded much in terms of mind-altering effects.[40] Whatever the case, there is relatively little note of hemp's psychoactive or other medicinal properties in the European sphere across many centuries, even after the broad revival of herbal scholarship that took hold during the sixteenth century.

In *The Herball, or Generall Historie of Plants* (1597), John Gerard summarized the properties of hemp relayed by Galen and Dioscorides many centuries earlier and recommended the pressed flesh of the seed for "yellow iaunders [jaundice], when the disease first appeereth," and to treat the gall bladder and to disperse and concoct (his terms) the choler.[41] A handful of decades later, John Parkinson's *Theatrum Botanicum* greatly elaborated on hemp's medicinal properties, listing first a series of ancient warnings dating to Galen—hemp seed is bad for digestion, "hurtful to the head & stomack," "breedeth ill blood and juyce in the body," and "dryeth up the natural seede of procreation"—and then recommending it for diarrhea and colic pain. The leaves, prepared by frying, are said to be good to staunch bleeding and to expel internal parasites; the juice of the leaf, Parkinson suggested, as had Dioscorides, for "wormes in the eares"; and the boiled root, for inflammation, gout, joint pain, and "hard tumours."[42] While the list of hemp's medicinal applications in Europe grew longer during the seventeenth century, it did not include (with the exception of its analgesic properties) the mind-altering effects so well known in South Asia and the Muslim world.

In an era of active trade across the Middle East and South Asia, the eleventh through sixteenth centuries saw considerable exchange of technology between nations, including the methods of hashish production for spiritual or recreational purposes.[43] Adherents to the mystic Sufi Muslim sect might have brought *hashish* eating to Egypt from Syria during the twelfth century; by the thirteenth century, *hashish* users were said to gather in Cairo's Garden of Cafour.[44] When European explorers reached exotic destinations in the sixteenth century, they discovered uses for cannabis unknown at home. For example, the Italian Prospero Alpini (1553–1617) visited Egypt and noted the ways local people used *hashish* to intoxicate themselves as Europeans did with alcohol.[45] A few decades earlier, the Portuguese physician Garcia de Orta (ca. 1501–1568) had written of his medicinal plant discoveries in India, describing *bangue* (*bhang*) as an herb capable of relieving anxiety, inducing sleep, and lessening sexual inhibitions.[46]

It wasn't until Napoleon's army rampaged through North Africa during the late eighteenth and early nineteenth centuries that medicinal-recreational cannabis became established in Europe. French soldiers returning from Egypt shared their *hashish* eating and smoking experiences, and before too long, this exotic amusement had spread among the avant-garde of society.[47] Part of the fascination and enjoyment of *hashish* must have been in the novel method of smoking the Europeans witnessed in North Africa.

Arabs had adopted an Indian water pipe, which came to them via the Persians and Turks, and used it to smoke *hashish* and other herbs (figure 9.7).[48] Using this device, the smoke is drawn through water before being inhaled, which moistens the vapor and softens its harshness. The enjoyment of such a means of smoking came through the sociable atmosphere of the water pipe garden, the pleasantly flavored smoke, and the intoxicating effects of the *hashish* (figure 9.8). Parisian literary circles became seduced by the vivid water-pipe experience and established the Club des hachichins in the 1840s, embracing artistic masters such as the poet Charles Baudelaire (1821–1867) and the novelist and playwright Alexandre Dumas (1802–1870).[49] Along with absinthe and opium, *hashish* inspired a generation of European intelligentsia in their exploration of the human mind and its creative outlets.

The new forms of expression emerging from the Paris artists' colonies were revolutionary, containing rich descriptions of surreal sensations and dramatic changes in mood and insight. These effects fascinated the French psychiatrist Jacques-Joseph Moreau (1804–1884), who viewed *hashish*

medicinal values. In a series of striking trials conducted during the mid-1840s, the Irish physician William O'Shaughnessy (1809–1889), working in India, produced an alcohol-based extract of cannabis resin for administration to animals and humans.[51] He documented its usefulness as an anticonvulsant for conditions such as tetanus and delirium tremens, an antidiarrheal for

FIGURE 9.8 The Shisha Hookah Bar, Washington, D.C. The water pipe historically has been used to smoke marihuana, opium, tobacco, and other herbs.

the treatment of cholera, and a palliative measure for rabies. As for the mental side effects, O'Shaughnessy considered them rather useful for patients facing frightening, life-threatening illnesses. "The changed state of mind it produces is truly wonderful," he wrote. "From the appalling terror which generally predominates, the patient soon passes into a state of cheerfulness, often of boisterous mirth, and soon sinks into a happy sleep."[52]

During the second half of the nineteenth century, physicians in Britain experimented with *hashish* imported from the South Asian colonies, publishing their findings and gaining a place for cannabis in the national pharmacopeias and pharmaceutical recipes of Europe and the United States (figure 9.9).

FIGURE 9.7 A man smoking a water pipe. (Photograph by Osgood Company [ca. 1901]; Library of Congress, Prints and Photographs Division, LC-USZ62-49951)

inebriation as a tool to study mental disorders. In an early expression of psychopharmacology, he wrote: "It seems there are two modes of existence. . . . The first one results from our communication with the external world, with the universe. . . . The second one is but the reflection of the first in the self, fed from its own distinct internal sources."[50] The world of dreams, mental illness, and drug-induced states of consciousness, according to the doctor, was the point of connection between the external and internal lives. More recently, neuroanatomists have located particular regions of the brain affected at a chemical level by medicinal plants such as cannabis, addressing to some degree the mysteries of Moreau's "modes of existence."

At about the same time that the social circles of poets and thinkers were using cannabis in the pursuit of inspiration, physicians were reassessing its

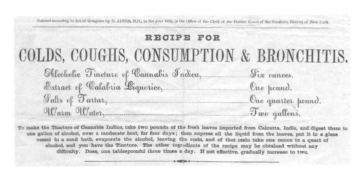

FIGURE 9.9 "Recipe for Colds, Coughs, Consumption & Bronchitis": a cannabis-containing medication to treat respiratory conditions, published by H. James, a doctor in New York, 1856. (Collection of George Glastris)

While cannabis was available in various medicinal preparations worldwide by the end of the nineteenth century and considered a useful drug, physicians were sometimes wary of using it, in light of its effects on the psyche. For example, *The Dispensatory of the United States of America* of 1892 recognized that "in morbid states of the system, it has been found to cause sleep, to allay spasm, to compose nervous disquietude, and to relieve pain" and listed it "as a decided aphrodisiac" and appetite enhancer. Yet doctors were cautioned of its "powerful narcotic" action and risk of "delirious hallucinations . . . drowsiness and stupor," "alarming effects," according to the *Dispensatory*.[53] Furthermore, despite considerable efforts, chemists had not succeeded in isolating the chemicals responsible for cannabis's effects, and the quality and efficacy of hemp tinctures was highly variable.[54] In an era when the biochemical extraction of the active principles of medicinal plants contributed to a drug's perceived legitimacy, many cannabis-containing concoctions came across as patent medicines, and its side effects hardly endeared it among the conservative medical set.[55] As a result of these factors and others, the reception of cannabis was mixed in mainstream Western medical practice, and it never became firmly established as a therapeutic agent.[56]

HEMP IN THE UNITED STATES

Hemp has a long history of cultivation in North America, dating back to colonial times, when it was grown extensively for rope, canvas, cloth, and paper.[57] It diminished as a fiber crop when cotton became more plentiful in the mid-nineteenth century, demand for sailcloth and rigging declined, and other fibers replaced hemp for making rope and canvas.[58] By the mid-twentieth century, hemp cultivation had essentially ended in the United States.

Toward the middle and end of the nineteenth century, the custom of smoking *hashish* among European artists spread to the United States, where literary types took pleasure in its effects.[59] In addition to the cosmopolitan use of cannabis among the educated classes of the cities, the late nineteenth and early twentieth centuries saw the rise of marihuana smoking among the disadvantaged tier of U.S. society.

Because cannabis grows as a weed, it became an inexpensive drug of choice among poor American and new immigrant communities both in cities and in the rural and deprived southern states.[60] Among these groups, it was frequently smoked not through elegant glass and decorated metal water pipes but as cigarettes rolled with inexpensive tobacco. In time, American artists and musicians gravitated toward hemp as a tool of social connection and inspiration, such that jazz performers of the 1920s considered "tea smoking" part of their revolutionary new culture.

From New Orleans to Chicago and New York, jazz music's new rhythm and structure attracted countless devotees and gave rise to a lifestyle that embraced marihuana. In "tea pads" scattered across Harlem in New York City, smokers—some white, most black—gathered to perform and listen to music. At dance halls, patrons and musicians smoked as new cannabis-inspired tunes caught hold, ultimately entering the American songbook on Broadway and performed by A-list artists such as Cab Calloway (1907–1994), Louis Armstrong (1901–1971), and Benny Goodman (1909–1986).[61]

To many journalists and the conservative elements in government, the seeming rise in marihuana use, particularly among immigrants and people of color, was perceived as a threat. Late-nineteenth-century immigration from southern Europe brought large numbers of the working poor to the American seacoasts and inland metropolitan regions, and after about 1910 Mexicans began moving into the southern states in increasing numbers. A general antiforeigner sentiment and directed racism converged in the vilification of marihuana, marihuana smoking, and the people who supposedly smoked it. (Indeed, the term "marihuana" was introduced into English during the late nineteenth century, ringing of Mexican or Mexican American slang.) The commissioner of the Federal Bureau of Narcotics, the antidrug crusader Harry Anslinger (1892–1975), argued on no linguistic basis that the name marihuana drew from a Nahuatl (Aztec) root meaning a "prisoner taken captive by the plant."[62]

Criminalization of Cannabis

Government officials and journalists reported in stark language the alleged risks of marihuana both to the user and to society. In 1915, for example, an American newspaper headline wondered in a sinister font whether cannabis smoking might explain the perceived character flaws in the people living south of the border: IS THE MEXICAN NATION "LOCOED" BY A PECULIAR WEED?[63] Other accounts elaborated on the tendency of cannabis to incite the user to violence and give him unusual strength, a common observation, apparently, among Mexicans. For instance, a Texas police captain wrote that under the sway of marihuana, Mexicans turn "very violent, especially when they become angry and will attack an officer even if a gun is drawn on him. They seem to have no fear. I have also noted that when under the influence of this weed they have enormous strength."[64] A newspaper report from 1920 described a knife-wielding "crazed Mexican" who turned himself on a crowd, a "seemingly demented man," who had to be "beaten into submission by police and citizens." The accessory to the crime, according to the article, was "marihuana, or 'loco weed.'"[65]

Public discourse over marihuana became increasingly targeted through the 1920s, and by the 1930s the plant was associated with extremes of violence, promiscuity, and ethnic contagion. The commissioner of public safety of the city of New Orleans warned of a "menace to public safety," including "unpremeditated and premeditated crime." One of the most troubling risks of marihuana, according to this official, was "the general lowering of morals and restraint which follows the continued use of this vicious dissipation."[66] In painting a scene of a cannabis party targeting high-school students, an antimarihuana activist warned that the drug induced young people to "the wildest sexuality" and sharpened his point by including "ordinary intercourse and several forms of perversion . . . girl to girl, man to man, woman to woman" among the threats to social propriety.[67]

The connection among marihuana, crime, and ethnicity was nearly always explicit during this era.

Who was responsible for the distribution of cannabis in New Orleans during the 1910s and 1920s? "Mexicans, Italians, Spanish-Americans and drifters from ships," according to the New Orleans official.[68] "The idle and irresponsible classes of America," wrote a professor of pharmacology.[69] In newspaper and magazine articles and on film, cannabis was increasingly portrayed as a drug of foreign origin with a unique capacity to incite the user to extreme violence and to corrupt the innocence of American youth.

The federal government played an important role in the demonization of cannabis, as the Bureau of Narcotics fed stories to the press that effectively illustrated its evils.[70] This agenda resulted in a proliferation during the 1930s of print pieces with titles such as "The Menace of Marihuana" (American Mercury), "Tea for a Viper" (New Yorker), and "Sex Crazing Drug Menace" (Physical Culture) and motion pictures such as Assassin of Youth and Marihuana (figure 9.10).[71] In the film Reefer Madness (1936), one of the best-known examples of this genre of antimarihuana dramas, several clean-cut white suburban high-school students are seduced into a seedy, amoral underworld by drug dealers.[72] Over the course of their marihuana encounters, their privileged existence quickly unravels into a morass of theft, vehicular hit-and-run, sexual promiscuity, murder, rape, and suicide.

The tactics of this era generated a public protest against hemp, which was ultimately regulated with the passage of the federal Marihuana Tax Act of 1937.[73] According to this legislation, anyone possessing or dealing commercially in hemp was required to register with the government and pay a tax through the purchase of tax stamps (figure 9.11).[74] The tax stamps were sold to only those individuals possessing cannabis, essentially requiring self-incrimination in any attempt to comply with the law. The Tax Act was overturned by the Supreme Court in 1969 for this reason, but Congress drafted and passed the Comprehensive Drug Abuse Prevention and Control Act of 1970, which includes the Controlled Substances Act, explicitly to criminalize marihuana as a schedule I controlled substance. The Controlled Substances

FIGURE 9.10 An advertising poster for the film *Marihuana* (1936). (Library of Congress, Prints and Photographs Division, LC-DIG-ppmsc-04768)

FIGURE 9.11 Marihuana tax stamps. (Smithsonian Institution, National Postal Museum, 1998.2013.4555.2.1-10)

Act and the earlier Marihuana Tax Act applied to all forms of cannabis extract, resin, and herb, with exceptions for only processed fiber and hemp seed, if sterilized before sale. Rarely, the cultivation for fiber production of varieties low in psychoactive potential has been permitted. For example, World War II cut off American supplies of imported fibers, and the United States funded hemp-fiber production between 1942 and the end of the war.[75]

Despite the federal regulatory status, hemp remained popular as a recreational drug in illicit use throughout the twentieth century, and clandestine growers trained the plant to thrive in the underground economy. From the 1970s onward, American hemp cultivation increasingly moved indoors, with specially selected varieties that performed well in greenhouses and interior grow rooms.[76] Under the pressures of directed breeding and intensive yield-centered techniques, cannabis has been manipulated to grow larger, denser buds rich in glandular leaf hairs, producing ever-increasing levels of THC. Aided by biochemical assays as well as subjective experiences, producers can select varieties that induce a range of subtle effects on mood and sensation. In the quest to produce a uniform harvest of potent material, many cultivators have preserved desirable female plants as essentially immortal vegetative specimens. Grown under particular controlled conditions, the plants do not flower and can be used to produce cuttings for plants that will flower (generate buds) in the commercial setting. In this way, the cultivator can harvest an abundant crop of *sinsemilla* buds several times a year.[77]

While breeders made advances in the production of potent marihuana in controlled, indoor settings, many users sought the herb for treating illness: to ease pain, alleviate anxiety, increase appetite, and suppress vomiting, for example. Dropped from the *United States Pharmacopeia and National Formulary* in 1941, strongly restricted by the federal government, and the subject of very little scientific research, marihuana's possible utility in therapy was only rarely considered during the second half of the twentieth century.[78] Recently, a number of state ballot initiatives have lifted state penalties for the possession and consumption of limited amounts of cannabis. In some cases, the new state measures permit the cultivation, manufacture, possession, and use of marihuana for what are deemed "medicinal" purposes. In other states, restrictions were lifted for "medicinal" or "recreational" uses. Despite these moves by states and the District of Columbia, the federal government maintains that marihuana remains prohibited nationally, placing cannabis in a knotty legal predicament. The federal government has discretion over which cannabis-related crimes to prosecute while holding the position that marihuana is a schedule I controlled substance, with a high potential for abuse, no accepted medical use, and no accepted safe use.

The state actions permitting the use of medical marihuana have carved an unusual exception to the standard practices of science-based pharmaceutics. Whereas drugs regulated by the Food and Drug Administration are subjected to appropriate clinical trials to demonstrate safety and efficacy, measures allowing physicians to prescribe and vendors to dispense marihuana as a pharmaceutical agent ignore the long-established role of the federal government to monitor the quality of pharmaceuticals.[79] In contrast to FDA-regulated medications, medical marihuana preparations are not subjected to federal oversight of manufacturing practices and content, or reporting of effectiveness or side effects, and patients face a diverse array of products with varying levels of potency.[80]

MECHANISM OF ACTION

The primary active principle of cannabis, THC, was isolated by the Israeli biochemist Raphael Mechoulam (b. 1930) and his co-workers in 1964. Unlike many psychoactive plant chemicals, THC and other biologically active marihuana compounds are soluble in fat rather than water. Active agents include a family of structures, of which THC is largely responsible for the psychoactive properties. Other members of this family of chemicals, called cannabinoids, include constituents such as cannabidiol (CBD) and about sixty other molecules (figure 9.12).[81] *Hashish* contains approximately 10 to 20 percent THC by weight, the buds and small leaves around 5 to 8 percent, and other plant parts 2 to 5 percent. Recent advances in cultivation and selection have resulted in an increase in the THC content in *hashish* made from certain elite varieties.[82]

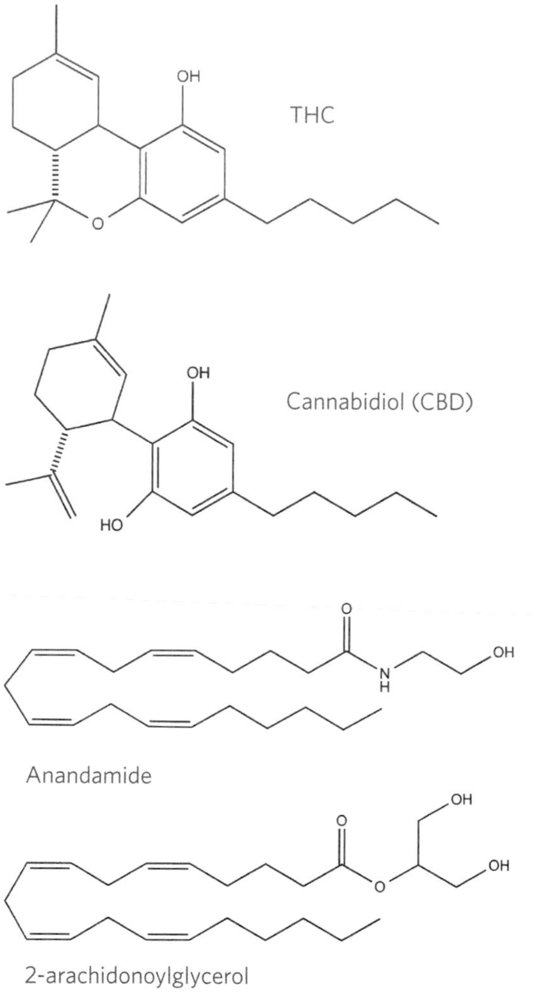

While hemp seeds have health value for their nutritive content, they are not naturally psychoactive because they contain negligible levels of THC.[83] Whatever minuscule amount of THC might be present in the prepared seeds is generally thought to be the result of contact with the resin in the flower parts carried over through harvesting.[84]

Throughout history, the most common methods of cannabis consumption have been via the oral route and in smoke. (Intravenous injection has not been practical because hemp's active principles are not water soluble.) It is also possible to apply cannabis-containing ointments or lotions to the skin. The route of administration influences the effects on the body.

In a marihuana cigarette containing about 30 milligrams of THC, approximately 30 percent of the active principle is destroyed by heat before inhalation, the remainder being taken up with more or less efficiency, depending on smoking habits such as depth and pace of draw and loss to sidestream smoke.[85] Uptake of the fat-soluble chemical is rapid, and levels of THC peak within a few minutes of smoking. In contrast to smoked cannabis, THC that enters the body via ingestion is broken down to some degree by stomach enzymes and is

FIGURE 9.12 Cannabinoid-signaling molecules: the cannabis active principles THC and cannabidiol (CBD); the endocannabinoids anandamide and 2-arachidonoylglycerol.

more slowly absorbed, reaching peak concentration in the blood after four to six hours.[86] (Unlike when taken up through the lungs, ingested THC is converted in the liver to 11-hydroxy-THC, a more potent and longer-lasting metabolite).[87] The concentration of CBD and perhaps other cannabinoids is also relevant to marihuana's pharmacological effects; these molecules are thought to prolong the effects of THC and may have specific physiological effects of their own.[88] Some herbalists consider marihuana to be a "synergistic medicine" composed of dozens of active principles that together generate therapeutic effects.[89] Regardless of the mode of consumption, cannabinoids become stored in fat deposits throughout the body and brain and are slowly released at a very low level over a period of days to weeks without behavioral effects.[90]

The cell receptors that perceive THC and other cannabinoids are present in many regions of the brain as well as in cells of the immune and reproductive systems. The brain receptors are particularly densely situated in the basal ganglia, cerebral cortex, hippocampus, cerebellum, and spinal cord, which may account for THC's effects on the sense of time; the experience of distortions in taste, color, and sound processing; and the ability to concentrate and form memories.[91] Cannabinoid receptors located in regions of the limbic system likely give rise to the sense of reward associated with cannabis.[92] However, cannabinoid receptors are absent in the brain stem, which may explain why THC does not affect respiration and is relatively nonlethal at high doses. THC mediates analgesia and euphoria through extensive connections to the opioid-signaling network and interferes with hunger regulation by acting as an appetite signal. It also suppresses vomiting.[93]

THC and related chemicals from cannabis mimic the brain's own signaling molecules, called the endocannabinoids (for endogenous cannabinoids), a group of chemicals that bear little resemblance to THC but nonetheless activate the same receptors (see figure 9.12). Endocannabinoid receptors on presynaptic neurons recognize endocannabinoids released by postsynaptic neurons in response to sustained signaling by glutamate,

acetylcholine, and other neurotransmitters.[94] As a result, endocannabinoids modulate a wide range of excitatory and inhibitory signals affecting numerous physiological activities.

Therefore, endocannabinoids play a role in regulating those processes also affected by THC consumption: the perception of pain, anxiety, and hunger, among others. The role of the endocannabinoids in sensory experiences (discernment of space and time) is less clear. In this regard, THC may have particularly strong effects because its dose in cannabis consumption is frequently higher than the endogenous molecules. Chronic marihuana users experience tolerance and a range of withdrawal symptoms on discontinuation of use, including anxiety, decreased food intake, and pain.[95]

With such diverse roles in memory, pain, anxiety, hunger, and other aspects of health, some cannabinoids and synthetic analogs have been developed for use as medicines in the modern era. For example, the synthetic THC drug dronabinol (sold as Marinol) and the synthetic THC analog nabilone (sold as Cesamet) are approved to be taken orally as appetite stimulants for use in patients with AIDS and antiemetics for cancer patients undergoing radiation treatment and chemotherapy.[96] There is great interest in developing new preparations of cannabis, including naturally derived active principles, for oral and inhaled delivery, as some patients wish to avoid the side effects of ingested THC and believe the combination of THC with other active principles to be more effective. As one step in this direction, the drug Sativex, a combination of natural THC and CBD in a one-to-one ratio delivered to the lungs via an inhaler, has been demonstrated effective for use by multiple sclerosis patients against neuropathic pain and spasticity.[97] The cannabinoid CBD itself is the target of much research because it appears to exert anticonvulsive, sedative, and anti-inflammatory effects without altering the patient's mood or sense of time and space.[98]

Clinical studies are addressing the efficacy and safety of THC and THC analogs for a wide array of conditions, including chronic pain and glaucoma.[99] At the same time, the laws of many jurisdictions are making cannabis more readily available to those

PHARMACOLOGICAL EFFECTS OF THC

Analgesia

Relaxation

Lightening of mood

Changes in perception

Disruption of short-term memory

Reduction of nausea

Increase in appetite

Euphoria

who wish to take it for "recreational" or "medical" purposes. As modern research on the use of medicinal marihuana is relatively recent, it is important to allow the scientific method to guide the development of therapies, taking sampling, placebo effects, and controlled route of administration into account.

Hemp has long been valued as a medicinal plant and a source of a versatile fiber throughout Asia and Europe. It was associated with spiritual and healing properties in ancient texts and has survived to modern times as a mystical, recreational drug. Europeans rediscovered its effects during the nineteenth century, when artists drew

inspiration from its calming and mind-opening influence. Although illegal in the United States according to federal law, many states have passed measures permitting its use for medical or recreational purposes, giving hemp a complex regulatory status. Meanwhile, researchers have made strides to understand how its assortment of active principles function singly or in concert to influence the mind and body in diverse ways. Since ancient times, people have documented hemp's social, spiritual, and therapeutic properties. As our modern society debates cannabis use, we are discussing and engaging the same physiological effects experienced by communities thousands of years ago.

Chapter 10

Coffee
Coffea spp.

A traditional Turkish coffee service in Ürgüp, Turkey.

Coffee grows as a shrub or tree up to 5 meters in height (although it is usually pruned aggressively in plantation), with simple oval-shaped deep-green leaves and multiple flowers clustered at leaf bases on its branches.[1] After fertilization, grape-size fruits develop containing two, or more rarely one, seeds (figure 10.1). Coffee plants arrive at maturity after about five years and continue to produce for another thirty-five years or more.[2] The fruits (known as "berries") are green when immature and turn to a deep red when ripe. The medicinal value of coffee is highest in the seeds (known as "beans").[3] The coffee plant originated in Africa, in the region that is now the highlands of Ethiopia, which is where people discovered its medicinal properties.[4]

Until the twentieth century, most coffee production consisted of the species *Coffea arabica*, a variety growing best in the higher elevations (1000–2000 meters) and producing a beverage with a mild, complex flavor. In cultivation, *C. arabica* tends to be susceptible to pests, which reduces its yield. During the European conquest of Central Africa in the mid-nineteenth century, Westerners noticed natives chewing on the leaves of a novel species of coffee, *C. canephora*.[5] *C. canephora* comes from the equatorial lowland (under 700 meters) forests of Africa and is thought to produce a bolder and harsher flavor than *C. arabica*.[6] It also exhibits greater pest resistance than its highland cousin.

FIGURE 10.1 Coffee (*C. arabica*): (*top*) plant; (*bottom*) mature berries. ([*top*] Photograph by Mark W. Skinner; USDA-NRCS PLANTS Database)

ORIGINS OF USE IN AFRICA AND ARABIA

One of many legends relates that coffee was discovered by a goatherd named Kaldi, who lost his flock one day, only to rediscover the animals frolicking, dancing, and playing in a mountain forest. The source of their bizarrely energetic behavior, he reasoned, were the green leaves and red berries of the plants they had eaten. The next day, he again lost his goats to the coffee tree and decided to see for himself the effects of these plants. He chewed on some leaves and, although they were bitter, felt a pleasant stimulation through his body. He next tried one berry, then two. He enjoyed the slightly sweet pulp of the fruit and licked the seeds, which were covered with a tasty mucilage. Before too long, Kaldi joined his herd in gleeful, exuberant romping. Thus coffee became Ethiopia's native stimulant, a source of energy and inspiration for dance and art.[7]

While the colorful myth of coffee's discovery has not been borne out by archaeological or textual evidence, the plant's fruits or seeds must certainly have been enjoyed by Ethiopians in the distant past, entered into commerce across the Red Sea to Yemen in the medieval era, and become known to the Arab and greater Islamic world (figure 10.2). As these early steps in the dissemination of coffee are far from clear, so too are the ways that coffee was first appreciated in Africa and the Middle East. In Ethiopia, people might have prepared the leaves and berries by steeping them in boiled water to make a coffee-flavored drink or

FIGURE 10.2 "Arbre du Café dessiné en Arabie sur le Naturel." (Engraving from Jean de la Roque, *Voyage de l'Arabie heureuse* [1716]; Library of Congress, Rare Book and Special Collections Division, LC-USZ62-90572)

FIGURE 10.3 Men preparing coffee in Gaza, 1870. (Photograph by Felix Bonfils)

ground the beans and mixed them with animal fat to eat as a snack. At some point, they also developed a coffee wine, made by fermenting the berry pulp, and a sweet drink of roasted coffee pulp in water, a tradition still observed in contemporary Ethiopia.[8]

The first written references to a plant identifiable as coffee indicate that its fruits and seeds were used medicinally as early as the ninth and tenth centuries.[9] According to the Persian scholar Muhammad ibn Zakariya Razi (Rhazes, ca. 865–925), a plant called *bunn* and a drink made from it, called *buncham*, have hot and dry properties (in humoral medical terms) and are "very good for the stomach."[10] Later, the Persian physician Abu Ali al-Husayn ibn Abd Allah ibn Sina (Avicenna, 980–1037) wrote of *bunn* and *buncham*, declaring,

"It fortifies the members, cleans the skin, and dries up the humidities that are under it, and gives an excellent smell to all the body."[11] Perhaps through a combination of commerce and the dissemination of medical scholarship, awareness of the coffee plant increased, and by the fifteenth century, someone (in Ethiopia or Yemen) had tried to roast and grind the beans to brew a dark beverage (figure 10.3).[12]

The practice of brewing coffee—the dark, bitter, stimulating beverage—seems to have become established most firmly in Yemen, where an Arab author of the mid-sixteenth century relates that "it drove away fatigue and lethargy, and brought to the body a certain sprightliness and vigor."[13] The Arabs called the coffee drink *qahwa*, the origin of the French *café* and the English "coffee."[14]

FIGURE 10.4 Outdoor coffeehouses in the Ottoman style: (*left*) men smoking and drinking coffee at a coffeehouse near Cairo, Egypt; (*right*) a coffeehouse in Istanbul, Turkey. ([*left*] Louis Haghe, lithograph after David Roberts, *The Coffee-Shop of Cairo* [1849]; Wellcome Library, London, V0019182)

Arab adherents of the Sufi Muslim sect were among the first ritual coffee drinkers; they used the beverage to keep themselves awake for midnight prayers.[15] While at first coffee was limited to spiritual-religious application (for its power over drowsiness), wealthy families eventually introduced coffee into their social gatherings, and merchants established coffeehouses for public use. Widespread commerce in the Arab and Muslim world during the late fifteenth and early sixteenth centuries brought coffee and coffeehouses from Yemen to Cairo, Damascus, and, in time, Istanbul.[16] With the expansion of the coffee beverage followed the growth of an intellectually stimulating coffeehouse culture. In the cities of Arabia, Persia, North Africa, and the growing Ottoman Empire, the coffeehouse evolved into a place of free political thought, social bonding, and avant-garde discourse (figure 10.4). Such progressive thinking, however, frequently irritated the governing religious and political authorities.

COFFEEHOUSE CULTURE AND THE WORLD COFFEE ECONOMY

In 1511, the governor of the Arabian city of Mecca grew wary of the unorthodox ideas emanating from his jurisdiction's coffeehouses, and determined that coffee—like wine—was forbidden by the Qu'ran.[17] Consequently, he closed all of Mecca's coffeehouses, a ban that lasted only until his coffee-drinking superior, the sultan of Cairo, heard about it and reversed the decree. (The official Cairo decree banned coffeehouses but permitted coffee drinking. In any case, efforts against coffee were only sporadically enforced, and coffee drinking continued both in secret and in the open.)[18] Still, numerous regional governments and religious groups during this period placed restrictions on coffee and coffeehouses based on public safety, medical, and theological grounds, with little long-term effect.[19] The lure of coffee remained strong, as the drink provided a medium around which people formed social and business connections, with a stimulating sense of well-being as a desirable side effect.

During the late sixteenth century, the coffeehouse culture blossomed in the large and multiethnic Ottoman Turkish city of Istanbul, and Turks grew to dominate coffee cultivation and model the coffee lifestyle.[20] The English philosopher Sir Francis Bacon (1561–1626) described an Ottoman coffeehouse in his posthumously published *Sylva Sylvarum* (1627), which illustrates coffee's perceived roles in social life and physical health:

They have in Turkey, a Drinke called Coffa, made of a Berry of the same Name, as Blacke as Soot, and of a Strong Sent, but not Aromaticall; Which they take, beaten into Powder in Water, as Hot as they can Drinke it; And they take it, and sit in their Coffa-Houses, which are like our Tavernes. This Drinke comforteth the Braine, and Heart, and helpeth Digestion.[21]

The connection of coffee with Turkey is also important, as it was through the Turkish-controlled Middle East that African coffee first emerged as a world commodity.

Most coffee destined for Europe was shipped out of the Yemeni Red Sea port of Mocha to Suez and overland by camel to Alexandria on the Mediterranean for purchase by Venetian and French agents.[22] While European travelers and traders grew fond of coffee and coffeehouses, the Turks, who exerted authority over most of Arabia and North Africa at that time, maintained their monopoly on coffee cultivation by fiercely guarding against any export of coffee plants and steeping or partially roasting all beans destined for export to prevent subsequent germination.[23]

Yet by the seventeenth century, coffee growing and drinking had spread, with smugglers establishing plantations of Yemen-derived coffee in India and throughout the Dutch merchant colonial system. Cultivation was initiated in Java and the other East Indian islands of Sumatra, Celebes, Timor, and Bali beginning in the 1690s. In the eighteenth century, the French began growing coffee in their holdings in Martinique (1720), Haiti (1725), and Guadeloupe (1726), and the British and Spanish set up plantations in Jamaica (1730), Cuba (1748), and Puerto Rico (1755). Coffee also found its way to Portuguese Brazil (1727) via Suriname and French Guiana.[24] Much of this vast diaspora of coffee originated with a handful of coffee plants smuggled out of the port of Mocha to the Dutch plantation in Java, two locations now firmly associated with coffee in modern parlance.[25] Because this widespread dispersal came from a small number of parent plants, genetic diversity among much of the coffee (C. arabica) of the world is low, resulting in an increased susceptibility to a number of bacterial and fungal plant diseases.[26]

Coffeehouses flourished in the British domain and continental Europe during the seventeenth century. London and Oxford saw their first coffeehouses open in the 1650s.[27] By 1663, more than eighty coffeehouses had opened in England, a number that increased to over 2000 by 1700.[28] The first American coffeehouse was established in Boston in 1689, followed by one in New York in 1696.[29] In Austria, many Viennese had their first taste of coffee following the retreat in 1683 of the Ottoman Turkish military, which had held an unsuccessful siege of the capital. The Turks left behind a considerable quantity of coffee beans, which the Viennese developed into a uniquely styled coffee.[30] In Italy, Venice saw its first coffeehouse open in 1683, the first of many.[31]

The coffeehouse provided a venue for lively discussion and a brush with the exotic, a combination that certainly attracted the most progressive of society, including writers, artists, and statesmen. In the late 1680s, an Italian immigrant to France, François Procope, founded a coffeehouse, the Café Procope, across from the storied Comédie-Française (figure 10.5).[32] The establishment drew the likes of the philosopher Jean-Jacques Rousseau, the playwright Voltaire, and the American statesmen Benjamin Franklin and Thomas Jefferson over the course of the following century.

In England, coffeehouses became important venues for social intercourse. Unlike alehouses, where drinking turned boisterous and uncivilized, coffeehouses allowed patrons to engage in erudite discussions—spurring some to call them "penny universities"—and to pursue business agreements.[33] For example, the mammoth British insurance firm Lloyd's of London originated as a coffeehouse where merchant mariners met to obtain insurance coverage for their cargoes.[34]

By the late eighteenth century, much of the world's coffee was no longer grown in the Middle East. Instead, the bulk of coffee originated in extensive plantations under European control in Asia, South America, and the Caribbean. To harvest and process so much coffee, Europeans required a large workforce of cheap laborers. The plantations in the French colony of Haiti supplied half of the world's coffee in 1788 and were tended by a huge

FIGURE 10.5 European coffeehouses were establishments of civil social interaction, as was Café Procope in eighteenth-century Paris. (Wellcome Library, London, V0014353)

number of African slaves living in the most inhumane of conditions. The slaves of Haiti led a long, bloody, but ultimately successful revolt against the French in 1791, and in the process they destroyed the hated plantations and murdered their former masters.[35] The Dutch filled the global coffee shortage with exports from its Java plantations, tended by enslaved Javanese. During the mid-nineteenth century, Portuguese landowning nobles in Brazil enhanced production with new slash-and-burn plantations established at the expense of tropical rain forest and worked by millions of African and native slaves.[36] In Guatemala and Nicaragua, coffee was cultivated during the nineteenth century through the forced labor of Mayan men, women, and children.[37] All this increased production fed the intense and growing coffee habits of Americans and Europeans.

MEDICINE OR POISON?

While coffee became a premier social drink in Europe, its medicinal use was alternatively upheld and questioned by various sources. To Arab physicians of the sixteenth century, there was a disagreement on the subject of its humoral properties: some scholars claimed that the seed and fruit were cold and dry; others held that the seed and fruit were hot and dry or that the seed was cold while the fruit was hot.[38] Coffee was generally thought to be good to dry up phlegmatic coughs and colds (by those who held to the hot-dry interpretation), a potent diuretic (and therefore good for the kidneys), and an effective hunger suppressor, but in excess it was blamed for causing hemorrhoids and headaches. The diuretic property of coffee was not universally appreciated as a benefit. For example, the author of a medical treatise of the era warns that an increased level of urinary excretion weakens the body. Other benefits of coffee documented by Arab authorities during this period include the treatment of "boiling (or bubbling) of the blood" and prevention of "smallpox, measles, and bloody skin eruptions."[39] The contention surrounding coffee's health effects spread with the beverage itself to Europe and around the coffee-drinking world.

In 1610, the British poet George Sandys (1577–1644) visited Istanbul and noted that the Turks sat "chatting most of the day" over cups of coffee, which he found "black as soote, and tasting not much unlike it."[40] He also found anecdotal evidence of coffee's medicinal value: "[I]t helpeth, as they say, digestion, and procureth alacrity [energy and liveliness]."[41] In 1632, an English writer claimed that coffee could "mend" the temperament, with the capacity to "expell feare and sorrow, and to exhilarate the mind."[42] By the mid-seventeenth century, coffee found its way formally into the European pharmacopeia, being described by the herbalist John Parkinson (1567–1650) as strengthening the stomach, improving digestion, and treating tumors and blockages of the liver and spleen.[43] To some, coffee's possible health benefits placed it squarely in the category of panacea, as witnessed by the wide-ranging descriptions offered by especially ardent promoters. For example, London's Rainbow Coffee-House on Fleet Street recommended coffee for conditions ranging from sore eyes to scurvy to the king's evil (scrofula, a bacterial infection).[44] The London coffee merchant Pasqua Rosee claimed that coffee could aid digestion, treat headaches and dropsy, prevent miscarriages, and cure a host of other ailments.[45]

Meanwhile, a number of voices emerged to warn of coffee's possible deleterious effects. For example, the German physician Simon Paulli (1603–1680), following a line of argument common in his era, wrote that coffee "surprisingly effeminates both the Minds and the Bodies of Persians," to the degree that men become infertile,[46] while a French physician claimed in the 1770s that coffee was responsible for nymphomania.[47] Although the seventeenth-century physician Thomas Willis (1621–1675) recommended coffee for its stimulating properties, he also warned that it could cause headaches, vertigo, and heart palpitations and could harm the joints.[48] After the opening of a coffeehouse in Marseille in the late seventeenth century, a group of local physicians protested, claiming that the drink would "burn up the blood," induce tremors, cause impotence, and lead to leanness.[49] Through most of the seventeenth and eighteenth centuries, there was no consensus on whether coffee was a healthful drink or a toxic substance.

While the perceptions of the medicinal value of coffee were mixed for several centuries after its introduction into Western commerce, coffee in the nineteenth century faced stronger criticism for its perceived negative health effects. Some of these attacks were deserved: unscrupulous wholesalers sometimes adulterated cheap unroasted coffee with lead- and arsenic-based coloring agents to produce the uniform yellow or green hue attractive to buyers. In 1884, a headline in the *New York Times* warned, POISON IN EVERY CUP OF COFFEE.[50] During the late nineteenth century, the health lifestyle entrepreneur C. W. Post (1854–1914) railed against the American coffee habit, suggesting in ubiquitous advertisements that people switch to his grain-based coffee substitute, Postum, and breakfast cereal, Grape-Nuts, for a more hearty diet. "You can recover from any ordinary disease," he said, "by discontinuing coffee and poor food, and using Postum Food Coffee." Post's attacks on coffee continued: "Coffee frequently produces indigestion and causes functional disturbances of the nervous system," and "Coffee is an alkaloid poison and a certain disintegrator of brain tissues."[51] While Post made his claims in the absence of any medical authority, a number of doctors joined Post's cause and warned of the myriad dangers of coffee. Indeed, coffee's perceived wholesomeness suffered during the first two decades of the twentieth century.

Despite the efforts of health evangelist businessmen such as Post, the prohibition of alcohol in the United States during the 1920s boosted coffee's place as the national social drink.[52] Coffeehouses again proliferated, and home consumption of the beverage increased. By 1923, American demand accounted for half of the world's coffee.[53] During the 1940s, the Allied war effort combated cold and fatigue with coffee, both in the battle zones and on the home front. After the war, coffee maintained its place in homes, restaurants, and offices as a social and economic fuel for the post–World War II age. While mainstream American tastes had grown accustomed to mild-flavored blends of beans and a brewing process that recirculated coffee over spent grounds, some spirited entrepreneurs saw an opportunity to redefine coffee and the coffeehouse.

THE NEW AMERICAN COFFEEHOUSE

During the decades following the 1950s, a number of coffee merchants began to specialize in selling select blends of coffee and developing coffeehouses that brewed elite beans into strong, highly flavorful drinks. Alfred Peet (1920–2007), a San Francisco–based coffee importer, established Peet's Coffee & Tea in Berkeley in 1966.[54] A few years later, three Seattle entrepreneurs visited Alfred Peet and trained to select, roast, grind, and brew coffee in his shop. They returned to Seattle and opened their own coffeehouse, Starbucks, in 1971.[55] The renewed interest in American coffeehouses combined an emphasis on quality coffee beans and strongly flavored coffee served in an environment conducive to social gathering. Serving much the same role in modern times as the coffeehouses of seventeenth-century Europe, the establishments of the twentieth and twenty-first centuries helped engage people in conversation over stimulating cups. During the 1990s and 2000s, the Starbucks operation grew enormously, generating a worldwide footprint of American coffeehouse culture with more than 20,000 locations.[56] Indeed, the explosion of coffeehouses on the American model, including corporate and independent firms, based in the United States or internationally, is in essence a recapitulation of manifested practices of earlier generations and other places, from Mecca and Istanbul to London and Paris.

THE PROCESSING OF COFFEE

Much of the success of the premium or specialty coffee shops is attributable to the selection and processing of the beans, which has a key role in the ultimate flavor of the drink. Before 1900, nearly all coffee consisted of the species *C. arabica*, which grows best in cool tropical highlands that receive steady rain throughout the year.[57] The beans contain about 1 to 1.5 percent of the stimulant compound caffeine and generate a highly flavorful brew.[58] However, *C. arabica* yields are sensitive to frost, and the plants are susceptible to the fungal pathogen coffee rust (*Hemileia vastatrix*).[59] *C. canephora*, a variety brought into cultivation from the Congo in 1898, grows acceptably in a wider geographic range and produces beans with slightly more caffeine than *C. arabica* and a harsher flavor.[60] It is more resilient against fungi than *C. arabica* and thus is grown widely and used in blending, particularly for cheaper or instant coffees.[61]

Most coffee berries grow with two seeds (beans), although a fraction of them (perhaps 5 or 10 percent) have a single seed. These rare, single-seeded coffee fruits are known as peaberries and are thought to produce a particularly richly flavored coffee.[62] The coffee berries are harvested when their color turns from green to red.[63] To remove the seeds from the berries, producers historically used one of two methods. The dry method consists of stripping berries from the trees and allowing them to dry on tarps under the sun. After frequent spreading and turning, while protecting from dew and any other moisture, the fruits shrivel and harden. The dried husks can be removed by pounding on them, yielding the seeds. Today, much of the drying and husk removal is done by machine. The wet method, developed in the West Indies and used throughout Central America, requires harvesting the ripe berries, removing much of the pulp, and allowing any remaining material adhering to the bean to soak and ferment in large tanks of water for up to forty-eight hours. Then, the softened tissues are washed off with running water and the beans dried under the sun or by machine.[64] The wet process is thought to impart a more nuanced flavor to the seeds but has also been responsible for a tremendous amount of water pollution: waste coffee pulp was discharged into streams and rivers, ultimately choking these waterways of oxygen as the organic material decomposed. However, recent ecologically conscious practices, such as retaining coffee pulp as natural compost, have improved the quality of water in places such as Guatemala and added value to former waste products.[65]

Raw coffee beans are known as green beans, although their color can range from green to yellow, white, or tan (figure 10.6). Coffee beans may be blended from different sources to produce a mix of flavors or characteristics deemed desirable for the consumer.

FIGURE 10.6 Green coffee beans (*top*) and green peaberries (*bottom*).

TYPICAL CAFFEINE CONTENT OF COMMON FOODS AND DRUGS

Source	Caffeine Content
BEVERAGES	
Brewed coffee	74–83 mg/5-oz. cup
Decaffeinated coffee	2–3 mg/5-oz. cup
Tea	24–30 mg/5-oz. cup
Cocoa	4–5 mg/5-oz. cup
Regular or diet colas	26–58 mg/12-oz. serving
Caffeine-free colas	0 mg
CHOCOLATES	
Milk or sweet chocolate	6–20 mg/oz.
Baking chocolate	35–60 mg/oz.
ANALGESIC DRUGS	
Anacin, regular strength	32 mg/tablet
Excedrin, extra strength	65 mg/tablet
OVER-THE-COUNTER STIMULANTS	
No Doz	100 mg/tablet
Vivarin	200 mg/tablet

Source: Jerrold Meyer and Linda Quenzer, *Psychopharmacology: Drugs, the Brain, and Behavior,* 3rd ed. (Sunderland, Mass.: Sinauer, 2013).

The roasting process also can add new flavors to the beans. A lighter roast is generally preferred in the United States, through which the beans are subjected to a temperature of 212 to 218° Celsius until they turn a uniform brown color. During the roasting process, the starches in the coffee bean turn to sugars at about 207°C and caramelize above 212°. At 238°, the caramelized sugars begin to burn. Therefore, the lighter roasts bring out more sweetness in the coffee; a darker roast imparts less mild flavors. A Vienna roast takes place at temperatures around 240°, and the French roast occurs at 250°.[66] Once roasted and ground to a fine powder, coffee has a very short shelf life before losing robustness and aroma, which is why most drinkers prefer a fresh grind.

MECHANISM OF ACTION

The stimulating property of coffee derives from its active principle, caffeine. First isolated from coffee in 1819 by the German chemist Friedlieb Runge (1795–1867), caffeine is responsible for many of the behavioral and physiological effects of coffee, tea, and other medicinal plants (figure 10.7).[67] Caffeine usually enters the bloodstream within thirty to sixty minutes of oral consumption, beginning through the stomach and continuing through the small intestine.[68] In low doses (100 milligrams, for example), caffeine produces subjective effects such as enhancement of alertness and reduction in tension as well as mild euphoria and improved sociability and self-confidence. At higher doses (200 milligrams or more), some people experience tension and anxiety. At very high doses (twelve or more cups of coffee

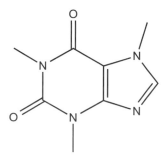

FIGURE 10.7 Caffeine.

per day; 1.5 grams of caffeine), caffeine causes agitation, tremors, and insomnia. The lethal dose of caffeine is about 10 grams (approximately 100 cups of coffee).[69]

Caffeine likely acts by blocking the receptors for the neurotransmitter-like chemical adenosine, which is normally involved in promoting calm and sleepiness.[70] Therefore, as an adenosine antagonist, caffeine hinders the ability of the brain to feel sleepy, resulting in alertness, increased cognitive performance, and anxiety.[71] Adenosine inhibits dopamine signaling, and since one type of adenosine receptor is especially concentrated in the striatum, a part of the brain rich in dopaminergic neurons, the blockade of adenosine function by caffeine is thought to stimulate a low level of dopamine signaling in this region, giving a mild sense of well-being without intensely activating the reward center of the limbic system.[72]

Caffeine consumption results in increased heart rate, blood pressure, and respiration both through antagonism of the adenosine system and through the stimulation of epinephrine release.[73] It dilates the coronary arteries but has the opposite effect on blood vessels in the brain, which it constricts, thus in some people relieving headaches. Caffeine is frequently added to over-the-counter analgesics containing aspirin or acetaminophen as it enhances their efficacy against pain.[74] Caffeine also relaxes the bronchial passages and increases urine output.[75]

Among heavy users, caffeine consumption during pregnancy is associated with an increased risk of spontaneous abortion and stillbirth. While caffeine is not a teratogen, antagonism of the adenosine system in the developing fetus is suspected of causing harm. However, recent investigations of any relationship between caffeine intake and perinatal outcomes have generated contradictory findings.[76] Unfortunately, many of the studies on caffeine and fetal health have been conducted improperly, with biased sample populations and poor controls. Still, many researchers recommend moderating caffeine use during pregnancy as a precaution.[77] In light of the concern over potentially damaging effects of caffeine during pregnancy, it is interesting that caffeine and closely related compounds can be employed therapeutically to treat apnea in prematurely born infants.[78]

Humans develop tolerance to caffeine and can become physically dependent.[79] Tolerance manifests as a reduced level of stimulation over time following ingestion of the same amount of caffeine and includes a lower effect of caffeine on tension, anxiety, blood pressure, heart rate, and urine production.[80] As a result, some users increase their caffeine use to produce the same physiological outcome. Among regular consumers, cessation of caffeine produces an array of typical withdrawal symptoms, including headache, irritability, fatigue, and poor mood, which usually last no more than a day or two.[81]

PHARMACOLOGICAL EFFECTS OF CAFFEINE

Increase in mental alertness
Wakefulness
Agitation
Slight increase in heart rate
Dilation of coronary arteries
Constriction of cerebral arteries
Bronchodilation
Increase in urine production

From its origins in eastern Africa to a place in the daily rituals of millions of people worldwide, coffee has melded into human life. Economies are built on its commerce, and workers, artists, and students alike are fueled by its stimulating effects. Few would argue that coffee—charged with its potent constituent, caffeine—is dispensable in the modern age. A beverage that ties together social groups, improves mental activities, and promotes scholarship and discourse is a remarkable type of medicine.

Chapter 11

Tea
Camellia sinensis

A terraced garden of young tea plants in Anxi County, Fujian, China. (Photograph by Winnie Lee)

The tea plant is a shrub or tree that bears simple, glossy green leaves and white-petaled flowers with multiple yellow stamens in the center (figure 11.1). Tea fruits are pale green, round, about the size of a small plum, and contain seeds that can be pressed to yield a commercially useful oil. Untended, tea can take the form of a tree reaching 10 to 15 meters in height; however, most tea in cultivation is pruned to a height around 1.5 meters. Tea plants require a few years of growth to become resilient enough to withstand harvesting, and plants can live for decades or centuries. Typically, the youngest leaves and leaf buds are plucked and processed into tea (figure 11.2). Successful plantations (called "gardens" or "estates") can exist from sea level to 2000 meters in elevation, although the plants do best in tropical to subtropical mountain regions where there is regular rainfall and no frost (figure 11.3).[1] The genus *Camellia* houses a handful of species native to Asia, including some planted as ornamental shrubs for their evergreen leaves and showy white, pink, or red flowers. The tea plant (*C. sinensis*) originated in the area of Southeast Asia where today's China, India, and Burma (Myanmar) converge, and it is now cultivated in two varieties and their hybrids: China tea (var. *sinensis*) has thinner, narrower leaves and produces a more delicately flavored beverage than Assam tea (var. *assamica*).[2]

FIGURE 11.1 Tea: (*top*) plant of the China tea variety; (*bottom*) flower.

ORIGINS OF USE IN CHINA

By steeping tea leaves and other ingredients in boiled water, people in China produced a flavored, slightly sweet, stimulating beverage that was probably used medicinally before the time of the Han dynasty (206 B.C.E.–220 C.E.).[3] The earliest account of tea preparation comes from the area now occupied by the borderlands between the central-western Chinese provinces of Sichuan, Hunan, and Hubei: it describes brewing roasted, powdered tea leaves with rice flour (*Oryza sativa*), ginger (*Zingiber officinale*), scallion (*Allium* spp.), and orange or tangerine peel (*Citrus* spp.).[4] Through the sixth century, tea leaves often were steamed after plucking and pressed into dense cakes, which were baked and strung together on ropes through holes punched in their center.[5] To brew the tea, the consumer chipped off and ground a portion of the hardened cake and steeped the resulting powder in water. As people developed new styles of tea preparation over many centuries, the new forms did not fully supplant the older ones, and as a result, today's regional tea cultures share some features with ancient practices. For example, tea pressed into cakes or bricks is still a preferred form in parts

FIGURE 11.2 A tea bud and young leaves in Nilgiris District, India.

FIGURE 11.3 A tea garden in Hangzhou, China.

of China, and grinding tea into a powder before steeping persists in Japan.

Tea diffused more deeply into the lives of the Chinese during the Tang dynasty (618–907), when it was widely adopted by Buddhist monks, who used it as an aid to meditation and study. As a medicinal substance of a certain rarity and value, tea entered the temple as a spiritual offering and found its way to the royal courts as tribute. It was also sold in markets and consumed by people of all walks of life.[6] In an atmosphere of widely celebrated tea culture, the scholar Lu Yu (733–804) composed *The Classic of Tea*, a thorough account of tea cultivation and preparation in Tang-era China.[7] In it, Lu describes what he considers the proper style of tea preparation, from the roasting and grinding of the tea cake, to the use of specialized utensils, to the technique for boiling the

salted water, to the correct way to raise a froth in the beverage. (The foam is the "essence" of the tea, according to Lu.)[8] As for additives popular during his time, such as scallion, ginger, jujube (*Ziziphus jujuba*), tangerine peel, dogwood berries (*Cornus* spp.), and mint leaves (*Mentha* spp.), Lu's purism is unambiguous. "Such preparations are the swill of gutters and ditches," he declared.[9] As for its medicinal properties, Lu described tea as cooling, useful "when feeling hot, thirsty, depressed, suffering from headache, eye-ache, fatigue of the four limbs, or pains in the joints."[10]

During the Song dynasty (960–1279), tea cakes were crushed and ground into a fine powder before being whisked into a thick froth. The tea was prepared by pouring unsalted, boiled water into cups individually, before frothing, rather than boiling the powder in a single vessel of salted water, as during the Tang.[11] During the Ming (1368–1644) and Qing (1644–1911) dynasties, loose-leaf tea became popular, and processing techniques proliferated, resulting in many new flavors and aromas from the steeped tea leaves.[12] As for additional condiments, they dwindled as improvements in tea handling produced ever more nuanced and flavorful brews. Still, tea scented with various types of flower petals became widely known, some of which remain common in modern China.[13]

As tea consumption expanded over the centuries, it wove itself into daily life. "Everyone, high and low, all drink tea, especially the farmers. Tea shops at the market places are numerous; merchants and travelers frequently exchange silk and lustrings [a type of glossy silk fabric] for tea," wrote a government official in northern China in 1206.[14] Teahouses were constructed in Chinese cities and towns, providing a venue for preparing and consuming tea in the company of friends and family, and also in the countryside, at particular scenic spots and near temples and shrines. Some teahouses were settings for quiet contemplation, and others were animated establishments of male entertainment; some teahouses served tea alone, and others offered tea, alcohol, and a complement of snacks and meals. The thirteenth-century author Wu Zimu describes certain recreational establishments in operation in the Northern Song capital of Kaifeng, where the "lanterns of tea shops burned all through the night." In the Southern Song capital of Hangzhou, some teahouses provided lessons in playing musical instruments, served as venues for chatting with acquaintances, and offered other respectable forms of entertainment. In contrast, Wu also noted that "on the main street there are several teahouses with prostitutes on the second floor," but these are "not places where a gentleman would set his foot."[15]

The evolution of tea in China can be seen in light of its many roles: as a medicine, an aid to religious practice, a tribute passed from subject to sovereign, a medium of social bonding, and a substance around which material culture has been built. Over many centuries, tea passed along major trade routes to caravan crossroads and maritime ports, into the hands of foreign traders and deep into the domestic landscape. Tea served as a crucial commodity in the important horse trade that supplied the Chinese with Tibetan steeds and that lasted, nearly uninterrupted, from the eleventh through the seventeenth century.[16] In Mongolia, where the compact and convenient brick tea suited the nomadic lifestyle, tea was accepted as a common currency, used to pay taxes and exchange for goods, a practice noted by European travelers as recently as the early twentieth century.[17]

THE SPREAD OF TEA: ALONG THE SILK ROAD AND BEYOND

It is likely that the first people to disseminate tea outside China were two Japanese Buddhist monks, who in 804 traveled to China to study Sanskrit and meditation at temples in Xi'an and Mount Tai. They returned to Japan having learned to drink tea and planted the first tea trees near Kyoto. "When I am not busy presiding over rituals," one of them wrote, "I study the language of India with tea by my side."[18] Later, in the twelfth century, the monk Myōan Eisai (1141–1215) returned to Japan from China, where he had studied Zen Buddhism, and promulgated tea's benefits for spiritual and physical health. Since tea in China at that time was customarily prepared by frothing powdered leaves in boiled water with a whisk, it is this form that Eisai advanced in his homeland.

Through contact with Mongols, Tibetans, and Persian and Arab traders along the Silk Road, the practice of tea drinking extended also to Central and western Asia, where unique tea customs conformed to local sensibilities and settings. For example, in Tibet and Mongolia, brick tea was preferred, sometimes with added salt and yak butter (Tibet) or cow, goat, or horse milk (Mongolia).[19] By the seventeenth century, tea had reached Persia, where, according to a European observer, "[they] boil it [the tea] till the water hath got a bitterish taste and blackish color and add thereto fennel, aniseed, or cloves, and sugar."[20] Tea was an attractive social beverage in the Muslim world, where alcohol was prohibited, and it spread rapidly through the major centers of Islamic life.[21] During the seventeenth century, tea also reached Russia, where it was warmly welcomed as a counteragent to the national beverage, vodka.[22] Extensive and lucrative trading routes were established that transported vast quantities of tea from northern and central China throughout Asia, a heritage evidenced by a shared name for tea along the route (northern Chinese 茶, *cha*; Hindi *chai*; Russian чай, *chai*; Persian *chai*; Turkish *çay*).[23]

Tea was unknown in Europe until the mid-sixteenth century, when travelers to Persia returned bearing stories of a Chinese herb with medicinal properties. In 1610, a Portuguese writer who had spent time in India described "a little herb . . . proclaimed to be very beneficial, and prophylactic of those disorders which Chinese gluttony might provoke."[24] Just a few years later, Portuguese and Dutch traders arrived in southeastern coastal China and Taiwan with a commission to barter for tea. Important trading posts were established along the Chinese seaboard and elsewhere in Pacific Asia. In this way, the maritime route of tea commerce was established, one that supplied Europeans for centuries.[25] In an interesting counterpart to the language built around the overland Silk Road from northern China, the word for tea in various European languages derives from the southeastern Chinese Min dialect (in Fujian 茶, pronounced *tay*, Dutch and German *thee*; French *thé*; Spanish *té*).[26]

Tea arrived in Britain by the 1650s and suited customers' tastes at a time when another imported stimulant, coffee, was making its own important entry into the world market. By the 1730s, the coffee-drinking fashion had nearly subsided in Britain, and the tea custom became firmly established. While coffee was brought to English and Scottish ports at great cost via the Ottoman Empire, British merchant mariners provided Chinese tea in abundance, which ultimately reduced its cost and secured its rank as beverage of choice. England imported a mere 100 pounds of tea in 1680, but in 1700 was bringing in over 1 million pounds annually. By 1780, England imported 14 million pounds, which demonstrates the rapid commitment of its people to an Asian leaf produced half a world away.[27] Meanwhile, in France, the opposite phenomenon occurred: strong imports of coffee and cacao together with weak imports of tea nearly eliminated tea from French beverage culture.[28]

Tea was likewise a mainstay of the American colonial lifestyle, and numerous importers provided the growing cities and villages with leaves to brew. The British East India Company served a primary role in transporting tea from China and distributing it throughout Britain and its colonies. In the rebellious American colonies, local tea smugglers and merchants sought to circumvent the overwhelming influence of the East India Company in colonial life by selling tea at a lower price than the East India Company could, skirting the taxes levied on the import and sale of all tea by the British Crown. As a result of this action, the East India Company suffered losses in the American colonial market. In 1773, the British government granted the East India Company the exclusive right to sell tea tax free in the colonies while still taxing American tea importers and sellers. As the duties on tea were significant, this tax law allowed the British company to undercut American merchants' tea prices.

Responding to this perceived injustice, in December 1773 a band of Boston revolutionaries protested the tea-tax policy by storming three British merchant vessels and dumping more than 300 casks containing nearly 100,000 pounds of East India Company tea into Boston Harbor (figure 11.4). One of the most incendiary events of the American Revolution, the Boston Tea Party helped set a defiant American attitude and presaged the wholesale rejection of tea by the newly minted American nation as a British beverage.[29] America's history following the colonial period is generally that of a coffee-drinking people, with tea taking a more minor role in beverage culture.[30]

THE PROCESSING OF TEA

Much of the variation in the flavor, color, and medicinal properties of tea derives from alternative methods of processing the leaves.[31] Tea leaves are harvested every one to two weeks during the growing season, which allows the desired young leaves and leaf buds to regrow from the stem for

FIGURE 11.4 Nathaniel Currier, *The Destruction of Tea at Boston Harbor* (1846). (Hand-colored lithograph; Library of Congress, Prints and Photographs Division, LC-USZC4-523)

periodic plucking.[32] The shape, size, color, and chemical composition of the fresh tea leaf are influenced by climate, soil type, and variety of plant, but nearly all differences in tea quality can be attributed to decisions made during harvesting and processing. For example, tea leaves that are hand plucked to include just a single leaf bud—a labor-intensive procedure—are considered to be of a very fine quality and delicacy. Plucking regimens that include one or two young leaves and the terminal bud are of incrementally decreasing quality. Hand-plucked shoots with older leaves are considered of lower quality still. Finally, harvesting leaves by machine is less discriminating and can yield tea of a relatively poor standard. In addition to the style of plucking, the processing of fresh leaves can influence the attributes of the final product.

Nearly all tea comes from young, vibrant green leaves (called the "flush") of the tea plant, and differences in color, flavor, and aroma accrue during processing (figure 11.5).[33] A minimally processed tea, white tea, consists of the youngest single buds

and immature leaves of the plant, at a stage when they are coated with fine hairs. The hairy texture of the leaves often gives them a silvery, shiny appearance. In white-tea manufacturing, the leaves and buds are simply harvested and quickly dried. When brewed in boiled water, they produce a pale yellow color and mildly sweet, vegetal flavor.

To produce green tea, the leaves and buds are plucked and quickly steamed or pan-fried to prevent the breakdown of cellular components, rolled to crush open the leaves, and dried either naturally or by machine.[34] Green tea, when brewed, produces an olive-green to slightly brown tea with a sweet, grassy, and mildly bitter flavor. In China and Japan, green tea is typically brewed from whole leaves; in Japan, it is also common to use powdered green tea (*matcha*).

Fresh leaves can also be harvested, lightly dried (withered), and then partially fermented by crushing them and allowing them to rest in a warm, humid chamber for several hours. Although the tea trade uses the term "fermentation," the biological process is not the same as the one of the same name

FIGURE 11.5 Different methods of tea processing result in the various styles of tea: (*clockwise from top left*) early-harvest white tea; unfermented green tea; mildly fermented oolong tea; highly fermented black tea; postfermented puer tea; roasted green tea, a Japanese specialty.

used in bread making and viticulture. The fermentation process in tea allows enzymes (biological catalysts) inside the leaves to begin degrading complex chemicals in the fresh leaf and produce new flavors via oxidation. After a certain amount of time, sometimes with intermediate rolling steps, a tea maker arrests fermentation by heating and drying the leaves. As a result of this process, called semifermentation, the leaves take on a greenish-brown color and brew to produce the oolong (*wu long* [black dragon])-style tea. Oolong teas have a complex flavor and aroma when brewed, often yielding a medium-brown color liquid. China and Taiwan manufacture numerous styles of oolong tea.

Alternatively, fresh leaves can be withered and then rolled and allowed to ferment for many hours in a warm, humid chamber to a dark, near-black hue before drying. The brewed tea is a reddish-brown color, with complex fruity, earthy, and sometimes astringent flavors. This style of highly fermented tea is known as black tea.[35] Black tea is produced in large quantity in China, India, Sri Lanka, East Africa, and the Middle East.

In ancient China, tea makers compressed tea into bricks or cakes to preserve its freshness and allow for more convenient transport by caravan. This style of tea processing is still prevalent in the southwestern Chinese provinces of Sichuan and Yunnan, among others. Tea bricks can be made of poor-quality black tea compressed with rice paste as a binding agent. However, many of the more prized forms of bricks or cakes are produced from carefully selected tea leaves that are

PROCESSING OF TEA LEAVES

Style	Method	Notable Producing Regions
White tea	Youngest leaves early in the growing season; quickly dried	Fujian (China)
Green tea	Leaves steamed or pan-fried, sometimes rolled, and dried	Hangzhou (China) Lake Tai, Jiangsu (China) Fukuoka (Japan)
Oolong tea	Leaves withered, rolled or crushed, and fermented for several hours	Fujian (China) Taiwan
Black tea	Leaves withered, rolled or crushed, and fermented for 12–24 hours	Anhui (China) Darjeeling (India) Assam (India) Nilgiris (India) Kenya Sri Lanka Rize (Turkey)
Puer tea	Leaves withered, heated, rolled, often compressed into bricks, and post-fermented for months or years	Yunnan (China)

steam pressed into thick discs. The tea cakes then undergo a lengthy postfermentation process in which they are matured in a controlled environment, sometimes for many years, during which time further oxidation occurs and microorganisms are allowed to subtly change the chemistry of the product. This form of tea, particularly famous in Yunnan, is called puer (also written "pu-erh" and "pu'er") and is esteemed for its ascribed medicinal properties.[36] The long postfermentation process develops complex flavors in the tea.[37] Puer teas have rich chemistries and mellow flavors generally lacking in bitterness and astringency. Unlike most other teas, which must be protected from the air and consumed within months to avoid stale and weak flavors, the flavors of properly stored puer tea improve with age, and thus fifty- and hundred-year-old puer bricks can reach high values in the marketplace (figure 11.6).[38]

In addition to the initial processing, the flavors of tea can be modified by roasting, as is common practice for some green and oolong teas. Also, as pioneered by the early Chinese, numerous contemporary tea cultures prefer tea flavored with additives. While white, green, oolong, black, and puer teas contain only leaves of the tea plant (*C. sinensis*), some styles of tea include mixtures of plant

FIGURE 11.6 A variety of postfermented teas prepared loose and compressed into cakes, bricks, and other shapes for sale in a shop in Guangzhou, China.

FIGURE 11.7 A worker packaging crush-tear-curl tea at a tea factory in Nilgiris District, India.

species to develop particular tastes and aromas. For example, in Britain, Earl Grey is a popular form that contains black tea with the oil of bergamot (*Citrus aurantium* ssp. *bergamia*), an Italian citrus tree. Jasmine tea, a northern China favorite, is usually produced by steeping a combination of green tea and flowers of the jasmine (*Jasminum* spp.) shrub.[39] The tea brewed is lightly sweet and fragrant. In Japan, green tea can be mixed with toasted rice and steeped in boiled water to make a flavorful drink called *genmaicha*. In contrast to these examples, herbal teas do not contain *C. sinensis* leaves and therefore are not true teas but tisanes. Some such herbal teas are made from chrysanthemum (*Chrysanthemum* spp.), chamomile (*Matricaria chamomilla*), hibiscus (*Hibiscus sabdariffa*), and rose (*Rosa* spp.).

In the present day, the tea industry is one of constant innovation, where new flavors and technologies compete to sate the thirst of discriminating and often thrifty consumers. One important transformation in tea consumption was the introduction of the tea bag, a fine cloth or paper sac in which tea leaves are sealed, making the steeping process more convenient. Invented around the turn of the twentieth century and first marketed by the British tea company Tetley in 1935, tea bags now account for nearly all the tea consumed in the West.[40] Rather than loose-leaf tea, most of today's tea is produced by an industrial technique developed in twentieth-century India and now deployed throughout the tea-growing world: a method called crush-tear-curl, frequently abbreviated CTC.[41] In CTC processing, the fresh tea leaves, sometimes harvested by machine to reduce labor costs, are withered and then fed through a series of high-speed roller wheels that shred them into small pieces. In this way, they oxidize quickly, require no further processing, and are easily sorted and packed into tea bags or sold in bulk. Nearly all CTC product is a darkly colored and strongly (some say harshly) flavored tea that can be drunk with sugar or milk (figure 11.7).

THE CULTURE OF TEA

The worldwide spread of tea has given rise to numerous traditions that now help define nations. As societies integrated tea into their community practices, unique tea preferences evolved, colored by lore and equipped with a specialized suite of implements used to prepare and consume tea. In China and Japan, where tea has a particularly long history, the preparation of this beverage took on a special importance. During the Tang dynasty, Lu Yu's *The Classic of Tea* conveyed the ways that tea connoisseurs paid close attention to the selection of water, leaves, and utensils to prepare tea, documenting the twenty-four accoutrements that he believed were key to tea preparation.[42] Yet the various tea rituals and implements demonstrate tremendous adaptability, which has yielded countless variations of tea service, tea ceremony, and accompanying tea wares.

The ritualistic preparation and drinking of tea serves a valuable social role in forming bonds between participants. For example, tea has long been used to signify respect and cordiality in China, whether in the form of a tribute good from vassal to feudal lord in an earlier era or as a token of hospitality during a visit to someone's residence, a practice that exists to this day. In China, guests to a home or an office are nearly always offered tea as a gesture of welcome. During a tea ceremony, a focus is placed on the precise movements of the tea master and on the aesthetics of the beverage produced, such that all guests are tied together in a singular activity. Over many years, the tea ceremony, at various levels of formality, has come to signify the formation of business and political links as well as family affiliation (figure 11.8). Indeed, an important event marking many Chinese weddings is the celebratory offering of tea between the new in-law families.

FIGURE 11.8 A tea-shop worker conducting a tea service in Fujian, China. (Photograph by Winnie Lee)

In Japan, tea preparation rituals became elevated forms of spiritual practice after being disseminated widely by Zen Buddhist monks during the twelfth and thirteenth centuries. Driven by the notion that *suki* (devotion to art) can lead to self-transcendence, religious scholars took up the tea ceremony as a way to regulate emotion, improve mindfulness, and ultimately achieve enlightenment.[43] The resulting Japanese tea ceremony—governed by intense discipline, precise movements, and an elaborate code of conduct—coupled the consumption of a stimulating, rich, frothed powdered tea with a meditative, spiritually significant physical pursuit. In response to the perceived overemphasis on ritual rather than the tea itself, an eighteenth-century Zen monk created a simpler tea ceremony called *sencha* (boiled tea), in which attention to the water, leaves, and brewed tea prevailed. The form of bright-green, flavorful, delicate tea used in this ceremony remains treasured today, and despite the monk's intended focus on the beverage to the exclusion of ceremonial pomp, *sencha* service gave way in time to its own specialized performance, replete with specially made implements and service ware (figure 11.9).[44]

While the diverse forms of ceremonial tea preparation have come to employ a tremendous variety of ritual performances and intangible social meaning, they have also generated an important industry in the fabrication of specialized tea service items. These products of human craft include bowls, cups, kettles, and pots of diverse designs and materials.

Where tea has spread outside East Asia, it has become incorporated into the social lives of people and been adapted to suit local tastes. Although tea plants were extant in the far northeastern reaches of India for many centuries, it was not until the British introduced large-scale tea cultivation in the nineteenth century that the Indians developed a widespread indigenous tea culture.[45] Today, tea is both an important source of export revenue and a national drink. Tea served in India is usually black tea prepared sweetened and spiced (figure 11.10). The recipe for Indian tea is variable, containing black-tea leaves boiled in water or in water and milk, usually with sugar or honey added. The spices frequently include cardamom (*Elettaria cardamomum*), cinnamon (*Cinnamomum* spp.), clove (*Syzygium aromaticum*), black pepper (*Piper nigrum*), orange (*Citrus* spp.) peel, ginger, nutmeg (*Myristica fragrans*), and star anise (*Illicium verum*).[46] Known in India as *masala chai*, this style of spiced tea has also become popular in the West. As India's productive industry sent out large shipments of flavorful black tea to the seaports of the United Kingdom, the custom of tea drinking be-

FIGURE 11.10 A streetside tea maker preparing *masala chai* for his eager customers in Kolkata, India. (Photograph by Winnie Lee)

came firmly seated in the lives of the British people.

In Britain, tea has become both a drink and a meal, in that the tea-drinking habit developed during the eighteenth century generated a daily event in which families and co-workers could rest and refresh.[47] The British traditionally consume black tea, often flavored with bergamot or other

FIGURE 11.9 A Japanese woman with her teaware. (Photograph by Herbert George Ponting [1905]; Library of Congress, Prints and Photographs Division, LC-USZ62-51044)

additives, poured into porcelain teacups with a small amount of milk.[48] The proteins in the milk bind to complex tannin compounds in the brewed tea, which both softens the astringency of the drink and reduces the amount of dark residue left on the inside surface of the teacup. Tea, often taken with sugar, provided a stimulating and energy-rich boost that helped fuel the increasingly industrial, and industrious, British lifestyle. The custom of "afternoon tea" in Britain is a midafternoon light meal accompanied by sandwiches or sweets; "high tea" is served in the early evening and in earlier times replaced a formal dinner, particularly among working families.[49] In many parts of the former British Empire, the word "tea" refers as much to a meal as to a drink. Indeed, during routinely long cricket matches, mealtime tea breaks allow players, officials, and spectators to dine and rest before resuming play.

While Turkey had a leading role in the spread of coffee, within the past two centuries tea has become a national drink, thanks in part to the successful cultivation of tea in Turkey beginning in the nineteenth century. Turks generally drink black tea brewed strong with no additional flavorings, adding one or two cubes of sugar. Customary tea service is in small glass cups narrow in the middle and with flared tops, roughly in the shape of a tulip flower (figure 11.11).[50] In Russia, black tea is preferred, and it is usually served with sugar alongside sweets such as chocolates and jams. Now a defining characteristic of Russian hospitality, the *samovar* (self-boiling) is a device present in many homes, a combined tea kettle and stove in which tea is kept heated all day, ready to welcome a guest with a soothing, stimulating beverage of kindness.

FIGURE 11.11 Turkish tea served in a traditional tulip-shaped glass.

Indeed, in these places and everywhere tea traveled, it integrated itself into local custom and spawned arts and artifacts that help to define regions.

TEA IN THE TRADITIONAL MEDICINES OF ASIA AND EUROPE

In early China, tea was valued as an herbal drug of particular versatility. "Tea lightens the body and changes the bones," wrote a Daoist master in the sixth century.[51] In various accounts, tea was said to cure sores and ulcers, treat breathing troubles, reduce phlegm, and improve digestion.[52] In twelfth-century Japan, the Buddhist monk Eisai wrote that the body requires a balance of bitter, sour, pungent, sweet, and salty flavors to harmonize the organs. Yet the Japanese diet dangerously lacked bitter ingredients. "Our country is full of sickly looking, skinny people, and this is simply because we do not drink tea," Eisai explained. "Whenever one is in poor spirits, one should drink tea. This will put the heart in good order and dispel all illness."[53] Later, the Chinese herbalist Li Shizhen (1518–1593) composed a detailed account of tea's medicinal properties in the framework of traditional medical scholarship. "Tea is bitter and cold, a *yin* among *yin*," he declared; tea dampens the body's internal fire so that "a clear *qi* can rise."[54] When Europeans learned of tea, they included it in their herbal pharmacy and set about explaining its properties.

The first reports of tea emerged in Europe during the sixteenth century, and firsthand accounts by missionaries and traders elaborated extensively on its purported medicinal effects. The Italian missionary to China Matteo Ricci (1552–1610) wrote that tea "is positively wholesome for many ailments if used often."[55] The French missionary Alexandre de Rhodes (1591–1660) agreed: "One of the things contributing to the great health of these peoples [the Chinese], who frequently reach extreme old age, is tay, which is commonly used throughout the Orient."[56] By 1641, the Dutch doctor Nikolas Dirx (1593–1674) had studied tea in great enough depth to exalt it in prose:

Nothing is comparable to this plant. Those who use it are for that reason, alone, exempt from all maladies and reach an extreme old age. Not only does it procure great vigor for their bodies, but it preserves them from gravel [kidney stones] and gallstones, headaches, colds, ophthalmia, catarrh, asthma, sluggishness of the stomach and intestinal troubles. It has the additional merit of preventing sleep and facilitating vigils, which makes it a great help to persons desiring to spend their nights writing or meditating.[57]

Among the first purveyors of tea in Europe were apothecaries and physicians, who praised its ability to regulate humors similar to bloodletting or laxatives—but less painfully.[58]

Entrepreneurs also lauded tea's myriad benefits in terms familiar to patrons accustomed to considering the harmony of their body's fluids in the Galenic framework of health. In 1660, for example, the London coffeehouse owner Thomas Garway promoted tea as a potent remedy in an advertisement. First, he spelled out that tea's "quality is moderately hot, proper for Winter or Summer. The drink is declared to be most wholesome, preserving in perfect health until extreme Old Age."[59] According to Garway, tea "prevents and cures Agues, Surfets and Fevers, by infusing a fit quantity of the Leaf, thereby provoking a most gentle Vomit and breathing of the Pores."[60]

As men of medicine and commerce alike advanced the cause of tea in Europe, it transformed from an exotic, foreign herb to a domestic necessity. During the eighteenth and nineteenth centuries, Europeans increasingly obtained tea from grocers rather than apothecaries and consumed it at home rather than in coffee houses.[61] Countless advocates heralded the plant's perceived health benefits, although a small number of detractors warned against it, such as the Scottish-French physician Daniel Duncan (1649–1735), whose enigmatic moralistic concern was tea's "voluptuousness." This property, he cautioned, "creates in us an Aversion to good Things . . . and an inclination to bad Things."[62] Nevertheless, tea effectively became part of the daily routines of many Europeans, and as

early as 1712 a widely circulated *materia medica*, *A Compleat History of Druggs*, summarized tea's effects, in the terminology of the day, as lightening the spirits, suppressing "vapours," driving away drowsiness, improving digestion, and cleansing the blood, among others. But, importantly, this pharmacy manual noted: "The Leaf is more used for Pleasure in the Liquor we call Tea, than for any medicinal Purpose."[63]

MECHANISM OF ACTION

The tea leaf and the beverage made from it contain a tremendous diversity of chemicals producing flavor, aroma, and other bioactive effects. Beyond the sheer complexity of compounds present in any tea, there is a great variability in the characteristics generated by growth conditions, harvesting, processing, and storage. Consequently, the best studies of the health-related properties of tea must take these numerous factors into account. Additionally, it is important to recognize that the chemicals present in the tea leaf and brewed liquid undergo changes in the human body that may further alter their structure and properties. Therefore, when evaluating the possible health effects of tea, researchers performing experiments in laboratory dishes and using test animals are cautious not to extend their findings to potential human outcomes without proper evidence. In short, biomedical research on the physiological properties of tea has produced a large body of data, but much more work needs to be done to account for the diversity of tea varieties and constituents.

In addition to caffeine, which tea leaves accrue to 2 to 4 percent of their weight, tea contains a number of other chemically related methylxanthines (figure 11.12), including theobromine (0.1–0.2 percent) and theophylline (around 0.05 percent).[64] Together, these compounds account for tea's stimulating properties, with a principal role ascribed to caffeine. Theophylline relaxes the lung's bronchial passages, a characteristic that makes purified theophylline one of the more potent asthma medications available.[65] Theophylline also contributes to tea's diuretic effects.[66] Beyond the roles

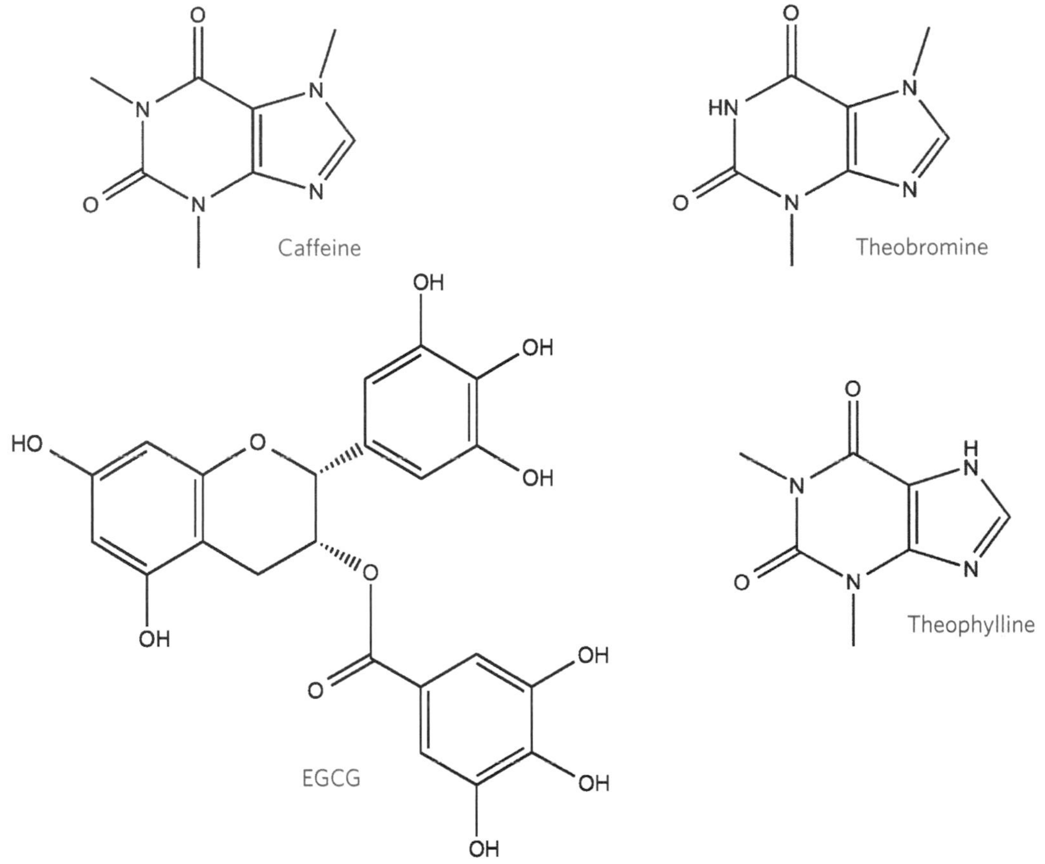

FIGURE 11.12 Active principles of tea: the stimulants caffeine, theobromine, and theophylline; the antioxidant epigallocatechin-3-gallate (EGCG).

of these chemicals, tea produces a variety of other compounds that are demonstrated to contribute to its physiological properties. Furthermore, tea contains theanine, a type of amino acid that accumulates to 1 to 3 percent of tea's dry weight and is thought to produce a calming effect via the brain's glutamate neurotransmitter system.[67]

Many of tea's purported medicinal effects derive from its chemicals that can reduce the oxidation state of cells and tissues of the body. Oxidation, a process by which sun exposure, toxins, or everyday "wear-and-tear" reduce the effectiveness of cellular biochemistry, is associated with diseases such as cancer and heart disease as well as the process of natural aging. Tea, like many plants, is rich in antioxidant chemicals that may counter the natural aging and disease-development processes.[68] The antioxidant properties of tea are generally thought to be strongest in minimally processed green and white teas.[69] During the processing of tea leaves into oolong or black-tea forms, the antioxidant chemicals in the leaves become oxidized themselves via natural enzymatic activity and exposure to air and heat and therefore less effective antioxidants.[70] Because of this, many scientists studying the medicinal properties of tea use green-tea liquid, powder, or chemical extract as a focus of investigation.

The chief antioxidants in (green) tea are a family of molecules called catechins, of which epigallocatechin-3-gallate (EGCG [see figure 11.12]) and epigallocatechin (EGC) are the most potent. Green-tea catechins have been demonstrated in test-tube experiments and small-scale clinical trials to inhibit the proliferation of cancer cells or reduce the incidence of a number of cancers and heart disease. In laboratory studies, EGCG was shown to reduce inflammation, which is associated with heart disease and immune conditions and treat stomach ulcers caused by the ulcer-associated bacterium *Helicobacter pylori*.[71] The scientific data, however, lack large-scale clinical trials that might determine unequivocally whether the catechins themselves are responsible for any therapeutic effects of tea or whether other compounds are involved. It is worth recognizing also that the antioxidant properties of the tea beverage might not directly extend to the cells of the human body because exposure to the harsh environment of the stomach and intestine, metabolism by the liver, and passage into the body's circulation can alter tea's chemical functionalities. Furthermore, there are concerns that highly concentrated forms of green-tea extract, such as those sold as dietary supplements, might be toxic to the liver in some people, particularly if taken on an empty stomach.[72]

In addition to their antioxidant properties, tea's complex polyphenolic compounds can bind to and coat cell surfaces and the teeth, protecting them from pathogens. With regard to dental health, there is some evidence that tea inhibits the growth of harmful bacteria through several mechanisms that change the chemical environment of the mouth, ultimately guarding the teeth against decay.[73] The antimicrobial nature of tea polyphenols has also

been harnessed in an ointment preparation sold as Veregen, effective against genital warts caused by human papillomavirus.[74]

In the more highly processed teas, especially those with a longer fermentation, the tannin composition is particularly rich. As research into tea's properties continues, its health benefits and risks can be better understood, ultimately yielding insight into the wisdom of ancient practices and guiding the development of new therapies from a storied leaf.

From the mountain forests of southern

Asia to homes around the world, the tea plant has convinced hundreds of millions of people to brew its simple green foliage into a stimulating medicinal beverage. Over the millennia, diverse cultures developed methods to process tea leaves into a rainbow of colored drinks and serve them with all sorts of additives. Tea is a medium of social bonding through Asia and Europe, and beyond, and it has changed the course of history in nations continents apart. As scientific investigation identifies the chemical basis of its medicinal properties, tea will certainly keep its place among the most useful and valuable plants in the centuries to come.

Chapter 12

Cacao

Theobroma cacao

An Ecuadorian postal stamp depicting a cacao pod, ca. 1930. (Smithsonian Institution, National Postal Museum, 2011.2005.181)

The cacao plant grows as a tree, a member of the lower story of the tropical rain forest. It can reach a height of 5 to 10 meters and produces simple oval-shaped, pointed, medium-green leaves and small (1 centimeter in diameter) pinkish-white flowers directly on its trunk and mature branches (figure 12.1).[1] The flowers give rise to slowly developing football-shaped pods, each maturing to a green, yellow, or red color and containing twenty to forty almond-size seeds surrounded by a mucilaginous pulp (figure 12.2). Cacao grows best in hot and humid tropical regions with consistent rainfall and a short or no dry season. It is often found in the shade of nitrogen-fixing legume trees[2] or in sunken *cenotes* (large, deep pits), where moisture and an environment protected from harsh sun and wind exist.[3]

ORIGIN AND USE AMONG INDIGENOUS MESOAMERICANS

Cacao likely originated in the upper Orinoco and Amazon River basins in the modern-day South American countries of Colombia, Venezuela, and Brazil, where it was first harvested around 7000 to 10,000 years ago for the sweet and flavorful pulp shrouding the seeds, called "beans."[4] From South America, cacao was brought to Mesoamerica and spread among the numerous cultural groups in the region.[5] Several species in the genus *Theobroma* are harvested for medicine and food, including *T. cacao* (the widely planted domesticated species) and *T. bicolor* (which grows semiwild).[6] By the time of European contact in the sixteenth century, cultivated cacao was seen only in Central America and southern North America.

A growing body of archaeological evidence establishes that beverages containing cacao were consumed as early as 1000 B.C.E., and it is likely that ancient peoples used cacao pulp to produce a fermented drink, a practice that exists today among indigenous peoples.[7] At some point, people learned to dry and grind the beans to produce a more flavorful beverage, facilitating its spread among the societies of the region. In time, cacao took on further cultural significance throughout Mesoamerica, along nutritional, economic, artistic, and

religious dimensions.[8] As the Maya civilization flourished in its Classic (200–900) and Postclassic (900–1200) periods, cacao was considered a path to the divine, a tree that connected the sky, earth, and underworld, allowing spirits to pass between all realms.[9] The power of cacao resided in both its physical properties and its metaphorical associations.

Early recipes called for water; a paste of ground, sometimes roasted, cacao beans; and various additives such as maize (*Zea mays*), chili peppers (*Capsicum annuum*), vanilla (*Vanilla planifolia*), and other spices to produce a rich, fatty, bitter, spicy, stimulating chocolate drink. There is no doubt that the biologically active chemicals in the cacao and its additives affected consumers' physiology in ancient times, as they do now.[10] By shaking the cacao beverage in a gourd, stirring vigorously, or pouring between vessels, ancient people worked up a stiff foam they considered particularly precious: they believed it to represent the breath of the gods.[11] Prepared chocolate also frequently contained *achiote* (*Bixa orellana* seeds, also known as annatto), which turned the

FIGURE 12.1 Cacao: (*top*) tree grows to a medium height in the jungle and can be highly branched; (*center*) flowers arise directly from the trunk and branches; (*bottom*) pods turn from green to yellow or red when ripe.

FIGURE 12.2 Cacao beans.

resulting liquid red and thereby evoked the blood of sacrifice and religious tribute.[12] It is the revered status of cacao that Linnaeus recognized by naming the genus *Theobroma*: "food of the gods."[13]

Maya, Mixtec, and Aztec Mesoamerican Indians, among others, cultivated cacao and integrated it into an elaborate culture of trade, tribute, worship, and ritual. From an early time, people probably used cacao to mark important events in social ceremonies, such as marriages, funerals, and rites of passage.[14] Maya vases often depict cups of frothy chocolate at the feet of rulers, and numerous artifacts show cacao beverages and seeds as royal tribute (figure 12.3).[15] Cacao is also found among the burial items of the elite Maya and is represented in Mixtec funeral ritual documents.[16] The detailed palace scenes on some Maya artifacts show sacks of cacao offered at the ruler's throne, and Aztec scribes recorded that cacao was given as tribute by provincial subjects to the emperor in Tenochtitlán.[17] An Aztec text demonstrates the preparation of the celebrated, foamy cacao beverage by a woman of the royal court (figure 12.4). Cacao has thus been long associated with spirituality, wealth, and power.

Cacao beans also circulated as currency among pre-Columbian Mesoamericans, used to purchase trade goods such as salt, bird feathers, animal hides, and precious stones. An early Spanish chronicler of mid-sixteenth-century Nicaragua relates that at market, a rabbit could be purchased

FIGURE 12.3 A Mayan ruler on a throne receiving what is probably a vessel of chocolate, depicted on a polychrome vessel, 600–900. (Vessel decoration rollout © Justin Kerr, K6418, http://www.mayavase.com)

FIGURE 12.4 An Aztec woman preparing chocolate by pouring the liquid chocolate from a height to make it frothy. (Illustration from *Codex Tudela* [1553]; Museo de América, Madrid)

drink the crushed roots of the *tlatlacotic* [a purgative root of unknown genus and species] in hot water, that he may vomit. A few days later let the roots and flowers of the *yollo-xochitl* [flowers of *Magnolia glauca* or *M. mexicana*; in Nahuatl, "heart flower"] and *cacua-xochitl* [flowers of *Theobroma cacao*] be crushed in water, and let him drink the liquor before eating, wherewith the evil humor in the chest will be largely driven out.

A cacao beverage is also listed as a restorative treatment "for relieving the fatigue of those administering the government, and discharging public offices."[24]

In the folk medicine of indigenous Mesoamerica, cacao figures heavily in supernatural healing, such as in concoctions administered to patients in shaman-led rituals to appease the spirits causing illnesses such as fevers and seizures.[25] The connection between cacao and social health is clear in contemporary Highland Maya indigenous ceremonies in which cacao is offered at ancestral grave sites to appeal for plentiful food and fertility, and among the Ch'orti' of eastern Guatemala, cacao is prepared to feed the gods responsible for rain and ensure a good harvest.[26] People throughout the region use cacao seeds, flowers, and leaves for health conditions as diverse as low weight, skin irritation, parasitic infection, and abdominal pain.[27] Cacao's role as a sacred, precious, medicinal plant remains at the heart of Mesoamerican life.

for ten cacao beans, a horse or mule for fifty beans, and a slave for a hundred.[18] Prostitutes sold their services for as little as ten cacao beans.[19] There is even evidence that cacao beans were counterfeited, as detailed clay replicas have been unearthed at an ancient Maya site, and early Spanish visitors noted that some cacao beans used as money had been carefully hollowed out and packed with dirt, probably to pass off as a more valuable intact bean in trade.[20] In addition to its use as currency, cacao served important roles in health.

As a sacred plant, cacao was thought to have powers unlike those of other trees in the forest. To the Maya, cacao pods were permeated with a life-giving force, rich in sap, akin to blood.[21] So revered was the cacao plant that the Aztecs believed that urinating on a cacao plant or flower resulted in retribution from the heavens in the form of a skin infection called *xixiotiz*.[22] Chocolate, a crimson metaphor for blood, was thought to be a remedy for illnesses involving the loss of blood.[23] In the Aztec herbal of 1552, a treatment "against stupidity of mind" is prescribed, by which the patient is to

CACAO IN EUROPE

During the Spanish conquest of the early sixteenth century, Europeans tasted cacao for the first time. While some found the beverages produced to be bitter and unpleasant, a page of the Spanish explorer Hernán Cortés (1485–1547) lauded the Indian beverage. "He who has drunk one cup," he wrote, "can travel a whole day without any further food, especially in very hot climates."[28] The Spaniards assimilated the natives' use of cacao as currency by making it the means by which the subjugated paid tribute to their new European lords. Indigenous people were required to pay tribute based on the size of the parcel farmed and generally out-of-date census data, meaning that during a time of widespread European-originated epidemics, widows were forced to pay tribute for their deceased husbands and children for their ill parents. The bishop of Guatemala objected to the unfair tribute system in 1603, but it continued nonetheless.[29]

When Cortés visited the Aztec emperor Moctezuma, he was impressed by the importance given to the frothy, bitter cacao drink enjoyed by the indigenous nobility. As the New World conquerors prepared shiploads of exotic animals, precious metals, and jewels to present to the Spanish throne, they may have included cacao in the cornucopia of American treasure. Although the precise date that cacao first appeared at the Spanish court is unknown, shipments of significant quantity were arriving in Seville in the mid-1580s, and by the early seventeenth century, cacao was prepared in Madrid at the royal palace.[30] Initially, the Spaniards found this crude and bitter substance unpleasant to the palate and added sugar and cinnamon (*Cinnamomum* spp.) to the cacao paste.[31] In this form, cacao became a trendy and popular nonalcoholic stimulating drink for society's elite. The Spanish experimented prolifically with this new drink, adding dozens of flavorings from anise (*Pimpinella anisum*) to almonds (*Prunus dulcis*) and hazelnuts (*Corylus avellana*) to modify the flavor and enhance the enjoyment of their liquid cacao.[32]

By the seventeenth century, the cacao plantations in Central America were losing productivity, owing to a depletion of the land and a labor pool decimated by plague. To expand cacao farming and increase yield, the Spanish introduced the plant into the Caribbean islands, Ecuador, Colombia, and (by way of the Portuguese) Brazil. They also brought to the New World large numbers of African slaves to replace the dwindling indigenous plantation workers.[33] Beginning in the seventeenth century, cacao gradually came under cultivation throughout the tropical band of European colonies worldwide, eventually spanning from Africa to India to Southeast Asia. Wherever cacao was grown, local people were harnessed to maintain plantations and carry out the laborious harvesting and processing operations.[34] With burgeoning production and widespread commerce, knowledge of cacao spread throughout Europe during the seventeenth and eighteenth centuries. This phenomenon, coupled with the highly productive (and equally exploitative) sugar plantations in the Caribbean and Brazil, produced conditions ripe for a substantial expansion of the appetite for cacao in Spain, Italy, France, and England.[35]

Sweetened hot chocolate (still in a liquid form, as solid chocolate bars did not come about until the mid-nineteenth century) and hot chocolate mixed with coffee became preferred luxury drinks among the wealthy of Europe. The British politician Samuel Pepys (1633–1703) wrote in his diary in 1664 that he was "off to a coffee house to drink jocolatte, very good."[36] The British are thought to have been the first to add milk to chocolate when in 1727, Nicholas Sanders made a milky hot chocolate, and Joseph Fry established the first English chocolate factory three years later. English chocolatiers continue to excel at the production of milk chocolates, with firms such as Cadbury enjoying international renown.[37]

The American statesman Thomas Jefferson (1743–1826) noted the "superiority of chocolate, both for health and nourishment," and preferred it to tea and coffee.[38] In the early United States, the first cacao-processing firm was established in 1780 as Walter Baker and Company in Massachusetts; the company still produces chocolate under the name Baker's Chocolate (although now headquartered in Delaware and part of a larger food conglomerate).[39] The United States produced a number of leading chocolate manufacturers through its history, many of which have gained worldwide recognition. For example, in 1894, Milton Hershey established a confectionary in Pennsylvania; Frank Clarence Mars started a candy business in the state of Washington in 1911.[40] Today, American chocolate manufacturers are among the most productive in the world.

CACAO IN EUROPEAN MEDICINE

The Europeans who first encountered cacao in the sixteenth century adhered to a traditional medical system that assigned qualities of hot and cold, wet and dry, to all medicinal substances.[41] These properties, they believed, could be harnessed to promote a balance of hot, cold, wet, and dry substances in the body and, thereby, health. Interestingly, the Aztecs also considered the universe in terms of opposing forces or principles such as hot and cold, and wet and dry.[42] Therefore, in addition to demonstrating the elaborate ritual symbolism and spiritual connection of cacao, the Aztecs probably taught the Spanish that cacao had cool and moist properties.[43] In this way, cacao entered the Western pharmacopeia.

The Europeans speculated that cacao's cool property should make it useful for treating fevers, especially in hot climates.[44] As an energy-rich and stimulating beverage, chocolate attracted many consumers in seventeenth-century Europe, who extolled its health-related virtues. For example, a Spanish physician in 1631 recommended chocolate to improve fertility in women, aid digestion, and cure chronic coughing, jaundice, inflammation, and other illnesses.[45] In England, an entrepreneur took out a newspaper advertisement in 1657 touting chocolate in vague but promising terms: "It cures and preserves the body of many diseases."[46] The Englishman Henry Stubbe (1632–1676) was even more emphatic in his praise of chocolate, relaying that it "doth more speedily and readily refresh and invigorate the bodily strength than any other sustenance whatever."[47] A French noblewoman, visiting Spain in 1680, wrote that "there is nothing better [than chocolate] for the health."[48] Over the course of a few decades, chocolate spread across Europe and Britain as a therapeutic, social beverage. But some observers considered cacao to be an unhealthy habit, particularly if consumed in excess or at inappropriate times. It seems that the compulsion to drink chocolate had presented itself frequently enough for a doctor in Seville to warn as early as 1624 that it should not be consumed more than two or three times a week and that if taken in the evening, it led to insomnia.[49]

THE PROCESSING OF CACAO

In the traditional method of preparing cacao, practiced in some parts of contemporary Mexico and Central America, cacao pods are harvested and the beans removed, still covered in their mucilaginous pulp. The pulp and beans are left partially to decompose by the action of natural yeasts and bacteria (in the trade, called "fermentation") for approximately three to six days, which tempers some of the acidity in the beans and builds flavor.[50] The beans are then dried under the sun.[51] After roasting, the beans' thin shells are removed and the meat of the seed (called the nib) is ground by hand on a heated stone grinding table, yielding a thick paste that can be pressed into small bricks or mixed with water and spices to produce chocolate.[52] Through the eighteenth century, Europeans experimented with roasting temperatures and used mechanical grinders but otherwise made very few

THE LANGUAGE OF CACAO

Cacao	The seeds or ground seeds of the cacao plant
Cacao beans	Cacao seeds
Cacao nibs	Broken bits of cacao seeds
Chocolate	A liquid or solid cacao-based food
Chocolate liquor	An oily paste of roasted, ground cacao beans
Baking chocolate	Solidified chocolate liquor
Cocoa powder	Ground cacao from which much of the fat has been pressed out
Cocoa butter	The fat removed from cacao beans
Milk chocolate	Milk solids, sugar, cocoa powder, and cocoa butter
White chocolate	Cocoa butter, milk solids, and sugar

modifications to the ancient process they learned from native Mesoamericans.[53] By the nineteenth century, however, English, Swiss, and Dutch innovators began to introduce new technologies that expanded the range of chocolate forms and flavors.

The physical properties of cacao and its manufactured products give rise to a particular vocabulary to describe its multiple forms and industrial techniques of preparation, which are as much chemistry as art. Cacao beans possess a high fat content, making cacao pastes and drinks greasy. When cacao is roasted and ground, it yields a dark, oily paste known as chocolate liquor. When solidified into squares, it becomes what is now known as baking chocolate, a thick and bitter substance unpleasant to eat or drink. In 1828, the Dutch chocolate makers Caspar and Coenraad van Houten developed a process to press much of the fat from roasted cacao beans, producing a cake that was amenable to grinding into a less oily cocoa powder.[54] The fat that is removed forms a dense, pale, lightly flavored substance called cocoa butter. Cocoa butter is a versatile oil, solid at room temperature, used in numerous cosmetic products (figure 12.5). In 1847, the Fry family pressed cocoa powder, cocoa butter, and sugar into chocolate bars.[55] Around 1875, the Swiss chocolatier Daniel Peter, with the assistance of Henri Nestlé, pioneered adding milk solids to sugar, cocoa powder, and cocoa butter to produce milk chocolate.[56]

To improve the solubility and mixing characteristics of cocoa powder, Dutch chocolate makers experimented with adding potassium and sodium carbonates, which are alkaline chemicals that neutralize organic acids. This process, called dutching, darkens the cocoa, improves its ability to mix with water and milk, and makes the flavor milder.[57] Today, about 90 percent of all cocoa is dutched. After processing the cacao into a melted chocolate blend, most manufacturers subject the liquid to conching, in which it is mixed and ground for many hours, improving flavor and smoothness.[58] During the nineteenth and twentieth centuries, confectioners refined the taste, texture, and aroma of chocolate products by improving mechanical techniques and introducing new additives.[59] They

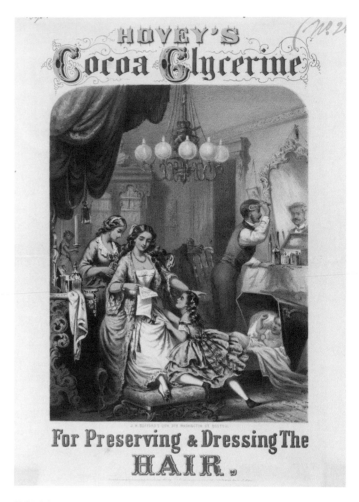

FIGURE 12.5 An unusual medical-cosmetic use of cacao in the nineteenth century: a hair pomade, ca. 1860. (Library of Congress, Prints and Photographs Division, LC-DIG-pga-05565)

also developed white chocolate, which is simply cocoa butter with sugar and other flavorings. Since it lacks cacao solids, white chocolate is not considered true chocolate.[60]

MECHANISM OF ACTION

In its most commonly consumed forms, cacao contains a mixture of nutrients, including fats, proteins, carbohydrates, and other components. The substantial fat (40–50 percent in beans), digestible carbohydrate (12–14 percent in beans), and protein (14–18 percent in beans) portion of cacao account for its reputation as a calorie-rich comestible.[61] Cacao also contains a variety of polyphenolic compounds with antioxidant properties, such as catechins, that have been speculated to reduce cancer risk and improve cardiovascular health. The body of literature on the clinical efficacy of cacao for these purposes is limited.[62] A small-scale study suggested that dark chocolate may benefit circulatory system health by reducing blood clotting, a risk factor for stroke.[63] However, few large-scale clinical trials using appropriate controls have demonstrated chocolate's effectiveness to treat cardiovascular disease or other ailments, so health claims are premature.[64]

In contrast to the nutritional components of cacao with less well-understood medicinal properties, cacao produces a complex assortment of chemicals that are demonstrated active in human physiology. In addition to caffeine, which is present at approximately 0.1 percent in cacao beans, the related alkaloid theobromine amounts to about 2 percent in cacao beans (figure 12.6).[65] Together, these chemicals exert a weak stimulant effect and increase heart rate.[66] They are mild bronchodilators and diuretics.[67]

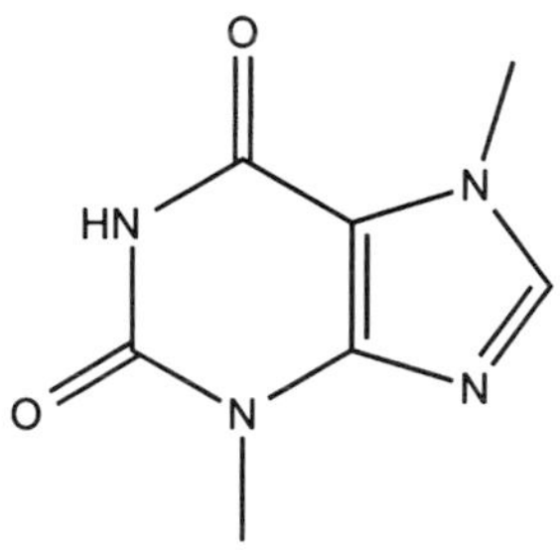

FIGURE 12.6 Theobromine.

The pleasure of eating chocolate may derive in part from the roles of the methylxanthine compounds theobromine and caffeine, which produce enjoyable sensations when taken in stimulating beverages such as tea and coffee.[68] Part of the positive experience may also owe to the high fat and sugar content in chocolate confections. The combination of fat and sugar, with or without the chocolate stimulants, provides a certain satisfaction qualitatively different from that of coffee or tea drinking. As for the euphoria that accompanies chocolate indulgence, the one that drives some people to profess an addiction to chocolate (chocoholism), future investigations may address that phenomenon. Chocolate, either through sweetness and fattiness or in combination with its potent assortment of methylxanthines, probably triggers the release of serotonin and dopamine. Neurotransmitters of this type are responsible for sensations of reward and pleasure, and when chocolate induces their actions, this neurochemistry may produce the satisfaction so widely known.

In Mesoamerica many centuries ago, people produced a blood-red frothy cacao drink as a tribute to their gods. Since that time, the land and people of the tropics have been exploited to profit from chocolate's worldwide appeal, and millions have drawn their livelihood from its cultivation, preparation, and sale. Cacao is at the center of dozens of national economies, deeply reliant on the world's demand for its captivating flavor and powerful stimulant compounds. From such an odd plant, whose trunk-borne flowers give rise to strange fruits, derives one of the most satisfying psychoactive foods on earth.

Chapter 13

Tobacco
Nicotiana tabacum

Tobacco leaves and flowers adorn the capital of a column in the U.S. Capitol, Washington, D.C. (Photograph by Winnie Lee)

Wild and cultivated tobaccos are leafy annual plants generally not reaching a height of more than 1 or 2 meters.[1] They often produce broad leaves, and their growing stems terminate in tubular yellow, cream, or pink flowers with fused petals and flared petal tips (figure 13.1). Originating in the tropical and subtropical Americas, the genus *Nicotiana* contains dozens of species historically distributed in the New World, Australia, the South Pacific, and Africa. The center of diversity for domesticated varieties is in the tropical Andes of South America, where the plant has probably been farmed for 5000 to 7000 years.[2] Tobacco was dispersed widely by humans in pre-Columbian times: the small-leafed cultivated species (*N. rustica*) was planted throughout northern South America, Central America, the Caribbean, and southwestern and eastern North America; the large-leafed cultivated species (*N. tabacum*) was grown in Central America and northern South America. Tobacco relatives of the same genus are used ornamentally, selected for their vibrant flower colors, pleasant bouquet, and tolerance to many pests. Through history, diverse peoples have used the various wild and domesticated *Nicotiana* species for medicinal, ritual, and social purposes, and they all contain psychoactive chemicals.[3] The most widely cultivated variety today is *N. tabacum*.

FIGURE 13.1 Tobacco: (*top*) plants; (*bottom*) flowers.

EARLY USE IN THE AMERICAS

Tobacco was valued by many nations of the Americas, and its uses were documented by Europeans as they made initial contact with indigenous peoples of the New World. Its role in the traditional practices of native societies has also been described by anthropologists in more recent times. The evidence points to tobacco consumption as diverse in form and nearly universal among the indigenous groups of the Americas (figure 13.2).[4] People used tobacco by chewing or sucking on the leaves, drinking tobacco leaf juice or water in which leaves were steeped, licking a tobacco leaf extract, snorting ground tobacco into the nose, taking as a liquid infusion or smoke via enema, applying as a leaf paste onto the body, and by leaf smoke taken into the lungs or blown onto the body, among other variations of consumption.[5] In some cases, tobacco was taken on its own; in others, it was combined with various additives. For example, tobacco was on occasion chewed with salt, earth, or lime (alkali) derived from crushed seashells or ashes. Tobacco prepared for chewing, smoking, drinking, and other means of consumption was sometimes mixed with other medicinal plant materials, including hallucinogenic substances.[6] Importantly, tobacco consumption among Native Americans cannot easily be described as solely medicinal, recreational, social, ritual, or spiritual because for many peoples these elements were interrelated.

Many indigenous people considered tobacco to have supernatural origins and powers.[7] For example, the Winnebago Indians of the southern coast of Lake Michigan held that tobacco was a gift to them from Earthmaker, and they offered it as a tribute to the spirits for blessings. Among other groups, tobacco played important roles in genesis stories. The Yecuana of Venezuela believed that women came from "clay over which tobacco smoke was blown," and the legends of the Yaqui of Mexico held that an ancient woman had been transformed into tobacco.[8] In the diverse cosmologies of the Native Americas, the interdependence of the spiritual and physical worlds is a common feature, and the mediator of contact between these domains was frequently a shaman. In the pursuit of their spiritual responsibilities, shamans took tobacco as a source of supernatural power.

There are numerous accounts of the use of tobacco by spiritual leaders to provide for the

FIGURE 13.2 Tobacco in indigenous American medical and social life: (*top*) an illustration of indigenous tobacco use in sixteenth-century Haiti; a shaman is tending to a patient (lying on the hammock) while other patients are smoking a cigar (standing) or intoxicated by the smoke (lying on the ground); (*bottom left*) a Mayan man smoking a cigar, depicted on an earthenware vessel, 600–900; (*bottom center*) a Piegan (Blackfeet) medicine man with a decorated ceremonial pipe; (*bottom right*) a Sioux man smoking a pipe. ([*top*]) "Modo che tengono i medici nel medicare gl'infermi" (How doctors heal the sick), woodcut from Girolamo Benzoni, *La historia del mondo nuovo* [1565]; Library of Congress, Prints and Photographs Division, LC-USZ62-71989; [*bottom left*] Museo Popol Vuh, Guatemala City; vessel decoration photograph © Justin Kerr, K3386, http://www.mayavase.com; [*bottom center*] photograph by Edward S. Curtis [1910]; Library of Congress, Prints and Photographs Division, LC-USZ62-117604; [*bottom right*] photograph by Edward S. Curtis [1907]; Library of Congress, Prints and Photographs Division, LC-USZ62-96972)

well-being of their communities through ritual offerings. For example, the shamans of the Huron Indians of the northern Great Lakes region instructed farmers to burn tobacco in their fields to honor the sky deity and ensure good weather for a bountiful harvest.[9] On the Ucayali River of Amazonian Peru, weather shamans "blow tobacco smoke to divert thunderstorms."[10] Along the northwestern California coast, the World Renewal priests of the Yurok and Karok drank tobacco juice and offered tobacco to commemorate the yearly transformation and rebirth associated with the seasons, a ceremony without which they believed the world would cease to exist.[11] In addition to its significance in such offerings, tobacco played a role in the curing practices of indigenous shamans.

Taurepan religious practitioners of Venezuela, for instance, drank tobacco juice to sustain contact with the spirit world and thereby extend their ability to treat illnesses, and shamans of the Caribs gave it to their patients to heal them.[12] The shamans of the Mundurucú in the Brazilian Amazon blew tobacco smoke on their patients to remove their illnesses.[13] Indeed, the use of tobacco smoke for identifying the source of illness and for curing was common among diverse peoples of the Americas. In numerous accounts from throughout the Americas, shamans are said to have blown tobacco smoke to determine the location of evil materials or spirits within the patient and to prepare the body for treatment. The blown tobacco smoke then served to drive out or appease the maleficent entities. Alternatively, the smoke directed the shaman where to excise the source of illness through a straw.[14]

Tobacco consumption in the spiritual-medical context was widespread in the Americas, frequently associated with altered states of reality induced by potent tobacco preparations and sometimes enhanced by the addition of other herbs. These effects, and tobacco's connection to the divine, are consistent with the plant's role in allowing shamans to communicate with the spirit world on behalf of his or her patients and community. In other instances, tobacco's perceived spiritual-medical power was harnessed in a variety of forms, by trained healers and common people alike, to treat diverse external and internal conditions. The spiritual and medical uses of tobacco in the traditional Mesoamerican context cannot be dissociated, an inherent duality common in tobacco-related beliefs and practices among numerous indigenous American groups.[15]

Among the Aztecs, tobacco was applied as a paste to the skin to kill parasites, and it was taken internally by steeping it in boiled water to make an herbal tea. The tobacco leaves were also often chewed or smoked. The Aztec herbal of 1552 lists drinking a beverage of "the intoxicating plant we call *piciyetl* [*N. rustica*]" as a treatment "for recurrent disease."[16] Mesoamericans employed tobacco for a variety of illnesses, including eye and ear ailments, toothaches, wounds, rashes, gout, and swelling of the belly.[17] In eastern and southeastern North America, the Cherokee used tobacco as an analgesic against cramps and headaches, an anticonvulsant, a diuretic, a treatment for dropsy, a remedy for snakebite, and a medicine for dizziness. The Montauk, Micmac, and Mohegan nations, among others, used tobacco as ear medicine and as a dental anesthetic, the smoke blown toward the sore site or the leaves applied directly.[18]

Tobacco has also served in the social-bonding customs of indigenous Americans. The people of the Blackfoot nation of Alberta and Montana smoked wild tobacco (*N. quadrivalvis*) leaves mixed with red osier dogwood (*Cornus sericea*) at ritual gatherings. In accounts of such rituals, the entire community gathered inside a large tipi, the women and children in a designated section and the men seated in order of wealth and community standing. Men passed the medicine pipe clockwise around the tipi, drew from the pipe precisely four times without inhaling, and blew the smoke upward. At elite, all-night Big Smoke ceremonies, only the wealthiest men of the community participated, and they regaled one another with a prescribed number of songs to vie for prestige among their peers. The Apache, Tewa, and other peoples of the southwestern United States smoked tobacco at medicine ceremonies, and the Paiute smoked it to ward off colds during ritual bathing and prayer.[19]

Among eastern North American tribes, political or economic agreements were formalized by the ritual exchange of tobacco. In many cases, this took the form of the communal smoking of a pipe, known as a calumet.[20] Such ceremonies took place to signify friendship, alliances, and treaties between peoples.[21]

The diverse cultural uses of tobacco across the Americas gave rise to countless technologies for the ingestion, inhalation, and insufflation of tobacco products. Some indigenous peoples of South America drank tobacco juice or a tobacco infusion out of a hollowed gourd, through either the mouth or the nose.[22] Throughout the Americas, people fashioned pipes out of stone, bone, clay, reed, or wood, through which they took smoke into their mouths or lungs.[23] Numerous groups rolled dried tobacco leaves in a wrapper made of tobacco, leaves of a different plant, or tree bark. People then lit the ends of these cigars and took the smoke into their mouths or lungs.[24] Also common was the use of finely ground tobacco (snuff), sometimes mixed with other materials, taken by tube onto the nasal mucosa.[25] Insufflators, as these tubes are known, were often made of bone and took various forms to accommodate one or both nostrils, for self-administration or administration by another. There are also examples of the use of rectally administered tobacco smoke or liquid, introduced by an enema made of materials such as animal bladder, rubber, bone, and reed.[26] The apparatus of tobacco consumption served in the ritual, social, and medical use of the herb in various forms. The European explorers who arrived in the New World learned from the indigenous peoples and adopted and adapted their customs.

TOBACCO AS A COLONIAL COMMODITY

When Christopher Columbus (1451–1506) first set foot on the island he called San Salvador (in the Bahamas) on October 15, 1492, the local people offered him tobacco leaves. Two weeks later, while exploring Cuba, his crew members observed "men and women with a firebrand of weeds in their hands to take in the fragrant smoke to which they are accustomed."[27] To Europeans searching for precious metals, exotic spices, and souls to convert to Christianity, the dried leaves held in such high esteem by the Caribbean islanders were at first no

more than a curiosity. The earliest Spanish writers described what they saw among the indigenous peoples of the Caribbean and Mesoamerica: people inhaling the smoke of tobacco, chewing tobacco with crushed seashells, and breathing powdered tobacco into their noses through Y-shaped tubes.[28] At first a novel New World botanical specimen among many, tobacco did not initially garner much attention by scholars in Madrid, Paris, or elsewhere. As tobacco's potential medicinal properties captured the Europeans' interest, however, it was transported by sea and land to the major trading posts of the world. The early explorers brought back to Europe samples of the leaves and seeds, and by the mid-sixteenth century, tobacco was grown in Spanish, Portuguese, Dutch, and French botanic gardens. By the end of the sixteenth century, tobacco was known throughout Europe and had been taken to Russia, Japan, China, India, and parts of Africa.[29] In less than 100 years, tobacco was propagated around the globe.

First the Spanish and later the Portuguese, English, Dutch, and French colonists cultivated tobacco both because they recognized its utility in trade with the indigenous Americans and because they themselves became accustomed to its physiological effects. In the journal of Columbus's voyage, tobacco smoke is said to reduce pain and fatigue among the indigenous people of Cuba, "so they say that they do not feel weariness."[30] A French chronicler noted in 1555 that tobacco smoke suppressed hunger and thirst among the natives living along what is now the coast of Brazil.[31] Europeans, and the African slaves they brought to the New World, ultimately tried tobacco for many of the same purposes: one account from the 1530s tells of Spaniards smoking tobacco to treat the pain associated with syphilis and describes African slaves smoking to relieve exhaustion.[32] Those explorers who smoked found it difficult to give it up, lamenting "that it was not within their hands to stop taking it."[33] Within a few decades of their arrival, a great number of colonists—settlers, sailors, and slaves—were hooked on tobacco.

During the sixteenth century, the Spanish established tobacco plantations in their Caribbean, Central and South American, and Philippine colonies, and the Portuguese grew tobacco in Brazil.

The French cultivated tobacco in their short-lived Florida colony. By the turn of the seventeenth century, tobacco was among the first crops grown in the English Jamestown colony.[34] The expansion of tobacco growing was driven by increasing demand in Europe and enabled by the plant's tolerance of relatively diverse climatic and soil conditions.[35] Therefore, settlers were able quickly to increase production of a highly adaptable cash crop.

THE PROCESSING OF TOBACCO

The Europeans, no doubt taking cues from inhabitants of the New World, innovated multiple means to consume tobacco.[36] Following the harvest of the mature leaves and their slow drying (or curing) in a shaded and airy place, in the sun, or over a slow-burning wood fire, tobacco is processed into a number of products.[37] Cigars, as modified by Europeans, contain shredded leaves wrapped into long, thin rods within a large tobacco leaf. Cigars are lit and smoked generally without inhaling deeply, instead allowing the smoke to linger in the mouth. Cigarettes, probably developed in Spanish America during the seventeenth century, contain tobacco leaf fragments rolled in paper, and the smoke is inhaled into the lungs.[38] Other types of smoking tobacco, sometimes mixed with fragrant plant oils, are slowly burned in the bowls of pipes or water pipes. Snuff is finely ground, scented tobacco sniffed into the nose, a form considered particularly sophisticated in much of Europe during the seventeenth and eighteenth centuries. Chewing tobacco consists of moist, cured tobacco-leaf pieces that are slowly chewed or held between the lips or cheek and gums.

HEALING HERB OR HEALTH HAZARD?

When Europeans first described tobacco use among American Indians, a common theme in their commentary is the plant's intoxicating effects. For example, a Spanish account from the mid-sixteenth century explains the consequences of tobacco chewing this way: "He who sells *picietl* [*N. rustica*] crushes the leaves first, mixing them with lime, and he rubs the mixture well between his hands. Placed in the mouth it produces dizziness and stupefies."[39] As an inebriant associated with native shamanic rituals, tobacco came to be a concern of some Christian explorers and clergy, who questioned its spiritual wholesomeness. A Dominican cleric in the Caribbean during the first third of the sixteenth century remarked that tobacco snuff "takes away the senses";[40] another visitor to Mexico railed against tobacco in 1541, calling it "a wicked and pestiferous poison from the devil."[41] These admonitions, however, did not prevent colonists from learning to use tobacco.

As colonists and sailors in the New World took up the tobacco habit during the early sixteenth century, botanists and physicians in European centers of learning debated the novel herb's effects on health.[42] In a French agricultural treatise of 1567, the author relays a story of a young boy whose cancerous growth on the nose was completely cured by applying tobacco pulp and a woman whose breast tumor was similarly eliminated, along with tales of treating ulcers, infections, and asthma.[43] The elaborate description of New World flora and fauna written by Nicolás Monardes (1493–1588) in 1571 claims that tobacco can be applied externally for skin ailments, muscle stiffness, and stomach pain; drunk as a liquid infusion for chest congestion, internal parasites, and bad breath; and inhaled as smoke to treat asthma, among other uses.[44] By 1576, tobacco was lauded as a panacea in a Belgian herbal, "effective for sores, wounds, ailments of the chest and wasting of the lungs."[45] As European scholars grew more familiar with tobacco's properties, they wrote tracts recommending a degree of moderation in tobacco use and outlined some of its risks. For example, an English guide to tobacco published in 1595 acknowledged the usefulness of tobacco smoke for headaches and arthritis but noted that it should not be taken unless necessary.[46] In a book on health published in London in 1600, the author suggests tobacco smoke to treat migraines and toothaches but warned that it was harmful to the brain and liver if taken after a meal.[47] By the early seventeenth century, numerous accounts of certain ill effects of tobacco had appeared. Among the most striking is the tract composed by King James I (1566–1625) of England.

FIGURE 13.3 Europeans took up smoking tobacco through small pipes: (*left*) an English smoker; (*right*) tobacco use in the Netherlands, depicted in David Teniers, *Peasants in a Tavern* (1633). ([*left*]) Woodcut from Anthony Chute, *Tabaco* [1595]; National Library of Medicine, 156492; [*right*] National Gallery of Art, Washington, D.C., 1991.140.1)

In *A Counterblaste to Tobacco* (1604), he famously describes smoking as "a custome lothsome to the eye, hatefull to the nose, harmefull to the braine, dangerous to the lungs, and in the blacke stinking fume thereof, neerest resembling the horrible Stigian smoake of the pit that is bottomelesse."[48]

Such concerns notwithstanding, tobacco captured people with particular efficiency; they probably were drawn more to the pleasure it provided than to any specific therapies it performed. Tobacco smoking crossed the Atlantic and attracted countless new habitués in Britain and throughout continental Europe (figure 13.3). As the value of leaves increased, it was the Jamestown settlement in Virginia—established in 1607 in honor of the English king who had written *Counterblaste*—where tobacco plantations first bolstered the North American economy.[49] The plant's leaves sold in England at luxury-good prices,[50] and the growth of Virginia tobacco

prompted the importation of the first African slaves to His Majesty's colonies to tend it.[51]

In the centuries following its entry into global trade, tobacco played numerous social roles. It retained its traditional value in staving off hunger and thirst among laborers in the New and Old Worlds alike and remained profitable as a material of social bonding, enjoyed in taverns and coffeehouses, royal courts and houses of worship, smoked, snuffed, and chewed. Despite the potential for harm noted by sixteenth- and early-seventeenth-century commentators, European and American physicians conjured dozens of new medical uses for the tobacco plant. Medical uses of tobacco, taken internally or externally, during the seventeenth through nineteenth centuries included treatment for ringworm, earache, constipation, and arthritis; healing of wounds; and prevention of the plague.[52]

Although the possible risks of tobacco use became better understood in the nineteenth century, and a number of anti-tobacco societies formed to carry on the public debate over tobacco, some physicians and laypeople promoted tobacco's health benefits well into the twentieth century, sustaining a public denial of the negative consequences of the widespread tobacco habit.[53] During times of military conflict, tobacco smoking has long been a reprieve for tired soldiers, and thus during the Crimean War (1853–1856), the Civil War (1861–1865), World War I (1914–1918), World War II (1939–1945), and more recent conflicts, tobacco drew a large cadre of weary souls who grew dependent on the mild euphoria it provided and the sense of sharpened awareness and enhanced reflexes it brought about. The innovation of manufactured cigarettes in Europe and the United States from the 1840s through the 1860s made tobacco especially convenient and cheaper

to produce and sell, enhancing its appeal for extensive distribution. Combined with effective branding and marketing schemes, the cigarette came to dominate the tobacco industry, now accounting for more than four-fifths of all leaf harvested.[54]

Tobacco companies drew from a long list of cultural and medical cues as they constructed effective advertising strategies for their products, an effort that greatly enhanced the spread of tobacco use—whether chewed, smoked, or insufflated—during the nineteenth and twentieth centuries. Importantly, a number of tobacco makers offered products whose health benefits were touted by doctors for improved digestion, as weight-loss aids, for their respiratory advantages, and so forth. This phenomenon of physician endorsement must certainly have convinced legions of Americans that their smoking habits were wholesome (figure 13.4).

FIGURE 13.4 Advertising tobacco in the United States: (left) manufacturers promoted tobacco through stereotyped images of indigenous Americans, such as In this advertising label for Indian Girl chewing tobacco, 1874; (right) tobacco products were endorsed by physicians through the mid-twentieth century, as in this advertisement for Lucky Strike cigarettes, manufactured by the American Tobacco Company, 1930. ([left] Library of Congress, Prints and Photographs Division, LC-DIG-ds-00874)

Advertisements notwithstanding, physicians noted a rising incidence of lung cancer among smokers during the 1920s and 1930s, and by the 1950s near-conclusive evidence had mounted that tobacco use contributed to cancer and respiratory and heart diseases.[55] In July 1957, the U.S. Public Health Service released its first warning that tobacco increases lung cancer risks; the Royal College of Physicians in 1962 and the U.S. Surgeon General in 1964 released reports stating the causal link between smoking and lung cancer, among other deleterious health effects.[56] As a result of these findings and numerous others, the U.S. government required labels on cigarette packs to alert smokers of the danger. As evidence has mounted, the labels have become more assertive in tone. In 1966, the mandated message was: "Caution: Cigarette Smoking May Be Hazardous to Your Health." By 1970, the phrasing became more forceful: "Warning: The Surgeon General Has Determined That Cigarette Smoking Is Dangerous to Your Health." In 2003, the World Health Organization issued guidelines for the labeling of tobacco products that has resulted in graphic designs and warning labels that occupy, in some cases, an entire side of the packaging. Several newer variations of these labels warn of the multiple negative consequences of tobacco use, including photographs of grotesquely impacted bodies. In nearly fifty countries and jurisdictions, the labels make a more concise statement, translated into many local languages, with one common message: "Smoking Kills."[57]

The unhealthful components of tobacco include products of curing and combustion as well as a large number of chemical compounds inherent to *Nicotiana* or added during the processing, flavoring, and texturing of the tobacco leaf. The chief psychoactive compound in tobacco, nicotine, has some adverse health consequences; however, many of the numerous pulmonary, cardiovascular, and carcinogenic effects can be attributed to other compounds in the product.[58] While tobacco has demonstrated negative effects in all its consumer forms, the delivery of tobacco through smoking conveys particular toxicity. Tobacco is thought to result in the premature deaths of 4.3 million people worldwide annually.[59]

ISOLATION OF TOBACCO'S ACTIVE PRINCIPLE

The chemical in tobacco responsible for its physiological effects is nicotine, an alkaloid first isolated by the German chemists Wilhelm Posselt and Karl Reimann in 1828 and synthesized around the turn of the twentieth century (figure 13.5).[60] Nicotine in its pure form is a clear, oily liquid, alkaline in nature. As an aerosol in smoke inhaled into the lungs, ingested, or applied to the skin and mucous membranes, it passes readily into the bloodstream. Although pure nicotine is a base,

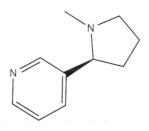

FIGURE 13.5 Nicotine.

it can exist in a slightly different structural arrangement in a more acidic (lower pH) environment. In the tobacco plant, nicotine is present in both an acidic form, in which it is joined to other molecules and therefore unavailable for release into a person's body, and a free-base form that is more prevalent in alkaline conditions. Therefore, traditional American Indian preparations of tobacco using crushed seashells, chalk, or ashes—all alkaline substances—converted nicotine into a more readily absorbed basic chemical form.[61]

Since nicotine most effectively crosses the mucous membranes under alkaline conditions, tobacco processing can substantially affect the product's physiological effects. Cigar and pipe tobacco are usually processed to be alkaline, and nicotine in their smoke can enter the body through the mouth. Since cigarette tobacco is usually processed to be slightly acidic, the large surface area of the alveoli in the lungs enhances the otherwise inefficient uptake of nicotine into the body.[62] Tobacco blending can modify the acid-base chemistry of cigarette smoke and alter the way nicotine affects smokers. Among the major types of dried leaves, sun-cured leaves are harvested and dried under the sun, air-cured leaves are dried in the shade, and flue-cured leaves are dried in a heated chamber.[63] Each process lends

PHARMACOLOGICAL EFFECTS OF NICOTINE

Reduction in muscle tone
Increase in attentiveness
Reduction in appetite
Increase in heart rate
Increase in blood pressure
Reduction in urine production
Mild euphoria

a particular aroma profile and chemistry to the smoked product, and many cigarette makers in the United States and Europe mix tobacco in a proprietary blend. In recent decades, the percentage of flue-cured tobacco in cigarette blends has decreased.[64] Since flue-cured tobacco is slightly acidic and air- and sun-cured tobaccos are neutral to slightly alkaline, the industry has moved to a type of cigarette that produces a greater dose of free nicotine in its smoke.

Typical tobacco leaves contain between 0.5 and 3 percent nicotine by weight, a figure that varies substantially depending on variety of plant, location on the plant, growing and harvest conditions, and curing technique.[65] The tobacco in a single cigarette contains, depending on the manufacture, approximately 10 milligrams of nicotine.[66] Less than 10 percent of the nicotine in a cigarette is inhaled and enters the bloodstream via the lungs, mouth, or stomach. The remainder is exhaled without absorption, destroyed by burning, or lost to uninhaled smoke.

MECHANISM OF ACTION

Nicotine exerts its effects by binding as an agonist to nicotinic acetylcholine receptors in the central and peripheral nervous system. In the periphery, nicotine gives rise to increased activity of the gastrointestinal tract and elevated blood pressure and heart rate, in part via the release of epinephrine from the adrenal glands. In the brain, nicotinic receptors are located at the presynaptic terminals

of dopaminergic, cholinergic, and glutamatergic neurons. Binding of nicotine causes the release of those neurotransmitters, which accounts for diverse effects on elevating mood and producing euphoria. Nicotine also reduces skeletal muscle tone, contributing to a sense of relaxation. Connections to the opioid and cannabinoid neurotransmitter systems, not yet fully investigated, may underlie some of the pleasure and analgesia resulting from tobacco use. Nicotine decreases appetite and increases psychomotor ability, attention, and memory consolidation. Nicotine is a demonstrated antidepressant, and co-occurrence of nicotine use and depression is an indication that people might self-medicate to reduce depressive symptoms.[67]

In addition to nicotine, which contributes to cardiovascular disease through an increased workload on the heart and blood vessels, tobacco, especially when smoked, contains numerous processing and combustion byproducts that cause cancer.[68] (Nicotine itself is toxic but not carcinogenic.) As awareness of the risks of consuming tobacco has grown in recent years, companies whose fortunes have long depended on the sale of nicotine products have diversified to include the production and marketing of nicotine delivery by other means, including aerosol, lozenge, and "dissolvable tobacco."[69] Indeed, many of the electronic cigarettes (generating doses of vaporized nicotine), nicotine patches, and nicotine chewing gums for smoking cessation are produced by "Big Tobacco."

When Europeans learned of the

tobacco plant from American Indians, they tried to put its combined stimulating and relaxing properties into a medical and social context. Some thought that consumption of the plant was a dignified expression of high class; others found the practice unsavory. Some viewed tobacco as a healthful herb; others saw it as a dangerous poison. In modern times, countries are pulled between the economic value of the tobacco crop and the obligation to protect people from its risks. Such has been the history of this divisive and important psychoactive plant, whose nicotine-laden leaves have gripped so many.

Chapter 14

Popular Herbs

Myriad seeds, stems, leaves, roots, and fruits—such as these for sale in a market in Kurseong, India—have a long history in the world's culinary and medicinal traditions.

"Let food be your medicine and medicine be your food," goes the oft-quoted capsule of advice attributed to the Hippocratic writers of ancient Greece.[1]

HERBAL DIETARY SUPPLEMENTS AND MEDICALLY ACTIVE FOODS

In many traditional medical systems, attention to diet is a core practice of healthful living, and illness might be rectified by changes in cuisine. A great number of herbs considered nutritive, flavorful, and wholesome have entered the pharmacopeia through the kitchen. Sometimes such plants, selected long ago, were ascribed various medicinal virtues and integrated into local conceptions of health. People gave these leaves, roots, fruits, and grains properties: they designated them as "warming" or "moistening," said they were good sources of energy, remembered they could make a person feel better if taken in moderation or worse if taken in excess. People established rules and procedures to prepare herbs properly, to mix them with the right types of ingredients and in the correct proportions, to taste a certain way, to have certain effects on the body. The experiences of generations of food preparers yielded such knowledge. Codified in the medical texts and preserved in folklore, beliefs in the health-related properties of edible plants are grounded in the observations of countless food providers and the people they helped feed.

As peoples met across the great routes of human movement, they shared knowledge and entered into commerce that extended plants into new territories and under new caretakers. People began again to taste, observe, and speculate, taking account of what they had heard and assigning to each plant a set of properties that made sense to them. They further disseminated these plants across vast distances and down through the generations, sharing and evolving what they had learned: this herb is good for earaches, this one is binding and "drying," this one is best in the winter. In this way, so many medicinal and culinary plants have come to the present, still serving their roles at the dinner table, and almost all awaiting the attention of biochemists and physiologists.

The inquisitive human spirit that gave rise to the diverse culinary traditions of the world, each incorporating so many flavorful, colorful, nutritious medicinal plants, is also evident in the countless roots, leaves, and seeds that made up the herbal teas, ointments, and other preparations people used to treat their illnesses and promote health. Generations of village elders, midwives, and physicians yielded a set of agents that, as part of and alongside cuisine, was believed to address the suite of maladies affecting a people. The traditional herbal pharmacy was enormous, and individuals relied on herbs both as part of a proper diet to maintain health and in directed medical treatment to deal with all types of ailments.

Drawing on numerous medical traditions, many in the West continue to look after their health through diet, giving attention to the timing, balance, and sourcing of their meals. Whether to treat particular acute complaints, address chronic concerns, or promote well-being in the broadest sense, people recognize that their choice of what to eat can influence their bodies in meaningful ways. They also choose to supplement their diets with readily consumed plant-based isolates and extracts, seeking health-related benefits and following an age-old practice of appealing to nature for help toward wellness. This chapter explores the lore and impact of several of the most popular herbal supplements and medically active foods. Some have come through thousands of years of Western herbal medical tradition; others have been adopted from the indigenous practices of Asia and North America. In a story common across so many herbal traditions, the cultural roles of these plants have changed over time, having been ascribed certain properties in antiquity and other properties that developed later, in new eras and locales. Today, these plants have been transformed yet again, shaped by new perceptions of their capabilities to modify health at the level of the body, tissue, cell, and molecule. Such herbs- encapsulated, extracted, and standardized—are modern products of the age-old pursuit of health through plants.

While previous chapters have addressed medicinal plants with unambiguous effects on human physiology (the strong stimulant properties of cocaine, from the coca plant, for example), many of the herbs in this chapter are proposed to have more subtle roles in health. These plants are thought to promote longevity, lighten the mood, and help resist infection, for instance, outcomes that can be harder to discern than those of products with more or less instantaneous effects. As a next large challenge for the study of herbal medicine, continued quantification and description of the properties of such popular supplements using objective measures will yield a great resource for practitioners and patients.

The oral and written records of traditional uses of plants provide much education to consumers and scholars, as generations of experience have delivered to modern researchers information about dosing, preparation, and particular uses of herbs. Human curiosity and government regulation demand also that medicinal products be subjected to scientific testing, to understand better the makeup and effects of plants and plant extracts on health. Such tests aim to characterize the chemical composition of herbal extracts to identify possible active principles, a task simpler for plant-derived agents with strong, immediate effects than for those with slower-acting effects. Some herbal compounds likely act as part of a mixture of chemicals, which calls for an analysis of some complexity. It is also important to recognize that experiments conducted in the laboratory using cell cultures or test animals (such as rodents) might not easily extend to the human being. Conducting clinical trials is a technically challenging and costly business, and many experiments investigating the possible effects of herbal supplements have been criticized for shortcomings of design. Ultimately, the most convincing data on the safety and efficacy of plant-based treatments will come from a consensus of scientifically sound experimental investigations that account well for dose, composition, mode of delivery, and specific, shared measures of health outcomes.

CRANBERRY

Vaccinium macrocarpon

The North American cranberry is a perennial evergreen vine adapted to grow in the wetlands of New England, the eastern maritime Canadian regions, and the northern Great Lakes area. The cranberry plant thrives in moist soil rich in organic material, where it creeps low to the ground, producing small, glossy, oval green leaves and bright-red fruits in late autumn.[2] The genus *Vaccinium* includes numerous small-fruited species mostly growing in the north temperate zone of the globe, including the American and Eurasian small cranberry (*V. oxycoccos*), Eurasian lingonberry (*V. vitis-idaea*), and North American blueberry (*V. corymbosum*).[3] Cranberry fruits have long been employed for their nutritive and medicinal properties (figure 14.1).

Among the indigenous North Americans, cranberry was a trade good and foodstuff eaten fresh or preserved by drying. Crushed berries were an ingredient in the American Indian food called pemmican, a mixture of cranberry, animal fat, and dried meat that kept well and provided a stable reserve of energy during lean times.[4] The first Europeans to tread the northern reaches of the New World may not have recognized the American cranberry as a relative of the smaller-fruited variety (*V. oxycoccos*) back home, but the latter was well known to botanists, who considered its juice useful against fevers and illnesses caused by the humor yellow bile.[5]

Early American accounts of cranberry demonstrate its culinary and medical roles, sometimes inspired by observations of indigenous American uses. For example, Roger Williams (ca. 1603–1683), the founder of New England's Providence Plantation, was introduced to cranberry by the Narragansetts and described it in 1643 as "another sharp, cooling Fruit growing in fresh Waters all the Winter, Excellent in conserve against Feavers." He also noted that they used cranberry and grain to "make a delicate dish which they cal *Sautáuthig*; which is as sweet to them as plum or spice cake to the English."[6] Later, in 1672, the English traveler

Figure 14.1 Cranberries: (*left*) fruit; (*right*) harvest. ([*right*] Photograph by Keith Weller; U.S. Department of Agriculture, K4418-6)

John Josselyn (1608–1675) remarked on cranberry's place in cuisine, writing that "the Indians and English use them much, boyling them with Sugar for Sauce to eat with their Meat." European explorers certainly experienced the effects of scurvy (now recognized as vitamin C deficiency) from a lack of fresh fruits and vegetables, and Josselyn noted its usefulness against that scourge as well as the "the fervour of hot Diseases."[7] Other early health-related applications of cranberry include the treatment of wounds, stomach concerns, liver problems, and infections.[8]

Cranberry became cultivated widely in North America during the nineteenth and twentieth centuries and is now marketed in several forms: as whole fruit, as prepared sauces, as juice, and in extracts in the form of powders or capsules. Cranberry is tart because of the presence of acidic and astringent chemicals, and it possesses several types of complex polyphenolic compounds, a mixture of substances that might account for its ascribed health-related properties.[9] Certainly, the effectiveness of cranberry against scurvy can be explained

by its vitamin C (ascorbic acid) content.[10] Cranberry also contains numerous compounds with antioxidant properties in addition to vitamin C.[11] Much current interest in cranberry is focused on its possible utility in preventing or treating urinary tract infections. Such infections result from the colonization of the urethra or bladder by bacteria such as the pathogenic strains of *Escherichia coli*.[12]

Although the mechanisms of cranberry's possible antibacterial properties have yet to be thoroughly defined, there seems to be a role for a particular form of polyphenolic compound called an A-type proanthocyanidin (figure 14.2).[13] In laboratory tests, these chemicals, which comprise a portion of a complex mixture, have been demonstrated to inhibit the adhesion of *E. coli* and other pathogenic bacteria to the type of epithelial cells that line the urinary tract.[14] Yet the extension of this laboratory finding to human health has been troubled by challenges of physiology and experimental design. For the former concern, investigators question whether the complex compounds present in cranberry juice or extract are absorbed from the

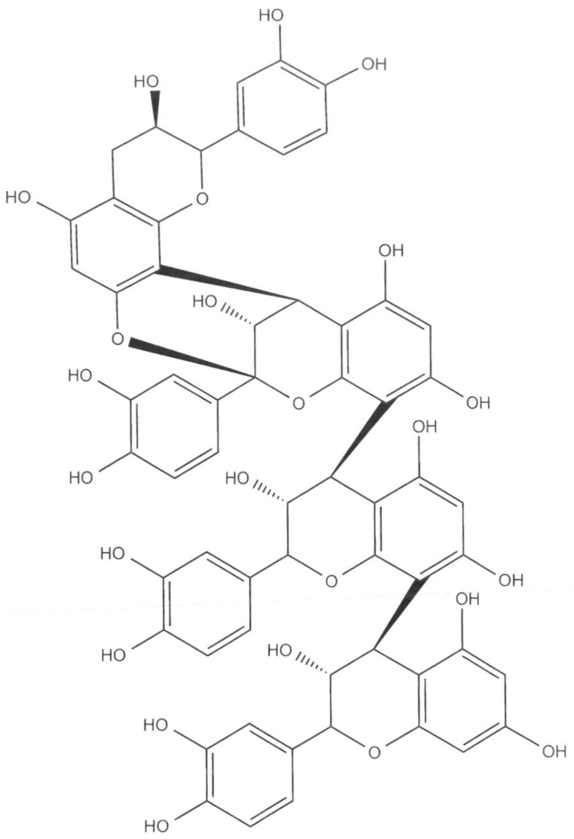

FIGURE 14.2 A-type proanthocyanidins are speculated to interfere with bacterial adhesion to the cells lining the urinary tract.

cranberry extracts in multiple forms, material not standardized to a common set of criteria.[16] It is therefore difficult to ascertain in such human studies whether cranberry or any of its possible active principles may play a role in preventing or treating urinary tract infections.

To date, there is little evidence supporting the use of cranberry to prevent urinary tract infections, nor have there been many recent studies on the effectiveness of cranberry to treat existing infections.[17] Yet the laboratory findings showing that certain cranberry constituents interfere with bacterial pathogenesis suggest that the fruit could yield useful medicinal products with demonstrable clinical activity in the future.

SOY

Glycine max

The soybean is an annual herbaceous legume that originated in eastern China.[18] It grows to a height around 1 to 1.5 meters and produces numerous small, white, pink, or purple flowers that give rise to pods, each containing two to four pea-size seeds (figure 14.3).[19] The plant has been cultivated for its leaves, which can be cooked and eaten or fed to animals as forage, and for its seeds and seed pods. The immature seed pods can be cooked and eaten, and the mature seeds can be prepared for human and animal consumption. The seeds are processed into numerous products, including soy meal, soy oil, soy milk (cooked, ground soybeans), and tofu (solidified soy proteins).

FIGURE 14.3 Soybean pods.

human digestive system and distributed efficiently. That is, if the A-type proanthocyanidins and related compounds are not taken up into the bloodstream from the intestine or are chemically broken down in the body, it is difficult to imagine their accumulation in the urine. Therefore, any extension of test tube or Petri dish findings to the whole organism would demand a better understanding of the metabolism of cranberry's constituents.[15] As to the matter of experimental design, recent studies of cranberry use in human populations have employed a suite of dissimilar scientific protocols and have tested commercial and proprietary

Standardized extracts

Mixtures of chemicals prepared from an herbal source in which the concentration of particular molecules of interest has been determined using appropriate laboratory techniques

Archaeological and literary evidence supports the idea that the soybean was domesticated sometime before about 3000 years ago. Soybeans were first grown in Europe and North America during

PROPERTIES OF THE POLYPHENOLIC COMPOUNDS

Structure	Polyphenols consist of rings of carbon atoms linked together in various ways
Molecular composition	Polyphenols contain carbon, oxygen, and hydrogen atoms; they do not contain nitrogen
Diversity	There are many thousands of types of polyphenols; hundreds can occur in any given plant tissue
Complexity	Some polyphenols are made up of dozens of units of simpler phenolic structures
Solubility	Polyphenols are often water soluble
Biological activity	Many polyphenols bind (chemically attach) to cell surfaces, which may explain antibacterial properties; some can bind to hormone receptors, enhancing or reducing natural hormone-related processes; some can bind to neurotransmitter receptors, mimicking or blocking their functions; many are also antioxidants
Vocabulary	Flavonoids, flavonols, isoflavones, anthocyanins, catechins, and tannins are all polyphenolic compounds

the eighteenth century and used primarily for animal feed. As a worldwide commodity, soybeans now account for a major part of agricultural production in the United States, Canada, Brazil, and China.[20] Economically important applications include food additives, such as the extract lecithin, which is used in the food industry to improve the texture of products containing both oil- and water-based ingredients, and soybean oils, which can be used for culinary purposes or for fuel as biodiesel.

In ancient China, soybean was both a food crop and a medicinal substance.[21] Classic texts of Chinese traditional medicine recommend soybean sprouts to clear bodily heat and treat painful obstructions of blood and *qi*, and cooked seeds as a remedy for illnesses characterized by irritability, headache, and restlessness. Soybean seed coats (the skin of the seed) were suggested as a poultice to treat sores and diaper rash.[22]

The past several decades have seen considerable interest in dietary soy as a possible preventive measure or therapy against cardiovascular disease, age-related bone loss, and the symptoms of menopause. Drawing on the observation that East Asian women appear to be less susceptible than Western women to postmenopausal bone loss, hot flashes, breast cancer, and other concerns of aging, investigators speculated that some aspect of the East Asian diet or lifestyle might be associated with a lower risk of these conditions.[23] Since the Japanese, Koreans, and Chinese, among others, consume larger quantities of soy products than do people in other nations, researchers questioned whether these foods might have properties related to women's health.

Menopause is accompanied by a decrease in the hormone estrogen, and it is possible that some components of soy might counteract this hormonal change, thereby reducing diverse menopause-associated health risks. Current studies often focus on a class of polyphenolic compounds called the isoflavones, chemicals that are in relatively high abundance in soybean seeds (figure 14.4).[24] Certain isoflavones have the property of binding to the cellular receptors for estrogen, thereby activating estrogen-dependent responses in the target tissues. This feature has earned them

Phytoestrogens

Chemicals from plants that act in the body like the female sex hormone estrogen

the name phytoestrogens: plant-derived chemicals that act like the steroid hormone estrogen. In addition to their estrogenic properties, isoflavones can interact with a suite of cellular enzymes, and they are antioxidants.[25]

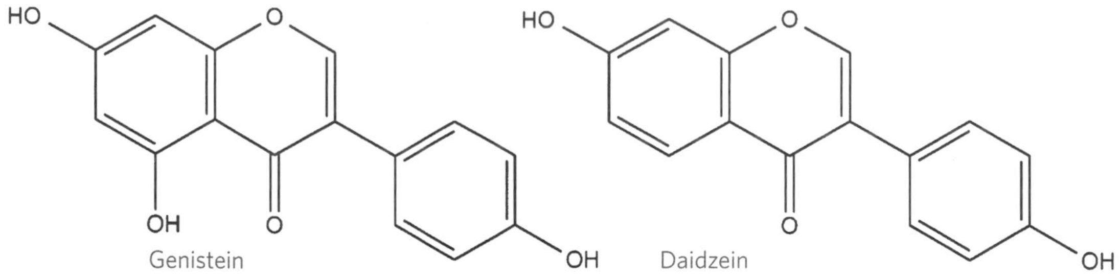

FIGURE 14.4 Soy isoflavones: genistein; daidzein.

Because isoflavones can mimic natural estrogens in laboratory tests, soy products have been examined against a wide range of menopausal symptoms in humans, from flushing and emotional changes to vaginal dryness and loss of bone density.[26] Most tests to date have been of a small scale and use differing sources and concentrations of soy products, rendering it relatively difficult to generalize findings.[27] Some statistically significant results have been obtained on the effect of soy products on bone density. In analyses of the results of several trials collected together, soybean or isoflavone supplementation appears to have moderately counteracted the loss of bone in postmenopausal women.[28]

While estrogen promotes bone density and staves off menopausal symptoms in women, it can also induce the proliferation of estrogen-sensitive cancer cells, leading some to question whether soy-isoflavone consumption might increase the risk of breast or ovarian cancers.[29] Although the mechanism of isoflavone binding to estrogen receptors could conceivably induce cancer in susceptible tissues, human trials have not demonstrated a strong relationship between soy consumption and markers of cancer proliferation.[30] Interestingly, some evidence may indicate a possible antiproliferative effect for isoflavones, suggesting that certain types or doses of these chemicals might reduce the growth of cancer. The dual pro- and anticancer cell–growth properties of certain isoflavones is speculated to be a function of their binding to alternative estrogen-receptor subtypes, with different, sometimes opposing roles in regulating cell proliferation. As there are so many factors influencing the development of cancers occurring in diverse tissues, it will take time for investigators to identify any specific components of soy that might play a role in this area.

Researchers have attempted to test associations between soy consumption and numerous aspects of health, including cardiovascular measures such as blood-lipid profile and blood pressure.[31] Such studies have generally yielded conflicting or equivocal results, with analysts complaining of small sample sizes and poor experimental designs.[32] Since much of the current understanding of soy and population health is built on epidemiological approaches with many possible variables and a relatively small set of controlled trials, there are few conclusive findings on the role of soy as a therapeutic food.

SAW PALMETTO

Serenoa repens

The saw palmetto is a low-growing palm abundant in its native range in eastern North America, along the coastal plains of South Carolina, Georgia, Florida, Alabama, and Mississippi.[33] The plant grows in the shrub layer of the forest ecosystem and as scrub on sandy coastal dunes, reaching a height of 1 to 2 meters, with branching stems close to the ground and fan-shaped divided green or blue-green leaves as large as 1 meter in diameter (figure 14.5). Numerous small flowers in the spring yield black grape-size fruits (called berries) in the autumn.[34] Saw palmetto has long been valued for its economically useful fibers, from which indigenous people made baskets and roof thatch, as well as for its slightly sweet, calorie-rich fruit.[35] However, early

FIGURE 14.5 A stand of saw palmetto.

European visitors to today's Florida did not share the local peoples' appreciation for saw palmetto berries, one seventeenth-century English writer describing their flavor as "rotten cheese steeped in tobacco."[36]

While saw palmetto was employed in traditional medicine among native North Americans for a wide variety of concerns, from respiratory to reproductive, it was not until the nineteenth century that saw palmetto came under the scrutiny of early biomedical science.[37] Literature from that time documents the ways that doctors and pharmacists experimented with saw palmetto berry juice, fluid extract, and oil, taken orally or as a suppository (figure 14.6). Medical entrepreneurs also concocted numerous patent medicines combining saw palmetto with sandalwood (*Santalum* spp.) oil, coca (*Erythroxylum coca*) leaf, cola (*Cola* spp.) nut, parsley (*Petroselinum crispum*), and other ingredients.[38] Late-nineteenth-century physicians recommended saw palmetto, in its various forms and preparations, for hemorrhoids, headaches, inflammation, bronchitis, gonorrhea, and a suite of other ailments of the male and female urogenital

FIGURE 14.6 An advertisement for Metto, a patent medicine that contained saw palmetto. (From *Ocala* [*Fla.*] *Evening Star*, July 27, 1901)

systems.[39] Much attention was paid to saw palmetto's effects on the sexual organs and on the breasts, for which doctors considered it "a great vitalizer, tending to increase their activity, to promote their secreting faculty, and add greatly to their size."[40]

After a series of curious trials on themselves and their patients, saw palmetto researchers at the turn of the twentieth century lauded its effects in increasing female bust size, improving male libido and the quality of the erection, and particularly treating the prostate gland.[41] While the prostate plays an important role in the male sexual response by secreting part of the seminal fluid, its position at the base of the bladder and tendency to enlarge with age can contribute to difficult and frequent urination in older men. The use of saw palmetto extract, doctors believed, reduced the size of the enlarged prostate and alleviated the urinary symptoms in their patients.

It is now recognized that certain fat-soluble chemicals (sterols and fatty acids) in the saw palmetto berry can interact with the normal pathways of testosterone and estrogen signaling, a finding that might help explain the plant's therapeutic properties.[42] Among several loci of action, saw palmetto fatty acids can inhibit the enzyme responsible for converting the steroid hormone testosterone to dihydrotestosterone, which is implicated in the (noncancerous) cell proliferation that increases the prostate's size in older men.[43] Numerous laboratory studies have demonstrated saw palmetto extract's ability to reduce prostate cell growth in response to normal hormonal stimuli.[44]

Controlled clinical studies of saw palmetto's effects on urinary symptoms associated with prostate enlargement have generated a rather large volume of literature, but evidence clearly demonstrating its therapeutic efficacy has not yet emerged. While some relatively small-scale trials using various types of extracts and methodologies have shown improved urinary symptom measures in patients treated with saw palmetto for durations of one to three months, the different experimental strategies make it difficult to generalize and compare outcomes.[45] A recent meta-analysis of large-scale trials, applying strict criteria for quality of experimental design, found no significant difference in urinary symptoms or prostate size of patients with enlarged prostate between groups taking saw palmetto extract at various common doses or those taking a placebo.[46] Despite laboratory results demonstrating saw palmetto extract's mode of action on the hormonal processes regulating cell physiology, clinical trials taken collectively failed to show a convincing effect, highlighting the importance of ongoing research to understand better how herbal drugs enter the human system, whether and how they reach their proposed target tissues, and what activities they retain when there. The paucity of conclusive evidence notwithstanding, saw palmetto extract has become widely employed in Europe and North America.[47]

While saw palmetto extracts are generally well tolerated, with few reported side effects, the commercial products available have been prepared to differing standards and vary tremendously in chemical composition.[48] Therefore, any physiological effects of a saw palmetto extract would depend on the particular dose, formulation, and processing employed in its manufacture.

Types of biomedical herbal research data

Laboratory studies test pharmaceutical effects on isolated cell components, cells, or tissues in a test tube or Petri dish (called *in vitro* research) or in captive animals (*in vivo* research).

Clinical trials administer chemical preparations to human volunteers and record various physiological measures, noting side effects.

Meta-analyses use defined selection criteria and advanced statistical methods to draw conclusions from the results of numerous aggregated research reports.

Hallmarks of good experimental design

- Large sample size to reduce risk of test subjects unrepresentative of general population
- Randomization of individuals receiving various experimental treatments
- Blinding of researchers to the identity of the treatment groups
- Credible placebo treatment to account for nonpharmacological effects

---

GARLIC

Allium sativum

A close relative of onion, chives, and leek, garlic has been grown for thousands of years, originating in Central Asia, disseminated throughout Eurasia and North Africa in ancient times, and now grown worldwide.[49] The culinary and medicinal parts are the fleshy leaf bases, which grow underground, surrounded by papery leaf sheaths (figure 14.7). Grasslike leaves can reach a height of 30 to 60 centimeters.[50] Although garlic, a perennial, occasionally produces flowers in starburst-like inflorescences at the end of long stems, it rarely sets seed. Therefore, most garlic propagation is vegetative, aided by the separation of individual vegetative buds and surrounding swollen leaf bases, the cloves.[51]

FIGURE 14.7 Garlic. (Woodcut from Hamsen Schönsperger, *Gart der Gesundheit* [1487]; Peter H. Raven Library, Missouri Botanical Garden, St. Louis)

Widely employed in cooking and medicine, garlic found its way into the customs of East Asia, South Asia, Mesopotamia, Egypt, and the Greco-Roman world.[52] At least as early as the Egyptian New Kingdom (from 1550 B.C.E.), garlic was a valued foodstuff, and, indeed, specimens of garlic bulbs accompanied numerous burials.[53] The Greek herbalist Pedanius Dioscorides (ca. 40–90) recommended garlic as a warm and drying herb that had the capability to relieve flatulence, expel intestinal worms, increase urine output, treat bald spots, clear coughs, and cure snakebites, among other uses.[54]

By the seventeenth century, an English medical text declared garlic "a remedy for all diseases and hurts," with warming properties particularly useful for "hydropick diseases, the Iaundise, Falling Sicknesse, Crampes, Convulsions, the Piles or hemorrhoides, and other cold diseases."[55] With such diverse applications across a long period of time, it is not surprising that modern researchers are engaged in testing garlic's efficacy against a wide spectrum of health concerns.

Garlic's medicinal and culinary properties derive in large part from an assortment of sulfur-containing molecules, some of them pungent, that are released when its tissue is crushed. Cells in the garlic bulb accumulate large amounts of the odorless chemical alliin.[56] As cells experience damage (such as when crushed or chopped), alliin reacts with an enzyme released from specialized subcellular compartments and is converted to the characteristically acrid chemical allicin (figure 14.8).[57] Allicin is unstable, and with heat or age it ultimately degrades into a diverse suite of potentially bioactive sulfur-containing compounds.[58] Because of the variety of allicin's degradation products, garlic's possible physiological effects depend on its fresh or aged state and method of preparation.[59]

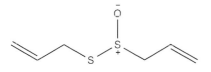

FIGURE 14.8 Allicin, a compound in freshly crushed garlic that gives rise to sulfur-containing chemicals with possible health-related effects.

The normal processes of cellular metabolism as well as the stresses of aging and illness can generate damaging (oxidizing) chemical byproducts in human cells, and some of garlic's sulfur-containing compounds are suggested to have antioxidant properties that counter these effects.[60] The antioxidant capacity of garlic appears to derive from the ability of the sulfur-containing compounds themselves both to neutralize oxidizing chemicals and to induce antioxidant gene expression in the body.[61] Such properties could recommend garlic for use against neurodegenerative disease and cancer, although large studies targeting the prevention or treatment of these conditions have not been performed. Garlic's antioxidant properties are credited in part for its potential role against cardiovascular disease, an area that has been investigated by biomedical researchers.

Mounting evidence indicates that garlic's complex assortment of bioactive chemicals, including the sulfur-containing compounds, have effects in lowering blood-lipid levels and reducing cholesterol, interfering with blood clotting, and bringing down blood pressure, all potentially beneficial outcomes in certain populations.[62] Laboratory assays, experiments in animals, and small-scale human trials have demonstrated garlic extract's capacity to reduce the oxidation of low-density lipoproteins in blood, which would limit the formation of arterial plaques and improve the blood-lipid profile.[63] While the results of trials investigating the effects of various preparations and dosing regimens of garlic on blood-lipid profile have been inconsistent, the collective data set indicates that garlic is probably efficacious in improving lipid parameters.[64] Furthermore, garlic in various forms has been demonstrated to reduce blood pressure in patients with hypertension.[65] The mechanism by which garlic is thought to reduce blood pressure has been partially elucidated: the allicin-derived sulfur-containing compounds are transformed in the blood and arteries to the chemical hydrogen sulfide, which acts as a signal to relax the muscles lining the arteries.[66] In addition to probable roles in improving blood-lipid profile and blood pressure, garlic is recognized to reduce blood clotting, a factor in cardiovascular disease and stroke.[67]

Garlic has an enduring reputation as a treatment for infectious disease, either by direct antimicrobial activity or by somehow boosting the immune system to deal better with pathogens.[68] While many laboratory tests have demonstrated the antibacterial and antiviral properties of garlic and its extracts, this activity is generally attributable to the unstable, oxidizing compound allicin.[69] Since allicin does not survive in the human system, it is not likely to be an effective antimicrobial in the body, except, perhaps, in topical applications. Although garlic has been suggested as an immune system modulator, studies of this possible function mostly remain preclinical. Several research centers have considered the use of garlic extracts as a preventive measure or treatment for infections such as the common cold, although to date the published experiments do not show an advantage of garlic over placebo treatment in this regard.[70]

In widespread use since ancient times and long associated with health and healing, garlic garnered esteem for treating diverse illnesses and adding its characteristic pungency to world cuisines. As garlic remains an important food item in diets around the globe, its role as a nutritive, flavorful ingredient will likely be further compounded with additional demonstrable medicinal properties in times to come.

GINKGO

Ginkgo biloba

The ginkgo, also known as the maidenhair tree, originated in East Asia and is now grown throughout the world's temperate zones. Ginkgo can reach a height of 35 meters or more, and some individuals are thought to be hundreds of years old.[71] The plant has fan-shaped medium-green leaves, sometimes divided into two portions, that turn bright yellow before falling in the autumn. As a dioecious plant, ginkgo has separate male trees that release sperm cells and females that grow small olive-size seeds covered in a fleshy tissue (figure 14.9).[72] Europeans first encountered ginkgo in East Asia during the sixteenth and seventeenth centuries and brought specimens back to the major botanic gardens by the

early eighteenth century. Today, ginkgo is grown as an ornamental and shade tree around the world, particularly desired in urban landscapes, where it has proved resilient against the inconsistent precipitation, temperature changes, pests, and pollution levels of modern cities. Nearly all ginkgo employed in landscaping is male, as the female flesh-covered seeds release an unpleasant odor. Since ginkgo's evolutionary origins predate the development of flowers and fruits, botanists classify it as a gymnosperm, along with the conifers and cycads. Ginkgo is unusual in being a monotypic taxon: the only extant species of its scientific genus, family, order, and class. The fossil record is rich in ginkgo relatives, however, leading some to dub *Ginkgo biloba* a "living fossil."[73]

Ginkgo found its way into East Asian medicine and cuisine at least 1000 years ago, and its seeds became part of the traditional pharmacopeia. In Chinese medicine, prepared ginkgo seeds are thought to restrain Lung *qi* and promote the movement of dampness, therefore effective for breathing disorders, cough, vaginal discharge, and urinary complaints.[74] (Lung is capitalized to distinguish the traditional Chinese physiological element from the Western anatomical structure.) The seeds, however, are toxic when fresh and in large quantity, and practitioners know to boil them well and prescribe them conservatively to avoid effects ranging from nausea to convulsions and death. Ginkgo seed is also commonly used as an ingredient in soups and porridges in contemporary East and Southeast Asia.[75]

FIGURE 14.9 Ginkgo: (*left*) leaves; (*right*) seeds.

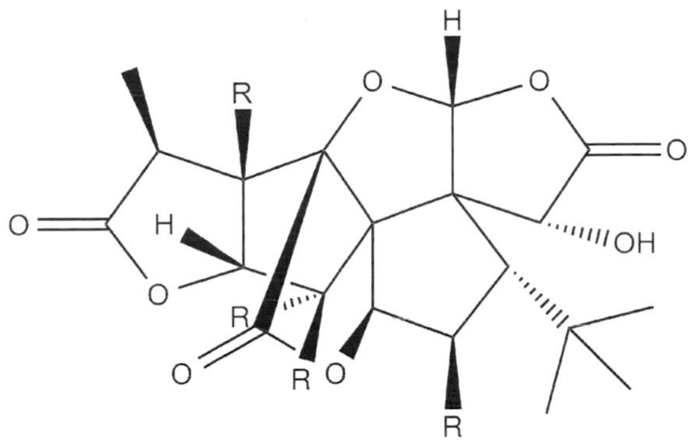

FIGURE 14.10 A ginkgolide, from ginkgo leaves.

In contrast to the Asian traditions, in which ginkgo seeds played a role in medicine and food, the Europeans took an interest in ginkgo leaves for possible health-related effects. By the twentieth century, German chemists had developed a process to purify some of the oil-soluble and water-soluble chemical constituents of the leaves as a standardized extract. Ginkgo leaf extracts contain a complex assortment of flavonoids, flavonoid glycosides, and terpenoid compounds called ginkgolides (figure 14.10). Many ongoing research projects focus on the possible physiological effects of standardized extracts, in principle rendering the laboratory and clinical studies more comparable.[76] Despite these efforts, many human trials using ginkgo leaf extracts have been criticized for inconsistent experimental design and dosing, and some commercial ginkgo extracts are unreliably prepared and labeled.[77] Ginkgo extract is speculated to benefit brain function and has been recommended to improve vision and blood flow in peripheral arteries.[78]

Although some specific activities of ginkgo extract have been assessed in the laboratory, many of ginkgo's possible physiological mechanisms remain unknown.[79] Among the most widely lauded applications of medicinal ginkgo is in the prevention and treatment of cognitive decline (such as accompanies aging and neurodegenerative diseases) and the enhancement of normal cognitive functions. To test these possible medical applications, several research groups have assessed a large body of experimental literature for evidence of efficacy. With regard to the use of ginkgo extract to treat cognitive decline, the

experimental evidence is mixed, with researchers at various sites using similar protocols but obtaining contradictory evidence. A meta-analysis concludes: "The evidence that *Ginkgo biloba* has predictable and clinically significant benefit for people with dementia or cognitive impairment is inconsistent and unreliable."[80] Recent large-scale studies on ginkgo extract as a preventive measure for age-related cognitive shortcomings—including declining memory, language ability, attention, and executive function—likewise failed to demonstrate significant differences over placebo control groups.[81]

Few human trials have investigated ginkgo's possible effects in the eyes, where its extract is speculated to prevent or treat age-related deterioration of the retina.[82] Ginkgo has also been suggested to treat tinnitus, the sensation of sound in the ears in the absence of any stimulus. A recent review of the literature revealed that ginkgo is generally ineffective for treatment of this condition.[83] Finally, work has been initiated to study the possible effect of ginkgo on the symptoms of peripheral artery disease, which causes patients leg pain and is associated with blockage of the arteries. An analysis of a number of relatively small controlled trials revealed no significant difference between ginkgo treatment and placebo on measures of peripheral artery disease.[84]

Despite centuries of use as a medicinal and culinary plant in Asia and decades of research as a therapy for a variety of age-related ailments, ginkgo's physiological activities remain enigmatic. Ongoing studies may add to the growing set of experimental outcomes questioning its utility as a drug or finally lend support to some of its long-ascribed properties.

MILK THISTLE

Silybum marianum

The milk thistle, indigenous to western and Mediterranean Europe, North Africa, and the Middle East, is an annual or a biennial herbaceous plant that now grows throughout Eurasia and is naturalized in southern Africa, North and

FIGURE 14.11 Milk thistle.

South America, and Australia. It grows a series of green, tapered oval leaves with spiky margins and whitish coloring along the veins, eventually sending up a number of flowering stalks to a height of 20 to 150 centimeters, each with a flowering head about 4 centimeters in diameter consisting of dozens of small, sharply pointed leaves (botanically bracts) surrounding a group of closely packed, pink-petaled flowers (figure 14.11). At maturity, the fruit case yields approximately 100 small, dry, seeded fruits, the medicinally important product.[85]

Milk thistle was well documented in the work of Greek and Roman scholars, and Theophrastus (ca. 371–287 B.C.E.), Pliny the Elder (23–79), and Dioscorides all mention its value for diverse medical concerns.[86] An abundant plant throughout western Europe during the Middle Ages and early modern period, milk thistle was frequently referenced in medical texts for its health-related properties. For example, an English herbal of 1657 recommends milk thistle as an herbal tea or in powdered form for agues, prevention and cure of the plague, to "open obstructions of the Liver and Spleen," to increase the production of urine, and to treat dropsy. The seeds were noted as particularly effective medicines and suggested "to cool the distemperature" of the liver and heart.[87] In time, medical practice focused on milk thistle's use for various ailments of the internal organs, and a medical text published in 1898 describes its efficacy to treat "chronic congestion of the kidneys, spleen and liver."[88]

During the twentieth century, biochemists extracted a diverse set of chemicals from milk thistle seeds, known as flavonolignans, dihydroflavones, and other phenolic compounds.[89]

Collectively, the mixture of chemicals in milk thistle extract is called silymarin, and it has been standardized for clinical and commercial purposes.[90] Silymarin exerts diverse effects on physiology, as demonstrated in laboratory tests. Silymarin components are antioxidants and have the property of stimulating cellular antioxidant proteins, thereby potentially reducing the effects of cellular damage from stress or age.[91] These compounds also interfere with inflammation-related signaling, modulate the cellular immune response, and seem to be able to suppress cell proliferation in Petri dish tests, possibly leading to applications in cancer treatment.[92] Much of the current research on milk thistle extract has centered on understanding its effects on the liver, where it is thought to be especially useful.[93]

Milk thistle and its extract silymarin have a reputation as a detoxifier and liver protectant, although such assertions are largely based on laboratory rather than clinical studies. Milk thistle's antioxidant, immunomodulatory, and anti-inflammatory properties were demonstrated to protect liver cells from hepatitis C virus infection, among other measures of liver health.[94] While orally administered silymarin suffers from poor absorption in the human system, its constituents can be chemically modified in the laboratory to improve uptake and, in particular, activity in the liver.[95] Despite the growing body of laboratory evidence for a number of liver-protective effects, challenges of dosing, formulation, and experimental design have rendered many human studies uninformative, and reliable clinical results for milk thistle have been slow to arrive.[96] The uncertainty over possible therapeutic benefits notwithstanding, it seems from the numerous studies that milk thistle extract in current dosing strategies generates few side effects (comparable to placebo) and therefore can be considered safe.[97] Ultimately, the results of large-scale, well-designed trials testing silymarin or other precisely measured forms of milk thistle extract to ameliorate the symptoms of cirrhotic, viral, and other types of liver disease will settle the open question whether the protective effects of the herb in the laboratory can be extended to human treatments.[98]

In addition to its potential role as a liver protectant and remedy for liver disease, milk thistle extract has been employed as an emergency measure in patients accidentally poisoned by eating *Amanita* mushrooms, whose toxins target the liver. Retrospective analyses and case studies generally support the idea that milk thistle extract was not harmful and, in the treating physicians' judgment, contributed to better patient outcomes.[99] Milk thistle has also been suggested as a galactagogue to increase milk production, but scientific research in this area is at an early stage, and possible mechanisms are unknown.[100]

PURPLE CONEFLOWER

Echinacea spp.

The purple coneflower is a perennial herbaceous plant native to central and eastern North America.[101] The genus is composed of several species, some of which are similar in appearance and partially overlapping in range. The species *Echinacea angustifolia* grows a series of oblong, lance-shaped leaves close to the ground and then produces flowering heads on stems 10 to 60 centimeters in length. The flowering heads are made up of dozens of small flowers, the outermost flowers growing long, thin, violet or pink petals, giving a flowering head about 10 centimeters in diameter (figure 14.12). The pale purple coneflower (*E. pallida*) is similar in appearance, with slightly taller stems and flowering heads with longer petals. *E. purpurea* has larger, more oval-shaped leaves.[102] Both the aboveground portion of the plant and the root have been used in medicine, and purple coneflower is a popular horticultural plant, growing well in temperate zone gardens.

FIGURE 14.12 A purple coneflower.

Numerous indigenous American groups employed purple coneflower in medicine and hygiene.[103] Among the Omaha and Ponca of present-day Nebraska and Oklahoma, the plant was called *mika-hi* (comb plant) because people used the mature flower head to brush the hair. Applications to treat illnesses were diverse: among Indians of the upper Missouri River region, the macerated root was used to treat snakebite and insect stings, the Dakota (Sioux) of the upper Great Plains used the root (probably applied to the skin) to treat bites and wounds, and the Lakota (Sioux) used it to ease toothache, bellyache, and other sorts of pain. Other groups in the central North American plains used *Echinacea* root for colic, colds, and sore throat and as a stimulant when traveling at night.[104] Indigenous American medicinal uses for purple coneflower were diverse. However, Europeans took very little note of its value in native health care, and only in the mid-nineteenth century did American physicians and botanists remark on its possible therapeutic utility.[105]

While mainstream medical practice in the West slowly shifted away from a humoral understanding of the body and became enamored with the advances in biochemistry emerging in Europe, a group of physicians and pharmacists in the United States developed a uniquely American school of medical thought during the nineteenth century. In this form of medicine, called eclectic medicine, local and indigenous botanic remedies and noninvasive therapies were favored over the violent purging, bloodletting, and strong drugs that sometimes characterized the patient's experience at the hands of conventionally trained physicians.[106] To the eclectic practitioners, native American herbs were appealing additions to the pharmacopeia, and they wrote extensively of the value of purple coneflower, one early adopter calling it a "blood purifier" and "antiseptic for internal and external use."[107] By 1915, eclectic physicians had documented dozens of

> **Eclectic medicine**
>
> A school of American medical practice that advocated the use of North American herbal treatments during the nineteenth and early twentieth centuries

therapeutic applications for *Echinacea* root extract, whether applied to the surface of the body, taken orally or rectally, or injected hypodermically. It was lauded as a sedative, cure for hair loss, and treatment for a diverse array of infections, including malaria, syphilis, spinal meningitis, and rabies.[108]

The American Indians and eclectic physicians primarily used *E. angustifolia* root (the latter extracting the root in alcohol and water) in their medical preparations, but by the 1930s, European practitioners transformed the use of the herb in two important ways. First, German and Swiss entrepreneurs brought back to Europe seeds of *E. purpurea* (instead of *E. angustifolia*), from which they prepared chemical isolates. Second, they made alcohol and water extracts from the aerial portions of the blooming plant rather than the roots.[109] Today, the diversity of purple coneflower preparations complicates their comparison in laboratory and clinical tests. Furthermore, the growing location and time of harvest influence the chemical profile of herbal extracts, rendering analysis of experimental outcomes somewhat challenging.[110]

Purple coneflower extracts contain an assortment of chemical constituents, including a volatile oil, fatty acid–derived alkamides, polyalkenes, polyalkynes, caffeic acid derivatives (including chicoric acid), flavonoids, polysaccharide carbohydrates, and a low concentration of alkaloids.[111] The relative concentration of these components differs among species and source (root versus aerial portion) of plant material.[112] Much of the current research on purple coneflower seeks to determine whether the herb can prevent or treat symptoms of upper-respiratory viral infections, such as those associated with the common cold and influenza. The particular chemical compound or combination of compounds responsible for *Echinacea*'s proposed biological activity have not been defined, but recent data indicate that commercial formulations of water-alcohol coneflower extracts can kill viruses, prevent their adhesion to human cells, and interfere with their ability to induce an inflammatory response in the laboratory setting.[113]

Clinical trials on purple coneflower and the common cold have been conducted by the dozens, but the results are mixed, with some studies showing effectiveness in reducing the incidence or severity of the illness and others showing no difference between the treatment and control groups. Among the largest studies performed to date, the investigators employed dissimilar dosing schemes, used diverse formulations of plant material, and examined different aspects of the progress of upper-respiratory infection.[114]

Once widely employed by indigenous people for infections and adopted by American physicians seeking native cures for difficult illnesses, the purple coneflower is among the most controversial of botanicals. Ongoing research in the laboratory and clinic should help define its possible bioactive components, and the standardization of experimental approaches would help generate a more robust dataset to allow a better understanding of its efficacy against viral challenges and perhaps also against those ailments for which it was recognized by generations of both indigenous and more recently arrived Americans.

BLACK COHOSH

Actaea racemosa

Black cohosh is a perennial herbaceous plant that grows in the understory of deciduous forests in the eastern part of North America, from southern Ontario through southern New England, Appalachia, and the Midwest.[115] The plant's leaves consist of three leaflets, each further divided into several pointed lobes, and grow both close to the ground and on upright stems 1 to 2.5 meters in height.[116] Flowers occur on wand-like stems, each one producing dozens of whitish flowers about 2 centimeters in diameter (figure 14.13). Fruits are small, with eight to ten seeds each.[117] The part of

FIGURE 14.13 Black cohosh.

the plant commonly used in medicine is the root and underground stem, or rhizome.

Diverse ethnobotanic accounts describe black cohosh root in the medical practices of indigenous eastern North American groups. For example, the Iroquois of the modern-day northeastern United States and Canada are said to have used it to promote the flow of milk in nursing women and in a bath to treat joint pain.[118] Among the Cherokee of the southeastern United States, black cohosh root was described to address a suite of health concerns—including colds and cough, constipation, backache, hives, and fatigue—as well as to help a baby sleep.[119] When the eclectic physicians of nineteenth-century America began compiling their list of herbal drugs, they included black cohosh, a "very active, powerful, and useful remedy."[120]

American physicians used black cohosh root for dozens of ailments, from syphilis to asthma to fevers. An extract of black cohosh root in alcohol was deemed to be effective against rheumatism and scrofula (generally a bacterial disease), among other ailments, and to have "an especial affinity for the uterus."[121] According to testimonials from the mid-nineteenth century, doctors employed black cohosh root, or its extract, to treat gynecological concerns such as sterility, irregular menstruation, vaginal discharge, prolapse of the uterus, and heavy menstrual bleeding.[122] By the late nineteenth century, black cohosh declined in use among mainstream medical practitioners in the United States but persisted in certain patent medicines (figure 14.14). It was also taken up to some degree by doctors in Germany.[123] Today, black cohosh is usually associated with treatments for the symptoms of menopause, such as hot flashes, depression, and discomfort.[124]

An assortment of chemicals has been isolated from black cohosh root, including various triterpenoids and some polyphenolic compounds (figure 14.15).[125] It is an ongoing challenge for biochemists and physiologists to determine whether an extract of black cohosh root has biological activities in the laboratory and clinic consistent with its perceived effects in menopause. One hypothesis is that black cohosh root compounds have estrogenic properties and therefore offset in some way the reduced estrogen hormone levels occurring at menopause. However, this estrogenic activity has not been consistently observed in the laboratory, and many recent reports show no such effect.[126] A more recent hypothesis posits that certain chemicals present in the black cohosh root extract are psychoactive, producing effects in the brain's perception of pain.[127]

In support of this notion, investigations have uncovered evidence that black cohosh extract

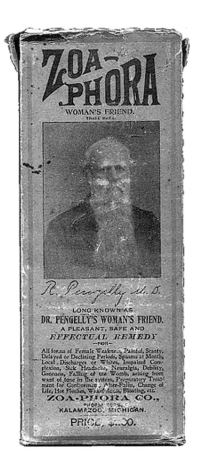

FIGURE 14.14 A package of Zoa-Phora, a patent medicine that contained black cohosh, early twentieth century. The label for this "woman's friend" indicates: "For all forms of female weakness, painful, scanty, delayed or declining periods, spasms at month, local discharges or whites, impaired complexion, sick headache, neuralgia, debility, goneness, falling of the womb, arising from want of tone in the system, preparatory treatment for confinement, after-pains, change of life, hot flushes, wakefulness, bloating, etc." (Smithsonian Institution, National Museum of American History, 1980.0698.068)

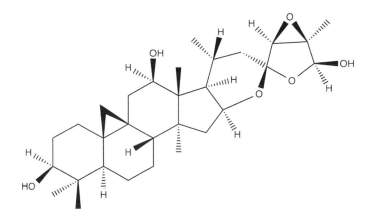

FIGURE 14.15 Acteol, one of the prospective active principles of black cohosh.

can act on the opioid receptors in the brain, possibly contributing to a calming and pain-reducing capacity of the herbal treatment.[128] Furthermore, it has chemicals that appear to be able to modulate the serotonin neurotransmitter signaling system, which may influence thermoregulation and mood in people taking black cohosh.[129] Taken together, the identification of components active in the central nervous system along with the paucity of evidence for estrogenic effects may help explain black cohosh's various cultural uses in treating many types of pain and discomfort, along with some of its uses in menopause.[130] While these observations might help address some long-standing questions of black cohosh's utility in therapy, the mechanisms of action are still not completely understood, and its entire chemical repertoire is not yet fully characterized.

The standard by which biomedical efficacy is judged is the well-designed clinical trial, and those testing black cohosh's utility against menopause symptoms have generated conflicting data. Recent clinical trials have used various preparations of black cohosh root extract, usually processed with ethanol or isopropanol as a solvent.[131] A review of clinical trials failed to show a positive effect of black cohosh extract on a series of menopausal symptoms but recommended further efforts to produce high-quality controlled studies to address this research question.[132] Ultimately, biomedical studies may shed further light on this plant of many virtues, so esteemed by indigenous Americans, eclectic physicians, and patients around the world.

ST. JOHN'S WORT

Hypericum perforatum

St. John's wort is an herbaceous perennial native to Europe that now grows both in cultivation and in the wild throughout the world's temperate zones.[133] The plant grows close to the ground, with stems rarely reaching 1 meter in height. Leaves are small, green, and oval, approximately 1 to 3 centimeters in length, with numerous translucent dots (that give the impression of perforations) and

FIGURE 14.16 St. John's wort.

occasional glandular black spots.[134] Flowers are bright yellow, with petals also bearing black spots. The flowers and aerial parts of the plant have been used in European medicine for at least 2000 years (figure 14.16).

Recognized in the works of the Greek and Roman writers, St. John's wort was listed by Dioscorides as "good for hip ailments" and able to draw out "much bilious matter and excrement." He also suggested it as a treatment for burns when applied to the injured skin.[135] The influential Greek physician Galen (129–ca. 216) considered the herb hot and dry, according to the humoral medical framework that he advanced.[136] To herbalists of the medieval period, the plant was associated with mystical powers in folk medicine, able to offer protection to those who used it, not by ingesting or applying it but simply by holding a sprig (figure 14.17).[137] In 1546, the German physician Hieronymus Bock (1498–1554) conveyed the popular wisdom in writing: "Many people carry these plants with them against evil spirits and thunderstorms."[138]

The controversial alchemist Paracelsus (1493–1541), who rejected classical humoral theory in favor of a folk and mystical medical practice, saw divine signs in St. John's wort's physical properties,

FIGURE 14.17 The use of St. John's wort to ward off demons, as shown in an illustration in an Italian herbal, ca. 1475–ca. 1525. (TR F Herbal, Special Collections, University of Vermont Library, fol. 27r; lower half: Erba ypericon)

according with his belief in the Doctrine of Signatures. The pores in the leaves and the red juice expressed when its flower petals are crushed demonstrated to him that the plant was placed on earth as a heavenly gift to heal the injuries of mortals.[139] "I declare to you that the holes that make the leaves so porous indicate that this plant is a help for all inward and outward openings in the skin—whatever should be driven out through the pores," Paracelsus explained, "and the putrefaction of its flowers into the form of blood, that is a sign that it is good for wounds." He also recommended St. John's wort stems and flowers, carried near the body or applied to the skin, to drive "phantasms out of people."[140]

While generations of physicians had considered St. John's wort valuable for ailments as diverse as broken bones and kidney stones, by the seventeenth century medical authorities also began to recommend the herb for "mental aberrations" such as "false imaginings, melancholy, fears, and corruption of the intellect."[141] The perception of St. John's wort's therapeutic value declined over time, as a popular British herbal from 1812 listed the herb only to treat urinary symptoms, wounds, and bruises; a French medical book of 1822 remarked at the lost status of an herb that had once been lauded for its "imaginary virtues" but by then had fallen into near oblivion.[142] Still, the author suggested

that it might have some utility for chest congestion and menstrual irregularity.[143]

Revived interest in St. John's wort in the medical community during the twentieth century prompted its testing for effects on depression, ability to heal wounds, and as a possible antiviral agent. St. John's wort foliage and flowers contain an array of potentially bioactive chemicals, including the pigments hypericin and related compounds, which accumulate to 0.05 to 0.3 percent and are responsible for the red color of the sap that exudes from crushed fresh herb, and the chemical hyperforin, which reaches 4 percent in the aerial portion of the plant (figure 14.18).[144] There are also polyphenolic chemicals that might be responsible for some physiological effects. Extracts are usually made in alcohol (either ethanol or methanol) and water and administered orally or topically.[145]

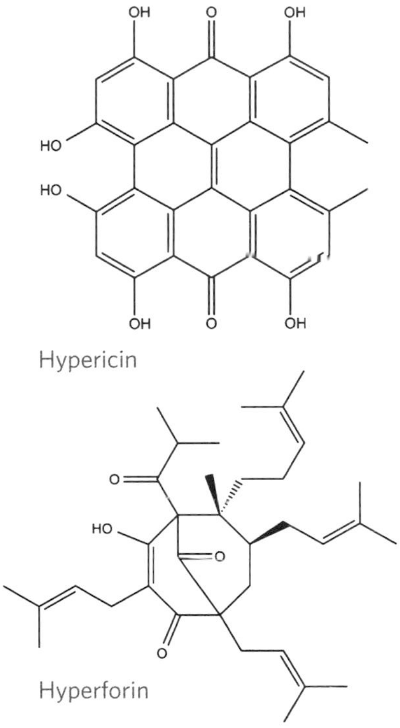

Hypericin

Hyperforin

FIGURE 14.18 Bioactive chemicals in St. John's wort: hypericin; hyperforin.

Laboratory tests suggest that hypericin and hyperforin have a wide range of interactions with components of various neurotransmitter signaling systems in the brain, although experiments demonstrating specific activities have not been uniformly replicated.[146] The collective evidence seems to indicate that the combination of constituents in St. John's wort extract has effects on the mood-related serotonin system, inhibiting reuptake of the neurotransmitter at the synapse and possibly increasing the abundance of its receptor. St. John's wort extract may also interact with the dopamine system, the acetylcholine system, the GABA system, and others, a clear impetus to further investigation of the many possible ways that St. John's wort might influence the mind.[147]

In the clinic, various St. John's wort extracts have been tested for their possible effect on depression versus placebo or synthetic antidepressant drugs. Statistical analysis of a body of literature demonstrated that patients taking the herb experienced a reduction of depressive symptoms superior to those taking a placebo preparation and on par with those taking the antidepressants. Furthermore, St. John's wort induced fewer side effects than the synthetic pharmaceutical treatment.[148]

Extracts of St. John's wort have also been investigated for their ability to aid in wound healing, and small-scale studies have indicated that a lotion made of the herb might be useful for this function. Antibacterial and antiviral activity has also been demonstrated in the laboratory.[149]

Appreciated by the ancients for its diverse roles in humoral, folk, and magical medicine, St. John's wort's use against depression is supported by a growing body of evidence. While medieval commentators noted the red juice issuing from crushed flowers and perceived it as a sign to use the plant to treat bloody sores, modern laboratory tests indicate its possible usefulness to improve wound healing and block infection, properties that can now be tested in the clinic. With further attention, examination of this plant's rich lore and complex assortment of chemical products might advance as-yet untested applications to human health.

GINSENG

Panax spp.

Ginseng is the name associated with several species native to East Asia and North America. The major Asian species, *Panax ginseng*, originates in northeastern China, Korea, and adjacent parts of Siberia and has been introduced into cultivation in Japan.[150] American ginseng (*P. quinquefolius*) comes from the eastern part of temperate North America, from southeastern Canada through the Appalachian Mountains and west to the Great Lakes.[151] Ginseng is a perennial herbaceous plant growing to a height of 50 centimeters and forming a small number of leaves composed of four or five leaflets, a cluster of inconspicuous pale-green flowers, and bright-red pea-size fruits containing a seed or two each.[152] The underground portion (root), which is harvested for commerce, consists of a stem and root system that can be as long as 30 centimeters or more, and 4 centimeters or more in diameter, although many specimens are much smaller. Ginseng can be collected in the wild, cultivated in wild settings, or grown in plantations.[153]

Ancient East Asian foragers must have noticed the contorted, branched form of the ginseng root and likened it to a human body, ascribing to it fantastic powers (figure 14.19).[154] In Chinese folk and scholarly medicine, ginseng was thought to be protective and therapeutic for the whole body. First described in the *Divine Husbandman's Classic of the Materia Medica* (first centuries C.E.) to "settle the ethereal soul and the corporeal soul, . . . expel pathogenic *qi*," and "strengthen the resolve," ginseng was believed to bolster all five organ systems of the Chinese medical framework.[155] Considered one of the most powerful herbs in the traditional East Asian pharmacopeia, ginseng is widely employed in tonic preparations as an herbal tea or in alcohol, soups, and medicinal formulas mixed with other herbs (figure 14.20). It is also a tribute and gift commodity, enveloped in luxurious packaging to accentuate the size and shape of the root.[156]

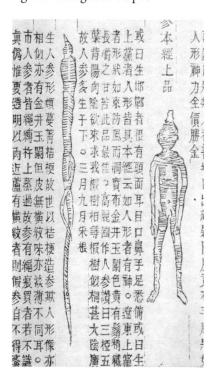

FIGURE 14.19 Ginseng in its regular (*left*) and anthropomorphic (*right*) forms, as depicted in a Chinese herbal. (Woodcut from Li Zongzi, *Origins of the Materia Medica* [1612])

FIGURE 14.20 Ginseng for sale at an herbal medicine market in Hong Kong.

Likewise in North America, indigenous groups harvested wild ginseng and used it in their medical practices. The Iroquois employed ginseng, for example, in a liquid infusion to prevent vomiting, as a blood medicine, for earaches, to improve the appetite, and applied to wounds and boils to help heal them. The Menominee of the upper Great Lakes viewed it as a "strengthener of mental powers," and the Cherokee of the Southeast used it "to relieve sharp pains in the breast."[157] (However, it is not possible to determine with certainty which indigenous uses of American ginseng predate contact with Westerners, since much of the ethnography related to native medical practices occurred decades or centuries after the ginseng trade was introduced by Europeans.)[158] Wherever it was discovered, ginseng found itself transformed into a drug deemed capable of procuring health in a profound way: whether treating illnesses directly or strengthening the body from inside, ginseng's purported powers are among the strongest of all the traditional medicines. Furthermore, in both East Asia and North America, ginseng was thought to be an aphrodisiac ("love medicine," according to ethnographers of indigenous North Americans) of remarkable potency.[159]

During the seventeenth century, Chinese exploitation of native ginseng stands brought the species close to extinction, sparking a spike in price and a search for new sources. In the early eighteenth century, a French missionary in China reported of ginseng's reputation as "a sovereign remedy for all weaknesses" and described its botanical features in a letter to the learned societies of Europe, which spurred a search for ginseng in the New World's largely unexplored northern colonies.[160] By 1716, French settlers located ginseng in Canada, and the British colonists found it in New England, New York, and Appalachia not long thereafter. The colonists generated a thriving trade in ginseng with the indigenous people of North America, who eagerly harvested the root on behalf of the Europeans. The European merchants then shipped it to China for sale at great profit.[161] Today's international ginseng market comprises a variety of cultivated and wild sources of ginseng, each with particular medicinal values (and therefore prices) assigned. For example, larger, older roots, especially those that closely resemble a human body, are prized over smaller, younger, less anthropomorphic examples.[162] The wild-harvested (or wild-crafted, cultivated in a forest setting with minimal artificial input) roots are considered more valuable than farmed ginseng. The high value of wild-harvested ginseng has contributed to a significant conservation problem in parts of East Asia and North America, where poaching has substantially reduced or eliminated the plant. To many East Asians, *P. ginseng* is believed to be more potent than *P. quinquefolius*. While generally thought to act similarly to the Chinese ginseng, American ginseng has attracted some controversy among traditional Chinese medical authorities. Some describe the herb as a strengthener of *qi*, though cooler in nature than the (warm) Chinese ginseng; others find it not to be strongly medicinal at all. Modern studies indicate that the chemical properties of Chinese and American ginseng are similar but distinguishable.[163]

Preparations of ginseng root in alcohol, water, tablets, and lozenges contain an assortment of about 100 different triterpene saponins called ginsenosides.[164] These components are collectively thought to be responsible for ginseng's adaptogenic properties. (Adaptogens are drugs that generally improve the body's ability to resist stresses, a traditional medical function not widely accepted

Triterpene saponins

Molecules made up of multiple carbon rings, often attached to a sugar component, having both water-soluble and fat-soluble aspects

by biomedical practitioners.)[165] Biomedical approaches to studying ginseng's therapeutic applications have focused on its possible role in modulating the immune system, affecting the pathways that govern diabetes, improving libido and sexual response, increasing mental function, and reducing nausea and stimulating appetite, among others.[166]

While laboratory studies and experiments in animals aiming to unravel the diverse effects of ginseng's likely active principles are plentiful, the extension of findings made on cell cultures and rodents to human beings is not always straightforward. Some clinical trials of ginseng's possible therapeutic effects have been criticized for poor experimental design, including small sample size, potential biases, and use of nonstandardized root extracts.[167] The base of evidence for ginseng's therapeutic effects in humans remains small and equivocal, but ongoing studies aim to generate data that might confirm or refute possible effects more convincingly.

HOREHOUND

Marrubium vulgare

Common horehound is a perennial herbaceous plant native to the eastern Mediterranean and grown widely in Europe, North Africa, and Central and South Asia since ancient times (figure 14.21). It is now naturalized in temperate North America, Australia, New Zealand, and South Africa.[168] Its small, roundish leaves grow opposite each other at nodes along stems 30 to 120 centimeters in height, terminating in spikes of clustered white flowers.[169] A member of the mint family, it is aromatic, producing a distinctive aroma and bitter constituents in the leaves and stem.

The Roman scholar Pliny, in his grand opus *Natural History*, introduced horehound as "a plant too well known to require any description," which speaks to its widespread cultivation in the ancient Mediterranean world. He lauded horehound's many uses, including a mixture of ground leaves and seeds for treating snakebites, pain in the torso, and coughs. He also recommended the stems boiled in water for "spitting of blood" and mixed with honey for "affections of the male organs." Made into an herbal tea with salt and vinegar, Pliny explained, horehound was useful as a laxative, beneficial as a promoter of menstruation, and effective in expelling the afterbirth, in total listing twenty-nine remedies incorporating the herb, against ailments as disparate as jaundice and hangnails.[170]

FIGURE 14.21 White horehound. (Illustration from Hieronymus Bock, *Kreüter Buch* [1546]; Peter H. Raven Library, Missouri Botanical Garden, St. Louis, Missouri)

Among the most useful herbs in the classic era, horehound's reputation followed it through to the seventeenth century, when the herbalist John Parkinson (1567–1650) echoed the advice of ancient authors in recommending it as "a remedy for those that are pursie [asthmatic], and short winded, for those that have a cough." He also listed it as an antidote to poison, a treatment for sores and ulcers, to improve eyesight, and to open "obstructions both of the liver and spleene." After an extensive summary of dozens of classical and contemporary applications of horehound, Parkinson encapsulated the prevailing, practical use of the herb: "There is a sirope made of Horehound to be had at the Apothecaries much used, and that to very good purpose for old coughes to rid the tough flegme."[171] While many of the varied uses of horehound failed to

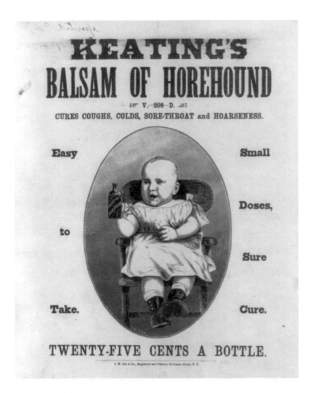

KEATING'S
BALSAM OF HOREHOUND

V.—208—D.

CURES COUGHS, COLDS, SORE-THROAT and HOARSENESS.

Easy   Small

to   Doses,

              Sure

Take.   Cure.

TWENTY-FIVE CENTS A BOTTLE.

FIGURE 14.22 An advertisement for Keating's Balsam of Horehound, a cough remedy, ca. 1870. (Library of Congress, Prints and Photographs Division, LC-USZ62-51231)

transition to the modern era, its preparation into a sweetened syrup or lozenge for respiratory complaints persisted well into the nineteenth century in mainstream medicine (figure 14.22). Although horehound is seldom employed today in formal medical practice, it is an ingredient in extant folk remedies such as Ricola, the Swiss expectorant cough drop.[172]

Horehound's purported medicinal properties are thought to derive from its chemical constituents, including bitter diterpenes such as marrubiin, which accumulate in the herb to 0.3 to 1 percent; some alkaloids; polyphenolics; and volatile oils.[173] While experimental evidence is largely restricted to the laboratory rather than trials involving patients, much of the work focusing on marrubiin and its derivatives has shown promise in potential analgesic and anti-inflammatory applications, which might relate to horehound's classical and folk use against cough and respiratory infections.[174] Some of the other chemical constituents, generally less well studied than marrubiin, may also

have physiological activity contributing to anti-inflammatory and antimicrobial properties, among others.[175] Time and dedicated clinical efforts will establish which of the numerous traditional uses of horehound might be explained by its complex chemistry.

VALERIAN

Valeriana officinalis

Valerian is an herbaceous perennial native to Europe and Asia, now naturalized in North America. While there are over 200 species in the genus *Valeriana*, a number of which have been used medicinally, the type with the longest history and most widespread interest is probably *V. officinalis*.[176] The plant consists of a stem that reaches 2 meters in height, with leaves composed of many narrow, oval green leaflets and clusters of tiny pink or white flowers. The underground portion of the plant, made up of rhizomes, stolons, and roots, is harvested for its ascribed health-related properties (figure 14.23).[177]

Documented as a medicinal plant as early as the writings of the Greek Hippocratic physicians and elaborated by authorities including Dioscorides and Galen, valerian has long been recognized for diverse therapeutic applications.[178] For example, in his *De materia medica* (ca. 60 C.E.), Dioscorides recommended that the root be prepared as an herbal tea to promote urination, treat pain in the sides, and "[draw] down the menses." He categorized the herb as warming and noted that its medicinal potency was revealed by the rather unpleasant smell of the dried root.[179] In 1597, the English physician John Gerard listed valerian root to treat urinary problems, jaundice, and "slight cuts, wounds, and small hurts." A liquid extract of the leaves, he wrote, was useful as a mouthwash or gargle to treat sore mouth and gums.[180] Several decades later, the popular herbalist Nicholas Culpeper outlined more than a dozen treatments involving valerian on its own and in combination with other substances, recommending it to cure coughs, expel phlegm, improve eyesight, and

FIGURE 14.23 Valerian: (*left*) flowers; (*right*) root.

remove thorns and splinters. In an era when periodic epidemics of infectious disease threatened the urban areas of Europe, Culpeper offered the helpful hint: "It is of special Vertue against the Plague," the foul odor of the root being repulsive to pestilence.[181]

Valerian's reputation gradually shifted toward a role in behavior and affect, and doctors recommended it for ailments (in nineteenth-century terms) such as hysteria, hypochondria, and chorea (tremors).[182] The English medical self-help book *The Working Man's Family Botanic Guide* (1852) called valerian "a nervine, and antispasmodic," useful to treat "nervous diseases."[183] By the twentieth century, valerian was less widely employed in medicine than previously but was known in informal medicine as a relaxant and sleep aid, particularly in Europe.[184]

The chemical profile of valerian root is highly variable, dependent on growing conditions and processing techniques. The constituents thought to be responsible for medicinal effects are volatile oils (0.2–2.8 percent in the root), including the

compound valerenic acid (figure 14.24).[185] Laboratory studies have demonstrated that valerenic acid and a closely related chemical from valerian can bind to, and enhance the activity of, a receptor for the neurotransmitter GABA.[186] Since GABA inhibits certain types of excitatory signals in the central nervous system, valerian extract is proposed to act as a calming agent. This hypothesis has been investigated in tests of anxiety in laboratory rats, and evidence is mounting that valerian extracts have such a property.[187] Clinical trials have sought evidence for anxiolytic effects among volunteers with anxiety or insomnia, and the results of such experiments are mixed. Reviewers of the clinical studies involving valerian have noted the diverse testing schemes, dosing, and

FIGURE 14.24 Valerenic acid.

source of valerian extract, along with other methodological concerns that (so far) obscure a conclusive interpretation of outcomes.[188]

With a long history in European medicine as a potent, though malodorous, herb, some of valerian's effect has now been described in neurochemical terms. Whether or not large-scale human trials of valerian extract yield definitive evidence of a calming effect in human beings, knowledge of valerenic acid's unique activities in the brain can shed new light on the basic circuitry of the nervous system, ultimately yielding a wealth of information for which modern science would have nature and generations of traditional medical practitioners to thank.

TURMERIC

Curcuma longa

Originating in South and Southeast Asia, turmeric is an herbaceous perennial now cultivated throughout the tropics. It grows to a height of about 1 meter, composed of numerous 50-centimeter-long flat, green, sword-shaped leaves and a spike of leaves and tubular pale-yellow flowers each about 5 centimeters in length. The underground portion, called the root, consists of a rhizome and roots that are used in cuisine, dyes, and medicine. Long valued in South Asia, the rhizome can grow to about 3 centimeters in diameter and 6 centimeters or more in length, and its orange-yellow flesh is eaten fresh, ground for use in flavoring dishes (turmeric being a key component of curries, for example), taken in medicinal food or herbal teas, and used as a pigment to color paper and textiles (figure 14.25).[189]

Widely esteemed for health-related properties in the traditional and folk medicines of Asia, its ancient use is evidenced in the written lore of India and China. At the inception of one of the major South Asian medical traditions, an ancient Sanskrit text suggests turmeric to treat the skin, slow graying hair, and—perhaps because of its color—ward

FIGURE 14.25 Turmeric root: (*left*) fresh; (*right*) dried and scraped; (*center*) ground.

off jaundice.[190] In China, turmeric first appeared in a seventh-century medical treatise, considered a warming herb capable of invigorating the blood and driving *qi* downward. Later, in 1596, the author Li Shizhen recommended turmeric to treat obstructions of *qi* and relieve pain.[191]

Turmeric entered the western European pantry during medieval times, considered a cheaper substitute for another precious imported spice: the brightly colored, subtly flavored female structures of the Near Eastern saffron crocus flower (*Crocus sativus*).[192] Turmeric was little employed in European medicine, where many considered it less potent than its botanical relative ginger (*Zingiber officinale*) and useful only as a dye.[193] Perhaps it is not surprising that Europeans found more value in turmeric as a pigment than as a medicine or spice because the combination of volatile oils and polyphenolic chemicals in the rhizome are liable to dissipate and become modified over time, which explains why freshly prepared turmeric is preferred by those seeking its flavor, aroma, and other health-related effects.[194]

The essential oil of turmeric makes up about 6 percent of the weight of the rhizome and contributes to the scent and flavor of the herb; a set of polyphenols (accumulating to 5 percent) collectively called curcumin is responsible for much of the color.[195] Curcumin is also thought to be responsible for many of turmeric's perceived medicinal effects. Considerable research has accrued in recent decades that shows curcumin to have numerous possible therapeutic targets.

In laboratory tests, curcumin polyphenols demonstrate antioxidant and anti-inflammatory properties, which might recommend them for age-related degenerative concerns and cancers.[196] Curcumin displays anti-inflammatory activity by suppressing a suite of enzymes that propagate pro-inflammatory signals and by blocking the expression of a key gene responsible for inflammation, cell proliferation, and other aspects of chronic disease.[197] The capacity to interfere with the inflammatory processes suggests curcumin as a possible agent against, for example, rheumatoid arthritis

and inflammatory bowel disease. It might also be of value in promoting wound healing and treating skin problems.[198] Furthermore, curcumin's inhibition of genes that regulate cell division make it an appealing candidate for the treatment of cancers, which are characterized by abnormal cell proliferation.[199] While active investigations in other areas seek to demonstrate whether curcumin might be a good candidate as a neuroprotectant, an antiviral, a cardiovascular protectant, and the like, few clinical trials have addressed its speculated therapeutic effects.[200]

One factor that may complicate curcumin's transition from laboratory to clinic is its hydrophobic nature (poor ability to dissolve in water). As a simple turmeric extract taken by mouth, very little of a curcumin dose passes into the bloodstream. There are some additives that might aid its uptake, but they are in the exploratory stages.[201] To render the active principles more effective, chemists have suggested altering the fundamental structure of the molecules into more water-soluble derivatives.[202] With the knowledge of thousands of years of traditional culinary and health practices, the experimental data of curcumin's prospects in the laboratory, a rational basis for modifying select compounds for increased activity, and a growing body of clinical literature, much more remains to be written on turmeric's medicinal legacy.

ALOE

Aloe spp.

The aloes comprise a group of hundreds of species of succulent plants adapted to survive in warm, dry conditions by storing water in thick, sword-shaped leaves, often protected by sharp, spiky leaf edges and tips (figure 14.26). The genus originated in Africa and diversified into shrub and tree forms now distributed throughout North and South Africa, Madagascar, the Arabian Peninsula, the Indian Ocean region, and the Mediterranean.[203] Many species have been employed in medicine, most commonly

FIGURE 14.26 Leaves of *Aloe vera*.

Aloe vera, a perennial cultivated variety that produces fifteen to thirty leaves that can range in size from 5 to 50 centimeters long and up to 10 centimeters wide at the base.[204] The pale-green leaves, sometimes bearing whitish spots and red streaks, and darkening with age, have been employed in medicine since ancient times.

Aloe was mentioned in medical writings of the Sumerians of ancient Mesopotamia (ca. 2100 B.C.E.) and the Egyptians (ca. 1550 B.C.E.).[205] When Pliny the Elder described aloe during the Roman period, he outlined an herb that he considered warming, "employed for numerous purposes, but principally as a purgative," and noted that it was one of the few available drugs that acted in this way while also being good for the stomach. He also recommended it, rubbed on the skin or made into an herbal tea, to allay headaches, and mixed with wine and worked onto the scalp to prevent baldness. Pliny further listed a series of external ailments treatable with applied aloe, including wounds, hemorrhoids, genital warts, and sores.[206] The Greek herbalist Dioscorides assured his readers that "it loosens the bowel and cleanses the stomach" when taken in water, is also useful for jaundice, and "is suitable for inflammations of the tonsils, for the gums, and for all conditions associated with the mouth."[207]

Aloe maintained currency in medieval European herbal practice and appeared in numerous manuscripts of the era. By the fifteenth and sixteenth centuries, reinvigorated trade throughout Europe and the Near East made aloe more widely available to the inhabitants of the cooler countries of northern Europe. John Gerard in 1597 noted that the British sometimes kept aloe as a houseplant, "hanged on the seelings and upper posts of dining

roomes," and characterized its juice as "good for many things." As a strong laxative, aloe was lauded by Gerard as "an enimie to all kinds of putrefactions, and defendeth the bodie from all manner of corruption," particularly helpful to expel intestinal worms. Like his Greco-Roman forbears centuries earlier, the English author listed several ways to employ aloe on the surface of the body, including to staunch bleeding and to heal sores—rather useful on the sensitive skin of the buttocks and "secret partes."[208] Doctors and folk healers long ago discovered two of the principal properties of aloe that have since been explained in biomedical terms: its capacity to stimulate intestinal contractions and to promote wound healing.

These physiological effects can be attributed to chemical compounds in two types of leaf extract (figure 14.27). One type of leaf product is the latex, a yellowish fluid that exudes just underneath the outermost layer of the leaf and is enriched in an assortment of molecules called aloin.[209] Intestinal bacteria metabolize aloin into its active form, aloe-emodin, which irritates the colon and causes muscular contractions. Aloe-emodin also interferes with the normal mechanism of water regulation in the intestine, increasing the water content in the fecal mass. These mechanisms render aloe latex an effective laxative, for which it was widely used until recently (figure 14.28). Aloe-emodin is also under investigation for a number of possible effects on cell proliferation and inflammation.[210] In 2002, the Food and Drug Administration banned the use of aloe latex in over-the-counter laxative preparations because of concerns about safety.[211]

In contrast to the latex, the clear, mucilaginous, juicy central portion of the leaf, called the leaf gel, is included in beverages and incorporated into lotions for external use. The active components of the gel appear to be the numerous complex carbohydrates that form a latticework inside and between the water-storing cells of leaf's interior.[212] Historically considered helpful in treating wounds, aloe leaf gel's carbohydrates (polysaccharides such as acemannan) and other components have been shown in laboratory tests to stimulate the immune

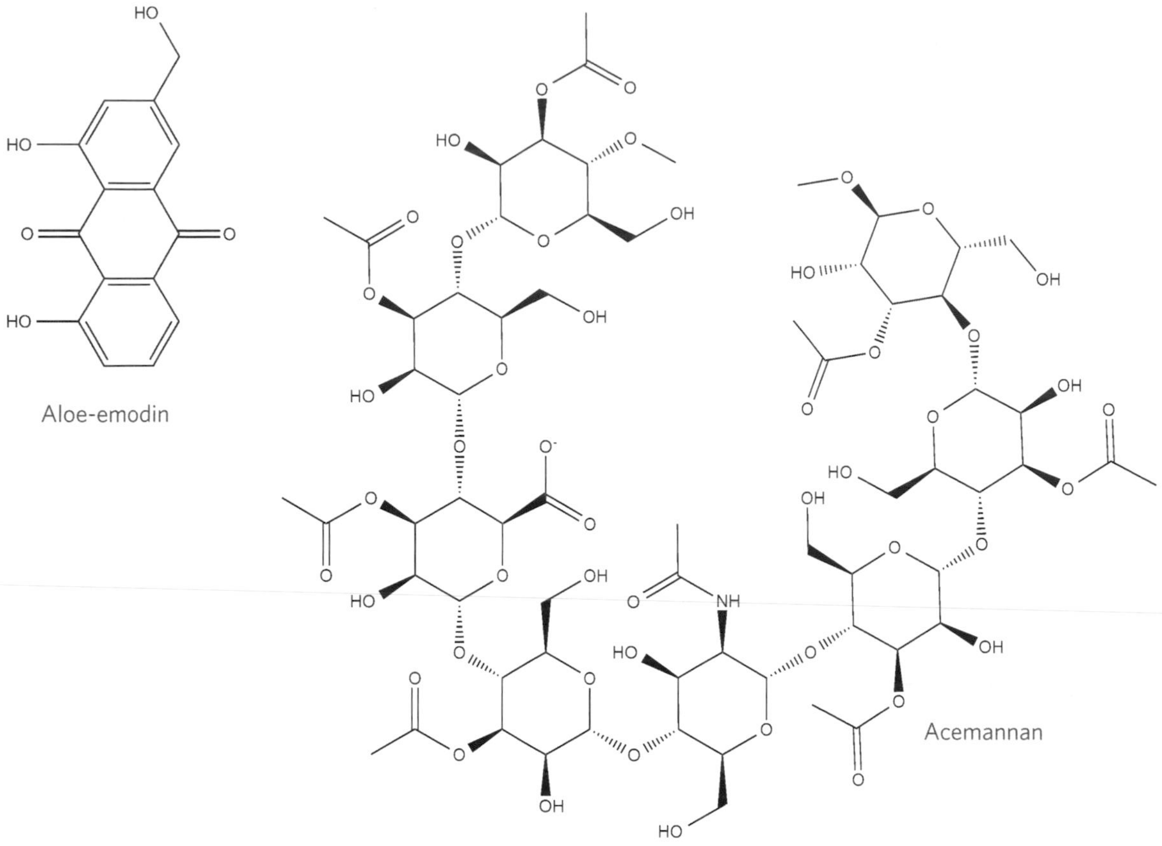

FIGURE 14.27 The active principles of aloe: aloe-emodin, a laxative compound generated in the intestine from aloin, a precursor in aloe latex; acemannan, one of the complex
carbohydrates in aloe leaf gel.

system, inhibit the growth of microbes, and thereby aid cuts, sores, and burns to heal more quickly.[213] Some people consider that aloe leaf gel, when taken orally, might reduce inflammation caused by ulcerative colitis and gastric ulcers, for example, and treat diabetes.[214] Since the methods by which aloe leaf gel is prepared can affect its chemical composition, and because clinical trials employ so many different testing protocols, the studies on human interventions using aloe have not generated consistent and easily compared data to demonstrate its therapeutic efficacy in wound healing and other applications.[215]

Clinical data yet to come should address whether the healing properties seen in the laboratory can be observed in humans: whether, for example, wounds heal more quickly when dressed with a preparation containing aloe gel, as compared with an inert placebo lotion. As aloe is already widely accepted as an ingredient in skin-care products, generally believed to soften skin and improve its health, there might also be value in its hydrating capacity, its texture, and the way it makes people feel to apply such an ancient herb to take care of themselves. Aloe's juice, Gerard wrote long ago, is "good for many things."

FIGURE 14.29 Ginger for sale at a street market in Hong Kong.

FIGURE 14.28 An advertisement for Beecham's Pills, a laxative that contained aloe. (From *Illustrated Sporting and Dramatic News*, March 3, 1888; National Library of Medicine, A028925)

GINGER

Zingiber officinale

Originating in South Asia and now cultivated throughout the tropics, ginger has long been valued in Asian cuisine and medicine and now also in Africa, the Caribbean, and elsewhere. The perennial plant reaches a height of about 1.5 meters, with numerous narrow, dark-green, sword-shaped leaves emerging alternately from leaf bases that sheath the stem and shorter inflorescence stems in the form of a spike of tubular orange and purple flowers.[216] The underground portion (gingerroot) is largely a branching rhizome of variable size that gives rise to

both roots and stems. The rhizome is harvested for its sweet, pungent spiciness and aroma, considered to be flavorful and healthy (figure 14.29).

The ancient Indian Sanskrit texts uphold ginger, endowed with special purifying properties, as "the great cure" and "the great medicine."[217] From its South Asian origin, ginger spread along the routes of commerce and took its place in East Asian and Mediterranean medicine.[218]

The Chinese employed ginger as a medicinal ingredient at least 2000 years ago, when gingerroot appeared in an early *materia medica*.[219] Classified as a hot herb, ginger has the capacity to promote *yang* in the body and disperse cold *qi*, according to traditional medical thought. In the Chinese pharmacological framework, it is often mixed with other herbs to modulate its effects and is used to treat abdominal pain and vomiting, cough, and some types of bleeding, among many other ailments.[220]

Ginger reached the Mediterranean region in antiquity, when traders operating between the Indian subcontinent and the Near East supplied the spice market with exotic products such as pepper

(*Piper nigrum*) and cardamom (*Elettaria cardamomum*). While the Greek herbalist Dioscorides did not quite know where ginger came from—he wrote that it grew in "Troglodytic Arabia"—he recommended it for its "warming and digestive properties." As he advised, "It gently softens the bowel and it is wholesome."[221]

Later European authorities elaborated on ginger's qualities, writers such as Gerard, whose herbal described dried gingerroot as hot and dry, in the humoral framework of medicine, but fresh or pickled ginger as hot and moist, with the particular attribute of "provoking venerie" as an aphrodisiac.[222] A French pharmaceutical encyclopedia of the late seventeenth century recorded that gingerroot was rarely used, except in powdered form as a flavoring called white spice. Northern Europeans, such as the Dutch and English, according to this source, consumed candied ginger or ginger marmalade from time to time as a warming agent, to improve digestion, and to prevent scurvy when at sea.[223] Ginger-containing beverages and syrups have persisted in the medical-culinary marketplace in recent centuries (figure 14.30).

FIGURE 14.30 An advertising poster for an elixir that contained ginger, ca. 1860. (Library of Congress, Prints and Photographs Division, LC-USZC2-291)

Gingerroot contains a complex assortment of chemical compounds that might act singly or in combination to produce its flavors and diverse medicinal attributes. Among the many chemicals extractable from the rhizome are an oleoresin (4–7.5 percent), composed of pungent chemicals (for example, phenolic compounds such gingerols and shogaols that accumulate on drying) and terpenoid volatile oils (1–3.3 percent).[224] Much of the current ginger research aims to test its capacity to reduce nausea and vomiting, such as accompanies gastrointestinal troubles, pregnancy, certain drug treatments, and motion sickness.[225] The mixture of ginger compounds being so flavorful and biting, it is a challenge to design proper placebo controls in such trials. Potential technical limitations aside, clinical evidence is gathering that ginger probably does reduce nausea and vomiting in diverse therapeutic settings.[226]

Gingerroot has also been suggested as an anti-inflammatory or analgesic herb, and clinical evidence for this application remains mixed.[227] With continued improvements in experimental design and a better understanding of potential biochemical mechanisms of action, ginger's ancient roles in health and cuisine may be extended in a new era of biomedical practice.

KAVA

Piper methysticum

From its Polynesian center of origin, kava was long ago dispersed by oceangoing peoples, eventually reaching islands across a vast portion of the Pacific, including parts of Micronesia, New Guinea, and much of Polynesia as far east as Hawai'i.[228] It is a perennial plant that produces many branched narrow stems bearing heart-shaped leaves and a tangled mass of medicinally useful roots underground (figure 14.31). Allowed to grow for five or

FIGURE 14.31 Kava.

more years, the plant can reach a height of up to 5 or 6 meters, but the roots are typically harvested from plants two to three years of age, at a height of around 2 to 2.5 meters.[229] Although kava occasionally grows small, spike-like flowering structures, the species is sterile and can be reproduced only through cuttings.[230] Therefore, it is thought to exist solely in cultivation.[231] Kava was domesticated perhaps 3000 years ago and now grows as dozens of varieties distinguished by coloration, growth habit, and medicinal properties.[232]

Kava's importance to social groups across the Pacific is recorded in legend and in the beliefs of people who employ the plant as a mediator of spiritual and community bonds. To many people of Oceania, kava is thought to be a gift of powerful ancestors or gods, sacred in origin and function.[233] The plant has most notably been incorporated into numerous traditional rituals, during which the roots are pounded or chewed, the resulting liquid collected and then consumed in customary tribal settings (figure 14.32).[234] Kava produces a sense

of relaxation and altered sensory experiences perceived as a means to communicate with ancestors.[235] It is also offered as a tribute to people of high social rank, shared between men to ease negotiations of a political or economic nature, and used to commemorate community events such as marriages, funerals, and initiation ceremonies.

In modern-day Pacific island cultures, kava is frequently consumed as a recreational beverage. For example, the islands of Vanuatu are dotted with small kava bars called *nakamals* where people gather, typically in the evenings, to consume freshly prepared kava.[236] In addition to kava's social and recreational roles, numerous medicinal uses have been recorded, including the treatment of female reproductive concerns, urinary problems, and many types of infections.[237]

European explorers of the eighteenth century and Christian missionaries of the nineteenth and twentieth centuries disapproved of the kava ritual and its accompanying inebriation. As they evangelized the peoples of the Pacific, they discouraged

FIGURE 14.32 Young women preparing kava in Samoa, 1916. (Photograph by A. J. Tattersall; Picture Collection, New York Public Library, Astor, Lenox and Tilden Foundations)

the consumption of kava, considering it an impediment to the moral development of the indigenous groups.[238] Despite the religious proscriptions imposed against kava in Oceania, the European and North American medical community of the early twentieth century brought the herb into its *materia medica*.[239] For example, the Parke-Davis pharmaceutical company recommended kava extracts to treat "gout, bronchitis, catarrhal affections" and "especially . . . acute gonorrhea." At a low dose, the drug was "a stimulant and tonic," according to the company experts, and at a larger dose, it produced "intoxication of a silent and drowsy nature accompanied by incoherent dreams."[240]

By the late twentieth century, kava production in many parts of the Pacific rebounded as self-governance developed and missionary zeal waned.[241] While kava's ritual and social uses remained important among local peoples, interest in kava rose particularly among Europeans and North Americans for its potential anxiety-reducing properties. Numerous companies imported large quantities of kava root for the manufacture of mood-altering supplements.[242]

Kava's effects on sensation, perception, and mood are attributed primarily to an assortment of fat-soluble compounds called kavalactones.[243] The chemicals exert a local anesthetic effect when absorbed through the skin and a general analgesic property when consumed orally.[244] In clinical trials, kava extracts have been shown to reduce anxiety among volunteer participants, an effect probably mediated by kavalactone binding to GABA neurotransmitter receptors in the brain.[245] Kavalactones also produce effects on the adrenergic neurotransmitter system.[246]

While kava has a long history of spiritual, social, and medicinal use in Oceania, along with a growing body of research indicating its mechanism of action and therapeutic efficacy, some questions have been raised about kava's environmental impacts and safety. With the revival of interest in kava on the Pacific islands during recent decades, farmers have expanded the cultivation of the plant at the expense of forested land, which threatens the islands' delicate ecosystems.[247] Furthermore, questions were raised about the safety of kava-containing products.

During the 1990s and early 2000s, Western regulators noted an association between kava use and liver damage among several dozen patients.[248] As a result, governments, including those of Great Britain, Germany, and Canada, banned the sale of kava products; in the United States, the FDA issued an advisory letter warning of kava's potential risks.[249] While the connection between kava constituents and liver damage remains unclear, some researchers speculate that differences in preparation method might account for possible health risks with imported kava. For example, it is possible that some of the dried kava material shipped overseas from Micronesian and Polynesian islands might not be of the same cultivated varieties as the kava consumed by locals and might include material other than the roots. Also, rather than preparing kava in a water-based drink, as is the Pacific custom, many Western companies extract the presumed active principles using organic solvents. These differences in preparation might account for differences in chemical profile, perhaps explaining an increased health risk.[250]

While researchers in the West continue to investigate the efficacy and safety of kava, it remains a plant of great cultural value in the Pacific. Many generations ago, people carried the plant from island to island, a living connection to a world of ancestors and gods, a mediator of social, spiritual, and physical health. Under the microscope of contemporary science, the herb's medical usefulness remains incompletely tested, and much more investigation awaits.

Thousands of years of tasting and sniffing, grinding and steeping, and trial and error have given humanity a set of medicinal plants that serve health in the form of spices, herbal teas, lotions, and foods. While some herbs declared their effectiveness early in human history, others came more recently into common use. Yet all have transformed the way people eat, drink, conduct

commerce, and view their health through diverse traditional medical systems. With the advent of a biochemical understanding of drug efficacy comes a new set of challenges. Of the many possible active principles in a root, leaf, stem, seed, or fruit, which ones might be responsible for physiological effects? Of the therapeutic properties gleaned from ancient texts, folk wisdom, and traditional practices, which might be supported by robust, controlled clinical trials? How should such plants and their extracts be regulated: as foods, dietary supplements, or drugs? The future of ancient medicinal plants in modern society expects that such questions be addressed in an integrated, multidisciplinary fashion.

Chapter 15

The Future of Medicinal Plants

A pharmacy technician at a hospital of Chinese medicine in Beijing preparing an herbal formula from powdered ingredients using a machine that automatically measures the correct quantities of each component and electronically guides her work. Red lights indicate which herbs to add to the prescribed mixture.

Medicinal plants have been a part of human life since the earliest times. To learn about this enduring relationship, modern-day scholars are unearthing new evidence of these ancient connections in nearly every imaginable discipline of study. Archaeologists are digging for signs of medicinal plant use at ancient sites of habitation to understand better when, how, and why such plants came to be used. Meanwhile, historians are examining original texts to decipher how these herbs were harvested, prepared, and valued in times long before our own, and linguists are reconstructing how such plant knowledge was communicated within and between societies. Anthropologists are documenting the ways people all over the world employ plants in health-related practices and how their beliefs give rise to culturally significant roles for plants. For their part, botanists are working out the hereditary relationships among plants, weaving genetic data and physical characteristics to recapture their evolutionary history and telling the story of those traits so useful to mankind.

For centuries, people have challenged their best technologies and analytic methods to understand the material basis of medicinal plant actions, and chemists continue to refine approaches to isolate and characterize the active principles of plants. Biochemists and pharmacologists are examining the synthesis and formulation of herbal extracts and mixtures while establishing the biological bases of such drugs' functions in cells, tissues, and living systems. Meanwhile, biomedical researchers are forging robust studies that aim to determine whether and how medicinal plants might treat disease and describe the basis of any side effects. Ongoing research into a large assortment of time-tested medicinal herbs owes its success to the knowledge built over centuries and transmitted to the present, along with new discoveries about their therapeutic potential. Likewise, as novel medicinal plants are being identified, they, too, become subject to a rich analysis from multiple perspectives.

Future advances in the study of medicinal plants will continue to follow the paths of multidisciplinarity, as several areas of investigation ultimately converge on the better understanding of how our medicinal plants came to be employed and how people can use them safely. The challenges ahead are substantial. The present era of population growth, rapid economic development in many parts of the world, unprecedented means of global communication and transportation, and environmental change gives impetus to document the many traditional medicines of the world and apply any new knowledge to address the most pressing of the world's health problems. As Western biomedicine accommodates the entry of herbs as therapeutic agents, it is increasingly important to pursue strong methodologies for assessing efficacy and risk. While some medicinal plants have already drawn much attention from researchers of various stripes, far more can be learned from the vast untapped potential of the plant kingdom. Collaboration across diverse disciplines, recognizing the value of traditional knowledge and the strengths of biomedical approaches, will yield a new array of plants with well-characterized effects on human health.

FOLLOWING ETHNOMEDICAL LEADS

It seems that nearly as soon as a people developed the ability to write, they began to document the ways they used plants as drugs. Ancient Egyptian medical papyri, Mesopotamian clay tablets, and Chinese texts on slips of bamboo all attest to the common human activity of communicating medical information.[1] Over the course of about 2000 years, through laboriously copied manuscripts and printed books, physicians and herbalists shared their knowledge in an ongoing conversation touching on theory and practice, folk wisdom and scholarly discourse, therapies and risks. These authors, now long passed, saw fit to contribute their accounts to a large corpus of medicinal plant knowledge. Their work tells of diverse ailments—some recognizable by today's medical framework, others not—and of the ways these people learned to treat them. There is no doubt that to these authors, plants were essential to health.

Ancient texts are occasionally criticized for relating medical ideas that have since been dismissed as irrelevant or unsafe by modern biomedical knowledge.[2] How could a medieval physician who advocates bleeding his patient's bad humors have anything to offer about herbal remedies? Why would an ancient Chinese author who recommends mercury-containing drugs (now known to be toxic) have worthwhile advice on medicinal plants? While it would be unrealistic to expect that an eleventh-century British herbal, for example, should be taken word for word as a guide to health in the modern day, it is still reasonable to consider that such a text was written to document the real ways that therapists tried to help their patients, who, in that place and time, suffered from real ailments and wanted to be healthy.[3]

Based on the premise that ancient texts record the ways that people prepared medicinal plants, assessed their benefits, and warned of their risks, such literature can be a great resource for modern-day investigations (figure 15.1).[4] At one level, simply compiling lists of medicines and therapeutic recipes into a large-scale database can allow associations and patterns to emerge that might be of interest to researchers. For example, are particular

FIGURE 15.1 Ancient texts can be used as a guide to selecting herbs to test using modern-day techniques. Authors long ago documented the numerous ways that people used plants for health and healing: (*left*) an apothecary's assistant preparing medicine as eminent scholars—including Dioscorides, Pliny the Elder, and perhaps Avicenna—converse; (*right*) an apothecary and a physician in a medicine shop. ([*left*] Woodcut from Hamsen Schönsperger, *Gart der Gesundheit* [1487]; Peter H. Raven Library, Missouri Botanical Garden, St. Louis; [*right*] illumination from Mattheus Platearius, *Circa instans* [early fourteenth century]; British Library, Sloane 1977, fol. 49v)

herbs mentioned in texts from diverse geographic or cultural settings for use against a certain type of ailment? It is possible that early physicians independently came upon an effective treatment that might be of interest to modern-day applications. Are certain herbs regularly paired together in ancient therapeutic recipes? Perhaps their formulation in this way is evidence of a beneficial combination that might be investigated today at the biochemical level.[5]

It is also possible to query such a database of ancient medicinal plant uses with hypotheses that can be addressed with a variety of laboratory and clinical approaches. For example, ancient accounts of laxative plants such as aloe (*Aloe* spp.) and senna (*Senna alexandrina*) can now be explained in terms of specific stimulatory chemical agents acting in the human intestine. Since the parasitic condition of malaria affected people across a wide expanse of the Old World, searching for mentions of malarial symptoms (intermittent fevers) in ancient texts might help find useful plants to supplement the natural and synthetic medications presently available.[6] Indeed, such an approach of mining classical Chinese texts yielded artemisinin, an effective antimalarial from sweet wormwood (*Artemisia annua*).[7] Ancient authors provide useful detail on both the possible benefits and the risks of medicinal plants. For example, the Greek herbalist Pedanius Dioscorides (ca. 40–90) carefully noted the toxicity of thorn apple (*Datura stramonium*, also known as jimsonweed [figure 15.2]), saying that a dose of 1 *drachma* (about 3.5 grams) of its root, drunk with wine, produces "not unpleasant fantasies," a quantity of 2 *drachmai* "drives a person out of his senses for up to three days," while a dose of 4 *drachmai* "kills."[8] The

FIGURE 15.2 Thorn apple. (Illustration from Joseph Roques, *Phytographie médicale* [1835]; BIU Santé, pharma_105443x04)

legacy of such medical documents includes meticulous descriptions of the therapeutic applications as well as warnings about the health-related properties of plants.

Yet the use of ancient texts to identify plants for biomedical testing has some limitations. For one, a great many of these sources remain largely unexplored, sharing their secrets with those few scholars with the ability to read their script, whether, for example, classical Chinese, Old English, Latin, or Greek. Such works are being translated by historians of medicine, opening up new avenues of research into the medicinal plants they document.[9] In many cases, however, the efforts of translation cannot span the chronological and cultural gap separating the modern scholar from the ancient author. So many Assyrian medical recipes excavated from Ashurbanipal's (685–627 B.C.E.) library contain references to plants that have eluded translation.[10] In the Egyptian papyri, as well, hide a large number of useful plants, for which there is no certain modern botanical equivalent.[11]

Descriptions of illness also depend on concepts of health prevalent in the time at which they were written. While today's "cancer" refers to an uncontrolled, malignant growth of tissue, ancient authors wrote of tumors, swellings, lesions, abscesses, and pain.[12] Are the latter symptoms indicative of the former ailment? If ancient texts discuss an herb's role in the humoral framework, how can such an understanding be adapted to a biomedical way of thinking? It might not be possible to achieve such a correspondence, and the physicians and patients who could clear up the uncertainty are no longer available for questioning. Despite these challenges, ancient texts have provided a rich base of evidence for the biochemical study of potentially useful medicinal plants and, in the hands of historians and linguists with skills to make these past contributions available to a large, modern-day audience, will give rise to many fruitful investigations.

While ancient medical texts can inform modern-day investigations of the physiological properties of herbal preparations, such documents generally record only the knowledge of those privileged to publish. In regions without a written tradition, ethnographers have learned from the expertise of practitioners by studying with them, living among their people, and describing how they employ plants in medicine.[13] Such studies can often take months or years, as generations of practical knowledge, embedded in a cultural setting, cannot be easily and quickly transmitted to a foreign visitor. Indeed, anthropologists who undertake these projects often try to integrate themselves into communities, learning prevalent languages and experiencing the pace and activities of local life as an insider.

This type of approach to documenting traditional knowledge of plants is worthwhile among people without a written heritage as well as in areas with a robust literature on herbal medicines, such as Europe and East Asia. Anthropologists are sensitive to the multiple levels of medicinal knowledge deployed in a complex society and know that many experienced individuals might not have the resources or inclination to commit their expertise to ink. For example, the plant-based treatments employed by midwives and folk-medicine experts might not be well represented in the medical literature of a region where a predominant, scholarly medical system produces the bulk of the written output. It is therefore important to learn about medicinal plants in all the cultural settings in which they are applied.

As economic development and transnational acculturation have come to touch even the remotest parts of the globe, many societies that had long relied on herbs in their traditional medical practices are adopting Western biomedical treatments. Furthermore, the move to new standards of hygiene, housing, religious practice, and employment has further eroded the close contact many groups maintained with medicinal plants. As tribal languages disappear and elders are no longer able to train their successors in the exercise of generations of accumulated knowledge about the local natural world, volumes of botanical wisdom vanish.[14] While researchers can do little to slow the evaporation of knowledge occurring as traditional societies give way to the forces of globalization, ethnobotanists are preserving as much as they are able, documenting what they learn, and making available to a broader community information that might soon disappear.[15] By cataloging the

diverse medical uses of plants among so many people and describing their cultural contexts, ethnobotanists are providing a rich resource for those working in many other disciplines. Such detailed accounts are useful to researchers looking for possible novel therapies.[16]

There are also challenges in communicating the outcomes of ethnobotanic projects to an audience working according to biomedical conventions. At one level, most societies employ illness concepts that do not align with the categories of Western biomedicine.[17] In many areas, people believe that supernatural entities play a role in health, and therefore people may resort to certain plants to appease offended spirits and heal their ailments. How can such properties be communicated usefully to investigators who have no equivalent disease concept? If a plant is said to have a heating or cooling nature, how can this information be used by biomedical researchers? Indeed, even in countries where Western medicine predominates, such as North America and Europe, most people who use herbs in their health-related practices converse in a vocabulary that makes sense to them, and rarely the precise language of formal medical training. Whether the outcomes of ethnobotanic study are presented using the terminology and conceptual framework of the traditional culture or "translated"—that is, converted judiciously into Western biomedical jargon—investigators must bear in mind the differences in the ways people conceptualize and treat their health.

Spanning the distance between local, culturally informed plant uses and the white-coat laboratory requires a language that accurately captures the former and is intelligible to the latter. While conveying the nuance of traditional illness concepts to researchers working in a world of scientifically defined disease agents is one aspect of this discourse, it is equally important to identify plants in a way that is useful to the broader community of researchers. With a multitude of local plant names, such a task is not as simple as asking an experienced informant to list the herbs that he engages in his practice. For example, the Lumbee Indians of North Carolina call the striped prince's pine (*Chimaphila maculata*) by the names rat's vein, pip, and lion's tongue, and botany manuals also list pipsissewa and spotted wintergreen among its common names.[18] Moreover, the common names of some plants might refer to more than one Linnaean species. For example, plants called snakeroot or chamomile are known botanically by several different scientific names.[19] Researchers relying on only vernacular names might easily mistake one plant for another.

To minimize such possible confusion, field workers conducting surveys of medicinal plant use record as much botanic information as possible, which usually includes a physical specimen of relevant plants. Together with local names, date, and precise geographic coordinates, several pressed examples of the stem, leaves, and flowers, for example, of a plant can be deposited in research herbaria as a permanent reference sample (a voucher specimen) of the species identified in the field (figure 15.3).[20] By documenting plants

encountered in this way, botanists can assign Linnaean binomials and help ensure an accurate record. In the age of DNA, samples of genetic material also help laboratory researchers determine the identity of plants encountered in traditional medical settings. In this way, publications utilizing ethnobotanic information can denote plants unambiguously and support the type of interdisciplinary collaboration that characterizes the field.[21]

The study of medicinal plants in ancient texts and in their many cultural settings supports a long-standing effort directed at the identification of potentially novel drugs. (At the same time, scholarly work on ancient medical sources and among modern-day people enhances humanity's knowledge of its history, traditions, and cultural diversity.) The leads uncovered in humankind's past experiences with medicinal plants and the countless traditions involving medicinal plants around the world support an agenda of bioprospecting on the premise that human experience has generated hypotheses to test in the laboratory. Such a strategy is hardly novel; after all, it was the pioneering work of the English physician William Withering (1741–1799) that characterized the pharmacological properties of a folk remedy involving foxglove (*Digitalis purpurea*), resulting ultimately in the identification of an active principle, digitalis, capable of treating heart disease.[22] Today, biomedical researchers are convinced that nature has many more undiscovered pharmaceuticals locked away in the molecular storerooms of plant cells.

The alternative to deriving medicines from natural sources is to synthesize chemically and test large numbers of molecules in the laboratory for effects on enzymes, cells, and tissues of therapeutic interest. This approach, called high-throughput screening, has certain benefits, in that researchers need not concern themselves with sourcing natural materials and

FIGURE 15.3 An herbarium specimen of thorn apple. (Courtesy of United States National Herbarium [US] 1689278)

High-throughput screening

The systematic, rapid testing of millions of (frequently synthetic) chemicals for biological activity using computer-aided experimental techniques

dealing with their complex, challenging chemistries. But high-throughput screening of synthetic molecules has a low "rate of return" in terms of prospective new drugs—a "hit rate" (that is, the number of potential novel drugs discovered per synthetic molecules screened) of less than 0.001 percent is estimated in high-throughput screens.[23] By deriving potentially active compounds from natural products and using chemistry to modify their structure and improve their activity, many pharmaceutical researchers are instead taking natural molecules as raw materials in the process of drug development, ultimately using some high-throughput technologies to identify the most active drug candidate molecules.[24]

Therefore, cultural and historical leads enrich the pool of candidate species that might be subjected to drug development programs. While such investigations might identify one or more active principles as candidate pharmaceutical agents, certain rational chemical modifications undertaken in the laboratory might yield even more potency or other desirable physiological properties.[25] For example, this approach has been used to generate the semisynthetic antimalarial agents chloroquine and mefloquine, more efficacious variations of the Peruvian fever tree's (*Cinchona* spp.) active principle, quinine. Likewise, as the *Plasmodium* parasites responsible for malaria are gradually developing resistance to the drug artemisinin, derived from the sweet wormwood (*Artemisia annua*), pharmaceutical researchers have modified its structure to produce active analogs, including the molecule OZ277 and dimeric artemisinin (figure 15.4).[26] Thus many modern-day drugs retain at their core the chemical signature of nature.

Useful medicinal plant knowledge comes from many sources, whether historical documents, folk wisdom, or formal traditional medical practices. It has been the work of countless scholars working in multiple disciplines to record, analyze, interpret, and ultimately disseminate a wealth of botanical information now available to a wide community, and there is more still to do. As the tools of chemistry can be applied to understand better the molecular constituents of herbs and their extracts, so are

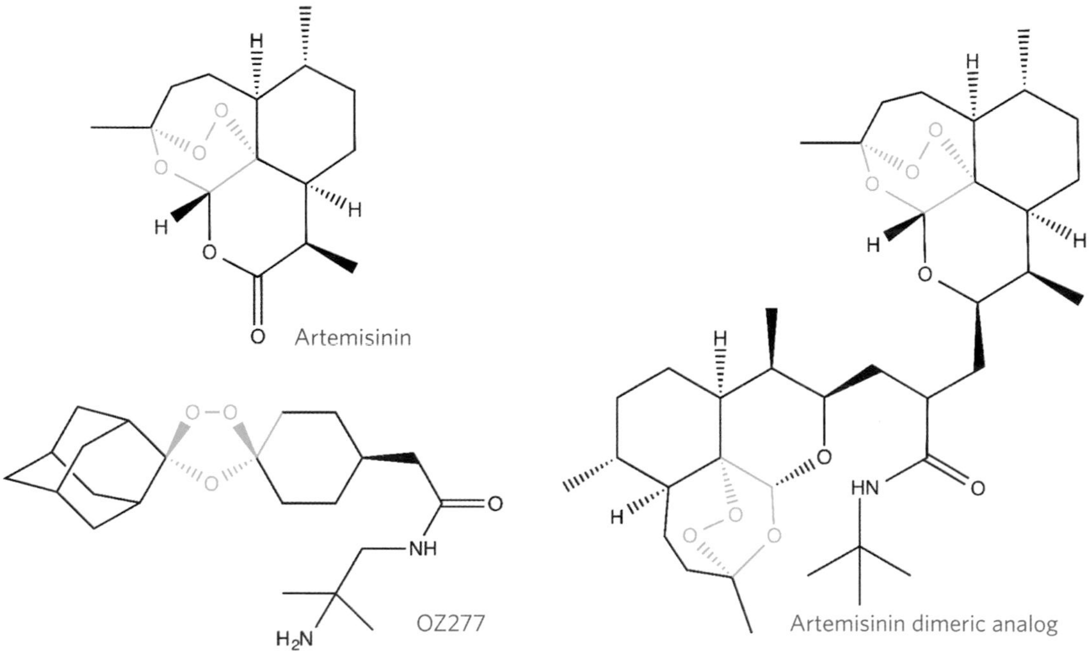

FIGURE 15.4 Natural products as a basis for synthetic drug development: artemisinin, an antimalarial compound from sweet wormwood; semisynthetic artemisinin dimeric analog; synthetic OZ277. The possible active sites are shown in green.

EVOLUTIONARY RELATIONSHIPS AMONG MEDICINAL PLANTS

Combining the nuanced analysis of plants' developmental patterns with new insight into their genetic heritage through DNA studies, botanists are constructing ever more accurate family trees (phylogenies) that place the hundreds of thousands of plant species among their most closely related kin. This increasingly detailed knowledge of plants' evolutionary histories has allowed medicinal researchers to examine hypotheses about the origins and geographic distribution of species with possible health-related effects.

Closely related plants often have similarities in structures such as flowers or leaves, an outward manifestation of shared ancestry. Related plants also frequently exhibit certain chemical similarities, such as the capacity to produce particular types of medicinal compounds. For instance, it was observed long ago that many plants of the nightshade family (Solanaceae) possess the biochemical pathways to produce medically active alkaloids. Nightshades such as tobacco (*Nicotiana* spp.), mandrake (*Mandragora officinarum*), angel's trumpet (*Brugmansia* spp.), and henbane (*Hyoscyamus niger*), although differing in growth habit and occurring in distinct parts of the world, have all been incorporated into traditional medical practices, presumably because of their potent psychoactive compounds.[1] Likewise, the globally distributed mint family (Lamiaceae) includes numerous medicinal and culinary plants, such as sage (*Salvia* spp.), horehound (*Marrubium vulgare*), lemon balm (*Melissa officinalis*), and the chaste tree (*Vitex agnus-castus*).

Although physically diversified to occupy the countless niches of the world's ecosystems, plants of a shared lineage often retain certain biochemical characteristics as part of their genetic heritage. It is therefore not surprising that human cultures have independently discovered the medicinal properties of plants separated by oceans yet bound together by their close ancestry. Some plant families appear to be greatly enriched in species that are used for health-related purposes in widely disparate places, a strong suggestion that certain species should be prioritized for further laboratory study of possible active principles.[2]

While the family trees of numerous medicinal plant species can provide hints to their possible chemical offerings, there are also some examples of convergent evolution, where particular medically relevant characteristics came about separately in species not closely related. For example, plants of numerous families can synthesize useful volatile-oil constituents such as eugenol, including clove (*Syzygium aromaticum* [myrtle family]), basil (*Ocimum basilicum* [mint family]), and bay laurel (*Laurus nobilis* [laurel family]). Similarly, many dozens of plant species across several families produce the alkaloid caffeine, including coffee (*Coffea* spp.), tea (*Camellia sinensis*), yerba mate (*Ilex paraguariensis*), and cola (*Cola* spp.).[3]

1. Tobacco produces nicotine; mandrake, angel's trumpet, and henbane produce tropane alkaloids such as atropine and scopolamine. The nightshades are also known for their diverse and useful terpenoid compounds, such as capsaicin, from chili pepper (*Capsicum annuum*), and lycopene, from tomato (*Solanum lycopersicum*).
2. Feng Zhu et al., "Clustered Patterns of Species Origins of Nature-Derived Drugs and Clues for Future Bioprospecting," *Proceedings of the National Academy of Sciences USA* 108 (2011): 12943–12948; C. Haris Saslis-Lagoudakis et al., "Phylogenies Reveal Predictive Power of Traditional Medicine in Bioprospecting," *Proceedings of the National Academy of Sciences USA* 109 (2012): 15835–15840.
3. Edward O. Kennedy, *Plants and the Human Brain* (Oxford: Oxford University Press, 2014), 98.

particular methodologies brought to bear on the questions of clinical efficacy and safety.

BUILDING A BASE OF EVIDENCE

Medicinal plants have served many roles in their diverse cultural and historical settings—roles that cannot be easily aligned with Western biomedical concepts of health. In contemporary China, for example, practitioners of Chinese traditional medicine believe that magnolia (*Magnolia officinalis*) bark is a warming herb that "regulates the *qi* and directs it downward."[27] In sixteenth-century England, the herbalist John Gerard (1545–1611?) recommended that the snapdragon (*Antirrhinum* spp.) be worn on the body as an amulet to protect the bearer against bewitchment and be soaked in water to treat the eyes, if they are tearing from "a hot cause."[28] In many parts of the world, people regularly take time to drink coffee (*Coffea* spp.)

or tea (*Camellia sinensis*) with friends and family, and it helps them affirm their social connections. Neither *qi*, nor bewitchment, nor hot causes of illness, nor social ritual have currency in the biomedical framework that attributes health to the coordinated interactions of molecules, cells, and structures in an anatomical machine.

Although Western biomedicine has become the dominant approach to health care in the past century, originating in Europe and North America and subsequently adopted in much of the world, many people still value their traditional and folk practices. In reality, Western medicine coexists with local, indigenous medicine throughout the world. The scholarly medical traditions of East Asia, Europe, India, and elsewhere present consistent theoretical bases for maintaining health and addressing illness, replete with their repertories of herbs and other treatments. There are plants thought to have properties relevant to a prevailing medical framework, such as the Greek humoral system or indigenous Chinese medicine. Incorporated into medical texts over centuries, such plants are taken for their perceived heating or cooling properties, for their dryness or *qi*-directing ability, and so forth. The collective experience of practitioners and patients helped assemble a set of herbs with qualities found useful in treating illnesses and promoting health according to the tenets of these diverse scholarly traditions.

Meanwhile, folk medical practices and religious medicine employ plants in their herbal teas, healing salves, and ritual concoctions. These medicines resulted from many years of trial and error, in which practitioners discovered techniques they thought worked and avoided those they suspected were dangerous. By noting the apparent effects of various herbal combinations and differences in preparation, they shaped their techniques to suit the needs of their patients and profession. In many parts of the world, the vast folk medical experience is passed orally between the generations, a great part of it disseminating extensive encounters with plant-based remedies.

The long history of herbal medicine and refinement of formulas across generations notwithstanding, until the development of the scientific

method, there was no way to determine objectively whether a given treatment was responsible for causing a particular physiological outcome. The scientific method, with its hypothesis-centered, experimental approach, is considered the most reliable way to examine therapeutic efficacy. Thus many investigators are subjecting medicinal plants and their extracts to laboratory assays and clinical trials to look for possible physiological activities. Yet such projects have their challenges. How should tests of medicinal plants be designed to yield the most useful results? How can effects described in traditional medical terms be examined using biomedical methodologies? What physiological mechanisms might explain the activities of plant-based treatments? The experimental sciences draw in ethnographic analyses, chemical methods, biological techniques, and clinical approaches collaboratively to address such questions.

Informed by historical and cultural knowledge, chemists and biologists since the nineteenth century have established clearly that active principles can be isolated from medicinal plant tissues and their effects on the human body rationalized in mechanistic terms. Morphine, they discovered, accumulates to a high level in poppy (*Papaver somniferum*) latex and is responsible for opium's effects on the body, largely by activating opioid receptors in the brain. They explained that caffeine is present in coffee, tea, and other plants and that it interacts with the adenosine receptor in the nervous system. It is now possible to describe precisely the effects of morphine on sleepiness and caffeine on alertness in ways that were not imaginable before the advances offered by chemical analysis and cell biology. Having isolated the active principles, modern-day pharmacists can formulate treatments of known, precise doses and predict more reliably the physiological outcomes. Yet many of these advances came without recourse to clinical trials, without applying the scientific method.

Because the effects of medicinal plants such as coca (*Erythroxylum coca*) and hemp (*Cannabis sativa*) are so quickly perceived by human subjects, early investigators did not doubt that these drugs produced the marked responses associated with their consumption. For many more herbs, however, the link of causality between medicinal plant use and physiological response has not been so clear. For example, if a patient regularly takes an extract of purple coneflower (*Echinacea* spp.) during the winter and does not catch a cold, how can one determine whether such an outcome is truly attributable to the herbal supplement? It is possible that the patient might also avoid illness without taking the purple coneflower treatment. The scientific method addresses such questions of therapeutic efficacy by comparing the outcomes of defined, potentially active treatments with the outcomes of parallel, pharmacologically inert treatments. In the clinical-trial setting, the more reliable of such experiments evaluate large numbers of subjects who have been randomly assigned to receive either the study drug or a placebo, and neither the patients nor the direct investigators are aware of their group placement.[29]

Herbal preparations are now regularly subjected to the robust scientific methodology that allows researchers to determine whether particular ascribed medicinal effects are borne out by data. While many trials are conducted to high standards of experimental design, some efforts at clinical testing could be improved.[30] At one level, investigators should choose their herbal treatments carefully and rationally. Various plant parts (root, shoot, leaf, flower, fruit, seed) can differ significantly in chemical makeup, and constituents can change across the growing season, in different geographic locations, and according to cultivation techniques.[31] Furthermore, the inherent genetic diversity of plants means that the assorted cultivars or regional varieties, called chemotypes, can differ in their distribution and levels of medically active compounds. Ultimately, research that carefully selects and describes the plant material under investigation is more easily compared with others and replicated. Furthermore, the form of plant-based treatment (whole herb, extract in solvent), route of delivery (capsule, injection, nasal spray, lotion, herbal tea), dosing scheme (once daily, twice daily, weekly), and trial duration influence the possible outcomes of the project and the ways that its results might

Chemotypes

Varieties of a single species that differ in chemical constituents

be contextualized. Placebo treatments should be credible in terms of flavor, smell, color, and other properties that might otherwise allow a patient to determine his or her group assignment. Additionally, the criteria by which patients are admitted to (or excluded from) the study can affect the observed results. At the level of outcomes, investigators must purposefully choose the ways they wish to measure the patients' possible response to treatment, using appropriate clinical and laboratory assays yielding quantitative data. In summary, there are many variables that demand an herbal researcher's attention in the conduct of clinical trials.

Quantitative data

Numerical experimental results (blood pressure, temperature, cell count, enzyme activity, and so on) that can be subjected to robust statistical analysis when collected from many test subjects

Many of these considerations are not unique to medicinal plants research but instead apply to clinical trials as a whole. To promote a strong and transparent methodology in clinical trials, many scientists have advocated the broad adoption by researchers of a basic set of experimental-design criteria. Since scientific journals are one of the primary outlets for reporting the results of clinical trials, the implementation of such standards by journals' editorial boards has influenced the way scientists design, conduct, and report their clinical trials.[32] Through improved standards of clinical trials and adherence to a set of global norms on the interpretation of experimental outcomes, herbal treatments are entering the realm of evidence-based medicine, the prevailing philosophy that demands that health-care decisions be grounded in the results of well-designed research yielding quantitative data. (That said, critics of evidence-based medicine point out that its emphasis on research findings diminishes the role of practitioner experience in health-related decisions. These critics also contend that forming clinical expectations from statistically derived trial data treats patients as uniform beings and ignores natural biological variation.)[33]

It is generally believed that larger clinical trials (having a greater number of subjects per treatment group) can produce more meaningful results than smaller ones, and statisticians have developed a methodology to combine the results of small trials into a larger, presumably more valid meta-analysis. Employing stringent inclusion criteria, researchers utilizing this approach search the scientific literature for trials involving a particular intervention on a therapeutic target of interest and select those that meet certain conditions.[34] For example, a project like this might collect all the available reports on trials of the extract of the South African geranium (*Pelargonium sidoides*, also known as umckaloabo) against acute respiratory infections.[35] Then the investigators might exclude those with poor design (according to established standards and their judgment) but otherwise include a variety of different plant preparations, data-collection strategies, and health-related endpoints in a wider analysis. For example, some of the included trials might test herbal tablets; others might test liquid preparations. Some might measure different aspects of respiratory disease—such as cough, sore throat, or fever—and track various outcomes, such as severity or duration, in ways that are difficult to compare. Meta-analyses can generate conclusions based on a much broader set of conditions and a larger number of subjects than the original trials, but the value of such projects depends to a great extent on the quality of the literature employed and the stringency of selection criteria. "A meta-analysis can show anything you want to show," one scientific observer noted. "Junk in, junk out."[36]

While clinical trials and their meta-analyses can lend support to the potential efficacy of herbal medicines, pharmacologists are establishing how the array of chemical constituents from plants interact with the body's systems. At the most basic level, assays can be performed in test tubes to look for certain properties in plant extracts: antioxidant activity, for example, or the ability to inhibit a metabolic enzyme. Such tests can help investigators establish the likely cellular targets of medicinal plant compounds and the way they might cause a physiological change—that is, the mechanism of action. By carefully establishing how specific plant-derived agents interact with molecular targets in the body, isolating active principles, and demonstrating clinical efficacy in human trials, medicinal plant researchers are advancing their work to take on some of the most challenging health problems.

Among the most important herbal medical discoveries of the twentieth century were a series of anticancer agents identified by the National Cancer Institute, many in massive screens looking for activity among thousands of plant extracts from all over the world. This project, which lasted from the 1960s to the 1990s, identified the vinca alkaloids vinblastine and vincristine from the Madagascar periwinkle (*Catharanthus roseus*) and paclitaxel (sold as Taxol) from the Pacific yew (*Taxus brevifolia*), among many others (figure 15.5).[37] The vinca alkaloids are commonly used in combination-chemotherapy regimens against leukemias, lymphomas, and breast and lung cancers, for example; paclitaxel is effective against breast, ovarian, non-small-cell lung cancer, and Kaposi's sarcoma (figure 15.6). These agents work by binding to the molecular machinery of cell division, preventing the uncontrolled cell proliferation that characterizes cancers.[38] Be-

FIGURE 15.5 Leaves of yew, a source of the cancer drug paclitaxel and its precursors.

cause the active constituents have been isolated, purified, and demonstrated effective by clinical trials, they were registered as drugs by the Food and Drug Administration and are available to patients by prescription. Since their discovery, chemists have learned to synthesize them from simpler chemical building blocks and make structural modifications to improve their effectiveness, ensuring that these agents—and many more such plant-derived cancer drugs—remain part of the armamentarium for a long time to come.

While the vinca alkaloids, paclitaxel, and other pharmacological agents can act as isolated

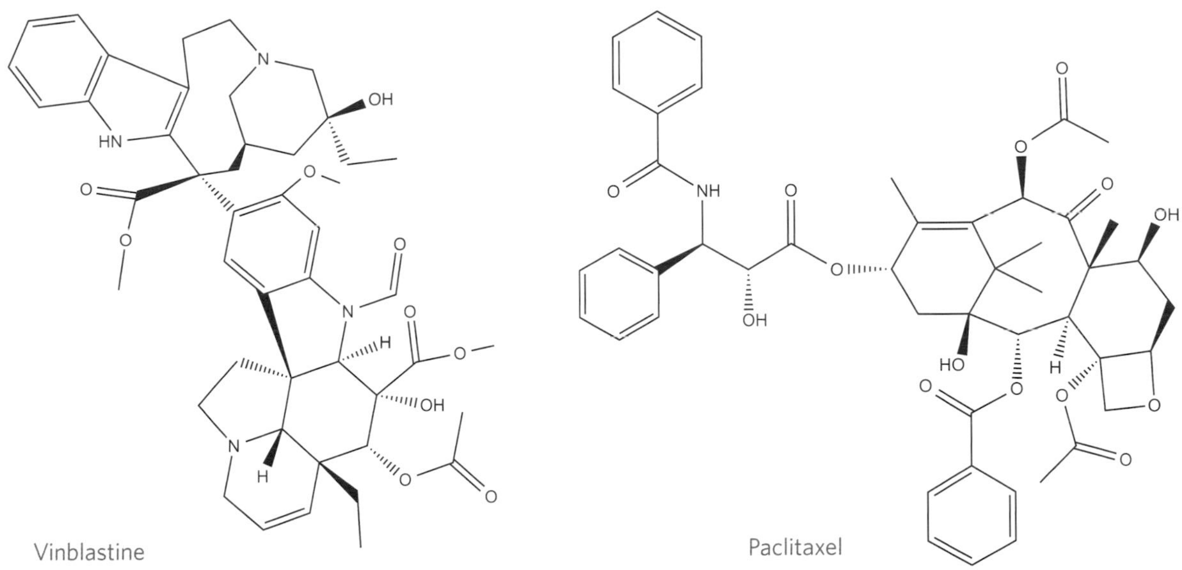

Vinblastine

Paclitaxel

FIGURE 15.6 Anticancer agents discovered in large-scale screens: vinblastine, from the Madagascar periwinkle; paclitaxel, from yew.

chemicals, it is also possible that combinations of active compounds might together have a physiological effect not present to the same degree in purified single compounds. This concept—"the mixture makes the medicine"—has led plant pharmaceutical development in a novel direction.[39] Rather than seek to identify solitary active principles for clinical trials, researchers can instead purify and thoroughly characterize an assortment of chemicals that together have an effect on health, demonstrable in clinical trials.[40] The first such botanical drug, a mixture of polyphenolic compounds from green tea effective against genital warts, received FDA approval in 2006.[41] (Unlike foods and dietary supplements, registered botanical drugs claim to treat disease.) The green tea–derived drug, called sinecatechin (sold as Veregen), is dispensed by a physician's prescription.

Another compelling example of this approach is the botanical drug crofelemer, from the South American tree *sangre de grado* (*Croton lechleri* [figure 15.7]). Also called *sangre de drago*, the tree produces a red latex that gives the plant its name: dragon's blood. Indigenous people in Ecuador, Colombia, Peru, Bolivia, and Brazil use its bark and stem exudate as a treatment for diarrhea and

dysentery, in a vaginal bath before and following childbirth, for treatment of intestinal and stomach ulcers, and to stop bleeding and heal wounds, among other applications.[42] The extract crofelemer is a mixture of polyphenolic compounds that has been found effective against diarrhea by blocking the release of water from the cells lining the intestine (figure 15.8).[43] The drug has been approved to treat patients suffering from loose stools as a result of HIV/AIDS antiretroviral therapy.[44]

FIGURE 15.7 The bark of *sangre de grado*, for sale at an Amazonian market, exudes a red latex when cut.

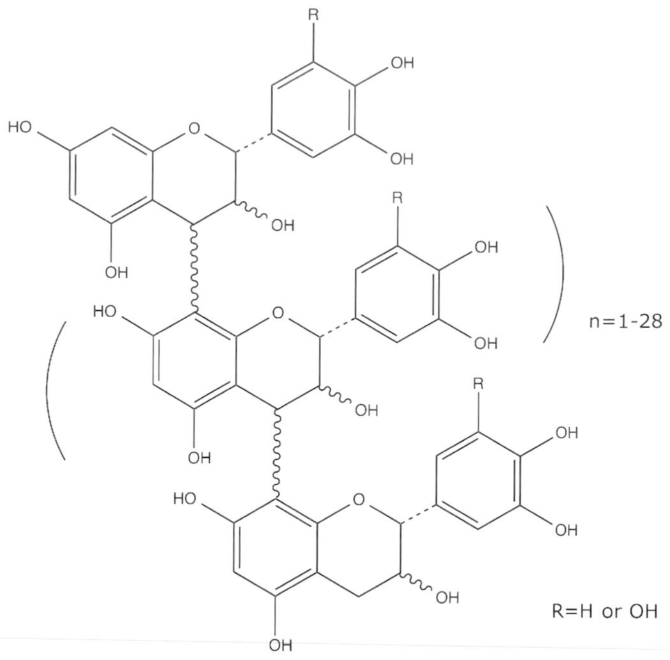

FIGURE 15.8 Crofelemer, an antidiarrheal botanical drug from the bark of *sangre de grado*. Crofelemer is a mixture of polyphenolic compounds.

Considering the sheer complexity of just a single plant's chemical makeup, consisting of thousands of distinct molecular structures, it is no wonder that many efforts at developing new herbal pharmaceuticals center on the isolation and characterization of a very small number of active constituents. Lone chemical constituents can be more simply characterized and tested than combinations of many compounds together in a single preparation. Accounting for the chemical differences among plant parts, variations of processing, the blending together of several different herbs, and other common herbal medical practices, the challenges to identify possible therapeutic compounds are formidable. Yet many plant researchers recognize that mixtures of chemicals might be more effective than single agents. This scenario, where two or more constituents produce a far greater effect when administered together than separately, is called pharmacological synergy.[45]

One form of synergy can exist when several distinct chemical compounds in a single medicinal plant work together to produce their physiological effects. For example, the combination of Δ9-tetrahydrocannabinol, other cannabinoids, and a variety of terpenoid compounds in hemp (*Cannabis sativa*) is thought to offer advantages over individual constituents in the treatment of pain, inflammation, anxiety, and a number of other conditions.[46] Synergies are also evident in herbal medicines formulated from more than one plant.

One such synergistic pharmacological relationship occurs in a medicinal beverage prepared by members of the Shipibo, Tukano, and other Amazonian groups for spiritual-healing rituals. The drink, called ayahuasca, is carefully concocted by experienced shamans and contains the ayahuasca (*Banisteriopsis caapi*) vine, chacruna (*Psychotria viridis*) leaves, and other plants. When ingested, the mixture gives rise to powerful visions of supernatural animals, spirits, and complex geometric patterns. The ingredients of the ayahuasca drink are now known to act together to produce these experiences: chacruna leaves contain the chemical dimethyltryptamine (DMT), which can activate receptors for the neurotransmitter serotonin in the brain, and the ayahuasca vine contains harmaline, which inhibits the enzyme monoamine oxidase in the digestive tract and thereby allows DMT to enter the bloodstream, cross the blood–brain barrier, and exert its mind-altering effects.[47] Indigenous peoples, then, discovered the combined activities of different medicinal plants long before chemists and pharmacologists came to understand their mechanisms of action.

Many herbal medical traditions employed combination therapies, mixtures of plants that were thought to work together. For example, indigenous Chinese medical practice relies on potentially thousands of herbal formulas, each composed of several (sometimes more than ten) medicinal plant ingredients (figure 15.9).[48] In ancient literature, centuries before the advent of Western chemistry, the Chinese described the art of formulas by assigning to each herb a role drawn from a model feudal society. The chief herb, also called the lord, was deemed responsible for the primary therapeutic effect; the

> **Synergy**
>
> A greater effect of a combination of drugs than would be predicted from their individual effects

FIGURE 15.9 A Chinese herbal prescription containing more than two dozen ingredients.

deputies or ministers assisted or enhanced the formula's efficacy. The assistants were believed to help treat accompanying symptoms of the patient's illness, counteract the harshness of one of the ingredients, and so forth. Finally, the envoys or couriers were thought to direct the formula to the proper channel or site of therapeutic activity.[49] It is possible that Chinese doctors long before the present identified synergistic herbal interactions that modern-day investigators can study at the molecular level.[50]

In indigenous South Asian medicine, mixtures of plant-derived ingredients abound in medical prescriptions and therapeutic food recipes, such as those that combine turmeric (*Curcuma longa*), the flavorful, yellow root that is powdered and added to numerous traditional sauces, with the ground-up fruits of the black pepper (*Piper nigrum*) or long pepper (*P. longum*) vine, a sharp, pungent spice.[51] Turmeric contains curcumin, a phenolic compound that is associated with anti-inflammatory and possible anticancer properties but is absorbed into the bloodstream rather inefficiently if taken alone.[52] Interestingly, the pepper compound piperine blocks certain enzymes in the intestine and liver that would ordinarily prevent the uptake and circulation of curcumin.[53] As a result, piperine enhances the ability of curcumin to reach the body's tissues, where it might exert physiological properties. Therefore, age-old South Asian recipes include two ingredients that together produce effects not possible with either isolated chemical agent.

Pharmacologists have described four aspects of human physiology affecting the ultimate ability of a systemic drug to act on its target tissues: absorption, distribution, metabolism, and excretion. It appears

Parameters affecting drug levels in the body	
Absorption	Uptake by the digestive tract, lungs, or other mucous membranes
Metabolism	Enzyme-mediated conversion into other molecules
Distribution	Transport in the body
Excretion	Removal from the body via the urine and feces

that many synergistic drug effects are attributable to the complex roles of various plant-derived chemicals in the way the body handles drugs introduced to it. By improving DMT's distribution into the brain, harmaline potentiates its psychoactive effect. By reducing barriers to curcumin's absorption and inhibiting its metabolism (breakdown), piperine allows it to achieve a higher concentration in the bloodstream and thereby reach tissues throughout the body. While the complexity of chemical compounds in single plants and multiherb mixtures remains a challenge in assessing herbal efficacy, traditional knowledge and modern-day pharmacology can shed new light on the many ways that ancient medicines affect human health.

SAFETY

Physicians, shamans, apothecaries, and other specialists in the properties of medicinal plants have concerned themselves with the challenge of distinguishing therapeutic effects from possible toxicities for thousands of years. The task of determining the safest ways to use medically active herbs while avoiding negative outcomes was once a matter of trial and error: knowledge of therapeutic effects and risks was gathered by observation. Doctors remembered their experiences and passed along new warnings to future generations. As an Amazonian herbalist remarked, "Many of our brothers and sisters died so that we can have these medicines."[54] Now that herbal preparations are increasingly subject to the techniques of biomedical science, it is possible to assess their safety with a new rigor and to examine an age-old problem in its many contemporary dimensions.

Classic texts and ethnographic sources are valuable resources in garnering time-tested methods for harvesting, preparing, and administering herbs safely. However, until the advent of biochemical pharmacology techniques and scientific, statistical methodology, there was no robust way for a physician to determine whether negative experiences—in modern terminology, adverse events—were caused by a treatment. Certainly, immediate reactions to a medicine, such as acute allergy or rapid gastrointestinal response, might have been connected to a medication, but in patients already suffering serious health problems, even such drug-related outcomes might not have been distinguishable from the normal course of illness. For toxicities revealing symptoms over the course of months or years, any link would have been nearly impossible to detect.

Species in the birthwort genus (*Aristolochia* spp.) were employed in numerous world medical traditions until their toxic properties were more fully recognized in the twentieth century (figure 15.10). The Greek herbalist Dioscorides relayed that the herb was "very helpful to women during childbirth," and one type, in an herbal tea, treated "asthma, hiccups, shivering, the spleen, ruptures, spasms, and pains in the side."[55] In China, several *Aristolochia* species figure in the classical pharmacy, including the root of *A. fangchi*, an

FIGURE 15.10 A South American species of birthwort.

herb first mentioned in the *Divine Husbandman's Classic of the Materia Medica* almost two millennia ago and thought to dispel wind, stop pain, clear heat, and promote urination.[56] Despite a long history of use, birthwort is now known to be harmful. The plant accumulates the chemical aristolochic acid, a genotoxin that is particularly damaging to the kidneys and can cause cancer of the upper urinary tract.[57] Numerous cases of kidney failure and cancer have been described in eastern Europe, where birthwort seeds were found to contaminate locally produced bread; western Europe, where patients at a weight-loss clinic were unintentionally treated with herbs including birthwort; and East Asia, where the herb is widely used in traditional medicine.[58] While many governments have banned the import or sale of birthwort, it is still employed in folk health practices in Asia, grows wild around the globe, and is available on the Internet to willing customers, without mention of its deadly nature.[59]

The example of birthwort highlights the ways that biomedical science can help identify dangerous medicinal herbs and establish the biological bases of their toxicity. Since many herbs are sold in the United States as dietary supplements, they are not routinely tested for safety in laboratory and clinical experiments before being placed on the market, as are pharmaceutical drugs. Under the Dietary Supplement Health and Education Act of 1994, which governs such products, many herbs and their extracts can be assumed safe for consumption and freely sold, leaving responsibility to the FDA to demonstrate any significant risk to the consumer. Since the early 2000s, following several deaths attributable to supplements containing jointfir (*Ephedra sinica*), the government and herbal-supplement industry have adopted more rigorous safety practices.[60] Manufacturers of herbal products are now required to submit reports of serious adverse events to the FDA. Reporting of serious adverse events includes those that result in death or are otherwise life threatening, cause birth defects, or require inpatient hospitalization, for example. The industry may voluntarily report mild or moderate adverse events, as may patients and practitioners. In addition, the industry must follow particular procedures regarding the processing, packaging, and storage of herbal supplements, called Current Good Manufacturing Processes.[61]

Some critics of the current state of affairs recommend that the FDA be given closer oversight of dietary-supplement safety, including the power to issue mandatory recalls if health concerns are suspected and required reporting of all adverse events.[62] The FDA considers that adverse events associated with dietary supplements are significantly underreported, impairing its ability to detect potentially harmful products. Since many consumers of dietary supplements in the United States do not view them as "drugs," they may be less apt to attribute side effects or other health problems to them. Many American patients neglect to share their herbal medicine use with physicians, and therefore clinicians cannot adequately discuss their health effects or report adverse events related to supplements.[63] Furthermore, misleading advertising and dietary-supplement labeling that may encourage consumers to misuse certain herbs also present health risks. For example, a recent government study uncovered numerous examples of deceptive marketing on Web sites and dangerous medical advice from salespeople of dietary supplements.[64] Therefore, it is likely that a combination of enhanced reporting of adverse events, improved communication between patients and health-care providers, and increased vigilance on the part of the FDA and other government agencies will help address herbal dietary supplement safety in the future.

In addition to the potential risks of herbs with undiscovered toxicity, there are concerns that some herbs may be misidentified, misrepresented, or adulterated during growth, harvesting, processing, or marketing. Because of the close physical resemblance of certain medicinal plants, it is possible that one species may be inadvertently collected in place of another, in either the field or the raw-herb market.[65] For example, the roots of the North American black cohosh (*Actaea racemosa*) are often sourced in the wild, a range shared in Appalachia with the closely related yellow cohosh (*A. podocarpa*) and other species. These plants' roots appear similar, especially when dry, and it is possible that they may be confused and the latter sold in place of the former. Furthermore, some samples of black cohosh root imported from China have been found to contain instead material from *A. cimicifuga* or *A. dahurica*.[66] Such substitutions present possible health risks because chemical composition can differ between species of the same genus.

Another example of mistaken identity occurred in a health product said to contain common plantain (*Plantago major*), an herbal bulk laxative.

When a Massachusetts woman became ill after taking the product, an investigation determined that the manufacturer probably had mistaken plantain for Grecian foxglove (*Digitalis lanata*), a plant of similar appearance but containing toxic cardiac glycosides.[67] Some cases of herbal misidentification occur because of similar appearance; others occur because of similar names. For example, over 100 cases of aristolochic acid–attributable kidney disease in Belgium during the 1990s have been traced to the mistaken substitution of the Chinese herb *hanfangji* (*Stephania tetrandra*) with *guangfangji* (*Aristolochia fangchi*) in an herbal weight-loss treatment.[68] A recent survey of bulk herb samples from American herbal retailers uncovered occasional substitution of look-alike plants for the labeled species, such as chamomile flowers of the genus *Anthemis* instead of *Matricaria chamomilla*, the former type associated with allergic reactions in some people.[69] Whether accidental or deliberate, the mislabeling of medicinal plants poses potential health risks, and care should be taken to monitor this aspect of herbal quality.[70]

In recent years, academic and government investigators have tested the accuracy of herbal product labels by screening commercially available whole herbs and extracts using sensitive DNA-based technologies.[71] These probes yielded data supporting the conclusion that some dietary supplements do not contain the advertised plant species and are instead composed of substituted plant material or fillers such as rice and wheat. While uncovering evidence of such inconsistencies calls into question the reliability of certain herbal product descriptions, it is worth noting that DNA-based techniques cannot adequately assess the origin of herbal extracts, whose source DNA is largely removed during manufacture. Instead, industry and university researchers recommend the testing of chemical content and microscopic analysis of plant material to assess the molecular makeup and botanical origin of commercial herbal products.[72] Improved and standardized methodologies for monitoring the content of herbal supplements will be useful tools in the determination of product identity and the oversight of quality.

In addition to these concerns, some herbal products have been found contaminated by various chemical agents. For example, pesticides occasionally make their way into processed herbs, generally at very low levels.[73] Heavy metals such as arsenic, lead, mercury, cadmium, and chromium have also been detected in some samples of raw herbs and products.[74] Heavy-metal contamination may arise from the environment in which plants are grown or through processing. In some herbal medical traditions, certain heavy metals are considered medicinal and therefore added to the preparations before sale.[75] It is also possible that herbs may be contaminated with microbes such as bacteria and yeast either before or after harvest.[76] Together, these possible sources of contamination and adulteration can be addressed by improved agricultural and manufacturing processes, the enactment of stringent regulations or the enforcement of existing tolerances, and adherence to recognized standards of quality.[77]

Just as mixtures of pharmaceuticals, including plant-derived chemicals, can act together in a positive, synergistic fashion, they can interact in a way that poses risks to human health.[78] Importantly, many such interactions cannot easily be predicted from laboratory studies that examine the effects of herbal extracts on isolated cells and tissues. On the contrary, by considering the complex pharmacological effects of medicinal plants, investigators are recognizing the role of the entire system on the absorption, distribution, metabolism, and excretion of their active constituents. As a result of such work, pharmacologists have discovered some risks associated with particular medicinal plants, whether taken in combination with other herbs or with synthetic drugs.

A widely publicized interaction occurs between grapefruit (*Citrus paradisi*) and many prescription (synthetic) drugs, discovered when patients who drank grapefruit juice experienced signs of medication overdose. It was found that furanocoumarins

Interaction

A situation in which one drug interferes with another, often in a harmful way

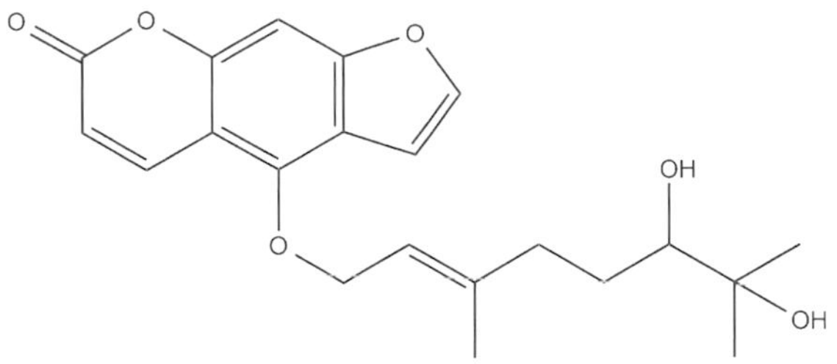

FIGURE 15.11 6′, 7′-dihydroxybergamottin, a furanocoumarin in grapefruit juice.

in grapefruit juice inhibited the activity of metabolic enzymes in cells lining the intestine, allowing a far greater amount of drug to absorb into the bloodstream (figure 15.11).[79] The enzymes blocked by grapefruit furanocoumarins are part of a large family called the cytochrome P450 (CYP) enzymes, expressed throughout the body and responsible for orchestrating much of the chemistry required for life. In the intestinal lining and liver, one of their principal roles is to break down foreign chemicals, such as herbal and synthetic drugs.

The effects of herbal products on CYP enzymes can lead to both elevated and reduced pharmacological activity. In contrast to grapefruit-juice furanocoumarins, which inhibit intestinal CYP, hyperforin from St. John's wort (*Hypericum perforatum*) induces the expression of certain CYP enzymes in the intestine.[80] In addition, hyperforin increases the activity of intestinal P-glycoprotein, a membrane channel that shuttles toxins from the body into the gastrointestinal tract.[81] Through these actions, hyperforin increases the capacity of the human body to both break down and excrete drugs, and this has resulted in serious adverse events in patients taking St. John's wort and prescription medication. For example, several instances of organ-transplant rejection have been attributed to concomitant use of St. John's wort and immunosuppressant drugs, and interactions are also known with heart medications such as foxglove-derived digoxin, blood lipid–lowering drugs such as atorvastatin, hormonal contraceptives, and many other pharmaceuticals. A recent analysis estimated that "more than 70% of all prescription medications are susceptible to [St. John's wort]–mediated interactions."[82] Although much work remains to be done

to identify possible herbal interactions involving CYP enzymes, several medicinal plants have been demonstrated to present such a risk, including the North American perennial goldenseal (*Hydrastis canadensis*) root and the medicinal berries of the East Asian *Schisandra* spp. vine.[83] In addition to the CYP enzymes and P-glycoproteins, many physiological factors might influence drug absorption, distribution, metabolism, and excretion, and therefore more research is warranted to establish the safe parameters of herbal treatment.[84]

SUSTAINABILITY AND ETHICS

More than 2500 years ago, the people of the Greek city-state of Cyrene on the Mediterranean coast of North Africa harvested a medicinal sap from a local, wild type of fennel (*Ferula* spp.) they called *silphion* (figure 15.12).[85] Aside from its great repute as a culinary flavoring, *silphion* was held to be a potent cough remedy and an effective oral contraceptive, among many health-related uses, so much so that it was one of the most valuable exports to cities throughout the dispersed Greek realm. Despite efforts to grow *silphion* elsewhere, to cultivate it for a more plentiful supply, the plant persisted only in the undomesticated state in a thin strip of land perhaps not much more than 175 kilometers in length, sandwiched between the inhospitable desert and the coastal farmlands in what is now Libya.[86] As demand grew and wild resources dwindled, regulations were established to limit collection, but ultimately they were ineffective against the intense profit motive.[87] The Roman natural historian Pliny the Elder (23–79) lamented, "Within the memory of the present generation, a single stalk is all that has ever been found [in Cyrene], and that was sent as a curiosity to the Emperor Nero."[88] Sometime in the first centuries of the new millennium, the esteemed medicinal plant *silphion* was harvested to extinction.

While some medicinal plants transition readily into cultivation, others resist domestication or require such an input of land, labor, and time that people find it disadvantageous to grow them. With *silphion* as an unfortunate precedent, the

FIGURE 15.12 A gold coin from Cyrene, depicting a horseman on the obverse (*left*) and the *silphion* plant on the reverse (*right*), ca. 331–322 B.C.E. (Numismatica Ars Classica, auction 54, lot 893)

pressure of intensive harvesting of wild herbs has threatened entire plant populations. For example, government-sanctioned and illicit ginseng (*Panax ginseng*) collection in northeastern China nearly eliminated the species during imperial times.[89] Likewise, North American ginseng (*P. quinquefolius*) has been threatened numerous times with local extinction because of both authorized digging and poaching in protected areas. Since wild ginseng is held to be more valuable than cultivated, diggers and merchants often take care to replenish the ginseng supply by replanting seeds where roots are removed, a conservation technique called wild crafting.[90] Numerous medicinal plants are sourced exclusively in the wild, rendering them susceptible to overharvesting and threatened by urban and agricultural development.[91] In addition to the careful stewardship of wild medicinal plants such as ginseng, black cohosh, and peyote (*Lophophora williamsii*), advances in biotechnology can be harnessed to protect native stands of valuable herbs.

When the bark of the Pacific yew was found to contain a potent anticancer compound, some conservationists were justifiably concerned that existing forests might not be able to furnish enough

material to meet the pharmaceutical needs. In time, these concerns were allayed with research demonstrating that chemical precursors could be isolated from leaves of a number of yew species, allowing a more sustainable approach to harvest and semisynthesis of the active drug agents.[92] For industrial-scale production, the process has been made more efficient still, as pharmaceutical scientists have developed a method to propagate large quantities of cells in vats of culture medium and extract the desired chemicals directly, thereby circumventing any need to collect from whole plants. In the developing world, however, the infrastructure and financial resources for such technological solutions might not exist. Therefore, drug manufacturers resort to harvesting local populations of plants. As a result of such practices, the population of the Central Asian yew (*Taxus contorta*) is threatened.[93]

While targeted exploitation of medicinal plants exerts significant pressure on many species, the global impact of changing climate, human population growth, and runaway development will include the loss of herbs known to be medicinal and those whose properties have not yet been (or may never be) discovered. Estimates of imminent extinction reach mind-boggling levels. The organization Plantlife projected that 15,000 of the world's 50,000 wild medicinal plants are at risk of extinction.[94] Thirty percent of all plant species are in danger of being lost, according to scientists at Britain's Royal Botanic Garden at Kew.[95] The rich biodiversity of the tropics is especially pressured because of the replacement of forests with farmland, the encroachment of human settlements and livestock, the introduction of nonnative invasive species, and new weather patterns. In the temperate regions as well, habitat loss and other human-attributable forces imperil countless species.[96]

Declining biodiversity erases the work of millions of years of plant evolution. Many unique and complex chemical profiles may never be known to science. Some of the large-scale screening projects that sample plants more or less randomly and subject their extracts to tests of possible biological activity have reached perhaps only 60,000 species.[97] Such screening programs, successful as they have been in identifying compounds such as paclitaxel, are never considered to rest as the final word on medicinal potential from the sampled plants. Indeed, plants sampled at different growth stages, specimens collected at different times of the year, and samples from different tissues are likely to present varying chemical repertoires. Taking into account also the many possible extraction techniques—using water, alcohol, organic solvents, and so on—any single species might render dozens of possible analytic units in terms of chemical profile. Furthermore, each isolate might be assayed in the laboratory against various models of cancer, virus infection, bacterial growth, diabetic conditions, and many more. In short, each lost plant species represents scores of experiments that can never be conducted and risks the permanent forfeiture of possible therapeutic agents against present and future diseases. Hundreds, and possibly thousands, of new drugs may never have a chance to be discovered.[98] Protecting biodiversity includes the defense of resources that may be found to have therapeutic value in the future.

The urgency to protect the earth's biological diversity is matched by the pressing need to preserve its cultural richness. As one indication of the loss of cultural diversity, it is estimated that 20 percent of the world's languages have lost their last remaining native speakers since 1970, and many other languages hold on to just a few.[99] About 90 percent of the world's estimated 6000 languages are unlikely to survive the twenty-first century.[100] Since human cultural diversity aligns geographically with plant and animal species richness, especially concentrated in the global band of the tropics, the decline of the world's languages threatens a loss of tremendous medicinal plant knowledge.[101] In regions with such biodiversity, indigenous peoples' languages have evolved to express distinctions among plants learned over many generations, distinctions not evident to outsiders. For example, the influential ethnobotanist Richard Evans Schultes (1915–2001) relates that in the Colombian Amazon, native people have thirty words to describe natural variations in the ayahuasca vine, differences imperceptible to a foreign botanist. The "intensely thorough indigenous familiarity of the intricately meticulous individuality of species in one of the world's richest floras," Schultes wrote, demonstrates "the often hidden biodiversities [in] the forests."[102]

The survival of indigenous languages is closely linked to the persistence and continued evolution of traditional knowledge about the environment and health. As the forces of acculturation advance, young people from tribal areas are increasingly drawn to a way of life in which information about useful plants and the skills to locate, cultivate, and prepare them are less salient. Seeking employment in multicultural towns and cities, adopting the predominant national and international languages, and frequenting clinics for health-related needs, recent generations are less likely to carry with them the traditional herbal knowledge of their ancestors. In many traditional societies, medical training consists of a young person's apprenticeship under a practitioner possessing hereditary secret formulas and herbal techniques handed down orally and elaborated over the course of centuries. Without interested students to instruct, traditional doctors are sometimes finding themselves the last of their ancient medical lineage.[103]

One approach to preserve endangered medicinal plant knowledge is to document as fully as possible the information and practices associated with the roles of plants in a society. To do this, outside researchers and local collaborators are collecting inventories of plant names and cultural uses by people serving various medical needs in the community. They are using photography to record where the plants grow and how they are prepared for medicine and capturing videos of herbalists at work. By having a community's long experiences with plants described and disseminated in academic publications, some aspects of its heritage can be made permanent against the forces of social change that might endanger it. Those projects perpetuating indigenous knowledge are more beneficial still. The recordings of expert plant specialists, donated to the source community, can help elders educate younger generations long after their passing. Maps, inventories, and guides prepared in collaboration with local people can serve as instructional aids. Academic, government, or industry outreach programs that generate revenue can be used to protect land and fund education in traditional herbal techniques among indigenous people.

Increasingly, programs to archive and prospect the traditional knowledge of indigenous people are undertaken within a framework of mutual benefit. Among the first to pursue this approach was the National Cancer Institute, which in the 1980s sought to collect tropical plant specimens for its large-scale bioassay screens. Before beginning its collection, it negotiated agreements with national governments to define the roles of foreign and domestic researchers and agencies in the collaboration and to establish an arrangement for the sharing of any revenues that might accrue from newly discovered products. The first such document was signed between the National Cancer Institute and the government of Madagascar in 1990, and numerous similar contracts have been developed since.[104] One year earlier, an ethnobotanist working in a Samoan village established a compact with its chief and elders whereby the village

received funds to support development of schools and protect the environment, even if the collaboration yielded no profitable drugs. A subsequent agreement to develop a novel potential antiviral compound identified from an indigenous hepatitis treatment stipulated the sharing of royalties among the government of Samoa, the village where the material had been collected, and the families of the village healers who had offered their knowledge.[105]

According to the Convention on Biological Diversity, an international agreement widely adopted after its signing in 1992, parties must uphold common standards toward the specified aims of conserving biological diversity, sustainable use of its components, and fair and equitable sharing of benefits.[106] As nations assert sovereign rights to their natural resources, any research that might generate benefits, such as the study of potential medicinal plants, is subject to prior informed consent. That is, researchers wishing to conduct investigations in a particular country must first obtain an agreement on the nature of the work and, should discoveries be made, the way benefits are to be distributed. In practice, compliance with this principle raises many questions. With whom should an agreement be made? Do national governments represent the interests of all their people, including those of diverse ethnic groups? What is to be done if a national government has not established a procedure for making such agreements? Within the community of interest, who should represent the group in such discussions, and how should disputes be resolved? Certainly, the definition of "benefit" is complex and variable, encompassing tangible and intangible manifestations that differ culturally and on multiple time scales.[107] For example, if some benefits of research are nonmonetary, how are they to be made equitable? As for the role of individuals with expertise in plant-based medicine, there is not yet consensus (and there might never be) on the appropriate sharing of benefit from indigenous common knowledge, secret knowledge, or community-held knowledge, for example.[108]

The convention and subsequent agreements provide a general framework for the conduct of responsible research and environmental steward-ship. With so many more possible therapies yet to be found in the richly diverse but gravely threat-ened ecosystems on the earth, collaborations with those people who hold herbal lore can cata-lyze great benefit at many levels.[109] Drawing on the expertise of a suite of academic disciplines, working among people and plants all over the world, medicinal plant researchers are improving our knowledge of herbs and their many roles in our planet's cultures. To identify medicinal com-pounds, characterize their actions and interac-tions, work extends from ancient archives to culturally diverse communities and to the labora-tory and clinic. Further study of medicinal plants will continue to follow the path of multidisci-plinarity, helping address some of our greatest health challenges.

Conclusion

Herbalists continue to serve as sources of medicines in many communities, such as at this street market in Chicago.

Humans are natural explorers, born curious, and driven to occupy new lands and provide for one another. As they've expanded across the surface of the earth, they have encountered novel plants they wished to investigate. Some plants they learned to eat, others to grind up, turn into teas, or even burn to smoke. People built lists of plants that they believed sustained them and made them healthy. As societies evolved, they distinguished themselves from one another in unique conceptions of the world around them. They developed special rites and rituals, traditions, language, art, literature—a tapestry of culture characteristic of the diversity that embodies the spark of the human species. Communities of people in all parts of the world thus lived parallel existences, sometimes in isolation and sometimes in communication. Often they shared worldviews, and often they did not.

These communities developed particular ways to think about health and practice medicine, an art that springs from a sense of how people ought to feel. In many parts of the world, these societies saw their health in physical, spiritual, and social terms. From the early days of humanity, medicinal plants played an important role in binding together communities in ritual, connecting people with the supernatural, and providing therapy over injury and ailment. By discovering plants with medicinal properties and seeking to ensure their abundance, humans gave these herbs great advantages in their evolutionary struggle to survive.

People collected their seeds and propagated them, determined the best conditions for cultivation, selected for healthy growth, learned to process them into useful products, and, in time, spread them around the world. Under the protection of civilization, the range of these plants grew to follow people wherever they went. When nations met, they exchanged medicinal plants, at times through peaceful trade, at times via conquest. Humans elevated these privileged plants—some to the status of gods, some to the value of gold, all as cherished companions of diverse cultures. Entered into global trade as commodities, cultivated across great expanses of land, extracted and chemically analyzed,

made objects of laws and regulations, such plants bear witness to the powers of human control and manipulation, made possible by our species' great skills in communication and technology.

These plants have achieved such a fate because they have compelled it. By nourishing us, intoxicating us, and putting us in touch with our deities, medicinal plants exert subtle but effective command of beings with the great capacity of cognition. These plants have engineered in themselves the ability to manipulate human neural pathways, to accumulate chemicals that can heal or kill us, and to thrive in diverse cultural and geographic settings. These plants have convinced humans to transport them around the world, to revere them, to battle over them, to lionize them in art and literature, to serve them in their conquest of us. The relationship between people and medicinal plants is thus one of mutual transformation. We cultivate them, but they have cultivated us, as well.

It is difficult to imagine a modern world without the medicinal plants that shaped it so profoundly. From them, people derived substances to allay the strains of their challenging lives and anesthetics to allow surgery and new treatments for disease. They learned to make herb extracts to protect them from contagious illness. They found potent stimulants, euphoriants, spiritual aids, and poisons. All these properties made the plants precious, and they generated fortunes for the merchants and nations that could effectively develop them as commodities. Medicinal plants made many people into slaves, fueled war and revolution, and enabled conquest. Medicinal plants inspired generations of artists, musicians, and writers, whose works represent the soul of humanity; they also extinguished the minds of countless others. Medicinal plants allowed scientists to understand their active principles, unravel the workings of the human body, discover signaling molecules and drug receptors, and improve the well-being of millions. The remarkable repertory of plant-derived biochemicals has the power to influence the human experience in many ways.

There is much more knowledge to be gained. There are many open questions on how to make the best use of herbs to prolong our health and treat illnesses. Methods to ensure safety and determine efficacy will refine our efforts to expand the modern pharmacopeia with ancient remedies. And there are many medicinal plants yet unknown to science. In the forests, in the fields, and in the handed-down lore of practitioners around the world are plants that make some of the most potent therapeutic chemicals on earth and molecules that will revolutionize our understanding of the brain and the body. The next chapter of the story of medicinal plants has not yet been written.

Notes

1. CONCEPTS OF ETHNOMEDICINE

1. Brenna M. Henn, L. L. Cavalli-Sforza, and Marcus W. Feldman, "The Great Human Expansion," *Proceedings of the National Academy of Sciences USA* 109 (2012): 17758–17764.

2. Mark D. Merlin, "Archaeological Evidence for the Tradition of Psychoactive Plant Use in the Old World," *Economic Botany* 57 (2003): 295–323. These dates are for findings of poppy seeds in Italy and hemp-cloth imprints on artifacts in China. Merlin reports earlier dates for a Neanderthal burial in Iraq in which pollen of an *Ephedra* species was identified (approximately 50,000 B.C.E.) and impressions of hemp rope on pottery in Taiwan (12,000 B.C.E.). Archaeological study of medicinal plant use is complicated by poor preservation of artifacts (including biological material) in many sites, potential contamination of ancient samples with more recent ones, and difficulty in determining the species identity of ancient plant material and in objects designed to represent them. In any case, the absence of artifacts in a region or during an ancient span of time is not an indication that medicinal plants were not used. Furthermore, plants such as poppy and hemp can be used for food and medicine (in the former case) and fiber, food, and medicine (in the latter case). A challenge in such work is that the identification of ancient plant remains does not prove their intended function or use. For an extensive discussion of these matters, with regard to the opium poppy and hemp, respectively, see Mark D. Merlin, *On the Trail of the Ancient Opium Poppy* (Rutherford, N.J.: Farleigh Dickinson University Press, 1984); and Robert C. Clarke and Mark D. Merlin, *Cannabis: Evolution and Ethnobotany* (Berkeley: University of California Press, 2013). Undoubtedly, many plant-derived dietary ingredients (grains, fruits, roots, and so on) were considered medicinal beyond their nutritional virtues. Still, investigators are drawn to document early uses of mind-altering plants, whose specific physiological actions imply a particular intention on the part of the ancestral human.

3. From Sanskrit *ayurveda*, which is alternatively translated as "science of life" and "knowledge of longevity." See Hari M. Sharma, "Contemporary Ayurveda," in *Fundamentals of Complementary and Alternative Medicine*, 4th ed., ed. Marc S. Micozzi (St. Louis: Saunders Elsevier, 2011), 495. The text in which this term originates is the *Atharva Veda*. Carlos Lopez places portions of the Atharva Veda as early as approximately 1000 B.C.E., in "Atharva Veda," in *Oxford Bibliographies: Hinduism* (November 21, 2012), http://www.oxfordbibliographies.com/view/document/obo-9780195399318/obo-9780195399318-0008.xml?rskey=Eag07N&result=12.

4. The *Yellow Emperor's Classic of Internal Medicine* is the first codified text of Chinese medicine, compiled from many sources and attributed to the wisdom of a mythical emperor. Paul U. Unschuld dates the text sources to around the third to first centuries B.C.E. in *Huang Di Nei Jing Su Wen: Nature, Knowledge, Imagery in an Ancient Chinese Medical Text* (Berkeley: University of California Press, 2003), 286. The book includes 224 drugs, of which 106 are of plant origin. Also from this period is a book of pharmaceuticals attributed to a godly author, the *Divine Husbandman's Classic of the Materia Medica*, which includes 365 drug prescriptions. See Paul U. Unschuld, *Medicine in China: A History of Pharmaceutics* (Berkeley: University of California Press, 1986), 17.

5. Cyril Bryan, *Ancient Egyptian Medicine: The Papyrus Ebers* (London: Bles, 1930), 28. The earliest Egyptian medical text, known as the Ebers papyrus, is generally dated to 1550 B.C.E. It was written in the ninth year of the reign of Amenhotep I (1534 B.C.E.), according to John F. Nunn, *Ancient Egyptian Medicine* (Norman: University of Oklahoma Press, 1996), 30–34.

6. The oral tradition of medicine outside Europe, Asia, and northern Africa is a rich repository of ethnobotanic wisdom. When Europeans first encountered the Aztec and Inca civilizations, they documented indigenous medical knowledge by asking natives to draw and write descriptions of their medicinal plants. Anthropologists and botanists continue to document medical practices disseminated by the oral tradition, recognizing the valuable resource of traditional, often unwritten, expertise.

7. The notion of balance is integral also to much contemporary medical discourse dealing with biochemical homeostasis, hormone regulation, pH balance, and the like.

In many traditional medical settings, the "balance" conducive to health is thought to be influenced by environmental, dietary, social, and spiritual factors in a way not articulated (and sometimes not accepted) by practitioners of mainstream biomedicine.

8. In Chinese medicine, "health is a physiological, emotional, and mental balance, maintained in all three spheres or not at all, for the three are inseparable" (Nathan Sivin, *Traditional Medicine in Contemporary China: A Partial Translation of Revised Outline of Chinese Medicine [1972] with an Introductory Study on Change in Present-Day and Early Medicine* [Ann Arbor: Center for Chinese Studies, University of Michigan, 1987], 4). Medical approaches that view the patient's physical, mental, social, and (sometimes) spiritual health together are called holistic. Holistic beliefs are ancient, but the term is a relatively new one.

9. The practice of medicine in China has evolved over thousands of years. It shares elements with (and probably gave rise to) similar conceptions of health in Japan, Korea, and Southeast Asia. This section addresses the cosmopolitan, widespread form of indigenous Chinese medicine as taught and practiced in China and elsewhere but not the many informal health-related practices, integrated Chinese-Western medicine, or the distinctive ethnic medicines of Asia's diverse populations.

10. According to Liu Yanchi, "Chinese medicine contends that disease develops when pathogenic factors disturb the equilibrium of *yin* and *yang* in the human body. Medicinal herbs help to increase body resistance, eliminate pathogenic factors, and restore a relative balance between *yin* and *yang*" (*The Essential Book of Traditional Chinese Medicine*, vol. 2, *Clinical Practice* [New York: Columbia University Press, 1988], 43).

11. The premodern repertory of Chinese pharmaceuticals was very large. The *Grand Materia Medica* (ca. 1596) listed 1,892 drugs, and the *Additions to the Grand Materia Medica* (ca. 1803) added another 716. See Sivin, *Traditional Medicine in Contemporary China*, 180.

12. Dan Bensky, Steven Clavey, and Erich Stöger, *Chinese Herbal Medicine: Materia Medica*, 3rd ed. (Seattle: Eastland, 2004), 830–832.

13. Liu, *Essential Book of Traditional Chinese Medicine*, 2:127; Bensky, Clavey, and Stöger, *Chinese Herbal Medicine*, 710–711, 822–823.

14. *Ayurvedic* medicine's earliest magico-religious Sanskrit texts date to around 2500 years ago; *siddha*'s origins are less clear. *Siddha* apparently took form beginning in the early centuries C.E., with a distinctive literature that emerged during the thirteenth through sixteenth centuries. See Kenneth G. Zysk, "Traditional Medicine of India: Ayurveda and Siddha," in *Fundamentals of Complementary and Alternative Medicine*, ed. Micozzi, 461–462. Another important form of South Asian medicine is *unani*, which came to the Indian subcontinent from the Mediterranean through Muslim scholars.

15. Zysk, "Traditional Medicine of India," 456.

16. According to Dagmar Benner, "Imbalance denotes both a change in quantity as well as the dislocation of a [*dosha*]" ("Traditional Indian Systems of Healing and Medicine: Ayurveda," in *Encyclopedia of Religion*, 2nd ed., ed. Lindsay Jones [New York: Macmillan, 2005], 3854). "The quantity of a [*dosha*] can both increase and decrease, which leads to various symptoms of disease" (3854).

17. Steven Kayne, "Indian Ayurvedic Medicine," in *Traditional Medicine: A Global Perspective*, ed. Steven B. Kayne (London: Pharmaceutical Press, 2010), 195–202.

18. Zysk, "Traditional Medicine of India," 460.

19. Zysk, "Traditional Medicine of India," 461–462.

20. The connection between matter, energy, and the zodiac was a widespread feature of Indian astrology. See Zysk, "Traditional Medicine of India," 462.

21. The five elements and three bodily constituents of *siddha* medicine are similar to those in *ayurvedic* medicine but occur in different proportions. See Zysk, "Traditional Medicine of India," 462–463.

22. According to Zysk, "The Siddha practitioners have a materia medica of at least 108 plants and plant products, some of which are imported from as far away as the Himalayas" ("Traditional Medicine of India," 466).

23. The lineage of many traditional practices was undoubtedly altered during the period of European colonization. The sources reference traditional activities and beliefs surmised from nineteenth- and twentieth-century anthropological study.

24. James W. Herrick summarizes Iroquois health concepts: "Bad health, illness, or disease must be understood in a general sense as a disturbance of balance in the power or life force of some person, place, thing, or event" (*Iroquois Medical Botany* [Syracuse, N.Y.: Syracuse University Press, 1997], 33).

25. John K. Crellin considers these to be "magico-religious/spiritual practices" ("Aboriginal/Traditional Medicine in North America: A Practical Approach for Practitioners," in *Traditional Medicine*, ed. Kayne, 55). The word "shaman" can refer to a person with special access to the spiritual world in many traditional settings.

26. Frances Densmore, "Uses of Plants by the Chippewa Indians," *Smithsonian Institution–Bureau of American Ethnology Annual Report* 44 (1928): 322.

27. Densmore, "Uses of Plants by the Chippewa Indians," 328.

28. According to Richard W. Voss, "Mental health and physical health were viewed as inseparable from spiritual and moral health. The good balance of the one's life in harmony with the woʼope, or natural law of creation, brought about wicozani, or good health, which was both individual and communal" ("Native American Healing," in *Fundamentals of Complementary and Alternative Medicine*, ed. Micozzi, 540).

29. Voss, "Native American Healing," 540.

30. Daniel E. Moerman identifies over 2800 medicinal plants used by North American native people in *Native American Ethnobotany* (Portland, Ore.: Timber Press, 1998), 11.

31. Moerman, *Native American Ethnobotany*, 101–102.

32. Moerman, *Native American Ethnobotany*, 607. The "stimulant" use is attributed to the Navajo and the "contraceptive" use to the Ramah.

33. Moerman, *Native American Ethnobotany*, 603.

34. The Chippewa medicine men were part of a *midewiwin*, a religious-medical sect whose teachings were secret.

35. The Aztecs, Toltecs, and other groups comprise a larger Nahuatl-speaking people that inhabit parts of modern-day Mexico and El Salvador.

36. Alfredo López Austin, *The Human Body and Ideology Concepts of the Ancient Nahuas*, trans. Thelma Ortiz de Montellano and Bernard Ortiz de Montellano (Salt Lake City: University of Utah Press, 1988), 1:52–53.

37. López Austin, *Ancient Nahuas*, 1:255–264.

38. López Austin, *Ancient Nahuas*, 1:204–222.

39. Bernard Ortiz de Montellano, *Aztec Medicine, Health, and Nutrition* (New Brunswick, N.J.: Rutgers University Press, 1990). According to López Austin, "Just as the force of the sacrificial victims could revitalize the gods, so their representatives on earth . . . participated in order to strengthen and prolong their own lives" (*Ancient Nahuas*, 1:377).

40. López Austin, *Ancient Nahuas*, 1:225–226.

41. Synonym *A. mexicana*. López Austin, *Ancient Nahuas*, 1:219.

42. Bernardino de Sahagún, *Florentine Codex: General History of the Things of New Spain*, book 11, *Earthly Things*, trans. Charles E. Dibble and Arthur J. O. Anderson (Santa Fe, N.Mex.: School of American Research, 1963), 131.

43. Francisco Hernández, *Historia natural de Nueva España*, vol. 1 (Mexico City: Universidad Nacional de México, 1959), cited in Ortiz de Montellano, *Aztec Medicine*, 141.

44. López Austin, *Ancient Nahuas*, 1:356, 358.

45. Kwasi Konadu, *Indigenous Medicine and Knowledge in African Society* (New York: Routledge, 2007), 66–68, 85–86.

46. Konadu, *Indigenous Medicine and Knowledge in African Society*, 52–55.

47. Konadu, *Indigenous Medicine and Knowledge in African Society*, xxi, 53.

48. Konadu, *Indigenous Medicine and Knowledge in African Society*, 95–96.

49. Mariana G. Hewson, "Indigenous Knowledge Systems: Southern African Healing," in *Fundamentals of Complementary and Alternative Medicine*, ed. Micozzi, 523. Another articulation of this idea is discussed by A. Moteetee and B.-E. Van Wyk, "The Basotho believe that good health is not only physical but also spiritual; therefore ancestral spirits (*balimo*) form an integral part of their lives. If a person experiences bad luck and misfortune, then this is most likely ascribed to the fact that they have neglected their ancestors and are therefore being punished" ("The Medical Ethnobotany of Lesotho: A Review," *Bothalia* 41 [2011]: 210).

50. Hewson, "Indigenous Knowledge Systems," 523.

51. Moteetee and Van Wyk, "Medical Ethnobotany of Lesotho."

52. Moteetee and Van Wyk, "Medical Ethnobotany of Lesotho," 210.

53. Moteetee and Van Wyk, "Medical Ethnobotany of Lesotho," 211.

54. Bruno Halioula and Bernard Ziskind, *Medicine in the Days of the Pharaohs*, trans. M. B. DeBevoise (Cambridge, Mass.: Harvard University Press, 2005), 8.

55. Halioula and Ziskind, *Medicine in the Days of the Pharaohs*, 8–10. Nunn notes that the earliest Egyptian medical documents (ca. nineteenth to sixteenth century B.C.E.) have less magical content than later ones, in *Ancient Egyptian Medicine*, 96.

56. Nunn, *Ancient Egyptian Medicine*, 24–41; Halioula and Ziskind, *Medicine in the Days of the Pharaohs*; Lise Manniche, *An Ancient Egyptian Herbal*, rev. ed. (London: British Museum Press, 2006); J. R. Campbell and A. R. David, "The Application of Archaeobotany, Phytogeography, and Phamacognosy to Confirm the Pharmacopeia of Ancient Egypt 1850–1200 B.C.," in *Pharmacy and Medicine in Ancient Egypt: Proceedings of the Conferences Held in Cairo (2007) and Manchester (2008)*, ed. Jenefer Cockitt and Rosalie David, BAR International Series 2141 (Oxford: Archaeopress, 2010), 20–29.

57. Halioula and Ziskind, *Medicine in the Days of the Pharaohs*, 31, 33. Nunn relates that identifying herbs in Egyptian hieroglyphic texts has been challenging, citing an estimate of 20 percent of the 160 plant products in papyri having reliable translations, in *Ancient Egyptian Medicine*, 151.

58. Translation of Ebers 25, in Nunn, *Ancient Egyptian Medicine*, 159.

59. The European lineage is set apart from other world medical traditions because it gave rise to the predominant form of healthcare in the United States, Europe, and much of the world, known as biomedicine. Its evolution provides context to the changing roles of medicinal plants highlighted in the text.

60. Vivian Nutton, "Medicine in the Greek World, 800–50 B.C.," in *The Western Medical Tradition 800 B.C. to 1800*, by Lawrence I. Conrad et al. (Cambridge: Cambridge University Press, 1995), 23. The authors of the Hippocratic Corpus lived around 420 to 350 B.C.E.

61. The notions of opposing qualities, elements, and humors might have originated in India. See Noga Arikha, *Passions and Tempers: A History of the Humours* (New York: HarperCollins, 2007), 4.

62. The Hippocratic authors wrote that health was a function of balance but differed on how imbalance led to illness: "The author of *On regimen* thought that the whole body was in a perpetual state of flux, and health consisted in keeping this flux within certain limits. By contrast, the author of *On the nature of man* argued that the body remained in a stable balance until something, external or internal, occurred to overturn it" (Nutton, "Medicine in the Greek World," 24).

63. Black bile was not considered a cardinal humor in the earliest Hippocratic writings. See Nutton, "Medicine in the Greek World," 24–25.

64. The humors "did not need to be observable to have explanatory power. In fact, none of the humors needed to be visible to exert their hold on the imagination, and to provide a credible, at times effective physiological account of the

unseen operations within the body" (Arikha, *Passions and Tempers*, 9). Yet physicians recognized that the humors had some connection to actual substances, fluids that on occasion emanated from the body. For example, blood was both a humor in the metaphorical sense and a red liquid that emerged from wounds: "Bile and phlegm were visible only when excreted during illness, and hence might be considered permanently dangerous [but] other fluids were more ambiguous" (Nutton, "Medicine in the Greek World," 24).

65. Arikha, *Passions and Tempers*, 13.

66. Arikha, *Passions and Tempers*, 40.

67. Excess of blood, stagnant blood, or bad blood were treated by slicing the patient (phlebotomy or venesection) or applying blood-sucking leeches. Bloodletting was common medical practice in Europe until the nineteenth century.

68. Jerry Stannard, "Hippocratic Pharmacology," *Bulletin of the History of Medicine* 35 (1961): 497–518.

69. The reference to black hellebore is from Pliny the Elder, *Natural History* 25.22; *The Natural History of Pliny*, trans. John Bostock and H. T. Riley (London: Bohn, 1856), 5:98–99. *Helleborus niger* is black hellebore, although the ancients may have referred to other *Helleborus* spp. or *Veratrum album*, known today as white hellebore or false hellebore. Both plants are medicinal and highly toxic in overdose; *V. album* is the more effective purgative.

70. This belief held sway for many centuries. Edward Shorter describes European medicine of a later era: "The aim of traditional therapeutics was getting the bowels open. . . . One treated fever with laxatives, getting those bad humours out by procuring an 'opening.'" ("Primary Care," in *The Cambridge History of Medicine*, ed. Roy Porter [Cambridge: Cambridge University Press, 2006], 108).

71. Stannard traces the first mention of scammony to the Hippocratic text *Diseases of Women*, in "Hippocratic Pharmacology," 503.

72. John Uri Lloyd, "History of the Vegetable Drugs of the Pharmacopeia of the United States," *Bulletin of the Lloyd Library of Botany, Pharmacy, and Materia Medica* 18, no. 4 (1911): 78–79; H. C. Wood, Joseph P. Remington, and Samuel P. Sadtler, *The Dispensatory of the United States of America*, 19th ed. (Philadelphia: Lippincott, 1907).

73. John Riddle, *Dioscorides on Pharmacy and Medicine* (Austin: University of Texas Press, 1985), xix, 11–13. Dioscorides assigned hot–cold and dry–moist properties to herbs, but generally did not reference the humors.

74. *De materia medica* is the better-known Latin name of a work written in Greek, sometime between 60 and 78 C.E. See Riddle, *Dioscorides*, 13–14. The number of medicinal substances described is subject to some discussion because the author may have made distinctions in naming that modern experts do not and vice versa. The five-book treatise contained over 1000 listings of plant, animal, and mineral-based pharmaceuticals; today's scholars itemize roughly 500 to 600 unique plant entries. See Jerry Stannard, "The Herbal as a Medical Document," *Bulletin of the History of Medicine* 43 (1969): 212–220; and Riddle, *Dioscorides*, xiii, xivn.2.

75. In *Medicine in China*, Unschuld considers *De materia medica* the world's first pharmacopeia. Although Chinese medical books of the era included descriptions of medicinal plants and their uses, they did not provide instructions for their standard preparation as in Dioscorides.

76. Susan P. Mattern, *The Prince of Medicine: Galen in the Roman Empire* (Oxford: Oxford University Press, 2013), 233–234.

77. A detailed account can be found in Owsei Temkin, *Galenism: Rise and Decline of a Medical Philosophy* (Ithaca, N.Y.: Cornell University Press, 1973).

78. A form of Galenic medicine is still practiced in the Muslim world under the name of *unani*.

79. Oswald Cockayne, ed., *Leechdoms, Wortcunning, and Starcraft of Early England* (London: Longman, Green, Longman, Roberts, and Green, 1863), 2:15, 113, 291; Audrey Meaney, "The Practice of Medicine in England About the Year 1000," *Social History of Medicine* 13, no. 2 (2000): 230.

80. Meaney, "Practice of Medicine in England," 222, 232–235.

81. Anne Van Arsdall, *Medieval Herbal Remedies: The Old English Herbarium and Anglo-Saxon Medicine* (New York: Routledge, 2002), 138–139. The original Old English, literally translated, diagnoses "evil night spirits."

82. Van Arsdall notes that at the time, "the moon was believed to cause 'lunacy,' and the Old English word for it is 'month-sickness,' months being measured in moons" (*Medieval Herbal Remedies*, 178).

83. Van Arsdall, *Medieval Herbal Remedies*, 138, 227. The herbal instructs the practitioner to pluck the plant while enunciating a special Latin incantation, helpfully provided.

84. Belief in the destructive potential of a malevolent gaze, or evil eye, is a pre-Christian idea that maintains currency in many parts of Europe and the Near East. A recent study of plants protective against the evil eye in Spain is José Antonio González et al., "Plant Remedies Against Witches and the Evil Eye in a Spanish 'Witches' Village," *Economic Botany* 66 (2012): 35–45.

85. Frederick Thomas Elworthy, *The Evil Eye: An Account of This Ancient and Widespread Superstition* (London: Murray, 1895).

86. Van Arsdall, *Medieval Herbal Remedies*, 152.

87. Percy Flemming, "The Medical Aspects of the Medieval Monastery in England," *Proceedings of the Royal Society of Medicine* 22 (1929): 771–782; Lucia Tongiorgio Tomasi, "The Renaissance Herbal," in *The Renaissance Herbal* (Bronx: New York Botanical Garden, 2013), 11.

88. Roy Porter, *The Greatest Benefit to Mankind: A Medical History of Humanity* (New York: Norton, 1997), 106–112.

89. Jerry Stannard, "The Theoretical Bases of Medieval Herbalism," *Medical Heritage* 1 (1985): 186–198.

90. Charles Webster, *Paracelsus: Medicine, Magic, and Mission at the End of Time* (New Haven, Conn.: Yale University Press, 2008), 15–19, 56–57. In addition to his unconventional beliefs about medicine and chemistry, Paracelsus wrote inflammatory tracts on theology.

91. Webster, *Paracelsus*, 150.

92. The Doctrine of Signatures touted by Paracelsus draws from a long line of folk practices in Europe. Nor is it

unique to Europe: many world folk practices have beliefs in treating disease with "similars" or "like-cures-like." Bradley C. Bennett considers the practice "ubiquitous" and argues that the Doctrine of Signatures serves as a mnemonic to help people in nonliterate societies teach and recall the uses of medicinal plants, in "Doctrine of Signatures: An Explanation of Medicinal Plant Discovery or Dissemination of Knowledge?" *Economic Botany* 61, no. 3 (2007): 246–255. See also Michael J. Balick, *Rodale's Twenty-First-Century Herbal: A Practical Guide for Healthy Living Using Nature's Most Powerful Plants* (New York: Rodale, 2014), 22–23.

93. There are also lungworts in the genus *Mertensia* and others.

94. R. C. A. Prior, *On the Popular Names of British Plants*, 2nd ed. (London: Williams and Norgate, 1870), 22, 136, 140. *A. clematitis* is a common European species of a widespread genus.

95. Mandrake was well known in antiquity and emerges in many ethnobotanic contexts through history.

96. These are John Gerard's (1545–1611?) *Herball or Generall Historie of Plantes* (1597, revised 1633); John Parkinson's (1567–1650) *Theatrum Botanicum*; and Nicholas Culpeper's (1616–1654) *The English Physitian* (1652). See Graeme Tobyn, Alison Denham, and Margaret Whiteleg, *The Western Herbal Tradition: Two Thousand Years of Medicinal Plant Knowledge* (Edinburgh: Churchill Livingstone/Elsevier, 2011), 13–15.

97. John Gerard, *The Herball, or Generall Historie of Plants* (London, 1597), 960.

98. For the significance of Linnaeus's contribution to simplifying nomenclature in a botanical community awash in impractical Latin names, see James L. Reveal, "What's in a Name: Identifying Plants in Pre-Linnaean Botanical Literature," in *Prospecting for Drugs in Ancient and Medieval European Texts: A Scientific Approach*, ed. Bart K. Holland (Amsterdam: Harwood, 1996), 57–90. Linnaeus's scheme for classifying biological specimens based on morphological features (in plants, flowers were deemed especially useful) came first, in 1735, and then his binomial was introduced in publications beginning in the 1740s, culminating in the grand *Species Plantarum* of 1753.

99. For a concise explanation of modern botanical classification, see Balick, *Twenty-First-Century Herbal*, 42–49. See also the Angiosperm Phylogeny Web site, http://www.mobot.org/MOBOT/research/APweb/.

100. Botanical convention expects that plants be identified by genus, species, and the authority (botanist or botanists) credited with the formal description of the organism in the scholarly literature. Therefore, the pumpkin is most properly identified as *Cucurbita pepo* L. The genus and species names are followed by L., which means Linnaeus was the first to describe this plant. Similarly, lavender (*Lavandula angustifolia* Mill.) bears the added name of the eighteenth-century botanist Philip Miller. This text will omit the authority in species names, although the precision of using the full botanic identifier is preferable in technical and scientific publications. See Bradley C. Bennett and Michael J. Balick, "Phytomedicine 101: Plant Taxonomy for Preclinical

and Clinical Medicinal Plant Researchers," *Journal of the Society for Integrative Oncology* 6 (2008): 150–157.

101. Systematists frequently reevaluate the taxonomic placement of species as new information about biological relationships become available in morphological and genetic studies. Up-to-date botanical name databases are maintained by the U.S. government (Integrated Taxonomic Information System, www.itis.gov), the Missouri Botanical Garden (www.tropicos.org), and the Plant List, an international consortium (www.theplantlist.org).

102. William Withering, *An Account of the Foxglove and Some of Its Medical Uses* (Birmingham, 1785).

103. Numerous plants were listed to treat dropsy in the medieval and renaissance herbals but probably not subjected to Withering's sort of curiosity.

104. Withering, *Account of the Foxglove*, 9.

105. Paul Hauptman and Ralph Kelly, "Digitalis," *Circulation* 99 (1999): 1265–1270.

106. Empirical evidence for a pharmaceutical's positive effects generally cannot be demonstrated without a carefully constructed experiment. Empirical evidence for negative outcomes might be deduced from folk experience. For example, surveying the hundreds of plant-derived treatments available to modern-day practitioners of indigenous medicine in the Peruvian rain forest, the Amazonian herbalist Justo Renquifo Guevara noted, "In the past, many of our brothers and sisters died so that we can have these medicines" (personal communication, 2009). He refers to the countless people who overdosed on medicinal plants or made poor combinations of herbs and whose experience served as warning to herbalists who followed them. Such lessons are passed along through the generations, helping shape a practice that seeks to maximize possible benefit and reduce risk.

107. Among the first analyses of this phenomenon was in Henry K. Beecher, "The Powerful Placebo," *Journal of the American Medical Association* 159 (1955): 98–109. The placebo effect in various clinical settings is reviewed in Howard Brody, "The Placebo Response: Recent Research and Implications for Family Medicine," *Journal of Family Practice* 49, no. 7 (2000): 649–654; F. Benedetti, "Placebo Analgesia," *Neurological Science* 27 (2006): S100–S102; and Damian G. Finniss et al., "Placebo Effects: Biological, Clinical and Ethical Advances," *Lancet* 375 (2010): 686–695. Despite the common vocabulary of the clinical research field that designates such treatments as "inactive," "mock," and "sham," the placebo effect is associated with quantifiable, replicable neurobiological-systemic changes in patients.

108. A shorthand for the therapeutic setting is the "doctor–patient relationship." For a reevaluation of the "placebo effect" phenomenon as a response to the "meaning of medicine," see Daniel E. Moerman, *Meaning, Medicine, and the "Placebo Effect"* (Cambridge: Cambridge University Press, 2002).

109. The mainstream or predominant health-care practice in North America and Europe, among other places, goes by many names among comparative medical scholars, including Western medicine, biomedicine, and orthodox medicine.

110. Chinese traditional medical training incorporates coursework in biomedical anatomy, pharmacology, and so forth.

111. Nutton, "Medicine in the Greek World," 24–25.

112. Zysk, "Traditional Medicine of India."

2. THE REGULATION OF DRUGS

1. The history of pharmacy includes several important early works that predate Pedanius Dioscorides's important first-century C.E. herbal, such as the Egyptian Ebers papyrus (1550 B.C.E.), the Chinese *Yellow Emperor's Classic of Internal Medicine* (300–100 B.C.E.), and the *Divine Husbandman's Classic of the Materia Medica* (early centuries C.E.), as well as writings by the Greek botanist Theophrastus (ca. 300 B.C.E.) and the Greek herbalist Krateus (first century B.C.E.). See Paul U. Unschuld, *Medicine in China: A History of Pharmaceutics* (Berkeley: University of California Press, 1986); and William E. Court, "Pharmacy from the Ancient World to 1100 A.D.," in *Making Medicines: A Brief History of Pharmacy and Pharmaceuticals*, ed. Stuart Anderson (London: Pharmaceutical Press, 2005), 21–36. Dioscorides's herbal is recognized for its systematic approach to pharmacy and its influence on future herbals in the European/Mediterranean context.

2. J. Vetter, "Poison Hemlock (*Conium maculatum* L.)," *Food and Chemical Toxicology* 42 (2004): 1374; Michael Wink and Ben-Erik van Wyk, *Mind-Altering and Poisonous Plants of the World* (Portland, Ore.: Timber Press, 2008), 94.

3. Will H. Blackwell. *Poisonous and Medicinal Plants* (Englewood Cliffs, N.J.: Prentice-Hall, 1990), 13–14; Enid Bloch, "Hemlock Poisoning and the Death of Socrates: Did Plato Tell the Truth?" in *The Death of Socrates*, by Emily Wilson (London: Profile, 2007), 258.

4. Blackwell, *Poisonous and Medicinal Plants*, 19; William Shakespeare, *Romeo and Juliet*, 5.1.

5. Bernard Ortiz de Montellano, "Empirical Aztec Medicine," *Science* 188, no. 4185 (1975): 216, referencing the *Codex Magliabechiano*,

6. Medical practice by religious authorities (for example, priests and monks) is a common theme in the history of health care in diverse geographical settings. For example, ancient Egyptian physicians practiced their art with priestly titles, in a medical system that invoked spirit possession as well as natural causation in illness. See Court, "Pharmacy," 25. During the European Dark Ages, classical medical thought was preserved by religious scholars in monasteries, who also added Christian elements to their principles of illness causation and treatment. The connection between religious calling and medical practice has long been apparent in Europe. As recently as the seventeenth century, as many as half of all London medical practitioners received their license to practice from the bishop of London or the dean of St. Paul's Cathedral. See Peter M. Worling, "Pharmacy in the Early Modern World, 1617 to 1841 A.D.," in *Making Medicines*, ed. Anderson, 62. Among many groups in the New World, health care falls under the purview of an individual specially skilled in mediating the religious-spiritual influences of health.

7. Penelope Ody, *The Complete Medicinal Herbal: A Practical Guide to the Healing Properties of Herbs, with More Than 250 Remedies for Common Ailments* (New York: Dorling Kindersley, 1993).

8. T. Van Andel et al., "The Medicinal Plant Trade in Suriname," *Ethnobotany Research and Applications* 5 (2007): 359.

9. Joseph Winter, "Traditional Uses of Tobacco by Native Americans," in *Tobacco Use by Native North Americans: Sacred Smoke and Silent Killer*, ed. Joseph Winter (Norman: University of Oklahoma Press, 2000), 9–58.

10. Nina L. Etkin explores the role of food and drink in social life from a biomedical and anthropological perspective in *Foods of Association: Biocultural Perspectives on Foods and Beverages That Mediate Sociability* (Tucson: University of Arizona Press, 2009). She calls such medically active, mostly plant-based foods and beverages "foods of association."

11. Stuart Anderson, "Researching and Writing the History of Pharmacy," in *Making Medicines*, ed. Anderson, 12; Peter G. Homan, "The Development of Community Pharmacy," in *Making Medicines*, ed. Anderson, 115–117. The pepperers and spicers of the Middle Ages gave rise over time to two new professions: the grocers and the apothecaries, the former specializing in foodstuffs and the latter in medicines. Joining the apothecaries in the pharmaceutical trade were the alchemists and druggists during the fifteenth and sixteenth centuries.

12. For a discussion of European witches, women medical practitioners, and their associated medicinal herbs, see John M. Riddle, *Goddesses, Elixirs, and Witches: Plants and Sexuality Throughout Human History* (New York: Palgrave Macmillan, 2010), 139–147. In addition to the roles noted in the text, it is important to note that female herbalists and wise women cultivated knowledge on drugs related to fertility, both male and female, and abortion.

13. Oversight of medical licensing was taken up by the American states beginning in the 1880s. See Edward Shorter, "Primary Care," in *The Cambridge History of Medicine*, ed. Roy Porter (Cambridge: Cambridge University Press, 2006), 110–111.

14. Samuel Thomson, *New Guide to Health or Botanic Family Physician*, 2nd ed. (Boston: House, 1825). Thomson's appeal lay in part in his "democratic approach to health." According to Thomson's teaching, "There was no need of doctors; every man could treat himself" (James Harvey Young, *The Toadstool Millionaires: A Social History of Patent Medicines in America Before Federal Regulation* [Princeton, N.J.: Princeton University Press, 1961], 54).

15. Thomson, *New Guide to Health*, 38–79; Young, *Toadstool Millionaires*, 44–57.

16. John S. Haller, *Medical Protestants: The Eclectics in American Medicine, 1825–1939* (Carbondale: Southern Illinois University Press, 1994).

17. According to Paul Gahlinger, "[The use of drugs] was considered a matter of personal choice and responsibility"

(*Illegal Drugs: A Complete Guide to Their History, Chemistry, Use, and Abuse* [New York: Plume, 2004], 57).

18. David F. Musto explains, referring to drugs of abuse: "Federal control over narcotic use and the prescription practices of the medical profession were thought in 1900 to be unconstitutional" (*The American Disease: Origins of Narcotic Control*, 3rd ed. [Oxford: Oxford University Press, 1999], 9). Any laws regulating medicinal substances were solely at the municipal or state level, and there were few.

19. See, especially, Young, *Toadstool Millionaires*; and Philip J. Hilts, *Protecting America's Health: The FDA, Business, and One Hundred Years of Regulation* (New York: Knopf, 2003), 46–49.

20. Federal Food and Drugs Act of 1906, P.L. 59-384, 34 Stat. 768, effective January 1, 1907.

21. Musto, *American Disease*, 6, 43.

22. Smoking Opium Exclusion Act, 35 Stat. 614, effective April 1, 1909; Harrison Narcotics Tax Act of 1914, P. L. 63-223, 38 Stat. 785, effective March 1, 1915. The Harrison Act permitted the sale of opium and coca products by prescription.

23. Musto, *American Disease*, 255.

24. Comprehensive Drug Abuse Prevention and Control Act of 1970, P.L. 91-513, 84 Stat. 1236, effective October 27, 1970.

25. A number of high-profile federal court cases worked through the system during the early 2000s, but more recently, the White House has taken a less aggressive stance toward such crimes. See James M. Cole, "Memorandum for All United States Attorneys: Guidance Regarding Marijuana Enforcement," Department of Justice, Office of the Deputy Attorney General, August 29, 2013. For the role of the Justice Department in choosing to prosecute violations of federal marihuana laws, see Ashley Southall and Jack Healey, "U.S. Won't Sue to Reverse States' Legalization of Marijuana," *New York Times*, August 29, 2013.

26. Musto, *American Disease*, 274–275.

27. *Kimbrough v. United States*, 06-6330, decided December 10, 2007. Congress responded by passing the Fair Sentencing Act of 2010, P.L. 111-220, effective November 1, 2011.

3. THE ACTIONS OF MEDICINAL PLANTS

1. Daniel E. Moerman, *Native American Ethnobotany* (Portland, Ore.: Timber Press, 1998), 502.

2. Daniel E. Moerman considers that many therapeutic responses derive from the cultural context in which disease is considered and medicine delivered. Rather than a "placebo response," Moerman calls this a response to the "meaning" of medicine, in *Meaning, Medicine, and the "Placebo Effect"* (Cambridge: Cambridge University Press, 2002).

3. Gerard J. Tortora and Bryan H. Derrickson, *Principles of Anatomy and Physiology*, 12th ed. (Hoboken, N.J.: Wiley, 2009), 922–923.

4. Many herbal medicines with activity in the digestive system cause the patient to vomit or defecate. These activities were integrated into existing conceptions of illness in which purgation removed bad influences from the body and helped restore health. See Edward Shorter, "Primary Care," in *The Cambridge History of Medicine*, ed. Roy Porter (Cambridge: Cambridge University Press, 2006), 103–135.

5. "Remedy to clear out the body and to get rid of the excrement in the body of a person" (Cyril Bryan, *Ancient Egyptian Medicine: The Papyrus Ebers* [London: Bles, 1930], 44).

6. George A. Burdock, Ioana G. Carabin, and James C. Griffiths, "Toxicology and Pharmacology of Sodium Ricinoleate," *Food and Chemical Toxicology* 44 (2006): 1689–1698; Sorin Tunaru et al., "Castor Oil Induces Laxation and Uterus Contraction via Ricinoleic Acid Activating Prostaglandin EP3 Receptors," *Proceedings of the National Academy of Sciences USA* 109 (2012): 9179–9184.

7. Holly J. Bayne, "FDA Issues Final Rule Banning Use of Aloe and Cascara Sagrada in OTC Drug Products," *HerbalGram* 56 (2002), 56; "Aloe Vera," National Center for Complementary and Integrative Health, Department of Health and Human Services, http://nccam.nih.gov/health/aloevera/.

8. Walter Lewis and Memory P. F. Elvin-Lewis, *Medical Botany: Plants Affecting Human Health* (Hoboken, N.J.: Wiley, 2003), 468.

9. Ben-Erik Van Wyk and Michael Wink, *Medicinal Plants of the World* (Portland, Ore.: Timber Press, 2004), 298.

10. These species names are accepted under the Integrated Taxonomic Information System but not by the Plant List (http://www.theplantlist.org), which accepts *Frangula alnus* and *F. purshiana*.

11. George A. Davis, *The Pharmacology of the Newer Materia Medica* (Detroit: Davis, 1892), 163–225.

12. John Davidson, *The Cascara Tree in British Columbia* (Victoria, B.C.: Ministry of Agriculture, 1942), 5–6, 21–23; Lewis and Elvin-Levis, *Medical Botany*, 470.

13. Bayne, "FDA Issues Final Rule," 56.

14. These plants' active principles are phenolic compounds called anthraquinones.

15. P. Simmonds, "Notes on Some Saps and Secretions Used in Pharmacy," *American Journal of Pharmacy* 67, no. 3 (1895): 7–15, and "Notes on Some Saps and Secretions Used in Pharmacy," *American Journal of Pharmacy* 67, no. 4 (1895): 7–11; John Uri Lloyd, "History of the Vegetable Drugs of the Pharmacopeia of the United States," *Bulletin of the Lloyd Library of Botany, Pharmacy and Materia Medica* 18, no. 4 (1911): 52.

16. John King and Robert S. Newton, *The Eclectic Dispensatory of the United States of America* (Cincinnati: Derby, 1852), 336–337.

17. Keith A. Sharkey and John L. Wallace, "Treatment of Disorders of Bowel Motility and Water Flux; Anti-Emetics; Agents Used in Biliary and Pancreatic Disease," in *Goodman and Gilman's The Pharmacological Basis of Therapeutics*, 12th ed., ed. Laurence L. Brunton, Bruce A. Chabner, and Björn C. Knollmann (New York: McGraw-Hill, 2011), 1337–1338.

18. Synonym: *Cephaelis ipecacuanha*. Hari Lorenzi and F. J. Abreu Matos, *Plantas medicinais no Brasil*, 2nd ed.

(Nova Odessa, Brazil: Instituto Plantarum de Estudos da Flora, 2008), 452–453.

19. M. R. Lee, "Ipecacuanha: The South American Vomiting Root," *Journal of the Royal College of Physicians, Edinburgh* 38 (2008): 355, 358.

20. Tortora and Derrickson, *Principles of Anatomy and Physiology*, 728.

21. Tortora and Derrickson, *Principles of Anatomy and Physiology*, 690–707.

22. Congestive heart failure has two clinical aspects that progress over time. One is pulmonary edema, fluid accumulation in the lungs. The other is peripheral edema, an increase in blood volume and fluid accumulation in the extremities. Either aspect can be evident first in the progression of the disease. See Tortora and Derrickson, *Principles of Anatomy and Physiology*, 742.

23. John Gerard, *The Herball, or Generall Historie of Plants* (London, 1597), 647.

24. Nicholas Culpeper, *The English Physitian Enlarged* (London: Streater, 1666), 109.

25. William Withering, *An Account of the Foxglove and Some of Its Medical Uses* (Birmingham, 1785).

26. Paul Hauptman and Ralph Kelly, "Digitalis," *Circulation* 99 (1999): 1265–1270; Bradley A. Maron and Thomas P. Rocco, "Pharmacotherapy of Congestive Heart Failure," in *Goodman and Gilman*, ed. Brunton, Chabner, and Knollmann, 802–803.

27. Maron and Rocco, "Pharmacotherapy of Congestive Heart Failure," 803. These include angiotensin-converting enzyme (ACE) inhibitors and β-adrenergic receptor antagonists (beta blockers).

28. Kevin J. Sampson and Robert S. Kass, "Anti-Arrhythmic Drugs," in *Goodman and Gilman*, ed. Brunton, Chabner, and Knollmann, 838.

29. Joseph Monachino, "*Rauvolfia serpentina*: Its History, Botany, and Medical Use," *Economic Botany* 8 (1954): 349–365; Rustom Jal Vakil, "*Rauwolfia serpentina* in the Treatment of High Blood Pressure: A Review of the Literature," *Circulation* 12 (1955): 220–229; World Health Organization, *WHO Monographs on Selected Medicinal Plants* (Geneva: World Health Organization, 1999), 1:221–230.

30. Reserpine binds irreversibly to vesicular catecholamine transporters in central and peripheral adrenergic neurons, which prevents norepinephrine and dopamine signaling.

31. Lewis and Elvin-Lewis, *Medical Botany*, 286–287.

32. Thomas Michel and Brian B. Hoffman, "Treatment of Myocardial Ischemia and Hypertension," in *Goodman and Gilman*, ed. Brunton, Chabner, and Knollmann, 776.

33. Thomas C. Westfall and David P. Westfall, "Adrenergic Agonists and Antagonists," in *Goodman and Gilman*, ed. Brunton, Chabner, and Knollmann, 300.

34. Tortora and Derrickson, *Principles of Anatomy and Physiology*, 874–889.

35. Elaine M. Marieb and Katja Hoehn, *Human Anatomy and Physiology*, 7th ed. (San Francisco: Cummings, 2007), 831–842.

36. Jan Timbrook, "Ethnobotany of Chumash Indians, California, Based on Collections by John P. Harrington," *Economic Botany* 44 (1990): 236–253.

37. N. Rajakumar and M. B. Shivanna, "Ethnomedicinal Application of Plants in the Eastern Region of Shimoga District, Karnataka, India," *Journal of Ethnopharmacology* 126 (2009): 64–73; Geone M. Corrêa and Antônio F. de C. Alcântara, "Chemical Constituents and Biological Activities of Species of *Justicia*—A Review," *Brazilian Journal of Pharmacognosy* 22 (2012): 220–238.

38. Lewis and Elvin-Lewis, *Medical Botany*, 491. Moerman's Native American Ethnobotany database (http://herb.umd.umich.edu) lists dozens of citations for the use of cherry bark in cough medicine among numerous tribes.

39. Among many indigenous uses of this species. See Paul B. Hamel and Mary U. Chiltoskey, *Cherokee Plants: Their Uses—A Four-Hundred-Year History* (Sylva, N.C.: Herald, 1975), 62; and J. T. Garrett, *The Cherokee Herbal: Native Plant Medicine from the Four Directions* (Rochester, Vt.: Bear, 2003), 249.

40. John Parkinson, *Theatrum Botanicum* (London: Cotes, 1640), 46.

41. Dan Bensky, Steven Clavey, and Erich Stöger consider the first citation of *ma huang* to be in the *Divine Husbandman's Classic of the Materia Medica* of the later Han period, in *Chinese Herbal Medicine: Materia Medica*, 3rd ed. (Seattle: Eastland, 2004), 4.

42. Bensky, Clavey, and Stöger, *Chinese Herbal Medicine*, 3–9.

43. Ehab A. Abourashed et al., "*Ephedra* in Perspective—A Current Review." *Phytotherapy Research* 17 (2003): 703–712.

44. M. K. Kaul, *Medicinal Plants of Kashmir and Ladakh: Temperate and Cold Arid Himalaya* (New Delhi: Indus, 1997), 117; Lewis and Elvin-Lewis, *Medical Botany*, 487. This is in contrast to R. N. Chopra and I. C. Chopra, who relay that *E. gerardiana* was not employed in traditional South Asian medicine (*Indigenous Drugs of India* [Calcutta: Art Press, 1933]), and I. I. Chaudhri, who claims it saw little use in the traditional medicine of modern-day Pakistan ("Pakistani Ephedra," *Economic Botany* 11 [1957]: 257–263).

45. Ephedrine activates β-adrenergic receptors in the lungs that relax the smooth muscle of the bronchial tree and also activates both α- and β-adrenergic receptors in the central and peripheral nervous system. Norepinephrine release is enhanced. See Thomas C. Westfall and David P. Westfall, "Neurotransmission: The Autonomic and Somatic Motor Nervous Systems," and "Adrenergic Agonists and Antagonists," both in *Goodman and Gilman*, ed. Brunton, Chabner, and Knollmann, 198, 300. The jointfir active principles are more informatively written l-ephedrine and d-pseudoephedrine, with prefixes that communicate to organic chemists the three-dimensional arrangement of atoms in the molecules.

46. Van Wyk and Wink, *Medicinal Plants of the World*, 134.

47. M. R. Lee, "The History of Ephedra (ma-huang)," *Journal of the Royal College of Physicians Edinburgh* 41 (2011): 78–84.

48. In severe asthma, the theophylline derivative aminophylline can be delivered intravenously.

49. Peter J. Barnes, "Theophylline: New Perspectives for an Old Drug," *American Journal of Respiratory and Critical Care Medicine* 167 (2003): 813–818; P. J. Barnes and R. A. Stockley, "COPD: Current Therapeutic Interventions and Future Approaches," *European Respiratory Journal* 25 (2005): 1084–1106; Mario Cazzola et al., "Pharmacology and Therapeutics of Bronchodilators," *Pharmacological Reviews* 64 (2012): 450–504.

50. Lewis and Elvin-Lewis, *Medical Botany*, 493.

51. Ronald Eccles, "Menthol: Effects on Nasal Sensation of Airflow and the Drive to Breathe," *Current Allergy and Asthma Reports* 3 (2003): 210–214; Diane Pappas and J. Owen Hendley, "The Common Cold and Decongestant Therapy," *Pediatrics in Review* 32 (2011): 47–54.

52. Tortora and Derrickson, *Principles of Anatomy and Physiology*, 1019–1021, 1049–1052.

53. Carole Rawcliffe, *Medicine and Society in Later Medieval England* (Stroud: Sutton, 1995), 46–52.

54. Bensky, Clavey, and Stöger, *Chinese Herbal Medicine*, 286.

55. Herbal medicines that increase urine volume are often termed aquaretics because they elevate the volume of urine without electrolyte loss. Eric Yarnell disputes that many herbal treatments fit this description and considers that most are properly described as diuretic, in "Botanical Medicines for the Urinary Tract," *World Journal of Urology* 20 (2002): 285–293. Robert F. Reilly and Edwin K. Jackson define diuretics as "drugs that increase the rate of urine flow" ("Regulation of Renal Function and Vascular Volume," in *Goodman and Gilman*, ed. Brunton, Chabner, and Knollmann, 677), and clinically useful diuretics also increase the rate of salt (usually NaCl) excretion.

56. Tortora and Derrickson, *Principles of Anatomy and Physiology*, 1045; Reilly and Jackson, "Regulation of Renal Function and Vascular Volume," 696–701.

57. Lewis and Elvin-Lewis list species in fifty-three genera with traditional uses as diuretics, in *Medical Botany*, 414–415. These include the horsetails *Equisetum arvense* and *E. robustum*, stinging nettle (*Urtica dioica*), and parsley (*Petroselinum crispum*). Moerman's Native American Ethnobotany database contains over 200 references to diuretic plant uses in North America alone. Few of these plants have clinically demonstrated diuretic or aquaretic activities.

58. Mark Blumenthal, Alicia Goldberg, and Josef Brinckmann, eds., *Herbal Medicine: Expanded Commission E Monographs* (Austin, Tex.: American Botanical Council, 2000); Katrin Schütz, Reinhold Carle, and Andreas Schieber, "*Taraxacum*: A Review on Its Phytochemical and Pharmacological Profile," *Journal of Ethnopharmacology* 107 (2006): 313–323; C. I. Wright et al., "Herbal Medicines as Diuretics: A Review of the Scientific Evidence," *Journal of Ethnopharmacology* 114 (2007): 1–31; Bensky, Clavey, and Stöger cite several East Asian *Taraxacum* spp. for this purpose in *Chinese Herbal Medicine*, 162–164. Although dandelion was widely used in North American Indian medicine, enhancing urine flow was not among its major applications. See Moerman, *Native American Ethnobotany*, 550–551.

59. Marta González-Castejón, Francesco Visioli, and Arantxa Rodriguez-Casado, "Diverse Biological Activities of Dandelion," *Nutrition Reviews* 70 (2012): 534–547.

60. Harmut Osswald and Jürgen Schnermann, "Methylxanthines and the Kidney," in *Methylxanthines: Handbook of Experimental Pharmacology*, ed. Bertil B. Fredholm (Berlin: Springer, 2011), 391–412.

61. Ruth G. Jepson, Gabrielle Williams, and Jonathan C. Craig conducted a review of trials in which cranberry or cranberry-lingonberry juice or capsules were tested versus placebo, water, or pharmaceutical treatment (depending on experimental design) for effect on the incidence of urinary tract infections. The data indicate that cranberry or cranberry-lingonberry do not significantly affect the occurrence of infections. The authors noted, however, substantial methodological problems in recent research that may have obscured detection of an effect of cranberry on urinary tract infection, if one indeed exists. See Ruth G. Jepson, Gabrielle Williams, and Jonathan C. Craig, "Cranberries for Preventing Urinary Tract Infections," *Cochrane Database of Systematic Reviews* (2012): CD001321. Mixed scientific evidence is common in clinical tests of herbal medicines. Different investigators choose their own preparations of plant material or commercially available preparations. Therefore, some researchers use concentrations of plant extracts and plant parts that are inconsistent, rendering the doses and bioactivity of separate studies incomparable. It is important to recognize that lack of uniform scientific evidence does not support or reject the efficacy of any herbal medicine since one must first consider the highly disparate testing conditions.

62. Elyad Davidson et al., "Prevention of Urinary Tract Infections with *Vaccinium* Products," *Phytotherapy Research* 28 (2014): 465–470.

63. Tortora and Derrickson, *Principles of Anatomy and Physiology*, 1082–1095.

64. Claus G. Roehrborn, "Benign Prostatic Hyperplasia: An Overview," *Review of Urology* 7 (2005): S3–S14.

65. Edwin M. Hale, *Saw Palmetto (Sabal Serrulata, Serenoa Serrulata): Its History, Botany, Chemistry, Pharmacology, Provings, Clinical Experience, and Therapeutic Applications* (Philadelphia: Boericke and Tafel, 1898).

66. J. Tacklind et al., "*Serenoa repens* for Benign Prostatic Hyperplasia," *Cochrane Database of Systematic Reviews* (2012): CD001423.

67. Fouad K. Habib, "*Serenoa repens*: The Scientific Basis for the Treatment of Benign Prostatic Hyperplasia," *European Urology Supplements* 8 (2009): 887–893.

68. K. M. Stewart, "The African Cherry (*Prunus africana*): Can Lessons Be Learned from an Overexploited Medicinal Tree?" *Journal of Ethnopharmacology* 89 (2003): 3–13; Kristine M. Stewart, "The African Cherry (*Prunus africana*): From Hoe-Handles to the International Herb Market," *Economic Botany* 57 (2003): 559–569. *P. africana* is at risk of becoming endangered because of overharvesting

of its bark in the wild to supply the international medicine market. See Ian Dawson, James Were, and Ard Lengkeek, "Conservation of *Prunus africana*, an Overexploited African Medicinal Tree," *Forest Genetic Resources* 28 (2000): 27–33.

69. Timothy Wilt and Areef Ishani, "*Pygeum africanum* for Benign Prostatic Hyperplasia," *Cochrane Database of Systematic Reviews* (2002): CD001423; World Health Organization, *WHO Monographs on Selected Medicinal Plants* (Geneva: World Health Organization, 2004), 2:246–258; Annie Hutchison et al., "The Efficacy of Drugs for the Treatment of LUTS/BPH: A Study in Six European Countries," *European Urology* 51 (2007): 207–216.

70. Tristan M. Nicholson and William A. Ricke, "Androgens and Estrogens in Benign Prostatic Hyperplasia: Past, Present, and Future," *Differentiation* 82 (2011): 184–199.

71. Tortora and Derrickson, *Principles of Anatomy and Physiology*, 1095–1116.

72. Bensky, Clavey, and Stöger relay that *ai ye* first appeared in the Chinese pharmacopeia in the *Miscellaneous Records of Famous Physicians* of Tao Hongjing (ca. 500 C.E.), in *Chinese Herbal Medicine*, 594–596.

73. Bryan, *Papyrus Ebers*, 83.

74. John M. Riddle, *Eve's Herbs: A History of Contraception and Abortion in the West* (Cambridge, Mass.: Harvard University Press, 1997), 46–47.

75. Parkinson, *Theatrum Botanicum*, 30.

76. Dieudonné Njamen et al., "Phytotherapy and Women's Reproductive Health: The Cameroonian Perspective," *Planta Medica* 79 (2013): 600–611; Ngeh J. Toyang and Rob Verpoorte, "A Review of the Medicinal Potentials of Plants of the Genus *Vernonia* (Asteraceae)," *Journal of Ethnopharmacology* 146 (2013): 681–723. A *Vernonia* species is used in the Gambia for a similar purpose. See Clare Madge, "Therapeutic Landscapes of the Jola, the Gambia, West Africa," *Health & Place* 4 (1998): 293–311. North American *Vernonia* species also have a reputation as uterine tonics or for female reproductive health. See John K. Crellin and Jane Philpott, *Herbal Medicine Past and Present*, vol. 2, *Monographs on Select Medicinal Plants* (Durham, N.C.: Duke University Press, 1990), 266.

77. For example, Pedanius Dioscorides of Anazarbus, *De materia medica*, trans. Lily Y. Beck (Hildesheim: Olms-Weidmann, 2005), 74. Both Dioscorides and Gerard (*Herball*, 1201), for example, recommend that chaste tree seed be taken with pennyroyal (the purported contraceptive or abortifacient plant) to promote menstruation. See also Riddle, *Eve's Herbs*, 57–58; and John M. Riddle, *Goddesses, Elixirs, and Witches: Plants and Sexuality Throughout Human History* (New York: Palgrave Macmillan, 2010), 123–126.

78. Riddle, *Eve's Herbs*, 57–58.

79. Gerard, *Herball*, 1201–1202.

80. Van Wyk and Wink, *Medicinal Plants of the World*, 343.

81. Rebecca L. Johnson et al., *National Geographic Guide to Medicinal Herbs: The World's Most Effective Healing Plants* (Washington, D.C.: National Geographic, 2010), 288–290.

82. M. Diana van Die et al., "*Vitex agnus-castus* Extracts for Female Reproductive Disorders: A Systematic Review of Clinical Trials," *Planta Medica* 79 (2013): 562–575.

83. W. Wuttke et al., "Chaste Tree (*Vitex agnus-castus*): Pharmacology and Clinical Indications," *Phytomedicine* 10 (2003): 348–357; D. E. Webster et al., "Activation of the μ-opiate Receptor by *Vitex agnus-castus* Methanol Extracts: Implication for Its Use in PMS," *Journal of Ethnopharmacology* 106 (2006): 216–221; Shao-Nong Chen et al., "Phytoconstituents from *Vitex agnus-castus* Fruits," *Fitoterapia* 82 (2011): 528–533.

84. Blumenthal, Goldberg, and Brinckmann, eds., *Herbal Medicine*.

85. A. Morales, "Yohimbine in Erectile Dysfunction: The Facts," *International Journal of Impotence Research* 12 (2000): S70–S74.

86. S. William Tam, Manuel Worcel, and Michael Wyllie, "Yohimbine: A Clinical Review," *Pharmacology & Therapeutics* 91 (2001): 215–243; Rany Shamloul, "Natural Aphrodisiacs," *Journal of Sexual Medicine* 7 (2010): 39–49.

87. Westfall and Westfall, "Adrenergic Agonists and Antagonists," 309; Christopher C. K. Ho and Hui Meng Tan, "Rise of Herbal and Traditional Medicine in Erectile Dysfunction Management," *Current Urology Reports* 12 (2011): 470–478. Yohimbine acts as an α_2-adrenergic receptor antagonist, increasing blood pressure and heart rate.

88. Bensky, Clavey, and Stöger, *Chinese Herbal Medicine*, 778–780.

89. Huiping Ma et al., "The Genus *Epimedium*: An Ethnopharmacological and Phytochemical Review," *Journal of Ethnopharmacology* 134 (2011): 519–541.

90. Ho and Tan, "Rise of Herbal and Traditional Medicine," 475.

91. Evidence is provided for both estrogenic and androgenic properties in P. Shen et al., "Taxonomic, Genetic, Chemical, and Estrogenic Characteristics of *Epimedium* Species," *Phytochemistry* 68 (2007): 1448–1458; and Ma et al., "Genus *Epimedium*." *Epimedium* spp. extracts have also been considered for the treatment of menopausal symptoms and many other conditions.

92. Icariin blocks the activity of phosphodiesterase enzymes, similarly to the synthetic drugs sildenafil (Viagra), vardenafil (Levitra), and tadalafil (Cialis).

93. Tortora and Derrickson, *Principles of Anatomy and Physiology*, 176–178, 302–303.

94. Lewis and Elvin-Lewis catalogued more than 100 species from over 40 taxonomic families contributing to curare among different tribes of modern-day South America, in *Medical Botany*, 238–240.

95. Norman G. Bisset relays that while some Southeast Asian and Central African *Strychnos* species produce the potent convulsant alkaloid strychnine, and although indigenous people use it in arrow poisons, strychnine is not the paralytic agent in South American arrow poisons, in "Arrow Poisons and Their Role in the Development of Medicinal Agents," in *Ethnobotany: Evolution of a Discipline*, ed. Richard Evans Schultes and Siri von Reis (Portland, Ore.: Dioscorides, 1995), 289–302. Instead, *Strychnos* spp. and

other plants contribute the nerve-blocking alkaloids of the curarine, toxiferine, and calabassine families. See also Geneviève Philippe et al., "About the Toxicity of Some *Strychnos* Species and Their Alkaloids," *Toxicon* 44 (2004): 405–416. To account for the three-dimensional arrangement of atoms, the active principle is more specifically conveyed as d-tubocurarine.

96. M. R. Lee, "Curare: The South American Arrow Poison," *Journal of the Royal College of Physicians Edinburgh* 35 (2005): 83–92; W. C. Bowman, "Neuromuscular Block," *British Journal of Pharmacology* 147 (2006): S277–S286.

97. Bowman, "Neuromuscular Block"; Ryan E. Hibbs and Alexander C. Zambon, "Agents Acting at the Neuromuscular Junction and Autonomic Ganglia," in *Goodman and Gilman*, ed. Brunton, Chabner, and Knollmann, 258–268.

98. Tortora and Derrickson, *Principles of Anatomy and Physiology*, 296.

99. Chongyun Liu, Angela Tseng, and Sue Yang, *Chinese Herbal Medicine: Modern Applications of Traditional Formulas* (Boca Raton, Fla.: CRC Press, 2005), 493.

100. C. K. Atal, O. P. Gupta, and S. H. Afaq, "*Commiphora mukul*: Source of Guggal in Indian Systems of Medicine," *Economic Botany* 29 (1975): 208–218; Shishir Shishodia et al., "The Guggul for Chronic Diseases: Ancient Medicine, Modern Targets," *Anticancer Research* 28 (2008): 3647–3664.

101. Jongbae Park and Edzard Ernst, "Ayurvedic Medicine for Rheumatoid Arthritis: A Systematic Review," *Seminars in Arthritis and Rheumatism* 34 (2005): 705–713; Dinesh Khanna et al., "Natural Products as a Gold Mine for Arthritis Treatment," *Current Opinion in Pharmacology* 7 (2007): 344–351.

102. Makoto Tominaga and Michael J. Caterina, "Thermosensation and Pain," *Journal of Neurobiology* 61 (2004): 3–12; Craig Burkhart, Dean Morrell, and Lowell Goldsmith, "Dermatological Pharmacology," in *Goodman and Gilman*, ed. Brunton, Chabner, and Knollmann, 1830.

103. P. Anand and K. Bley, "Topical Capsaicin for Pain Management: Therapeutic Potential and Mechanisms of Action of the New High-Concentration Capsaicin 8% Patch," *British Journal of Anaesthesia* 107 (2011): 490–502.

104. Lewis and Elvin-Lewis, *Medical Botany*, 247; Burkhart, Morrell, and Goldsmith, "Dermatological Pharmacology," 1830. Despite the well-resolved mechanism of action (Anand and Bley, "Topical Capsaicin for Pain Management"), it appears that externally applied capsaicin at normally prescribed doses is not effective in reducing pain deeper in the body, such as caused by arthritis. See Melainie Cameron and Sigrun Chrubasik, "Topical Herbal Therapies for Treating Osteoarthritis," *Cochrane Database of Systematic Reviews* (2013): CD010538. Anand and Bley discuss the possible efficacy of a capsaicin patch delivering the active principle at 8 percent, in "Topical Capsaicin for Pain Management."

105. Tortora and Derrickson, *Principles of Anatomy and Physiology*, 842–844.

106. Tortora and Derrickson, *Principles of Anatomy and Physiology*, 867.

107. Rawcliffe, *Medicine and Society*, 14–15.

108. Gerard, *Herball*, 462–463, 826–827.

109. Kelly Kindscher, "Ethnobotany of Purple Coneflower (*Echinacea angustifolia*, Asteraceae) and Other Echinacea Species," *Economic Botany* 43, no. 4 (1989): 498–507.

110. Moerman, *Native American Ethnobotany*, 205–206.

111. Christopher Hobbs, "Echinacea: A Literature Review. Botany, History, Chemistry, Pharmacology, Toxicology, and Clinical Uses," *HerbalGram* 30 (1994): 33; Francis Brinker, "*Echinacea* Differences Matter: Traditional Uses of *Echinacea angustifolia* Root Extracts Versus Modern Clinical Trials with *Echinacea purpurea* Fresh Plant Extracts," *HerbalGram* 97 (2013): 46–57.

112. Lewis and Elvin-Lewis, *Medical Botany*, 494.

113. S. N. Groom, T. Johns, and P. R. Oldfield, "The Potency of Immunomodulatory Herbs May Be Primarily Dependent upon Macrophage Activation," *Journal of Medicinal Food* 10 (2007): 73–79; S. M. Sharma et al., "Bactericidal and Anti-inflammatory Properties of a Standardized *Echinacea* Extract (Echinaforce): Dual Actions Against Respiratory Bacteria," *Phytomedicine* 17 (2010): 563–568; M. R. Ritchie et al., "Effects of Echinaforce® Treatment on *ex vivo*-Stimulated Blood Cells," *Phytomedicine* 18 (2011): 826–831; James B. Hudson, "Applications of the Phytomedicine *Echinacea purpurea* (Purple Coneflower) in Infectious Diseases," *Journal of Biomedicine and Biotechnology* (2012): Article ID 769896.

114. S. A. Shah et al., "Evaluation of *Echinacea* for the Prevention and Treatment of the Common Cold: A Meta-Analysis," *Lancet Infectious Disease* 7 (2007): 473–480. See also Ronald B. Turner et al., "An Evaluation of *Echinacea angustifolia* in Experimental Rhinovirus Infections," *New England Journal of Medicine* 353 (2005): 341–348; Bruce Barrett et al., "*Echinacea* for Treating the Common Cold: A Randomized Controlled Trial," *Annals of Internal Medicine* 153 (2010): 769–777; and M. Jawad et al., "Safety and Efficacy Profile of *Echinacea purpurea* to Prevent Common Cold Episodes: A Randomized, Double-Blind, Placebo-Controlled Trial," *Evidence-Based Complementary and Alternative Medicine* (2012): ID 841315.

115. Cristina Fiore et al., "A History of the Therapeutic Use of Liquorice in Europe," *Journal of Ethnopharmacology* 99 (2005): 317–324; Gayle Engels, "Licorice," *HerbalGram* 70 (2006): 1, 4–5; Marjan Nassiri Asl and Hossein Hosseinzadeh, "Review of Pharmacological Effects of *Glycyrrhiza* sp. and Its Bioactive Compounds," *Phytotherapy Research* 22 (2008): 709–724.

116. Xiaoying Wang et al. call licorice a "guide drug" for its use in traditional medicine to improve the efficacy of other components in multiherb formulas, in "Liquorice, a Unique 'Guide Drug' of Traditional Chinese Medicine: A Review of Its Role in Drug Interactions," *Journal of Ethnopharmacology* 150 (2013): 781–790.

117. Fiore et al., "History"; C. Fiore et al., "Antiviral Effects of Glycyrrhiza Species," *Phytotherapy Research* 22 (2008): 141–148. Tzu-Chien Kao, Chi-Hao Wu, and Gow-Chin Yen review the laboratory evidence for possible roles of licorice constituents against a range of disease conditions, in

"Bioactivity and Potential Health Benefits of Licorice," *Journal of Agricultural and Food Chemistry* 62 (2014): 542–553.

118. There is some laboratory evidence that certain medicinal plants considered to support general resilience may mediate changes in the expression of particular gene products in the brain associated with stress-lowering activity. See Alexander Panossian et al., "Adaptogens Exert a Stress-Protective Effect by Modulation of Expression of Molecular Chaperones," *Phytomedicine* 16 (2009): 617–622; and Alexzander Asea et al., "Evaluation of Molecular Chaperons Hsp72 and Neuropeptide Y as Characteristic Markers of Adaptogenic Activity of Plant Extracts," *Phytomedicine* 20 (2013): 1323–1329.

119. I. I. Brekhman and I. V. Dardymov define an adaptogen as a substance capable of enhancing the "state of non-specific resistance" ("New Substances of Plant Origin Which Increase Nonspecific Resistance," *Annual Review of Pharmacology* 9 [1969]: 419–430).

120. Van Wyk and Wink, *Medicinal Plants of the World*, 38–39.

121. Lewis and Elvin-Lewis, *Medical Botany*, 294.

122. Joseph M. Vinetz et al., "Chemotherapy of Malaria," in *Goodman and Gilman*, ed. Brunton, Chabner, and Knollmann, 1406.

123. Bensky, Clavey, and Stöger, *Chinese Herbal Medicine*, 219–221; Elisabeth Hsu, "*Qing hao* 青蒿 (*Herba Artemisiae annuae*) in the Chinese *Materia Medica*," in *Plants, Health, and Healing: On the Interface of Ethnobotany and Medical Anthropology*, ed. Elisabeth Hsu and Stephen Harris (New York: Berghahn, 2010), 86–90.

124. N. J. White, "*Qinghaosu* (Artemisinin): The Price of Success," *Science* 320 (2008): 330–334.

125. Nicholas J. White, "Artemisinin Resistance: The Clock Is Ticking," *Lancet* 376 (2010): 2051–2052; Vinetz et al., "Chemotherapy of Malaria," 1395–1396.

4. THE ACTIONS OF MEDICINAL PLANTS ON THE NERVOUS SYSTEM

1. Gerard J. Tortora and Bryan H. Derrickson, *Principles of Anatomy and Physiology*, 12th ed. (Hoboken, N.J.: Wiley, 2009), 508; Eric R. Kandel et al., *Principles of Neural Science*, 5th ed. (New York: McGraw-Hill, 2013), 8.

2. Tortora and Derrickson, *Principles of Anatomy and Physiology*, 508; Kandel et al., *Principles of Neural Science*, 8.

3. Kandel et al., *Principles of Neural Science*, 342.

4. Tortora and Derrickson, *Principles of Anatomy and Physiology*, 508; Kandel et al., *Principles of Neural Science*, 8.

5. Tortora and Derrickson, *Principles of Anatomy and Physiology*, 547; Joshua M. Galanter, Susannah B. Cornes, and Daniel H. Lowenstein, "Principles of Nervous System Physiology and Pharmacology," in *Principles of Pharmacology: The Pathophysiologic Basis of Drug Therapy*, 3rd ed., ed. David E. Golan et al. (Philadelphia: Lippincott Williams & Wilkins, 2012), 93–96.

6. Tortora and Derrickson state that neuron cell bodies range from 5 to 135 micrometers (μm), in *Principles of Anatomy and Physiology*, 419. Kandel et al. give a typical cell body diameter of 50 micrometers, an axon diameter between 0.2 and 20 micrometers, and an axon length of 0.1 millimeter to 2 meters or more, in *Principles of Neural Science*, 22).

7. Kandel et al., *Principles of Neural Science*, 22, 24. While a spinal motor neuron might typically receive about 10,000 contacts from other neurons, some cells in the cerebellum receive up to 1 million synaptic contacts. A single axon's branches might give rise to 1000 synapses with neighboring neurons.

8. Kandel et al., *Principles of Neural Science*, 175.

9. Kandel et al., *Principles of Neural Science*, 23. Some synapses are electrical rather than chemical.

10. Kandel et al., *Principles of Neural Science*, 213. In figure 4.4, glutamate is presented as glutamic acid.

11. Eric J. Nestler, Steven E. Hyman, and Robert C. Malenka, *Molecular Neuropharmacology: A Foundation for Clinical Neuroscience*, 2nd ed. (New York: McGraw-Hill, 2009), 130–133; Robert M. Julien, Claire Advokat, and Joseph E. Comaty, *A Primer of Drug Action*, 12th ed. (New York: Worth, 2011), 81.

12. Recent evidence also implicates GABA in some excitatory processes and in the generation of oscillatory electrical waveforms. See Nestler, Hyman, and Malenka, *Molecular Neuropharmacology*, 133.

13. For the role of fly agaric in early South Asian religion, see Walter Lewis and Memory P. F. Elvin-Lewis, *Medical Botany: Plants Affecting Human Health* (Hoboken, N.J.: Wiley, 2003), 706–707. For the use of fly agaric for ritual healing also in the New World, see Richard Evans Schultes, Albert Hofmann, and Christian Rätsch, *Plants of the Gods: Their Sacred, Healing, and Hallucinogenic Powers* (Rochester, Vt.: Healing Arts, 2001), 82–85. See also Christian Rätsch, *The Encyclopedia of Psychoactive Plants: Ethnopharmacology and Its Applications* (Rochester, Vt.: Park Street Press, 2005), 631–639.

14. George S. Bause, "From Fish Poison to Merck Picrotoxin," *Anesthesiology* 118 (2013): 1261–1263.

15. Nestler, Hyman, and Malenka, *Molecular Neuropharmacology*, 141.

16. Rätsch, *Encyclopedia of Psychoactive Plants*, 482–484.

17. John Gerard, *The Herball, or Generall Historie of Plants* (London, 1597), 1363.

18. Julien, Advokat, and Comaty, *Primer of Drug Action*, 72–74.

19. Elaine M. Marieb and Katja Hoehn, *Human Anatomy and Physiology*, 7th ed. (San Francisco: Cummings, 2007), 543–544.

20. Rätsch, *Encyclopedia of Psychoactive Plants*, 376–393.

21. Julien, Advokat, and Comaty, *Primer of Drug Action*, 376; Ryan E. Hibbs and Alexander C. Zambon, "Agents Acting at the Neuromuscular Junction and Autonomic Ganglia," in *Goodman and Gilman's The Pharmacological Basis of Therapeutics*, 12th ed., ed. Laurence L. Brunton, Bruce A. Chabner, and Björn C. Knollmann (New York: McGraw-Hill, 2011), 270–271.

22. Joan Heller Brown and Nora Laiken, "Muscarinic Receptor Agonists and Antagonists," in *Goodman and Gilman*, ed. Brunton, Chabner, and Knollmann, 225–230;

Alireza Atri, Michael S. Chang, and Gary R. Strichartz, "Cholinergic Pharmacology," in *Principles of Pharmacology*, ed. Golan et al., 124.

23. John Parkinson, *Theatrum Botanicum* (London: Cotes, 1640), 349.

24. Jeffrey D. Henderer and Christopher J. Rapuano, "Ocular Pharmacology," in *Goodman and Gilman*, ed. Brunton, Chabner, and Knollmann, 1785.

25. Anticholinergic drugs are more effective against chronic obstructive pulmonary disease than asthma. See Brown and Laiken, "Muscarinic Receptor Agonists and Antagonists," and Peter J. Barnes, "Pulmonary Pharmacology," both in *Goodman and Gilman*, ed. Brunton, Chabner, and Knollmann, 228, 1044–1045.

26. Norepinephrine, the closely related neurotransmitter epinephrine, and dopamine are called catecholamines in the neuroscience literature, and their activity catecholinergic. In addition to their roles in the central nervous system, norepinephrine and epinephrine are active in the periphery.

27. Nestler, Hyman, and Malenka, *Molecular Neuropharmacology*, 156–157.

28. Brian B. Hoffman and Freddie M. Williams, "Adrenergic Pharmacology," in *Principles of Pharmacology*, ed. Golan et al., 132.

29. Julien, Advokat, and Comaty, *Primer of Drug Action*, 286–287; Thomas C. Westfall and David P. Westfall, "Adrenergic Agonists and Antagonists," in *Goodman and Gilman*, ed. Brunton, Chabner, and Knollmann, 300.

30. Ehab A. Abourashed et al., "*Ephedra* in Perspective—A Current Review," *Phytotherapy Research* 17 (2003): 703–712; Rakesh Amin, "FDA Issues Final Rule Banning *Ephedra*," *HerbalGram* 62 (2004): 63, 64–67.

31. Yohimbine is selective for α_2-adrenergic receptors, and norepinephrine release stimulates cardiac β_1-adrenergic receptors and peripheral vascular α_1-adrenergic receptors. See Westfall and Westfall, "Adrenergic Agonists and Antagonists," 309; and Hoffman and Williams, "Adrenergic Pharmacology," 141.

32. Norepinephrine release and yohimbine's additional effects on the serotonin system may play improve a man's mood and outlook and thereby "benefit some patients with psychogenic erectile dysfunction" (Westfall and Westfall, "Adrenergic Agonists and Antagonists," 309). Nestler, Hyman, and Malenka report that yohimbine "induces fear and anxiety in laboratory animals and humans" (*Molecular Neuropharmacology*, 156).

33. Nestler, Hyman, and Malenka, *Molecular Neuropharmacology*, 159; Thomas Michel and Brian B. Hoffman, "Treatment of Myocardial Ischemia and Hypertension," in *Goodman and Gilman*, ed. Brunton, Chabner, and Knollmann, 776.

34. Nestler, Hyman, and Malenka, *Molecular Neuropharmacology*, 147.

35. David G. Standaert and Ryan R. Walsh, "Pharmacology of Dopaminergic Neurotransmission," in *Principles of Pharmacology*, ed. Golan et al., 191–193.

36. The severe depression affecting about 15 percent of patients taking reserpine for hypertension was a factor in the move to other blood pressure medications during the twentieth century. See Nestler, Hyman, and Malenka, *Molecular Neuropharmacology*, 159.

37. Nestler, Hyman, and Malenka, *Molecular Neuropharmacology*, 160.

38. Charles P. O'Brien, "Drug Addiction," in *Goodman and Gilman*, ed. Brunton, Chabner, and Knollmann, 662; Julien, Advokat, and Comaty, *Primer of Drug Action*, 403.

39. Nestler, Hyman, and Malenka, *Molecular Neuropharmacology*, 158.

40. Nestler, Hyman, and Malenka, *Molecular Neuropharmacology*, 158.

41. Schultes, Hofmann, and Rätsch, *Plants of the Gods*, 124–132; Rätsch, *Encyclopedia of Psychoactive Plants*, 456–457. Almost two dozen New World, Old World, and Australian plant species produce dimethyltryptamine. See Schultes, Hofmann, and Rätsch, *Plants of the Gods*, 138. Chacruna is a major source of dimethyltryptamine in the ayahuasca drink.

42. Julien, Advokat, and Comaty, *Primer of Drug Action*, 526–527.

43. The term "endorphin" is used to mean endogenous opioid peptide and refers to neuropeptides including β-endorphin, enkephalins, dynorphin, and related molecules. See Tony L. Yaksh and Mark S. Wallace, "Opioids, Analgesia, and Pain Management," in *Goodman and Gilman*, ed. Brunton, Chabner, and Knollmann, 481–484. Substance P also plays a role in transmitting pain signals from the peripheral nerves at the level of the spinal cord. See Kandel et al., *Principles of Neural Science*, 536–538.

44. Nestler, Hyman, and Malenka, *Molecular Neuropharmacology*, 284–285; Julien, Advokat, and Comaty, *Primer of Drug Action*, 318.

45. Yaksh and Wallace, "Opioids, Analgesia, and Pain Management," 492–496.

46. Julien, Advokat, and Comaty, *Primer of Drug Action*, 482–486. There are also endocannabinoid receptors on the surface of immune system cells, the heart, and in other tissues thought to be involved in pain and inflammation response.

47. Julien, Advokat, and Comaty, *Primer of Drug Action*, 489–494.

48. Kandel et al., *Principles of Neural Science*, 367.

49. William E. Fantegrossi, Aeneas C. Murnane, and Chad J. Reissig, "The Behavioral Pharmacology of Hallucinogens," *Biochemical Pharmacology* 75 (2008): 17–33; Julien, Advokat, and Comaty, *Primer of Drug Action*, 515–516.

50. Kandel et al., *Principles of Neural Science*, 1104–1107.

51. Julien, Advokat, and Comaty, *Primer of Drug Action*, 511–513.

52. Kandel et al. explain that the rewarding properties of opiates are attributable to both dopamine-dependent and independent mechanisms, in *Principles of Neural Science*, 1107.

53. Peter R. Martin, Sachin Patel, and Robert M. Swift, "Pharmacology of Drugs of Abuse," in *Principles of Pharmacology*, ed. Golan et al., 284–293; Kandel et al., *Principles of Neural Science*, 1109–1110. Diagnosis and treatment of

addiction has historically been a contentious matter, in part because it is a stigmatized condition fraught with value judgments and moral commentary and in part because its neural mechanisms have not been well understood. In the past, some clinicians used the term "dependence" in describing both the physical adaptations to drug use and compulsive drug taking, which are actually separable phenomena. See the critique in Charles P. O'Brien, Nora Volkow, and T.-K. Li, "What's in a Word? Addiction Versus Dependence in DSM-V," *American Journal of Psychiatry* 163 (2006): 764–765. The most recent *Diagnostic and Statistical Manual of the American Psychiatric Association* (DSM-5, 2013), a standard clinical guide, treats "maladaptive" drug taking as a "substance use disorder."

54. This is just one of many possible classification schemes for psychoactive substances. Among the most influential works on the effects of mind-altering drugs is by the Berlin pharmacologist Louis Lewin (1850–1929), in *Phantastica: Narcotic and Stimulating Drugs* (1924 [originally published in German]). Lewin lays out five categories: (1) *Euphorica*, "sedatives of mental activity"; (2) *Phantastica*, "hallucinating substances"; (3) *Inebrianta*, which cause first "mental excitation," then "a state of depression"; (4) *Hypnotica*, "sleep-producing agents"; and (5) *Excitantia*, "mental stimulants" (*Phantastica: A Classic Survey on the Use and Abuse of Mind-Altering Plants*, trans. P. H. A. Wirth [Rochester, Vt.: Park Street Press, 1998], 26). A review of hallucinogen history and pharmacology is David E Nichols, "Hallucinogens," *Pharmacology & Therapeutics* 101 (2004): 131–181. Nichols's definition encompasses several of the five categories: "Hallucinogens are drugs that are characterized principally by their ability to produce alterations in perceptual processes" ("Hallucinogens," in *Encyclopedia of Psychopharmacology*, ed. Ian P. Stolerman [Berlin: Springer, 2010], 572). In any case, when it comes to describing the effects of mind-altering drugs used in traditional medical-spiritual practices, the boundaries of "false" and "true" sensory experiences depend very much on perspective. As the ethnobotanist Richard Evans Schultes commented, "From the point of view of the indigenous culture, such distinctions [between categories of hallucinogens] are of course meaningless, since the drug-induced experience is regarded as neither "hallucination" nor "illusion" but as another form of reality—indeed, often the ultimate reality" ("Overview of Hallucinogens in the Western Hemisphere," in *Flesh of the Gods: The Ritual Use of Hallucinogens*, ed. Peter T. Furst [New York: Praeger, 1972], 4n.

55. According to Nestler, Hyman, and Malenka, "The functions of most small-molecule neurotransmitters have been discovered partly because of the availability of selective agonists and antagonists" (*Molecular Neuropharmacology*, 190).

56. Raphael Mechoulam and Lumir Hanus, "A Historical Overview of Chemical Research on Cannabinoids," *Chemistry and Physics of Lipids* 108 (2000): 1–13.

57. Nestler, Hyman, and Malenka, *Molecular Neuropharmacology*, 191.

58. Some scholars contend that the mind is an emergent property of human existence irreducible to the mechanical interactions of cells and biological molecules. This is not the view of many contemporary neuroscientists, who argue: "What we commonly call the mind is a set of operations carried out by the brain" (Kandel et al., *Principles of Neural Science*, 5–6).

5. POPPY

1. Michael Wink and Ben-Erik van Wyk, *Mind-Altering and Poisonous Plants of the World* (Portland, Ore.: Timber Press, 2008), 180. This chapter examines the opium poppy (*Papaver somniferum*), one of many poppy species.

2. For a review of the nutritive properties of the poppy seed, see James Duke, "Utilization of Papaver," *Economic Botany* 27 (1973): 390–400. For several culinary uses of poppy seed in the classical Mediterranean, see Paolo Nencini, "The Rules of Drug Taking: Wine and Poppy Derivatives in the Ancient World. VI. Poppies as a Source of Food and Drug," *Substance Use & Misuse* 32, no. 6 (1997): 757–766.

3. Daniel Zohary, Maria Hopf, and Ehud Weiss, *Domestication of Plants in the Old World*, 4th ed. (New York: Oxford University Press, 2012), 109–111. There are several dozen species belonging to the *Papaver* genus and other closely related genera with a distribution in the Old and New World. See James C. Carolan et al., "Phylogenetics of *Papaver* and Related Genera Based on DNA Sequences from ITS Nuclear Ribosomal DNA and Plastid *trnL* Intron and *trnL–F* Intergenic Spacers," *Annals of Botany* 98 (2006): 141–155. *P. somniferum* probably derives from the weedy *P. setigerum* and can be described as having two cultivars: *P. somniferum* ssp. *somniferum*, the opium variety, and *P. somniferum* ssp. *hortensis*, grown for oilseed. Péter Tétényi lists *P. somniferum* ssp. *setigerum* as a variety and gives evidence for a hybrid source of the domesticated species, in "Opium Poppy (*Papaver somniferum*): Botany and Horticulture," *Horticultural Reviews* 19 (1997): 374, 376–368. The geographic origin of cultivated poppy remains unclear. Most experts place it in either the Mediterranean region or western Asia. See Mark D. Merlin, *On the Trail of the Ancient Opium Poppy* (Rutherford, N.J.: Farleigh Dickinson University Press, 1984), 87; and Tétényi, "Opium Poppy." More recently, other authors have summarized evidence for a western Mediterranean domestication. See Zohary, Hopf, and Weiss, *Domestication of Plants*, 111.

4. Mark D. Merlin, "Archaeological Evidence for the Tradition of Psychoactive Plant Use in the Old World," *Economic Botany* 57 (2003): 295–323.

5. J. R. Campbell and A. R. David claim that identification of the opium poppy in the Ebers papyrus (782) is "improbable" and instead suggest a reference to the corn poppy (*P. rhoeas*), in "The Application of Archaeobotany, Phytogeography, and Pharmacognosy to Confirm the Pharmacopeia of Ancient Egypt 1850–1200 B.C.," in *Pharmacy and Medicine in Ancient Egypt: Proceedings of the Conferences Held in Cairo (2007) and Manchester (2008)*, ed.

Jenefer Cockitt and Rosalie David, BAR International Series 2141 (Oxford: Archaeopress, 2010), 28.

6. Cyril Bryan translates the herbal ingredient as "poppy pod," in *Ancient Egyptian Medicine: The Papyrus Ebers* (London: Bles, 1930), 162; however, it is rendered as "poppy seeds" in John F. Nunn, *Ancient Egyptian Medicine* (Norman: University of Oklahoma Press, 1996), 153–156; and Lise Manniche, *An Ancient Egyptian Herbal*, rev. ed. (London: British Museum Press, 2006), 138–140. Both Nunn and Manniche express some doubt that the substance in question is from the opium poppy.

7. P. G. Kritikos and S. P. Papadaki, "The History of the Poppy and of Opium and Their Expansion in Antiquity in the Eastern Mediterranean Area," *Bulletin on Narcotics* 3 (1967): 17–38; John Scarborough, "Theophrastus on Herbals and Herbal Remedies," *Journal of the History of Biology* 11 (1978): 353–385; Jenö Bernáth, "Introduction," in *Poppy: The Genus Papaver*, ed. Jenö Bernáth (Amsterdam: Harwood, 1998), 1–2.

8. Pedanius Dioscorides of Anazarbus, *De materia medica*, trans. Lily Y. Beck (Hildesheim: Olms-Weidmann, 2005), 274.

9. Mark Grant, *Galen on Food and Diet* (London: Routledge, 2000), 106.

10. Dioscorides, *De materia medica*, 274.

11. Duke, "Utilization of Papaver," 390–394.

12. Poppy seeds washed in hot water and ground into a powder have very low levels of morphine and codeine. However, when contaminated by capsule material during harvesting, the opiate content can rise to pharmacologically relevant concentrations. Morphine and codeine levels in poppy seeds are low but variable, and growing conditions and variety of plant can also influence alkaloid accumulation. See Dirk W. Lachenmeier, Constanze Sproll, and Frank Musshoff, "Poppy Seed Foods and Opiate Drug Testing—Where Are We Today?" *Therapeutic Drug Monitoring* 32 (2010): 11–18; and Constanze Sproll et al., "Guidelines for Reduction of Morphine in Poppy Seed Intended for Food Purposes," *European Food Research Technology* 226 (2007): 307–310.

13. With only minor differences in practice—scoring the capsules in the morning versus the evening, using horizontal versus vertical strokes, employing a single blade or multiple blades per cut, and so on—the procedure summarized by Dioscorides persists wherever poppy is grown. For example, in late-eighteenth- and early-nineteenth-century England, see the account of Dr. Young and others in Virginia Berridge, "Our Own Opium: Cultivation of the Opium Poppy in Britain, 1740–1823," *British Journal of Addiction* 72 (1977): 90–94. A more recent description of opium production in the Sinai Peninsula is in Joseph J. Hobbs, "Troubling Fields: The Opium Poppy in Egypt," *Geographical Review* 88 (1998): 64–85.

14. Dioscorides, *De materia medica*, 275.

15. Dioscorides, *De materia medica*, 274.

16. Pliny, *Natural History* 20.26; *The Natural History of Pliny*, trans. John Bostock and H. T. Riley (London: Bohn, 1856), 4:276. "Many other persons, too, have ended their lives in a similar way," Pliny continued.

17. The poison is presented as opium in Martin Booth, *Opium: A History* (New York: St. Martin's Press, 1996), 20. The murder is attributed to hemlock (*Conium maculatum*) in L. Cilliers and F. P. Retief, "Poisons, Poisoning, and the Drug Trade in Ancient Rome," *Akroterion* 45 (2000): 88–100.

18. Kritikos and Papadaki, "History of the Poppy and of Opium"; Nencini, "Rules of Drug Taking," 760.

19. Booth, *Opium*, 20; Merlin, "Archaeological Evidence," 310.

20. The scarcity of textual sources on the poppy in the centuries after the fall of the Roman Empire probably has much to do with the state of scholarly learning during this period and less to do with the abundance of the poppy. In fact, the *Old English Herbarium*, an early Middle English translation, dating perhaps to the eleventh century, of a Latin herbal from as early as the fourth century, lists the "white poppy" for "pain in the temples" and "sleeplessness" (Anne Van Arsdall, *Medieval Herbal Remedies: The Old English Herbarium and Anglo-Saxon Medicine* [New York: Routledge, 2002], 68, 101–104, 126). The existence of such a text demonstrates that people used the poppy among many plant-based medicines throughout these centuries and found it worthwhile to preserve, copy, and disseminate their knowledge in writing.

21. Friedrich A. Flückiger and Daniel Hanbury, *Pharmacographia: A History of the Principal Drugs of Vegetable Origin Met with in Great Britain and British India*, 2nd ed. (London: Macmillan, 1879), 43–44; Booth, *Opium*, 21–23.

22. Virginia Berridge and Griffith Edwards, *Opium and the People: Opiate Use in Nineteenth-Century England* (London: Allen Lane/St. Martin's Press, 1981), xxiii.

23. D. Fraser-Harris, "The Former Importance of Our Sea-Borne Trade in Drugs," *Canadian Medical Association Journal* 18 (1928): 468.

24. Paracelsus is quoted as saying: "I possess a secret remedy which I call laudanum and which is superior to all other heroic remedies" (Booth, *Opium*, 24). He is also credited with calling his opium pills "the stone of immortality." See Berridge and Edwards, *Opium and the People*, xxiii–xxiv. However, Philip Ball expresses doubt that the Paracelsus laudanum recipes contained opium, in *The Devil's Doctor: Paracelsus and the World of Renaissance Magic and Science* (New York: Farrar, Straus and Giroux, 2006), 181–183.

25. Recipes are provided in Oswald Croll, *Basilica Chymica* (Savoy: Paul Marcell, 1610), 254–258.

26. For example, the *U.S. Pharmacopeia* of 1851 gives the following recipe for laudanum (*Tinctura Opii*): 2.5 ounces of powdered opium in 2 ounces diluted alcohol, macerated 14 days, strained. An easy-to-swallow formulation called paregoric elixir included honey, camphor, benzoic acid, anise, and alcohol.

27. Quoted in Berridge and Edwards, *Opium and the People*, xxiv.

28. George Young, *A Treatise on Opium Founded upon Practical Observations* (London: Millar, 1753), v–vi.

29. Samuel Crumpe, *An Inquiry into the Nature and Properties of Opium* (London: Robinson, 1793), 217.

30. J. Edkins, *Opium: Historical Note or the Poppy in China* (Shanghai: Statistical Department of the Inspectorate General of Customs, 1889), 6–24; Timothy Brook and Bob Tadashi Wakabayashi, "Opium's History in China," in *Opium Regimes: China, Britain, and Japan, 1839–1952*, ed. Timothy Brook and Bob Tadashi Wakabayashi (Berkeley: University of California Press, 2000), 5.

31. Edkins, *Opium*, 6.

32. *Materia Medica of the Kaibao Period*, translated in Edkins, *Opium*, 7.

33. Quoted in Yangwen Zheng, *The Social Life of Opium in China* (New York: Cambridge University Press, 2005), 11.

34. Edkins, *Opium*, 11.

35. Translated in Edkins, *Opium*, 13.

36. Zheng, *Social Life of Opium*, 16. Peter Lee states that the term "poppy milk," used in texts of the Song dynasty, indicates knowledge of opium production as early as the twelfth century, in *Opium Culture: The Art and Ritual of the Chinese Tradition* (Rochester, Vt.: Park Street Press, 2006), 7. Edkins traces the first Chinese description of opium ("Opium is produced in Arabia from a poppy with a red flower") to the author Wang Xi, who died in 1488, in *Opium*, 15.

37. Zheng, *Social Life of Opium*, 12.

38. Translated in Zheng, *Social Life of Opium*, 19.

39. Control of the male response is an ancient Chinese medical practice. See Donald J. Harper, *Early Chinese Medical Literature: The Mawangdui Medical Manuscripts* (London: Kegan Paul, 1998), 137–138.

40. Lee, *Opium Culture*, 8–9.

41. Zheng, *Social Life of Opium*, 18.

42. Carol Benedict proposes that tobacco entered China either in direct commerce with European visitors or from Chinese mariners returning from markets in places such as Japan or the Philippines, in *Golden-Silk Smoke: A History of Tobacco in China, 1550–2000* (Berkeley: University of California Press, 2011), 15–21.

43. Timothy Brook, "Smoking in Imperial China," in *Smoke: A Global History of Smoking*, ed. Sander L. Gilman and Zhou Xun (London: Reaktion, 2004); Zheng, *Social Life of Opium*, 25–30.

44. Lee, *Opium Culture*, 6–18.

45. Zheng, *Social Life of Opium*, 165; Lee, *Opium Culture*, 182–190, 201–204.

46. Gregory Blue, "Opium for China: The British Connection," in *Opium Regimes*, ed. Brook and Wakabayashi, 32–36.

47. This is the narrative of Zheng, *Social Life of Opiume*.

48. Michael Greenberg, *British Trade and the Opening of China, 1800–1842* (Cambridge: Cambridge University Press, 1951), cited in Zheng, *Social Life of Opium*, 93. Each chest weighs about 63.5 kilograms. See Brook and Wakabayashi, "Opium's History in China," 6.

49. Jonathan D. Spence, *The Search for Modern China*, 3rd ed. (New York: Norton, 2013), 143–178.

50. Europeans and Americans benefited from the opium trade despite its supposed continued ban because the treaty ending the First Opium War in 1842 gave British consular officials oversight of the British end of the opium trade without Qing intervention. Furthermore, foreigners enjoyed extraterritoriality in China, rendering them immune from prosecution. See Blue, "Opium for China," 35–36.

51. In other words, "British initiated hostilities over the Qing authorities' seizure of a Chinese opium-running boat after the craft's British registration had expired" (Blue, "Opium for China," 36).

52. The Second Opium War was "settled" by the Treaty of Tianjin in 1858 and amendments put forward at the Convention of Beijing (1860). See Blue, "Opium for China," 36.

53. Furthermore, domestic opium production was made legal, and taxed, in 1890. See Blue, "Opium for China," 36–37.

54. Zheng, *Social Life of Opium*, 110–115; Alexander Des Forges, "Opium/Leisure/Shanghai: Urban Economies of Consumption," in *Opium Regimes*, ed. Brook and Wakabayashi, 170.

55. In the words of a foreign observer, the British commissioner of customs at Amoy (Xiamen) in 1870, "Opium-smoking appears here as elsewhere in China to be becoming yearly a more recognized habit, almost a necessity of the people" (quoted in George Hughes, "Amoy Trade Report for the Year 1870," in *Reports on Trade at the Treaty Ports in China for the Year 1870* [Shanghai: Customs Press, 1871], 87).

56. For Xiamen, Hughes classified these smokers as heavy (10 percent), moderate (30 percent), and light (60 percent) users, giving daily consumption figures, in "Amoy Trade Report," 87.

57. David T. Courtwright, *Dark Paradise: Opiate Addiction in the United States Before 1940* (Cambridge, Mass.: Harvard University Press, 1982), 67. See also comments from protestant missionaries in J. Hudson Taylor's report, in Wilbur Crafts and Mary Leitch, *Intoxicating Drinks and Drugs in All Lands and Times* (Washington, D.C.: International Reform Bureau, 1909); and International Opium Commission estimates, in R. Bin Wong, "Opium and Modern Chinese State-Making," in *Opium Regimes*, ed. Brook and Wakabayashi, 194.

58. The Reverend W. E. Soothill, quoted in Crafts and Leitch, *Intoxicating Drinks and Drugs*, 111.

59. One early report on the chemical analysis of opium is John Leigh, *An Experimental Investigation into the Properties of Opium and Its Effects on Living Subjects* (Edinburgh: Charles Elliot, 1786).

60. F. W. A. Sertürner, "Über das Morphium, eine neuesalzfahige Grundlage und die Makosaure, als Hauptbestandtheile des Opiums," *Annalen der Physik* 55 (1817): 56–89. Sertürner isolated the alkaloid morphine, described its physical properties, and administered the substance to himself and a handful of young volunteers to record its apparent effects.

61. Alternatively "morphium" or "morphia" in period literature.

62. Berridge and Edwards, *Opium and the People*, 136, 138; John Swann, "The Evolution of the American Pharmaceutical Industry," *Pharmacy in History* 37 (1995): 76–86.

63. For example, F. Magendie, *Formulaire pour la préparation et l'emploi de plusieurs nouveaux médicamens* (Paris: Méquignon-Marvis, 1821), 13–19; Martyn Paine, *A Therapeutical Arrangement of the Materia Medica* (New York: Langley, 1842), 199; and George W. Wood and Franklin Bache, *The Dispensatory of the United States of America*, 8th ed. (Philadelphia: Grigg, Elliot, 1849), 1033–1044.

64. The hypodermic syringe was invented for the administration of morphine. It was first brought to America in 1856. See Courtwright, *Dark Paradise*, 46. See also Roberts Bartholow, *A Manual of Hypodermic Medication* (Philadelphia: Lippincott, 1869), 13–18.

65. The use of hypodermic injections during the mid-nineteenth century is reviewed in David F. Musto, "Iatrogenic Addiction: The Problem, Its Definition and History," *Bulletin of the New York Academy of Medicine* 61, no. 8 (1985): 694–705.

66. Courtwright, *Dark Paradise*, 54–55.

67. An example is "Mrs. Winslow's Soothing Syrup for Children Teething," a patent medicine containing morphine and recommended "for complaints incident to children, such as griping in the bowels, dysentery, or diarrhoea" (quoted in Peter G. Homan, Briony Hudson, and Raymond Rowe, *Popular Medicines: An Illustrated History* [London: Pharmaceutical Press, 2008], 143–147).

68. Alonzo Calkins, *Opium and the Opium-Appetite* (Philadelphia: Lippincott, 1871), 48–49.

69. Courtwright has shown that in late-nineteenth-century America, the majority of opium and morphine addicts were native-born white women in small and medium-size towns, often in the South, in *Dark Paradise*, 36–42.

70. Courtwright discusses the physician's incentive to use morphine, in *Dark Paradise*, 47, 50–51. Berridge and Edwards discuss the acceptance of hypodermic morphine administration within the medical profession and in particular its vaunted superiority over opium eating, in *Opium for the People*, 140–141.

71. Tony L. Yaksh and Mark S. Wallace," Opioids, Analgesia, and Pain Management," in *Goodman and Gilman's The Pharmacological Basis of Therapeutics*, 12th ed., ed. Laurence L. Brunton, Bruce A. Chabner, and Björn C. Knollmann (New York: McGraw-Hill, 2011), 501.

72. Berridge and Edwards, *Opium for the People*, xix–xx; Booth, *Opium*, 77.

73. Booth, *Opium*, 78–79.

74. John Ellis lists *Papaver somniferum* among the crops whose cultivation should be encouraged in the southern colonies of British North America, "for the sake of obtaining the opium pure" ("A Catalogue of Such Foreign Plants as Are Worthy of Being Encouraged in the American Colonies, for the Purposes of Medicine, Agriculture, and Commerce," *Transactions of the American Philosophical Society* 1 [1769–1771]: 255–266).

75. Courtwright, *Dark Paradise*, 68–69.

76. Courtwright describes the role of the Chinatown as a "safety valve" against the pressures of Chinese indentured servitude, where Chinese laborers could indulge in opium smoking, prostitution, and gambling as an escape from the hardships of life in America, in *Dark Paradise*, 68.

77. Courtwright, *Dark Paradise*, 70–78. Opium smoking was romanticized in American and European art and literature. Jos Ten Berge treats the opium aesthetic of the late nineteenth and early twentieth centuries in "The *Belle Epoque* of Opium," in *Smoke*, ed. Gilman and Zhou, 108–117. Yet there was white hostility to opium dens in London's Chinatown, which was frequently articulated in racial alarmist terms, as in the United States, according to Berridge and Edwards, *Opium for the People*, 195–205. This theme is also explored in Barry Milligan, "The Opium Den in Victorian London," in *Smoke*, ed. Gilman and Zhou, 118–125. Opium dens were associated with gambling and prostitution, which may have contributed to their allure to both Chinese and non-Chinese patrons, and they provided a reliable shelter for travelers. As Courtwright explains, "An opium den (or "dive" or "joint") was . . . also a meeting place, a sanctuary, and a vagabonds' inn" (*Dark Paradise*, 73).

78. Courtwright, *Dark Paradise*, 79.

79. The Smoking Opium Exclusion Act, 35 Stat. 614, effectively banned the importation, possession, and use of smoking opium.

80. Hamilton Wright, *Report on the International Opium Commission and on the Opium Problem as Seen Within the United States and Its Possessions*, Senate Document No. 377, 61st Cong., 2nd sess. (Washington, D.C.: Government Printing Office, 1910), 45.

81. Morphine and its related chemicals can make up 25 percent of the dried opium latex. See Susanna Fürst and Sándor Hosztafi, "Pharmacology of Poppy Alkaloids," in *Poppy*, ed. Bernáth, 291–318. Alkaloid content is highly variable, dependent on variety, growing conditions, and time of day.

82. Yaksh and Wallace," Opioids, Analgesia, and Pain Management," 499.

83. Yaksh and Wallace," Opioids, Analgesia, and Pain Management," 494–497; Andrea Trescot, "Clinical Use of Opioids," in *Comprehensive Treatment of Chronic Pain by Medical, Interventional, and Integrative Approaches*, ed. Timothy R. Deer et al. (New York: Springer, 2013), 101–102.

84. Samuel C. Hughes, "Intraspinal Narcotics for Obstetric Analgesia," *Western Journal of Medicine* 162, no. 1 (1995): 54–55; Trescot, "Clinical Use of Opioids," 106. Morphine and related compounds can also be delivered to the spinal fluid (intrathecal).

85. Selina Read and Jill Eckert, "Opioids and the Law," in *Comprehensive Treatment of Chronic Pain*, ed. Deer et al., 135–144.

86. Andrea M. Trescot et al., "Opioid Pharmacology," *Pain Physician* 11 (2008): S133–S153.

87. Yaksh and Wallace," Opioids, Analgesia, and Pain Management," 503.

88. Yaksh and Wallace," Opioids, Analgesia, and Pain Management," 492.

89. Robert M. Julien, Claire Advokat, and Joseph E. Comaty, *A Primer of Drug Action*, 12th ed. (New York: Worth, 2011), 336–337.

90. Yaksh and Wallace," Opioids, Analgesia, and Pain Management," 496.

91. Julien, Advokat, and Comaty, *Primer of Drug Action*, 335; Yaksh and Wallace," Opioids, Analgesia, and Pain Management," 504–505. Lomotil is a combination of diphenoxylate, an opioid receptor agonist, and atropine, an acetylcholine receptor antagonist.

6. COCA

1. Bruce Bohm, Fred Ganders, and Timothy Plowman, "Biosystematics and Evolution of Cultivated Coca (*Erythroxylaceae*)," *Systematic Botany* 7, no. 2 (1982): 121–133. Emanuel Johnson et al. give the following geographic breakdown: *E. coca* var. *coca*, "eastern slopes of the Andes in areas of humid montane forests"; *E. coca* var. *ipadu*, "cultivated only sparingly in the western part of the Amazon Basin"; *E. novogranatense* var. *novogranatense*, "plantation crop in Colombia and . . . an ornamental throughout the tropics"; *E. novogranatense* var. *truxillense*, "mostly confined to the desert areas of northern Peru" ("Identification of *Erythroxylum* Taxa by AFLP DNA Analysis," *Phytochemistry* 64 [2003]: 187–197).

2. Steven B. Karch, *A Brief History of Cocaine*, 2nd ed. (Boca Raton, Fla.: CRC Press, 2005), 1–2. Based on archaeological evidence, Timothy Plowman concludes that coca chewing existed around 3000 B.C.E., making the practice 5000 years old, in "The Ethnobotany of Coca (*Erythroxylum* spp., Erythroxylaceae)," *Advances in Economic Botany* 1 (1984): 72–73.

3. Richard Martin, "The Role of Coca in the History, Religion, and Medicine of South American Indians," *Economic Botany* 24, no. 4 (1970): 424–425. Among the early Inca, according to the first Inca chronicler, Garcilaso de la Vega, in the early seventeenth century, coca was both a sacred plant ("They worshipped also the plant cuca, or coca, as the Spaniards called it") and a ritual offering ("The things offered to the Sun were of diverse sorts: the chief and principal sacrifice was that of lambs, but besides, they offered all sorts of cattle . . . all sorts of grain, the herb cuca . . . all which they burnt in the place of incense, rendering thanks and acknowledgments to the Sun, for having sustained and nourished all those things for the use and support of Mankind") (*The Royal Commentaries of Peru*, trans. Paul Ricaut [London: Christopher Wilkinson, 1688], 31, 119).

4. Martin, "Role of Coca," 430. Nicolás Monardes recorded that Andean peoples chewed coca and tobacco (probably *Nicotiana tabacum*) leaves together ("a thing that gives them great contentment") (*Joyfull Newes Out of the New-Found World*, trans. John Frampton [London, 1596], 102–103). This mode of consumption seems no longer to exist.

5. Michael Wink and Ben-Erik van Wyk give a concentration of all tropane alkaloids, including cocaine, of 0.1 to 1.4 percent, in *Mind-Altering and Poisonous Plants of the World* (Portland, Ore.: Timber Press, 2008), 125; Robert M.

Julien, Claire Advokat, and Joseph E. Comaty give a concentration of cocaine of 1 percent in leaves, in *A Primer of Drug Action*, 12th ed. (New York: Worth, 2011), 399; Christian Rätsch lists an alkaloid concentration of 0.5 to 2.5 percent in leaves, in *The Encyclopedia of Psychoactive Plants: Ethnopharmacology and Its Applications* (Rochester, Vt.: Park Street Press, 2005), 251; Plowman records cocaine concentrations in the range of 0.63 to 1.02 percent, depending on species, variety, and growing conditions, in "Ethnobotany of Coca," 91–92.

6. Plowman, "Ethnobotany of Coca," 95.

7. Karch, *Brief History of Cocaine*, 6.

8. Karch, *Brief History of Cocaine*, 7–8.

9. *Cédula* of October 18, 1569, quoted in W. Golden Mortimer, *Peru: History of Coca: "The Divine Plant" of the Incas* (New York: Vail, 1901), 108.

10. Monardes published his compendium of New World medicinal plants as two books in 1565 and in revised form in 1571 and 1574. For an overview of Monardes's career and role in mediating knowledge of New World plants, see Marcy Norton, *Sacred Gifts, Profane Pleasures: A History of Tobacco and Chocolate in the Atlantic World* (Ithaca, N.Y.: Cornell University Press, 2008), 110–117.

11. Garcilaso de la Vega quotes a now-lost sixteenth-century text by Father Blas Valera, in Mortimer, *Peru*, 159–160.

12. J. J. von Tschudi, *Travels in Peru During the Years 1838–1842*, trans. Thomasina Ross (London: Bogue, 1847), 457, 452.

13. Karch, *Brief History of Cocaine*, 23–25.

14. There are variations in cocaine extraction techniques using fresh or dried leaves and various types of solvents, acids, and bases. In illicit cocaine production, field laboratories are established to produce crude cocaine base or crude acid cocaine sulfate, which is transported to destination countries for further purification into cocaine base or hydrochloride. See Paul Gootenberg, *Andean Cocaine: The Making of a Global Drug* (Chapel Hill: University of North Carolina Press, 2008), 298–301; and Department of Justice, Office of the Inspector General, appendix B: "Production of Cocaine Hydrochloride and Cocaine Base," in "The CIA-Contra-Crack Cocaine Controversy:A Review of the Justice Department's Investigations and Prosecutions," http://www.justice.gov/oig/special/9712/appb.htm.

15. David T. Courtwright, "The Rise and Fall and Rise of Cocaine in the United States," in *Consuming Habits: Drugs in History and Anthropology*, ed. Jordan Goodman, Paul E. Lovejoy, and Andrew Sherratt (London: Routledge, 1995), 217–219.

16. Gootenberg, *Andean Cocaine*, 57.

17. Gootenberg, *Andean Cocaine*, 57.

18. William H. Helfand, "Mariani et le Vin de Coca," *Revue d'histoire de pharmacie* 247 (1980): 227–234. Mariani marketed his product as "the concentrated extractive of the leaf of *Erythroxylon* [-um] *coca*, and an excellent special quality of Bordeaux wine" (Mariani and Company, *Vin Mariani, Erythroxylon Coca: Its Uses in the Treatment of Disease*, 2nd ed. [Paris: Mariani, 1884], 48).

19. Joseph F. Spillane, *Cocaine: From Medical Marvel to Modern Menace in the United States, 1884–1920* (Baltimore: Johns Hopkins University Press, 2000), 79–85.

20. Cocaethylene metabolism is described in S. Casey Laizure et al., "Cocaethylene Metabolism and Interaction with Cocaine and Ethanol: Role of Carboxylesterases," *Drug Metabolism and Disposition* 31 (2003): 16–20.

21. Mark Pendergrast, *For God, Country, and Coca-Cola: The Definitive History of the Great American Soft Drink and the Company That Makes It*, 3rd ed. (New York: Basic Books, 2013), 18–22. Pemberton's French Wine Coca was created no later than 1884.

22. *Cola* spp. Also spelled kola, a West African plant that produces high levels of caffeine in its seeds.

23. Probably *Turnera diffusa*, a shrub native to Central and South America and used in traditional medicine as an aphrodisiac.

24. Advertisement, ca. 1885, in Pendergrast, *For God, Country, and Coca-Cola*, 23.

25. The decocainized coca leaf extract in the secret Coca-Cola formula is slyly labeled "Merchandise No. 5." See Gootenberg, *Andean Cocaine*, 62; and, especially, Pendergrast, *For God, Country, and Coca Cola*, app. 1.

26. This assumes about 50 milligrams per "line" of cocaine, a figure that must vary quite a bit, depending on the custom of the user and source of illicit cocaine. Julien, Advokat, and Comaty give the figure of 25 milligrams per "line," in *Primer of Drug Action*, 398; Karch gives 50 milligrams, in *Brief History of Cocaine*, 37.

27. Spillane, *Cocaine*, 84–87.

28. Sigmund Freud, *Über Coca*, in *Cocaine Papers*, ed. Robert Byck (New York: Stone Hill, 1974), 48–74.

29. The broad applicability of Koller's findings was predicted in the *British Medical Journal* as early as November 29, 1884.

30. Tammy P. Than and Jimmy Bartlett, "Local Anesthetics," in *Clinical Ocular Pharmacology*, 5th ed., ed. Jimmy Bartlett and Siret Jaanus (St. Louis: Butterworth-Heinemann, 2008), 87–88.

31. Gootenberg, *Andean Cocaine*, 110. Demand probably outstripped the ability of the pharmaceutical companies to provide.

32. Gootenberg, *Andean Cocaine*, 58–59.

33. Gootenberg puts global cocaine production in the year 1900 at 15 to 18 metric tons, in *Andean Cocaine*, 121.

34. David F. Musto, "America's First Cocaine Epidemic," *Wilson Quarterly* 13, no. 3 (1989): 59–64.

35. For example, Edward Huntington Williams, "Negro Cocaine 'Fiends' Are a New Southern Menace," *New York Times*, February 8, 1914.

36. Courtwright, "Rise and Fall and Rise of Cocaine," 211–213.

37. While cocaine is still used in certain surgeries as a local anesthetic, its use as a nerve block has given way to the drugs procaine (Novocain), introduced 1905; lidocaine, introduced 1948; and others. See Daniel Becker and Kenneth Reed, "Essentials of Local Anesthetic Pharmacology," *Anesthesia Progress* 53 (2006): 103. For topical anesthesia, benzocaine is used. For an outline of the mechanism of action of local anesthetic agents, see Joshua M. Schulman and Gary R. Strichartz, "Local Anesthetic Pharmacology," in *Principles of Pharmacology: The Pathophysiologic Basis of Drug Therapy*, 3rd ed., ed. David E. Golan et al. (Philadelphia: Lippincott Williams & Wilkins, 2012), 147–162.

38. William A. Catterall and Kenneth Mackie, "Local Anesthetics," in *Goodman and Gilman's The Pharmacological Basis of Therapeutics*, 12th ed., ed. Laurence L. Brunton, Bruce A. Chabner, and Björn C. Knollmann (New York: McGraw-Hill, 2011), 570.

39. Julien, Advokat, and Comaty, *Primer of Drug Action*, 403–405.

40. Julien, Advokat, and Comaty, *Primer of Drug Action*, 403–404; Peter R. Martin, Sachin Patel, and Robert M. Swift, "Pharmacology of Drugs of Abuse," in *Principles of Pharmacology*, ed. Golan et al., 284–309. Blockage of dopamine reuptake is effected at lower cocaine concentrations and norepinephrine and serotonin at higher concentrations. While cocaine affects neurotransmitter reuptake throughout the body, increase in synaptic dopamine levels in the midbrain and pons seem to be most strongly responsible for cocaine's effects via connections to the limbic system.

41. M. S. Gold, "Cocaine (and Crack): Clinical Aspects," in *Substance Abuse: A Comprehensive Textbook*, 3rd ed., ed. J. H. Lowinson et al. (Baltimore: Williams & Wilkins, 1997), 185, reprinted in Julien, Advokat, and Comaty, *Primer of Drug Action*, 400.

42. Julien, Advokat, and Comaty, *Primer of Drug Action*, 407.

43. Julien, Advokat, and Comaty, *Primer of Drug Action*, 406; Bryan G. Schwartz, Shereif Rezkalla, and Robert A. Kloner. "Cardiovascular Effects of Cocaine," *Circulation* 122 (2010): 2558–2569.

7. PEYOTE

1. Edward F. Anderson, "The Biogeography, Ecology, and Taxonomy of *Lophophora* (Cactaceae)," *Brittonia* 21 (1969): 299–310. The yellow flower color is assigned to *L. diffusa* in Edward F. Anderson, *Peyote: The Divine Cactus*, 2nd ed. (Tucson: University of Arizona Press, 1996), 167.

2. Peyote's range is "widely distributed on limestone soils of low hills and flatlands in the Chihuahuan Desert of Texas, northern, and central Mexico from 50 to 1800 m elevation" (Anderson, "Biogeography, Ecology, and Taxonomy of *Lophophora*," 304).

3. *L. williamsii* occurs from the Rio Grande region through much of the Chihuahuan Desert and into the Sierra Madre Oriental foothills; *L. diffusa* occurs in an isolated, restricted area of high desert in the state of Querétaro. The latter species does not produce the medicinal compound mescaline. See Anderson, *Peyote*, 168.

4. Weston La Barre, "Twenty Years of Peyote Studies," *Current Anthropology* 1, no. 1 (1960): 45–60.

5. Jan G. Bruhn and Peter A. G. M. De Smet, "Ceremonial Peyote Use and Its Antiquity in the Southern United States," *HerbalGram* 58 (2003): 30–33.

6. J. G. Bruhn et al., "Peyote Alkaloids: Identification in a Prehistoric Specimen of *Lophophora* from Coahuila, Mexico," *Science* 199 (1978): 1437–1438.

7. Hesham R. El-Seedi et al., "Prehistoric Peyote: Alkaloid Analysis and Radiocarbon Dating of Archaeological Specimens of *Lophophora* from Texas," *Journal of Ethnopharmacology* 101 (2005): 238–242.

8. Francisco Hernández, *De historia plantarum Novae Hispaniae*, 15:25, translated in J. S. Slotkin and David P. McAllester, "Menomini Peyotism, a Study of Individual Variation in a Primary Group with a Homogeneous Culture," *Transactions of the American Philosophical Society* 42, no. 4 (1952): 571. Hernández conducted his Mexican research expedition during the 1570s.

9. Bernardino de Sahagún, *Historia general de las cosas de la Nueva España*, 11:7, sec. 1, translated in Slotkin and McAllester, "Menomini Peyotism," 571. Sahagún conducted his ethnographic studies from the 1550s through the 1570s.

10. Inquisition document of June 29, 1620, translated in Irving Leonard, "Peyote and the Mexican Inquisition, 1620," *American Anthropologist* 44, no. 2 (1942): 324–326.

11. José Ortega, *Historia del Nayarit, Sonora, Sinaloa y ambas Californias*: "*raíz diabólica*" (1754), quoted in Weston La Barre, *The Peyote Cult* (1938; Hamden, Conn.: Archon, 1975), 10; and, at length, in Anderson, *Peyote*, 11–13. Earlier, in 1591, the physician Juan de Cárdenas had written that without some other type of witchcraft, peyote could not by itself initiate communion with the devil. It altered the mind ("taking the man out of his judgments"), induced an unpleasant sleep ("a sleep horrible and terrifying"), and caused visions ("imagine types of figures of monsters, bulls, tigers, lions, and ghosts"), but "it is completely false to say that the herb out of its virtue makes the devil appear" (*Problemas y secretos maravillosos de las indias*, trans. G. A. Alles, quoted in Anderson, *Peyote*, 7–8).

12. The Tarahumara call themselves Rarámuri, and the Huichol call themselves Wixáritari. The Cora (calling themselves Náayarite) also employ peyote ritually.

13. Carl Lumholtz, *Unknown Mexico* (London: Macmillan, 1903), 2:359–361.

14. Myerhoff also worked among the Huichol with the anthropologist Peter Furst.

15. Barbara G. Myerhoff, "The Deer-Maize-Peyote Symbol Complex Among the Huichol Indians of Mexico," *Anthropological Quarterly* 43 (1970): 64–78.

16. Myerhoff, "Deer-Maize-Peyote."

17. There is some evidence that Huichols occasionally administer peyote by rectal enema. See Anderson, *Peyote*, 20–21.

18. The *mara'akame* Ramón Medina Silva, quoted in Barbara G. Myerhoff, "Peyote and Huichol Worldview: The Structure of a Mystic Vision," in *Cannabis and Culture*, ed. Vera Rubin (The Hague: Mouton, 1975), 417–438. The *mara'akame* serves as the shaman, priest, healer, and leader to the Huichol. See also Barbara G. Myerhoff, *Peyote Hunt:*

The Sacred Journey of the Huichol Indians (Ithaca, N.Y.: Cornell University Press, 1974), 94.

19. Myerhoff, "Peyote and Huichol Worldview," 433.

20. Wirikuta is located in the Chihuahuan desert of San Luis Potosí, several weeks' walk (about 500 kilometers) from the present-day Huichol communities in the mountainous parts of Nayarit, Jalisco, Zacatecas, and Durango, although many Huichols now travel by motor vehicle.

21. Myerhoff, "Deer-Maize-Peyote," 74.

22. Myerhoff, *Peyote Hunt*, 44, 77, and "Peyote and Huichol Worldview," 420–421.

23. Myerhoff, "Peyote and Huichol Worldview," 425.

24. Evidence for knowledge of peyote among the Hopi, Taos, Isleta, Queres, and Caddo of modern-day Arizona, New Mexico, Texas, Oklahoma, Arkansas, and Louisiana before the nineteenth century is in J. S. Slotkin, "Peyotism, 1521–1891," *American Anthropologist* 57, no. 2 (1955): 202–230. In "Twenty Years of Peyote Studies," La Barre disputes the methodology by which Slotkin both identified peyote in written sources and assigned it to particular cultural groups. Of course, archaeological specimens of peyote show that people harvested it within its range thousands of years ago. See also Anderson, *Peyote*, 23–24.

25. La Barre, "Twenty Years of Peyote Studies"; Anderson, *Peyote*, 43; Thomas Constantine Maroukis, *The Peyote Road: Religious Freedom and the Native American Church* (Norman: University of Oklahoma Press), 22–24.

26. Slotkin and McAllester, "Menomini Peyotism," 571; La Barre, "Twenty Years of Peyote Studies."

27. Robert L. Bee, "Peyotism in North American Indian Groups," *Transactions of the Kansas Academy of Science* 68 (1965): 13–61.

28. Ruth Shonle, "Peyote, the Giver of Visions," *American Anthropologist* 27, no. 1 (1925): 53–75; La Barre, *Peyote Cult*, 122, and "Twenty Years of Peyote Studies," 50–51.

29. Anderson, *Peyote*, 44–47.

30. Slotkin and McAllester refer to the "Great Spirit's power in Peyote" among the Menominee of Wisconsin ("Menomini Peyotism," 569); Charles S. Brant quotes a Kiowa Apache informant: "God comes to the Indian through peyote" ("Peyotism Among the Kiowa-Apache and Neighboring Tribes," *Southwestern Journal of Anthropology* 6, no. 2 [1950]: 212–222).

31. Slotkin and McAllester, "Menomini Peyotism," 180.

32. La Barre, *Peyote Cult*, 27.

33. Bee, "Peyotism," 26.

34. James Collins, "A Descriptive Introduction to the Taos Peyote Ceremony," *Ethnology* 7 (1968): 427–449.

35. Variations on the form and quantity of peyote consumption abound in the literature. Among the Taos, Collins reports in "Taos Peyote Ceremony," peyote was ground to a powder and taken in small balls or ingested "green" (raw). Slotkin and McAllester record that the Menomonee usually chew dried buttons or take it as a peyote tea. Some women reported having taken just one or two buttons; a "heavy user" recalled having taken twenty-five to thirty buttons in a session. Another man recollected having taken sixty when he was sick. According to one of the Menomonee

men interviewed, "That medicine, the more you take it, you have that much more power, see?" ("Menomini Peyotism," 568, 594, 616, 645).

36. Slotkin and McAllester, "Menomini Peyotism," 568.

37. Slotkin and McAllester, "Menomini Peyotism," 570.

38. Shonle, "Peyote"; Slotkin and McAllester, "Menomini Peyotism"; Collins, "Taos Peyote Ceremony"; Anderson, *Peyote*, 62–64; Maroukis, *Peyote Road*, 30–31.

39. Anderson, *Peyote*, 63. According to Vincenzo Petrullo,

> Most peyotists strongly affirm the Christian elements as an important part of their religion. One of the most interesting claims is that "Peyote was sent to the Indians and that afterwards Jesus was sent to the Whites, with the same purpose. However, the Whites killed Jesus in their ignorance, and thus have only the cross left; whereas the Indians never killed Peyote, with the result that they still have him, and the material manifestation of Peyote is the plant." (*The Diabolic Root: A Study of Peyotism, the New Indian Religion, Among the Delawares* [Philadelphia: University of Pennsylvania Press, 1934], 142, quoted in Anderson, *Peyote*, 64)

40. Quoted in Maroukis, *Peyote Road*, 37–38. In 1922, the Peyote Church of Christ changed its name to the Native American Church of Winnebago, Nebraska.

41. Maroukis, *Peyote Road*, 57.

42. Bee, "Peyotism"; Maroukis, *Peyote Road*, 146–151. The Native American Church of North America now serves as an umbrella organization for numerous regional chapters. The central organization advocates on behalf of issues important to members but otherwise does not ordain worship leaders or establish doctrine. See Maroukis, *Peyote Road*, 209–210.

43. Bee, "Peyotism"; Richard Evans Schultes, Albert Hofmann, and Christian Rätsch, *Plants of the Gods: Their Sacred, Healing, and Hallucinogenic Powers* (Rochester, Vt.: Healing Arts, 2001), 152; Maroukis, *Peyote Road*, 5.

44. American Indian Religious Freedom Act: 95‑341, 92 Stat. 469, effective August 11, 1978; Religious Freedom Restoration Act: 103‑141, 107 Stat. 1488, effective November 16, 1993; American Indian Religious Freedom Act Amendments: 103‑344, 108 Stat. 3125, effective October 6, 1994.

45. Lumholtz, *Unknown Mexico*, 358.

46. George A. Davis, *The Pharmacology of the Newer Materia Medica* (Detroit: Davis, 1892), 37–38.

47. H. C. Wood, Joseph P. Remington, and Samuel P. Sadtler, *The Dispensatory of the United States of America*, 19th ed. (Philadelphia: Lippincott, 1907), 1387.

48. The prolific Berlin physician Louis Lewin (1850–1929) is responsible for the first detailed description of peyote's subjective effects.

49. Christian Rätsch reports concentrations of mescaline between 0.4 and 2.74 percent in fresh buttons, and up to 3.7 percent in dried buttons, in *The Encyclopedia of Psychoactive Plants: Ethnopharmacology and Its Applications* (Rochester, Vt.: Park Street Press, 2005), 334. Mescaline

content in cultivated peyote was found to be in the 1.27 to 4.83 percent range. See Masako Aragane et al., "Peyote Identification on the Basis of Differences in Morphology, Mescaline Content, and trnL/trnF Sequence Between *Lophophora williamsii* and *L. diffusa*," *Journal of Natural Medicine* 65 (2011): 103–110.

50. Wood, Remington, and Sadtler, *Dispensatory*, 1387.

51. Aldous Huxley, *The Doors of Perception* (New York: Harper, 1954), 16.

52. To Ginsberg's eye, peyote superimposed New York on San Francisco. "The first time he looked at San Francisco under the influence of peyote, he saw New York City as though it was right outside his window" (Jonah Raskin, *American Scream: Allen Ginsberg's "Howl" and the Making of the Beat Generation* [Berkeley: University of California Press, 2004], 131).

53. Raskin, *American Scream*, 138.

54. Five milligrams per kilogram of body weight is the "usual oral dose," according to Robert M. Julien, Claire Advokat, and Joseph E. Comaty, *A Primer of Drug Action*, 12th ed. (New York: Worth, 2011), 515.

55. David O. Kennedy, *Plants and the Human Brain* (Oxford: Oxford University Press, 2014), 110–111.

56. David E. Nichols, "Hallucinogens," in *Encyclopedia of Psychopharmacology*, ed. Ian P. Stolerman (Berlin: Springer, 2010), 572–577; Julien, Advokat, and Comaty, *Primer of Drug Action*, 515–516.

57. Kennedy, *Plants and the Human Brain*, 111.

8. WORMWOOD

1. Richard N. Mack, "Plant Naturalizations and Invasions in the Eastern United States: 1634–1860," *Annals of the Missouri Botanical Garden* 90 (2003): 77–90; Department of Agriculture, Agricultural Research Service, Germplasm Resources Information Network, http://www.ars-grin.gov.

2. *Ancient Egyptian Medicine: The Papyrus Ebers*, trans. Cyril P. Bryan (London: Bles, 1930), 40, 45, 62; Jad Adams, *Hideous Absinthe: A History of the Devil in a Bottle* (Madison: University of Wisconsin Press, 2004), 15. Many of the wormwood-containing Ebers remedies are compound preparations, including numerous medicinal substances such as poppy, juniper, and fennel.

3. Pliny the Elder, *Natural History*, 27.28; *The Natural History of Pliny*, trans. John Bostock and H. T. Riley (London: Bohn, 1856), 5:233.

4. Pliny the Elder, *Natural History*, 14.19; *The Natural History of Pliny*, trans. John Bostock and H. T. Riley (London: Bohn, 1855), 3:259, and Pliny the Elder, *Natural History*, 27:28; *Natural History of Pliny*, 5:233–235; .

5. *Natural History of Pliny*, 5:233–235.

6. Adams, *Hideous Absinthe*, 16; Graeme Tobyn, Alison Denham, and Margaret Whiteleg, *The Western Herbal Tradition: Two Thousand Years of Medicinal Plant Knowledge* (Edinburgh: Churchill Livingstone/Elsevier, 2011), 109. Works attributed to Hippocrates were probably written by many hands over a period of time, forming a corpus of

literature difficult to assign to any individual authors. See Roy Porter, "What Is Disease?" in *The Cambridge History of Medicine*, ed. Roy Porter (New York: Cambridge University Press, 2006), 78.

7. Pedanius Dioscorides of Anazarbus, *De materia medica*, trans. Lily Y. Beck (Hildesheim: Olms-Weidmann, 2005), 188–189.

8. Dioscorides, *De materia medica*, 189.

9. Quoted in Phil Baker, *The Book of Absinthe: A Cultural History* (New York: Grove Press, 2001), 102; and in Barnaby Conrad, *Absinthe: History in a Bottle* (San Francisco: Chronicle, 1988), 86.

10. Conrad, *Absinthe*, 86.

11. Quoted in Tobyn, Denham, and Whiteleg, *Western Herbal Tradition*, 105, 110.

12. John Gerard, *The Herball, or Generall Historie of Plants* (London, 1597), 936–938. The etymology of the term "wormwood" is disputed, some claiming that the "worm" syllable and medicinal properties against parasites is but a coincidence. For example, Ernest Weekley's etymological dictionary proposes that wormwood is a corruption of Anglo-Saxon *wermōd* or *weremōd*, whose two elements represent "man" and "courage" (*An Etymological Dictionary of Modern English* [London: Murray, 1921]). As for the knowledge of wormwood as a vermifuge, the classic texts are mixed in their treatment: Pliny recommends absinthe wormwood against internal parasites, but not Dioscorides or Galen. Medieval herbals do not uniformly address wormwood's potential against intestinal worms, and only in the eighteenth century do medical texts reach consensus affirming this property. See Tobyn, Denham, and Whiteleg, *Western Herbal Tradition*, 112–113.

13. Thomas Tusser, *Five Hundred Points of Good Husbandry* (1557).

14. Conrad, *Absinthe*, 86.

15. Quoted in Conrad, *Absinthe*, 85; and Adams, *Hideous Absinthe*, 18. The latter author attributes the quote to the alternatively titled Marquise de Sévigné.

16. Baker, *Book of Absinthe*, 101; Stephan A. Padosch, Dirk W. Lachenmeier, and Lars U. Kröner, "Absinthism: A Fictitious Nineteenth-Century Syndrome with Present Impact," *Substance Abuse Treatment, Prevention, and Policy* 1 (2006): 1–14.

17. Shakespeare, *Romeo and Juliet*, act 1, scene 3.

18. Quoted in Adams, *Hideous Absinthe*, 17. The original French rhymes: "L'absinthe confort les ners / Est bon aussi pour les vers."

19. Adams, *Hideous Absinthe*, 18–9.

20. Adams, *Hideous Absinthe*, 19.

21. Distillation appears to be an ancient practice in Europe and Asia, with a history of perhaps 2000 years or more. However, it became much more widespread and improved during the Renaissance, at least as evidenced by the numerous printed books on the subject. Adams attributes some of the earliest European writings on steam distillation to Hieronymus Brunschwig, in 1500 and 1512, in *Hideous Absinthe*, 18. Steam distillation reduces the temperature of the boiling solvent to minimize any heat-related breakdown of chemically sensitive components of the distillate. The application of distillation to medicine was inherent to Brunschwig's work, and the English translation of his magnum opus, *The Vertuose Boke of Distyllacyon of the Waters of All Maner of Herbes* (trans. 1527), is subtitled to the attention of surgeons, physicians, apothecaries, and all manner of people, so that they might "lerne to dystyll all maner of herbes to the profit, cure, & remedy of all maner [of] dysieases and infirmytees."

22. Brunschwig, *Vertuose Boke of Distyllacyon*, chap. 275.

23. Marie-Claude Delahaye, *L'absinthe: Son histoire* (Auvers-sur-Oise: Musée de l'Absinthe, 2001), 28–35; Adams, *Hideous Absinthe*, 21.

24. Adams, *Hideous Absinthe*, 22. The Pernod absinthe was notable in having a much higher strength of 120 proof, or 60 percent alcohol. See Baker, *Book of Absinthe*, 107.

25. Adams reports a "classical procedure" for producing absinthe:

> Dried wormwood, anise and fennel were steeped in a high concentration of ethyl alcohol (that is "drinking alcohol"). After a day, water was added and the concoction boiled with the distillate collected. The process was completed with a further extraction of wormwood and hyssop and lemon balm, which was filtered to give a clear, green liquid. Other ingredients of the era included tansy, angelica, dittany of Crete, juniper, star anise, mint, coriander and veronica. A surviving recipe calls for 3.5 kg of "Grande Absinthe" (*Artemisia absinthium*), 4 kg of green anise, 4 kg of fennel, 675 g of star anise, 1 kg of lemon balm and 1.5 kg of hyssop. (*Hideous Absinthe*, 22)

26. Baker, *Book of Absinthe*, 105.

27. Conrad, *Absinthe*, 90; Delahaye, *L'absinthe*, 84; Baker, *Book of Absinthe*, 106. The French invaded Algeria in 1830, and the fighting was most intense between 1844 and 1847.

28. Baker, *Book of Absinthe*, 106.

29. Conrad cites a pamphlet by proabsinthe doctors published in 1904, in *Absinthe*, 109, 111. While some medical authors of the period, such as M.-J.-F.-Alexandre Pougens, recommended absinthe for digestive, parasitic, and menstrual concerns (*L'art de conserver la santé, de vivre longtemps et heureusement* [Montpellier: Picot, 1825], 261–262), François Antoine Fabre documented ongoing research to employ absinthe in the treatment of "dyspepsia, hypochondria, obstructions of the liver and spleen, helminthiasis [parasitic worm infection], dropsy, intermittent fever, gout, amenorrhea, etc." (*Dictionnaire des dictionnaires de médecine, français et étranger* [Paris: Béthune et Plon, 1840], 39).

30. Conrad, *Absinthe*, 90–93.

31. Delahaye, *L'absinthe*, 49–61.

32. Adams links the proliferation of absinthes using cheaper industrial alcohol to the *Phylloxera* grapevine plague of the 1860s and 1870s, which greatly reduced the availability of wine alcohol, in *Hideous Absinthe*, 50–51.

33. Conrad, *Absinthe*, 96; Dirk W. Lachenmeier et al., "Chemical Composition of Vintage Preban Absinthe with

Special Reference to Thujone, Fenchone, Pinocamphone, Methanol, Copper, and Antimony Concentrations," *Journal of Agricultural and Food Chemistry* 56 (2008): 3073–3081. The authors speculated, however, that the cheapest absinthes may have been adulterated with methanol and other toxic substances but are less likely to have survived to undergo modern chemical testing.

34. Adams, *Hideous Absinthe*, 179.

35. For more modern work on the possible negative health effects or adulterants in absinthe, see Lachenmeier et al., "Chemical Composition of Vintage Preban Absinthe"; Padosch, Lachenmeier, and Kröner, "Absinthism"; and Adams, *Hideous Absinthe*, 244–249.

36. http://www.pernod-ricard.com.

37. Adams, *Hideous Absinthe*, 33–35.

38. Adams, *Hideous Absinthe*, 29, 38–39.

39. For example, Adams, *Hideous Absinthe*, 48–50, 55.

40. Charles Baudelaire, "Enivrez-vous," in *Le spleen de Paris* (1869).

41. Adams, *Hideous Absinthe*, 56.

42. V. Magnan, *On Alcoholism, the Various Forms of Alcoholic Delirium, and Their Treatment*, trans. W. S. Greenfield (London: Lewis, 1876), 22–32. It is worthwhile to note that Magnan's trials consisted of single doses of absinthe injected directly into an animal's (such as a dog) stomach or bloodstream, carried out without controls.

43. Conrad, *Absinthe*, 104–105; Karin M. Höld et al., "Alpha-Thujone (the Active Component of Absinthe): Gamma-Aminobutyric Acid Type A Receptor Modulation and Metabolic Detoxification," *Proceedings of the National Academy of Sciences USA* 97, no. 8 (2000): 3826–3831.

44. C. Cadéac and A. Meunier, "Nouvelle note sur l'étude physiologique de la liqueur d'absinthe," *Comptes-rendus de la Société de biologie* (1889): 633–638.

45. Defenders of absinthe both in the nineteenth century and today hold that toxic adulterants in low-grade absinthe are responsible for the perceived health risks of the wormwood drink. See Adams, *Hideous Absinthe*, 245.

46. The terms of the statutes differed regionally. The U.S. prohibition was codified by Food Inspection Decision 147 (1912), which restricted the importation and interstate commerce of absinthe. Thujone was banned as a food additive in 1972. See Adams, *Hideous Absinthe*, 225, 234.

47. Conrad, *Absinthe*, 125.

48. U.S. Code of Federal Regulations Title 21 § 172.510. The European Union has a standard of 35 ppm. See Lachenmeier et al., "Chemical Composition of Vintage Preban Absinthe," 3079–3080.

49. Adams, *Hideous Absinthe*, 243–244.

50. Pete Wells, "A Liquor of Legend Makes a Comeback," *New York Times*, December 5, 2007.

51. Lachenmeier et al., "Chemical Composition of Vintage Preban Absinthe."

52. Adams, *Hideous Absinthe*, 245.

53. Höld et al., "Alpha-Thujone."

54. Höld et al. report toxicity in mice injected into the body cavity, achieving 50 percent mortality at 45 mg/kg and 100 percent mortality at 60 mg/kg, in "Alpha-Thujone,"

3828. The National Toxicology Program (http://ntp.niehs.nih.gov) provides a wide range of toxicity in different animals and representing different routes of exposure. An oral LD_{50} dose (enough to kill 50 percent of test animals) of thujone (α and β forms) is 500 mg/kg in rats. A similar toxic level in rabbits of intravenous exposure is 0.031 mg/kg. In rats, the LD_{50} for injection into the body cavity is reported at 0.2 ml/kg. Clearly, the route of administration matters for this chemical.

55. Assuming data obtained in the mouse could be scaled to a 70-kilogram human being, and that typical modern absinthes contain about 5 mg/L α-thujone and 20 mg/L β-thujone, according to Lachenmeier et al. ("Chemical Composition of Vintage Preban Absinthe"), the LD_{50} dose of 45 mg/kg α-thujone (Höld et al., "Alpha-Thujone") is equivalent to more than *600 liters* of undiluted absinthe. Vintage absinthes did not much surpass modern absinthes in α-thujone level; Pernod samples from 1895 to 1905 had 6 to 8 mg/L, only slightly more than the recent product. See Lachenmeier et al., "Chemical Composition of Vintage Preban Absinthe."

56. Ben-Erik Van Wyk and Michael Wink, *Medicinal Plants of the World* (Portland, Ore.: Timber Press, 2004), 398.

57. Richard W. Olsen, "Absinthe and γ-aminobutyric Acid Receptors," *Proceedings of the National Academy of Sciences USA* 97, no. 9 (2000): 4417–4418.

9. HEMP

1. The hemp plant's growth habit is highly dependent on environment. In open, sunny areas with abundant nutrients, the plant can reach a height of 5 meters; in shaded and restricted locations, the plant can mature at 20 centimeters. See Robert C. Clarke and Mark D. Merlin, *Cannabis: Evolution and Ethnobotany* (Berkeley: University of California Press, 2013), 13.

2. Cannabis "seeds" are technically fruits: seeds surrounded by dried layers of maternal tissue. Cannabis "seeds" can be shelled, yielding the soft, edible seeds proper. Hemp seeds can be eaten directly, pressed for oil, or processed into a type of "hemp milk."

3. The subspecies designations relayed here are those accepted by the Integrated Taxonomic Information System (http://www.itis.gov). There has been considerable attention paid to the taxonomy and evolutionary history of *Cannabis* varieties. See Richard Evans Schultes et al., "Cannabis: An Example of Taxonomic Neglect," in *Cannabis and Culture*, ed. Vera Rubin (The Hague: Mouton, 1975), 21–38; and Karl W. Hillig and Paul G. Mahlberg, "A Chemotaxonomic Analysis of Cannabinoid Variation in *Cannabis* (*Cannabaceae*)," *American Journal of Botany* 91 (2004): 966–975. After reviewing a century of work on *Cannabis* systematics and incorporating the most recent genetic data, Clarke and Merlin (*Cannabis*, 17–18, 311–331) support the species designations *C. indica* (drug-type hemp, although *C. indica* ssp. *chinensis* is a fiber-type hemp) and *C. sativa* (fiber-type hemp), further circumscribed into several subspecies and

regional cultivars, including *C. indica* ssp. *chinensis*, which is a fiber-type hemp subspecies bred from a drug-type hemp ancestral lineage. There is also a Central Asian hemp called *C. ruderalis* that may be closely related to the progenitor of *C. indica* and *C. sativa*.

4. Clarke and Merlin, *Cannabis*, 6–8.

5. Clarke and Merlin, *Cannabis*, 26.

6. Clarke and Merlin collect citations to the archaeological evidence of early hemp use in Europe and Asia, in *Cannabis*, 64–65.

7. Clarke and Merlin, *Cannabis*, 64–65, 88–89.

8. For example, Richard Evans Schultes, Albert Hofmann, and Christian Rätsch, *Plants of the Gods: Their Sacred, Healing, and Hallucinogenic Powers* (Rochester, Vt.: Healing Arts, 2001); and Mark D. Merlin, "Archaeological Evidence for the Tradition of Psychoactive Plant Use in the Old World," *Economic Botany* 57 (2003): 295–323.

9. Clarke and Merlin, *Cannabis*, 214–218.

10. *Atharva Veda* XI.6.15, cited in Clarke and Merlin, *Cannabis*, 221, 243. The date of the *Atharva Veda* is uncertain. For a review and bibliography of recent scholarship, see Carlos Lopez, "Atharva Veda," *Oxford Bibliographies: Hinduism* (November 21, 2012), http://www.oxfordbiblio graphies.com/view/document/obo-9780195399318 /obo-9780195399318-0008.xml?rskey=Eago7N&result=12.

11. Herodotus, *The Histories* 4.75. Rather than "howl in joy" or "pleasure," James L. Butrica suggests a translation closer to "howl in mourning" ("The Medical Use of Cannabis Among the Greeks and Romans," *Journal of Cannabis Therapeutics* 2 [2002]: 54–55). This passage is further discussed in Christian Rätsch, *The Encyclopedia of Psychoactive Plants: Ethnopharmacology and Its Applications* (Rochester, Vt.: Park Street Press, 2005), 142–143, 146; and Clarke and Merlin, *Cannabis*, 214–215; among many others. Herodotus also mentions a certain practice among the Araxes, a nomadic Central Asian tribe, in which they use an unnamed plant as an intoxicant:

> They know (it is said) of trees [tall hemp plants?] bearing a fruit [flowering shoots?] whose effect is this: gathering in groups and kindling a fire, the people sit around it and throw the fruit into the flames; then the fumes of it as it burns make them drunk as the Greeks are with wine, and more and more drunk as more fruit is thrown on the fire, until at last they rise up to dance and even sing. (*Histories* 1.102, in *Herodotus*, vol. 1, trans. A. D. Godley (New York: Putnam, 1920)

12. Rätsch, *Encyclopedia of Psychoactive Plants*, 129–131.

13. Clarke and Merlin, *Cannabis*, 214–216, and references cited.

14. Raphael Mechoulam, "The Pharmacohistory of *Cannabis sativa*," in *Cannabinoids as Therapeutic Agents*, ed. Raphael Mechoulam (Boca Raton, Fla.: CRC Press, 1986), 3; Lise Manniche, *An Ancient Egyptian Herbal*, rev. ed. (London: British Museum Press, 2006), 88; Clarke and Merlin, *Cannabis*, 246. The obstetric use documented here (in the original text of the Ebers papyrus, simply "for mothers and children") might be interpreted as a medication to induce contractions. For a summary of evidence for medicinal hemp in ancient Egypt, see Ethan Russo, "History of Cannabis and Its Preparations in Saga, Science, and Sobriquet," *Chemistry & Biodiversity* 4 (2007): 1614–1648.

15. Mechoulam, "Pharmacohistory of *Cannabis sativa*," 2. The translation of the Assyrian cuneiform tablets from which these medicinal properties were drawn remains controversial. See Ethan Russo, "Hemp for Headache: An In-Depth Historical and Scientific Review of Cannabis in Migraine Treatment," *Journal of Cannabis Therapeutics* 1 (2001): 21–92.

16. The evolution of the character *ma* 麻 and its role in compound words and as a component of several other Chinese characters is discussed in Hui-Lin Li, "The Origin and Use of Cannabis in Eastern Asia: Their Linguistic-Cultural Implications," in *Cannabis and Culture*, ed. Rubin, 51–62. For the adoption of the character 麻 for fiber plants generally in ancient China, see Tsuen-Hsuin Tsien, "Raw Materials for Papermaking in China," *Journal of the American Oriental Society* 93, no. 4 (1973).

17. In "Origin and Use of Cannabis," Li cites sources from the second through tenth centuries C.E. See also Clarke and Merlin, *Cannabis*, 243.

18. Dan Bensky, Steven Clavey, and Erich Stöger, *Chinese Herbal Medicin: Materia Medica* (Seattle: Eastland, 2004), 245–247.

19. Mechoulam cites the *Divine Husbandman's Classic of the Materia Medica*, of the first centuries C.E., in "Pharmacohistory of *Cannabis sativa*," 9. Li renders the original passage: "ma-fên (the fruits of hemp) . . . if taken in excess will produce hallucinations (literally 'seeing devils'). If taken over a long term, it makes one communicate with spirits and lightens one's body" ("Origin and Use of Cannabis," 56). Later Chinese texts elaborate on this theme, distinguishing the toxic fruits from the nonpoisonous seeds.

20. In "Medical Use of Cannabis," Butrica demonstrates that hemp's role in Greco-Roman medicine is far from certain, as it appears in only some of the major medical writings of the era: for example, it is absent in Hippocrates (ca. 450–370 B.C.E.), A. Cornelius Celsus (ca. 25 B.C.E.–ca. 50 C.E.), and Soranus (second century C.E.).

21. Pliny the Elder, *Natural History*, 20.97; *The Natural History of Pliny*, trans. John Bostock and H. T. Riley (London: Bohn, 1856), 4:297–298. For discussion and an alternative translation, see Butrica, "Medical Use of Cannabis." Butrica argues that Pliny confuses the attributes of other plants with those of cannabis.

22. Pedanius Dioscorides of Anazarbus, *De materia medica*, trans. Lily Y. Beck (Hildesheim: Olms-Weidmann, 2005), 247.

23. Butrica translates Pliny's warning as "Its seed is said to extinguish men's semen"; Dioscorides's as "when consumed in quantity [it] extinguishes the semen," and Galen's as "if consumed in quantity, it dries up the semen" ("Medical Use of Cannabis," 57); Bostock and Riley translate Pliny's note of caution as "Hempseed, it is said, renders men impotent" (*Natural History of Pliny*, 4:298).

24. Mark Grant, *Galen on Food and Diet* (London: Routledge, 2000), 106–107.

25. Grant, *Galen on Food and Diet*, 107. Butrica translates the first portion of this passage: "It warms sufficiently, and for that reason it also intoxicates quickly" ("Medical Use of Cannabis," 62).

26. Hemp was not an important source of fiber in South Asia. See Clarke and Merlin, *Cannabis*, 98.

27. I. C. Chopra and R. N. Chopra, "The Use of *Cannabis* Drugs in India," *Bulletin on Narcotics* 9 (1957): 4–29; Clarke and Merlin, *Cannabis*, 225–226.

28. Clarke and Merlin, *Cannabis*, 228–229.

29. Chopra and Chopra, "Use of *Cannabis* Drugs in India"; Clarke and Merlin, *Cannabis*, 225–231.

30. Clarke and Merlin, *Cannabis*, 43, 227.

31. Clarke and Merlin, *Cannabis*, 238.

32. Chopra and Chopra, "Use of *Cannabis* Drugs in India"; Clarke and Merlin, *Cannabis*, 227.

33. Technically, hand-rubbed resin is called *charas*; resin sieved from dried flowering tops is called *hashish*.

34. Clarke and Merlin, *Cannabis*, 213.

35. Russo, "History of Cannabis," 1620. THC is, however, soluble in alcohol, including the ethanol of wine and distilled spirits.

36. Clarke and Merlin, *Cannabis*, 239.

37. Ethan Russo, "Cannabis Treatments in Obstetrics and Gynecology: A Historical Review," *Journal of Cannabis Therapeutics* 2 (2002): 5–35.

38. Sula Benet, "Early Diffusion and Folk Uses of Hemp," in *Cannabis and Culture*, ed. Rubin, 42–48.

39. The papal bull on Satanism is *Summis desiderantes affectibus*, issued on December 5, 1484. It does not specifically reference cannabis, but hemp is connected with the targeted folk medicine and occult spirituality by Ernest L. Abel, *Marihuana, the First Twelve Thousand Years* (New York: Plenum, 1980), 107–108; and Martin Booth, *Cannabis: A History* (New York: St. Martin's Press, 2003), 58.

40. Russo, "History of Cannabis"; Clarke and Merlin, *Cannabis*, 48–50.

41. John Gerard, *The Herball, or Generall Historie of Plants* (London, 1597), 573.

42. John Parkinson, *Theatrum Botanicum* (London: Cotes, 1640), 598.

43. Clarke and Merlin, *Cannabis*, 234.

44. Abel, *Marihuana*, 38–43.

45. Abel, *Marihuana*, 110.

46. Abel, *Marihuana*, 111; Booth, *Cannabis*, 60.

47. Abel, *Marihuana*, 148–151.

48. The water pipe is called a *hookah* (*huqqa*) in Hindi/Urdu, *nargila* or *qalyan* in Persian, *nargile* in Turkish, and *shisha* in Arabic. The device probably originated in the Indian subcontinent in the early sixteenth century, devised to smoke tobacco, and then applied to the consumption of other medicinal plants and mixtures. See Rudi Matthee, "Tobacco in Iran," in *Smoke: A Global History of Smoking*, ed. Sander L. Gilman and Zhou Xun (London: Reaktion, 2004), 58–60.

49. Abel, *Marihuana*, 154–163.

50. J. Moreau, *Du hachisch et de l'aliénation mentale: Études psychologiques* (Paris: Fortin, Masson, 1845), 41–42.

51. W. B. O'Shaughnessy, "On the Preparations of the Indian Hemp, or Gunjah," *Provincial Medical Journal*, January 28, 1843; February 4, 1843.

52. O'Shaughnessy, "On the Preparations of the Indian Hemp."

53. H. C. Wood, Joseph P. Remington, and Samuel P. Sadtler, *The Dispensatory of the United States of America*, 16th ed. (Philadelphia: Lippincott, 1892), 351. The entry for cannabis in the various editions of the *Dispensatory* is basically unchanged from 1854 through the century.

54. Abel, *Marihuana*, 183–184. Abel also reports *hashish* samples contaminated with opium, according to a nineteenth-century source (178).

55. Abel, *Marihuana*, 184–185; Booth, *Cannabis*, 94–97.

56. The experience of a British physician might well have been typical: "Indian hemp for some time back has been vaunted as a medicine of some therapeutic value in cases of dysmenorrhea; to me, however, its action seems so variable, and the preparation itself so unreliable, as to be hardly worthy of a place on our list of remedial agents at all" (James Oliver, "On the Action of *Cannabis indica*," *British Medical Journal* 1 [1883]: 905–906).

57. Abel, *Marihuana*, 78–95.

58. Charles Richard Dodge, *A Report on the Culture of Hemp and Jute in the United States* (Washington, D.C.: Government Printing Office, 1896), 7; Lyster H. Dewey, "Hemp," in *Yearbook of the United States Department of Agriculture* (Washington, D.C.: Government Printing Office, 1914), 284–285.

59. Abel, *Marihuana*, 171–178.

60. According to Abel, "Even when it was not against the law, marihuana was used by very few Americans. Those who used it were typically from minority groups like the Mexicans and the Negroes, and this made their drug preferences highly visible" (*Marihuana*, 200).

61. Abel cites numerous hemp-centered musical numbers, including the song "Smokin' Reefers," which was in the Broadway review *Flying Colors* (1932), in *Marihuana*, 219–220.

62. Booth, *Cannabis*, 148–149.

63. *Ogden (Utah) Standard*, September 25, 1915.

64. Richard J. Bonnie and Charles H. Whitebread, *The Marihuana Conviction: A History of Marihuana Prohibition in the United States* (Charlottesville: University Press of Virginia, 1974), 34.

65. *San Antonio Light*, December 12, 1920.

66. Frank R. Gomila and Madeline C. Gomila Lambou, "Present Status of the Marihuana Vice in the United States," in *Marihuana: America's New Drug Problem*, by Robert P. Walton (Philadelphia: Lippincott, 1938), 38–39.

67. Courtney Riley Cooper, *Here's to Crime* (Boston: Little, Brown, 1937), 338, quoted in Abel, *Marihuana*, 228.

68. Gomila and Gomila Lambou, "Present Status of the Marihuana Vice," 30.

69. Walton, *Marihuana*, 2.

70. Abel, *Marihuana*, 240–241.

71. Abel, *Marihuana*, 241; Booth, *Cannabis*, 151.

72. The film was originally titled *Tell Your Children*.

73. 75–238, 50 Stat. 551, effective October 1, 1937.

74. Abel, *Marihuana*, 243; David F. Musto, *The American Disease: Origins of Narcotic Control*, 3rd ed. (Oxford: Oxford University Press, 1999), 221–229.

75. Booth, *Cannabis*, 159–160.

76. Clarke and Merlin, *Cannabis*, 304–305.

77. Robert C. Clarke and David P. Watson, "*Cannabis* and Natural *Cannabis* Medicines," in *Marijuana and the Cannabinoids*, ed. Mahmoud A. ElSohly (Totowa, N.J.: Humana, 2007); Clarke and Merlin, *Cannabis*, 371.

78. Tod H. Mikuriya, "Marijuana in Medicine: Past, Present, and Future," *California Medicine* 110 (1969): 34–40.

79. Peter J. Cohen, "Medical Marijuana: The Conflict Between Scientific Evidence and Political Ideology (Part One)," *Journal of Pain & Palliative Care Pharmacotherapy* 23 (2009): 4–25, and "Medical Marijuana: The Conflict Between Scientific Evidence and Political Ideology (Part Two)," *Journal of Pain & Palliative Care Pharmacotherapy* 23 (2009): 120–140.

80. Jane E. Brody, "Tapping Medical Marijuana's Potential," *New York Times*, November 4, 2013.

81. Rudolf Brenneisen, "Chemistry and Analysis of Phytocannabinoids and Other *Cannabis* Constituents," in *Marijuana and the Cannabinoids*, ed. ElSohly, 17–50. The ratio of THC to CBD is highly variable and depends on the genetic makeup of the cannabis variety. In general, strains selected for fiber and seeds have low THC (less than 1 percent by weight) and slightly higher CBD (approximately 2 percent). Strains selected for medicinal use have higher THC (often 5–10 percent) and lower CBD (0–5 percent). In modern times, breeders have further manipulated the THC-to-CBD ratio to produce effects on the mind of a desired quality. See Clarke and Watson, "*Cannabis* and Natural *Cannabis* Medicines."

82. THC potency has trended sharply upward in recent years because of advances in breeding and indoor cultivation. Marihuana buds and *sinsemilla* can reach 30 percent THC; *hashish* can reach 50 percent or higher. See Clarke and Watson, "*Cannabis* and Natural *Cannabis* Medicines," 10–12; and Brenneisen, "Chemistry and Analysis of Phytocannabinoids," 25–27.

83. Hemp seed consists of about 25 percent protein and 30 percent oil, the latter of which is a rich source of essential fatty acids. See J. C. Callaway, "Hempseed as a Nutritional Resource: An Overview," *Euphytica* 140 (2004): 65–72.

84. In "Chemistry and Analysis of Phytocannabinoids," Brenneisen reports THC levels in the μg/g level, on the order of millionths of a percent by weight.

85. Marilyn A. Huestis and Michael L. Smith describe standardized studies in which a cigarette containing 1.75 percent THC (about 16 milligrams) is considered low dose and a cigarette containing 3.55 percent THC (about 30 milligrams) is considered high dose, in "Human Cannabinoid Pharmacokinetics and Interpretation of Cannabinoid Concentrations in Biological Fluids and Tissues," in *Marijuana and the Cannabinoids*, ed. ElSohly, 207. Robert M. Julien,

Claire Advokat, and Joseph E. Comaty estimate that a marihuana cigarette contains 75 milligrams of THC, of which 25 to 50 percent is available in the smoke, in *A Primer of Drug Action*, 12th ed. (New York: Worth, 2011), 486. The dose ultimately present in the bloodstream would be 5 to 10 milligrams.

86. Huestis and Smith, "Human Cannabinoid Pharmacokinetics," 207–210. Bioavailability of THC by the oral route is reported at 4 to 20 percent.

87. Clarke and Merlin, *Cannabis*, 239. Huestis and Smith report that 11-hydroxy-THC levels reach 50 to 100 percent of THC levels in the blood of those who ingest THC, compared with 10 percent for those who smoke it, in "Human Cannabinoid Pharmacokinetics," 209–210.

88. Clarke and Watson, "*Cannabis* and Natural *Cannabis* Medicines," 7–8, 12; Angelo A. Izzo et al., "Nonpsychotropic Plant Cannabinoids: New Therapeutic Opportunities from an Ancient Herb," *Trends in Pharmacological Sciences* 30 (2009): 515–527, Clarke and Merlin, *Cannabis*, 253.

89. Clarke and Merlin, *Cannabis*, 253. In contrast to this opinion, Brenneisen reports that THC alone and THC in cannabis preparations "produced similar, dose-dependent subjective effects" and "few reliable differences between the THC-only and whole-plant condition" ("Chemistry and Analysis of Phytocannabinoids," 39).

90. Huestis and Smith, "Human Cannabinoid Pharmacokinetics"; Julien, Advokat, and Comaty, *Primer of Drug Action*, 488.

91. While cannabis modestly alters the senses and mood, it does not cause "delusions, paranoia, hallucinations, confusion and disorientation," and the like except at very high doses, according to Julien, Advokat, and Comaty, *Primer of Drug Action*, 491.

92. Balapal S. Basavarajappa, "Neuropharmacology of the Endocannabinoid Signaling System—Molecular Mechanisms, Biological Actions, and Synaptic Plasticity," *Current Neuropharmacology* 5 (2007): 81–97; Billy R. Martin, "The Endocannabinoid System and the Therapeutic Potential of Cannabinoids," in *Marijuana and the Cannabinoids*, ed. ElSohly, 126–127. For a discussion of cannabis and reward, see David O. Kennedy, *Plants and the Human Brain* (Oxford: Oxford University Press, 2014), 233.

93. Martin, "Endocannabinoid System," 129–130.

94. Kennedy, *Plants and the Human Brain*, 230.

95. Martin, "Endocannabinoid System"; Julien, Advokat, and Comaty, *Primer of Drug Action*, 496–497.

96. Julien, Advokat, and Comaty, *Primer of Drug Action*, 498.

97. Clarke and Merlin, *Cannabis*, 253.

98. Julien, Advokat, and Comaty, *Primer of Drug Action*, 499; A. J. Hill et al., "Phytocannabinoids as Novel Therapeutic Agents in CNS Disorders," *Pharmacology & Therapeutics* 133 (2012): 79–97.

99. Jerrold Meyer and Linda Quenzer, *Psychopharmacology: Drugs, the Brain, and Behavior*, 3rd ed. (Sunderland, Mass.: Sinauer, 2013), 334–335; R. D. Hosking and J. P. Zajicek, "Therapeutic Potential of Cannabis in Pain Medicine," *British Journal of Anaesthesia* 101, no. 1 (2008): 59–68.

10. COFFEE

1. John K. Francis, "*Coffea arabica* L.," Department of Agriculture, Forest Service, International Institute of Tropical Forestry, http://www.fs.fed.us/global/iitf/pdf/shrubs /Coffea%20arabica.pdf.

2. Coffee trees begin to produce fruit after three to four years and reach maximum yield after six to eight years. See Francis, "*Coffea arabica.*" Coffee plants can live for more than a hundred years. The plants are productive for up to thirty years, according to Beryl Simpson and Molly Ogorzaly, *Economic Botany: Plants in Our World*, 3rd ed. (New York: McGraw-Hill, 2001), 318.

3. To some botanists, the coffee fruit is an accessory berry: two seeds enveloped in a fleshy pericarp, the fruit developing between the stem and the point of attachment of the petals. Other botanists consider the fruit to be a drupe, in that the seeds are surrounded by a fibrous endocarp and the remainder of the pericarp is fleshy. In the coffee trade, the "berry" or "cherry" refers to the fruit, and the "bean" refers to the seed. One exception to this nomenclature is the term "peaberry," which refers to a seed that grows single, rather than paired, in a particular berry. Some sources refer to the coffee "husk" (fleshy part of the fruit) or coffee "kernel" (seed).

4. Genetic evidence supports the hypothesis that the major *Coffea arabica* cultivars derived from wild populations growing in Ethiopia's southwestern highlands. See F. Anthony et al., "Genetic Diversity of Wild Coffee (*Coffea arabica* L.) Using Molecular Markers," *Euphytica* 118 (2001): 53–65.

5. Mark Pendergrast, *Uncommon Grounds: The History of Coffee and How It Transformed Our World*, 2nd ed. (New York: Basic Books, 2010), 42.

6. *C. canephora* is sometimes described by its promotional moniker *robusta*.

7. Pendergrast, *Uncommon Grounds*, 3–4; Ralph S. Hattox, *Coffee and Coffeehouses: The Origins of a Social Beverage in the Medieval Near East* (Seattle: University of Washington Press, 1985), 13.

8. Pendergrast, *Uncommon Grounds*, 4. Sources documenting the indigenous preparations of coffee in East Africa and Arabia are summarized in Bennett Alan Weinberg and Bonnie K. Bealer, *The World of Caffeine: The Science and Culture of the World's Most Popular Drug* (New York: Routledge, 2001), 4–5, 22–23.

9. Weinberg and Bealer, *World of Caffeine*, 5–6.

10. Quoted in William H. Ukers, *All About Coffee* (New York: Tea and Coffee Trade Journal Company, 1922), 12, from a translation by Philippe Sylvestre Dufour (1685). Although it is far from certain that texts as early as the eleventh century describe the properties of coffee (rather than some other medicinal plant), many authors on the early history of coffee accept the proposition. Consistent with the entries in ancient medical books, they consider the Arabic *bunn* and *buncham* as the coffee plant and a drink made from it. Weinberg and Bealer lend support for this interpretation by relaying that in Abyssinia, then

and now, *bunn* refers to the coffee berry and *buncham* the drink, in *World of Caffeine*, 5. The word *bunn* glosses both the drink coffee and the coffee bean in Arabic, while in Ethiopian dialects the words for coffee and coffee bean can variously be expressed as *buno*, *bun(n)ä*, and *bunna*, according to Alan S. Kaye, "The Etymology of 'Coffee': The Dark Brew," *Journal of the American Oriental Society* 106 (1986): 557–558. When the earliest Islamic authors referenced coffee, it was of an unroasted form, since roasting was an innovation of the fourteenth or fifteenth century, perhaps, according to Pendergrast, *Uncommon Grounds*, 5.

11. Quoted in Ukers, *All About Coffee*, 12.

12. There are records of a coffee drink prepared from the husk or dried fruit as well as roasted, ground seed in the Arab and Turkish domain into the eighteenth century. See Hattox, *Coffee and Coffeehouses*, 83–85.

13. Abd al-Qadir al-Jaziri (fl. 1558), referencing an earlier authority, quoted in Hattox, *Coffee and Coffeehouses*, 14.

14. The etymology of the Arabic word *qahwa* is speculated to be either the co-option of an earlier term, *qahwa* (wine), or a reference to the Ethiopian place-name Kaffa (alternatively spelled Kefa and Keffa), possibilities that are not mutually exclusive. Furthermore, the term *qahwa* seems also to have been used to describe a stimulating beverage made from the leaves of *qat* (*Catha edulis*), which implies that *qahwa* may have been a general term for a stimulating, hunger-suppressing, or intoxicating beverage. See Hattox, *Coffee and Coffeehouses*, 18–20. In "Etymology of 'Coffee,'" Kaye argues that the word *qahwa* draws its linguistic roots from adjectives related to the dark color of wine and brewed coffee. The term "coffee" entered English from the French *café*, which—like the Italian *caffè*, Dutch *koffie*, and German *Kaffee*—derives from the Turkish *kahve*, originally from the Arabic *qahwa*. See Weinberg and Bealer, *World of Caffeine*, 24.

15. The importance of Sufism in the dissemination of coffee is underlined by Hattox, *Coffee and Coffeehouses*, 24. As a sect with an "emphasis on the mystical reaching out for God," Sufism encouraged its adherents to pursue a "trancelike concentration on God" during nighttime communal worship services. Sufis were known to consume *hashish* and *qat* in aid of prayer, and in this context, any substance, including coffee, that could help them focus on their spiritual duties was easily accepted. As most Sufis led regular daytime lives, performed their trades, and traveled regionally, they spread coffee drinking around the Arab and Muslim world.

16. Fernando Vega, "The Rise of Coffee," *American Scientist* 96 (2008): 138–145; Hattox, *Coffee and Coffeehouses*, 26–28.

17. Hattox explains the complex sociopolitical milieu of the (ambivalent) official opposition to coffee, in *Coffee and Coffeehouses*, 30–38. Although there were religious and medical arguments to be made against coffee drinking—and such arguments have been documented in formal minutes of meetings over coffee's fate—the underlying concern

in such efforts is the perceived tavernlike atmosphere of the coffee house, where gambling, antiauthoritarian talk, and other disagreeable activities might occur.

18. Hattox, *Coffee and Coffeehouses*, 36.

19. Hattox describes short-lived restrictions on coffee in Mecca, Damascus, and Cairo during the early and mid-sixteenth century, in *Coffee and Coffeehouses*, 38–40.

20. The Ottoman Empire expanded to control much of Arabia during the sixteenth century, placing most of the world's coffee plantations and nearby ports under Turkish control. Coffee and the coffeehouse were introduced to Istanbul to an enthusiastic reception in 1555. See Hattox, *Coffee and Coffeehouses*, 77.

21. Quoted in Markman Ellis, *The Coffee House: A Cultural History* (London: Weidenfeld and Nicholson, 2004), 22.

22. Pendergrast, *Uncommon Grounds*, 7.

23. Pendergrast, *Uncommon Grounds*, 7.

24. Pendergrast, *Uncommon Grounds*, 7.

25. Vega, "Rise of Coffee"; Anthony et al., "Genetic Diversity of Wild Coffee."

26. Anthony et al., "Genetic Diversity of Wild Coffee," 63–64.

27. Ellis proposes a date for the first Oxford coffeehouse in the 1652 to 1654 era, in *Coffee House*, 29–33. He argues that the first coffeehouse in Christendom was probably Pasqua Rosee's, established in London as early as 1652 and no later than 1656.

28. Pendergrast, *Uncommon Grounds*, 12.

29. Ellis, *Coffee House*, 259.

30. Pendergrast, *Uncommon Grounds*, 9–10.

31. Pendergrast, *Uncommon Grounds*, 8; Weinberg and Bealer, *World of Caffeine*, 18.

32. Ellis, *Coffee House*, 80.

33. Pendergrast, *Uncommon Grounds*, 12. The authorities in seventeenth-century Britain, as in the Muslim world during the sixteenth century, suspected coffeehouse patrons of unwholesome activities and seditious discourse. As in Arabia a century earlier, official decrees were no match for peoples' enthusiasm for coffee. Thus when Charles II of England (r. 1660–1685) ordered the revocation all coffee-selling licenses in 1675, essentially banning coffeehouses, the resulting public displeasure—and threat to the revenue stream that coffee generated in taxes—convinced him to repeal the edict before it went into effect. See Ellis, *Coffee House*, 72–75, 86–105.

34. Ellis, *Coffee House*, 177–178.

35. Pendergrast, *Uncommon Grounds*, 17–18.

36. Pendergrast, *Uncommon Grounds*, 20–23.

37. Pendergrast, *Uncommon Grounds*, 31–32.

38. Scholarly Arab and Persian medicine was based on Galenic medicine transmitted through the Greeks. See Hattox, *Coffee and Coffeehouses*, 65.

39. Hattox, *Coffee and Coffeehouses*, 66–68.

40. Quoted in Ellis, *Coffee House*, 8.

41. Quoted in Pendergrast, *Uncommon Grounds*, 8. Ellis attributes the quote to Robert Burton, in *Coffee House*, 23.

42. Robert Burton, *The Anatomy of Melancholy* (1632), quoted in Ellis, *Coffee House*, 23.

43. John Parkinson, *Theatrum Botanicum* (London: Cotes, 1640), entry: *Arbor Buncom fructo suo Buna* (The Turkes berry drinke), quoted in Ellis, *Coffee House*, 23. Although widely cited, this entry did not appear in the edition of Parkinson's classic consulted for this work.

44. Ellis, *Coffee House*, 134.

45. Pendergrast, *Uncommon Grounds*, 12.

46. Simon Paulli, *Commentarius De Abusu Tabaci Americanorum Veteri, Et Herbae Thee Asiaticorum in Europe Novo* (1665), quoted in Ellis, *Coffee House*, 137. Paulli's critique accorded the same demasculinizing properties to tobacco, tea, and coffee.

47. Ellis, *Coffee House*, 137.

48. Quoted in Ellis, *Coffee House*, 142.

49. Ukers, *All About Coffee*, 33.

50. *New York Times*, May 3, 1884.

51. Pendergrast, *Uncommon Grounds*, 91–97.

52. The idea that coffee might replace alcohol as a social drink originated very early in the history of coffee, first in the Middle East and then in the West. In the seventeenth century, the Englishman James Howell wrote: "[T]his Coffee drink hath caused a greater sobriety among the Nations," declaring it a "wakeful and civil drink" (quoted in Ellis, *Coffee House*, 134). Much later, the Briton John Crawford (1852) wrote that "tea and coffee . . . have contributed materially to the sobriety, decency, and even morality, of the inhabitants of this country." In the United States, coffee was accepted as a temperance drink that easily substituted for alcohol as a facilitator of social bonding. On September 16, 1923, during Prohibition, the *New York Times* played on this beverage swap in an article with the headline "Coffee-Drunken New York," and in the article that followed, it described coffee's capacity to stimulate workers and artists in a stressful, dynamic era.

53. Pendergrast, *Uncommon Grounds*, 146.

54. Peet's Coffee & Tea, "It All Began on Vine Street," http://www.peets.com/about-us/our-history.

55. Pendergrast, *Uncommon Grounds*, 279–280.

56. Starbucks, "Company Information," http://www.starbucks.com/about-us/company-information.

57. Ukers, *All About Coffee*, 198.

58. Christian Rätsch gives caffeine concentration of 0.58 to 1.7 percent, in *The Encyclopedia of Psychoactive Plants: Ethnopharmacology and Its Applications* (Rochester, Vt.: Park Street Press, 2005), 176.

59. Ukers, *All About Coffee*, 148, 198.

60. Ukers, *All About Coffee*, 144–145; Pendergrast, *Uncommon Grounds*, 141; Weinberg and Bealer, *World of Caffeine*, 245–246. The West African *C. liberica*, which shares the characteristics of heat tolerance and pest resistance with the Central African varieties, is grown to a lesser extent than *C. arabica* and *C. canephora*. See International Coffee Organization, "Botanical Aspects," http://www.ico.org/botanical.asp.

61. Pendergrast, *Uncommon Grounds*, 238–239.

62. Pendergrast, *Uncommon Grounds*, xi. In the past, peaberries were considered waste because of their odd shape and irregular size. Among coffee producers and consumers, the aesthetic and economic value of peaberries is contested.

63. Some rare varieties are mature when yellow. See Pendergrast, *Uncommon Grounds*, 25.

64. Ukers, *All About Coffee*, 249–254; Pendergrast, *Uncommon Grounds*, 26, 34.

65. Pendergrast, *Uncommon Grounds*, 34, 366.

66. Tom Johnson, "Using Sight to Determine Degree of Roast," Sweet Maria's Coffee Library, http://www.sweetmarias.com/library/content/using-sight-determine-degree-roast.

67. Weinberg and Bealer, *World of Caffeine*, xix.

68. Jerrold Meyer and Linda Quenzer, *Psychopharmacology: Drugs, the Brain, and Behavior*, 3rd ed. (Sunderland, Mass.: Sinauer, 2013), 394.

69. Robert M. Julien, Claire Advokat, and Joseph E. Comaty, *A Primer of Drug Action*, 12th ed. (New York: Worth, 2011), 364.

70. Eric J. Nestler, Steven E. Hyman, and Robert C. Malenka, *Molecular Neuropharmacology: A Foundation for Clinical Neuroscience*, 2nd ed. (New York: McGraw-Hill, 2009), 202; Tarja Porkka-Heiskanen and Anna V. Kalinchuk, "Adenosine, Energy Metabolism, and Sleep Homeostasis," *Sleep Medicine Reviews* 15 (2011): 123–135.

71. Nestler, Hyman, and Malenka, *Molecular Neuropharmacology*, 204.

72. Julien, Advokat, and Comaty, *Primer of Drug Action*, 365–366; Nestler, Hyman, and Malenka, *Molecular Neuropharmacology*, 205.

73. Niels P. Riksen, Paul Smits, and Gerard A. Rongen, "The Cardiovascular Effects of Methylxanthines"; and Stephen L. Tilley, "Methylxanthines in Asthma," both in *Methylxanthines: Handbook of Experimental Pharmacology*, ed. Bertil B. Fredholm (Berlin: Springer, 2011), 413–437, 439–456.

74. Jana Sawynok, "Methylxanthines and Pain," in *Methylxanthines*, ed. Fredholm, 311–329.

75. Julien, Advokat, and Comaty, *Primer of Drug Action*, 365; Harmut Osswald and Jürgen Schnermann, "Methylxanthines and the Kidney," in *Methylxanthines*, ed. Fredholm, 391–412; Tilley, "Methylxanthines in Asthma," 439–456. The chemically similar methylxanthine alkaloid compounds caffeine, theophylline, and theobromine generally resemble one another in their physiologic actions but differ in their relative potency. For example, theophylline is a more effective bronchodilator and diuretic than caffeine.

76. In high-dose animal studies, caffeine caused fetal malformations, but there is no evidence for birth defects attributable to caffeine in human populations. The effects on fertility and miscarriage are called into question by confounding factors of experimental design. See Ulrika Ådén, "Methylxanthines During Pregnancy and Early Postnatal Life," in *Methylxanthines*, ed. Fredholm, 373–389. Ådén summarizes the results of epidemiological studies on high-dose caffeine consumption during pregnancy as "inconclusive" and states that "there is currently insufficient evidence for advising mothers to avoid caffeine during the last two thirds of pregnancy."

77. Julien, Advokat, and Comaty, *Primer of Drug Action*, 367–368. Alba Minelli and Ilaria Bellezza review the literature on caffeine and reproductive health, concluding that "clinical counselors can inform prepregnant/pregnant women who do not smoke or drink alcohol and who consume moderate amounts of caffeine (5–6 mg/kg per day, five cups [of coffee]) that they do not have an increase in reproductive risks or adverse effects" ("Methylxanthines and Reproduction," in *Methylxanthines*, ed. Fredholm, 366).

78. Ådén, "Methylxanthines During Pregnancy," 380.

79. The question of whether caffeine and related methylxanthines can produce dependence is a source of controversy in the field. Clinical definitions of dependence have been built around the effects of substances such as the opiates and cocaine that produce persistent desire, elicit strong behavioral changes, and have much longer withdrawal periods than caffeine. A widely accepted compromise is to call caffeine an "atypical drug of dependence." See John W. Daly and Bertil B. Fredholm, "Caffeine—An Atypical Drug of Dependence," *Drug and Alcohol Dependence* 51, no. 1 (1998): 199–206; and Micaela Morelli and Nicola Simola, "Methylxanthines and Drug Dependence: A Focus on Interactions with Substances of Abuse," in *Methylxanthines*, ed. Fredholm, 483–507.

80. Morelli and Simola, "Methylxanthines and Drug Dependence," 486.

81. Morelli and Simola, "Methylxanthines and Drug Dependence," 485.

11. TEA

1. William H. Ukers, *All About Tea* (New York: Tea and Coffee Trade Journal Company, 1935), 268–271.

2. The Integrated Taxonomic Information System (http://www.itis.gov) does not accept the distinction of Assam tea and China tea as formal varieties under *Camellia sinensis*. They are rather similar in growth habit and readily hybridize. But the Royal Botanic Gardens at Kew curate these forms as distinct taxonomic varieties. F. N. Wachira et al. support the formal classification of *C. sinensis* var. *sinensis* and *C. sinensis* var. *assamica* and explore some of the challenges in determining tea species and cultivars, in "The Tea Plants: Botanical Aspects," in *Tea in Health and Disease Prevention*, ed. Victor R. Preedy (London: Elsevier, 2013), 3–18. There is also an accepted third cultivar, *C. sinensis* var. *lasiocalyx* (Cambod tea), which is not in wide commerce.

3. Scholars disagree on the origins of tea cultivation and processing, but it was certainly well established by 168 C.E., when a county in Hunan Province was named Tuling (Tea Ridge). Part of the difficulty in ascertaining the earliest tea consumption is in the scarcity of artifacts unequivocally identifiable as used for tea (rather than for other culinary or medicinal herbs), and part is in the interpretation in ancient texts of one of the Chinese characters for "tea," which shifted from *tu* 荼 (referring to tea and possibly other plants) to *cha* 茶 (used only for tea and differing by a single stroke) during the eighth century. See Victor H. Mair and Erling Hoh, *The True History of Tea* (New York: Thames & Hudson, 2009), 29, 33, 262–268.

4. The account is from the third-century *Guangya* dictionary: "In the region between Jing and Ba [the area between modern eastern Sichuan and the western parts of Hunan and Hubei] the people pick the leaves and make a cake. If the leaves are old, rice paste is used in forming the cake. [People who] wish to brew the tea first roast [the cake] until it is a reddish color, pound it into a powder, put it into a ceramic container, and cover it with boiling water. They stew scallion [spring onion], ginger, and orange peel with it" (quoted in Mair and Hoh, *True History of Tea*, 33).

5. Mair and Hoh, *True History of Tea*, 36.

6. Mair and Hoh, *True History of Tea*, 42–43.

7. For translations, see Lu Yü, *The Classic of Tea: Origins and Rituals*, trans. Francis Ross Carpenter (Boston: Little, Brown, 1974); and Ukers, *All About Tea*, 15–22. For historical perspective, see Mair and Hoh, *True History of Tea*, 45–50.

8. Ukers, *All About Tea*, 15–22.

9. Mair and Hoh, *True History of Tea*, 49.

10. Ukers, *All About Tea*, 15. The term "cooling" refers to the traditional Chinese medical concept rather than to physical temperature.

11. Mair and Hoh, *True History of Tea*, 61–62. Some people probably steeped loose-leaf tea during this time as well.

12. Mair and Hoh, *True History of Tea*, 110–119.

13. Mair and Hoh, *True History of Tea*, 120.

14. Mair and Hoh, *True History of Tea*, 68.

15. Quoted in Mair and Hoh, *True History of Tea*, 65, 67.

16. Mair and Hoh, *True History of Tea*, 76–83.

17. Mair and Hoh, *True History of Tea*, 136.

18. The monk Kūkai, quoted in Mair and Hoh, *True History of Tea*, 44.

19. Mair and Hoh, *True History of Tea*, 129.

20. Adam Olearius, *Travels of the Ambassadors Sent by Frederic, Duke of Holstein, to the Great Duke of Muscovy and the King of Persia* (1647), trans. John Davies (1662), quoted in Mair and Hoh, *True History of Tea*, 155.

21. Mair and Hoh, *True History of Tea*, 151–163.

22. Mair and Hoh, *True History of Tea*, 139.

23. Frederick J. Simoons, *Food in China: A Cultural and Historical Inquiry* (Boca Raton, Fla.: CRC Press, 1991), 463n.

24. Pedro Teixeira, quoted in Mair and Hoh, *True History of Tea*, 166.

25. Much of the tea that first arrived in Europe probably shipped from Japan, where the Portuguese and Dutch had established footholds that proved temporary for the former and stable for the latter. See Mair and Hoh, *True History of Tea*, 171.

26. A valuable linguistic study of tea is in Mair and Hoh, *True History of Tea*, 262–268. Ukers calls attention to the British trading post at Amoy (Xiamen), Fujian, established after 1644, in the entry of tea into the English lexicon, in *All About Tea*, 38.

27. Beatrice Hohenegger discusses the sharp increase in tea consumption in Britain during the seventeenth and eighteenth centuries, in *Liquid Jade: The Story of Tea from East to West* (New York: St. Martin's Press, 2006), 128. There were apparently strong fluctuations in tea imports. Imports

totaled 100 pounds in 1664; 4,713 pounds in 1678; 12,070 pounds in 1685; 38,390 pounds in 1690; and 13,082 pounds in 1699. See Roy Moxham, *Tea: Addiction, Empire and Exploitation* (New York: Carroll and Graf, 2003), 21. Years of larger imports flooded the market and were followed by years of meager imports.

28. Mair and Hoh, *True History of Tea*, 177.

29. Ukers, *All About Tea*, 51–60.

30. Tea consumption in the British way continued in America after independence, and it evolved to include iced tea, a form popular from the nineteenth century to the modern day. See Mair and Hoh, *True History of Tea*, 206–209.

31. Mary Lou Heiss and Robert J. Heiss discuss numerous variations of tea processing and their influence on the brewed product's color, flavor, and aroma, in *The Story of Tea: A Cultural History and Drinking Guide* (Berkeley, Calif.: Ten Speed Press, 2007), 48–110. Processors delineate the categories of tea generally based on differences in manufacturing technique, starting with the plucked leaves and buds, but disagree on the number of classes. Experts usually recognize six categories based on a Chinese system (green, yellow, white, blue-green, red, and black), with numerous regional variations and subcategories. In common English-language usage, these categories go by green, yellow, white, oolong, black, and dark, although yellow tea is sometimes omitted because of its rarity. Dark tea, comprising teas produced in many Chinese regions, is sometimes named for one of its more popular subtypes, puer, although the manufacturing processes are distinct. See Heiss and Heiss, *Story of Tea*, 51; and Jinghong Zhang, *Puer Tea: Ancient Caravans and Urban Chic* (Seattle: University of Washington Press, 2014), 13.

32. The tea-growing season varies regionally. In cooler climates, such as China and northern India, the tea plant is productive during the spring, summer, and early autumn. In warmer places, such as southern India and Sri Lanka, tea grows throughout the year. In many cases, the youngest leaves and bud are preferred, but in some areas, mature leaves are harvested.

33. Ukers, *All About Tea*, 280–294.

34. There are numerous variations in green-tea processing that do not alter the fundamental condition that the leaves are prevented to oxidize before drying.

35. There is some confusion in the tea trade over the naming of fully fermented teas. In English, the term "black" is used to describe the color of the processed leaves; in Chinese and Japanese, the same tea is characterized by the reddish brew it produces—for example, *hong cha* 红茶 (red tea). Therefore, some Asian-derived black teas may be marketed as red tea in Western countries. In China, *hei cha* 黑茶 (black tea) refers to postfermented teas called dark tea in English. To further complicate these matters, the red-tea label is also used to describe an unrelated brewed drink made from the South African rooibos (*Aspalathus linearis*) plant, whose Afrikaans name translates as "red bush."

36. In Chinese folk medicine, puer tea has been associated with benefits for eye health, treatment and prevention of obesity, improvement in digestion, detoxification, and enhanced longevity, among other effects. See Huai-Yuan

Chen, Shoei-Yn Lin-Shiau, and Jen-Kun Lin, "Pu-erh Tea: Its Manufacturing and Health Benefits," in *Tea and Health Products: Chemistry and Health-Promoting Properties*, ed. Chi-Tang Ho, Jen-Kun Lin, and Fereidoon Shahidi (Boca Raton, Fla.: CRC Press, 2009), 9–16. They note, however, that little biomedical evidence exists to support these health claims.

37. Since the 1970s, some Chinese producers have simulated the long aging process by treating green-tea leaves with microbes during processing, rendering their tea market ready in a matter of months. See Zhang, *Puer Tea*, 12–13.

38. Zhang, *Puer Tea*, 12, 45.

39. Mair and Hoh, *True History of Tea*, 120.

40. Mair and Hoh, *True History of Tea*, 25.

41. Mair and Hoh, *True History of Tea*, 248. CTC alternatively denotes cut-twist-curl, the same process.

42. Mair and Hoh, *True History of Tea*, 49–50.

43. Mair and Hoh, *True History of Tea*, 85–87.

44. Mair and Hoh, *True History of Tea*, 108–109.

45. Ukers, *All About Tea*, 137–144. The British introduced both China tea and Assam tea to estate cultivation. Tea drinking among native South Asians expanded substantially during the twentieth century.

46. Mair and Hoh, *True History of Tea*, 249.

47. Woodruff Smith, "From Coffee-House to Parlour: The Consumption of Coffee, Tea, and Sugar in Northwestern England in the Seventeenth and Eighteenth Centuries," in *Consuming Habits: Drugs in History and Anthropology*, ed. Jordan Goodman, Paul E. Lovejoy, and Andrew Sherratt (London: Routledge, 1995), 142–157.

48. Drinking tea with dairy (milk or butter) is an ancient tradition in Tibet and Mongolia, and Europeans encountered milk tea at the Manchu court and in Canton in the mid-seventeenth century. The European milk-tea custom, however, is attributed to a Parisian salon hostess, Madame de la Sablière, who offered the concoction to her guests beginning in 1680. Soon thereafter, English aristocrats mimicked their French peers by adding milk to their tea. As to the question of whether to add milk to the cup before or after the tea, this matter is a contentious one in British tea culture and evokes arguments of class, taste, and propriety. See Mair and Hoh, *True History of Tea*, 175, 247.

49. Mair and Hoh, *True History of Tea*, 235–238.

50. The Turks shared their inclination toward tulip-shaped glasses with the Persians. See Mair and Hoh, *True History of Tea*, 162–163.

51. Quoted in Mair and Hoh, *True History of Tea*, 35.

52. Ukers, *All About Tea*, 20–21, citing Lu Yü's review of medical literature in *Classic of Tea*.

53. Quoted in Mair and Hoh, *True History of Tea*, 86.

54. Quoted in Mair and Hoh, *True History of Tea*, 121–122.

55. Quoted in Ukers, *All About Tea*, 25.

56. Quoted in Ukers, *All About Tea*, 27.

57. Quoted in Ukers, *All About Tea*, 32–33. Dirx published under the pseudonym Nikolas Tulp.

58. Ukers, *All About Tea*, 32.

59. Thomas Garway claimed that tea has a hot property in humoral terms, in contrast to the Chinese herbalist Li Shizhen's determination that tea is cooling.

60. Quoted in Ukers, *All About Tea*, 39.

61. Ukers, *All About Tea*, 46.

62. Daniel Duncan, *Wholesome Advice Against the Abuse of Hot Liquors, Particularly Coffee, Tea, Chocolate, Brandy, and Strong Waters* (London, 1706). Duncan's passion is directed against coffee, tea, and chocolate together.

63. Quoted in Bennett Alan Weinberg and Bonnie K. Bealer, *The World of Caffeine: The Science and Culture of the World's Most Popular Drug* (New York: Routledge, 2001), 113. The original source was first published by Pierre Pomet, *Histoire générale des drogues* (Paris: Jean-Baptiste Loyson and Augustin Pillon, 1694). It underwent numerous revisions over at least a half century, with coauthors Nicolas Lémery, a chemist, and Joseph Pitton de Tournefort, a botanist. The 1712 edition appears to be the first English version. The passage cited must be new to the English edition, or at least added since 1694, because Pomet declines to describe tea's medicinal properties, instead referring his audience to other authorities: "A l'égard de ses qualitez, je n'en diray rien, renvoyant le Lecteur aux Livres qu'en ont fait les Sieurs du Four & de Blegny" (144).

64. Ben-Erik Van Wyk and Michael Wink, *Medicinal Plants of the World* (Portland, Ore.: Timber Press, 2004), 75. The methylxanthine content in tea leaves and the brewed beverage is highly variable and depends on numerous factors, such as cultivar type, climate, harvesting technique, processing, storage, and steeping procedure. Caffeine is more concentrated in the leaf hairs (2.25 percent) and young leaves (first and second: 3.4 percent) than in older leaves (fifth and sixth: 1.5 percent) and stalk (between fifth and sixth: 0.5 percent), according to Mair and Hoh, *True History of Tea*, 26. Furthermore, the methylxanthines are water soluble and steep into the tea beverage more fully at higher temperature, with longer infusion times, and from finer leaf particles. There is also the question whether white, green, oolong, or black tea has more caffeine than the others, a problem rather challenging to address in light of the known factors influencing this criterion. Since the fermentation process employed in oolong- and black-tea manufacture does not alter the methylxanthine profile of the tea leaf, there is no difference in caffeine content attributable to tea-leaf processing.

65. Stephen L. Tilley, "Methylxanthines in Asthma," in *Methylxanthines: Handbook of Experimental Pharmacology*, ed. Bertil B. Fredholm (Berlin: Springer, 2011), 439–456. The use of theophylline against asthma is declining, particularly among wealthy nations, as there is some risk of theophylline toxicity, and specific adrenergic agonists have been demonstrated more effective.

66. Harmut Osswald and Jürgen Schnermann, "Methylxanthines and the Kidney," in *Methylxanthines*, ed. Fredholm, 391–412.

67. Liang Zhang et al., "L-Theanine from Green Tea: Transport and Effects on Health," in *Tea in Health and Disease Prevention*, ed. Preedy, 425–435; Xiaochun Wan et al., "Chemistry and Biological Properties of Theanine," in *Tea and Tea Products*, ed. Ho, Lin, and Shahidi, 255–273.

68. Many of tea's antioxidant chemicals are in the poly-phenolic structural class of compounds, a diverse group of chemicals involved in signaling, cellular energy metabolism, and pigmentation. Among them are the flavonoids, of which over 4000 have been identified in the plant kingdom. The flavonoids themselves are subdivided into groups based on shared structural features: the chalcones, flavanones, flavones, flavonols, isoflavones, anthocyanidins, and flavanols. While all the polyphenolics are potentially bioactive, the latter group, also called the catechins, has been the focus of intensive study in recent years for its strong antioxidant potential. See Ingrid A.-L. Persson, "Tea Flavanols: An Overview," in *Tea in Health and Disease Prevention*, ed. Preedy, 73–78.

69. F. N. Wachira et al., "Cultivar Type and Antioxidant Potency of Tea Product," in *Tea in Health and Disease Prevention*, ed. Preedy, 91–102.

70. Ankalo A. Shitandi, Francis Muigai Ngure, and Symon M. Mahungu demonstrate a sharp reduction in concentration of certain catechins during the fermentation process, attributable to both time and temperature, in "Tea Processing and Its Impact on Catechins, Theaflavin, and Thearubigin Formation," in *Tea in Health and Disease Prevention*, ed. R. Preedy, 193–206. However, the total polyphenol concentration is affected less strongly by fermentation, which is consistent with the observation that more fully oxidized black tea possesses its own suite of potentially bioactive polyphenolic compounds, including thearubigins and theaflavins. For a review of the current state of knowledge on thearubigin/theaflavin biochemistry, see Nikolai Kuhnert, "Chemistry and Biology of the Black Tea Thearubigins and of Tea Fermentation," in *Tea in Health and Disease Prevention*, ed. Preedy, 343–360. For a brief review of certain metabolic and genetic markers of oxidation in green- and black-tea studies, see Kaushik Das and Jharma Bhattacharttya, "Antioxidant Functions of Green and Black Tea," in *Tea in Health and Disease Prevention*, ed. Preedy, 521–528. Interestingly, the literature is not at consensus regarding the antioxidant capacity of green and black teas. Although the conventional wisdom still holds that green tea is more effective on this measure, as confirmed by test-tube studies, some animal experiments have yielded conflicting outcomes, showing black tea as either superior or equal to green tea in antioxidant capacity. Furthermore, certain tea polyphenolic extracts have been shown to have both antioxidant and prooxidant properties, depending on concentration and biological setting. In short, the question of tea's antioxidant properties is far from settled.

71. Joerg Gruenwald, Thomas Brendler, and Christof Jaenicke, eds., *PDR for Herbal Medicines*, 4th ed. (Montvale, N.J.: Thomson, 2007), 414–422; Elvira Gonzales de Mejia, Sirima Puangpraphant, and Rachel Eckhoff, "Tea and Inflammation"; and Calin Stoicov and JeanMarie Houghton, "Green Tea and Protection Against *Helicobacter* Infection," both in *Tea in Health and Disease Prevention*, ed. R. Preedy, 563–580, 593–602.

72. Dandapantula Sarma et al., "Safety of Green Tea Extracts," *Drug Safety* 31 (2008): 469–484.

73. Baruch Narotzki et al., "Green Tea: A Promising Natural Product in Oral Health," *Archives of Oral Biology* 57 (2012): 429–435; Puneet Goenka et al., "*Camellia sinensis* (Tea): Implications and Role in Preventing Dental Decay," *Pharmacognosy Review* 7, no. 14 (2013): 152–156.

74. Yukihiko Hara, "Tea Catechins and Their Applications as Supplements and Pharmaceutics," *Pharmacological Research* 64 (2011): 100–104. Veregen is a trade name for one such preparation. Another is Polyphenon E.

12. CACAO

1. Cameron L. McNeil, "Introduction: The Biology, Antiquity, and Modern Uses of the Chocolate Tree (*Theobroma cacao* L.)," in *Chocolate in Mesoamerica: A Cultural History of Cacao*, ed. Cameron L. McNeil (Gainesville: University Press of Florida, 2006), 4.

2. For example, the canté tree (*Gliricidia sepium*), known as *madre de cacao* in some parts of Mesoamerica. See Meredith L. Dreiss and Sharon Edgar Greenhill, *Chocolate: Pathway to the Gods* (Tucson: University of Arizona Press, 2008), 157.

3. Dreiss and Greenhill, *Chocolate*, 80, 157.

4. Allen M. Young, *The Chocolate Tree: A Natural History of Chocolate* (Gainesville: University Press of Florida, 2007), 1–4. To this day, Arawete and Asurini Indians in Amazonia eat the pulp of a related species, *T. speciosum*, and make a crude chocolate from its seeds. See Young, *Chocolate Tree*, 11.

5. For the scholarship on cacao's South American origins, see Sophie D. Coe and Michael D. Coe, *The True History of Chocolate*, 3rd ed. (New York: Thames & Hudson, 2013), 25–26. For a discussion of domestication hypotheses, see Nathaniel Bletter and Douglas C. Daly, "Cacao and Its Relatives in South America: An Overview of Taxonomy, Ecology, Biogeography, Chemistry, and Ethnobotany," in *Chocolate in Mesoamerica*, ed. McNeil, 35–37. Numerous species of *Theobroma* are said to have palatable pulp.

6. Bletter and Daly, "Cacao and Its Relatives in South America," 48–60. Just as *T. bicolor* and *T. cacao* have been described as distinct botanical species, the Maya in contemporary Mesoamerica recognize both as related forms of cacao, the former called *pataxte*. See Cameron L. McNeil, "Traditional Cacao Use in Modern Mesoamerica," in *Chocolate in Mesoamerica*, ed. McNeil, 341.

7. John S. Henderson et al., "Chemical and Archaeological Evidence for the Earliest Cacao Beverages," *Proceedings of the National Academy of Sciences USA* 104 (2007): 18937–18940; John S. Henderson and Rosemary A. Joyce, "Brewing Distinction: The Development of Cacao Beverage in Formative Mesoamerica," in *Chocolate in Mesoamerica*, ed. McNeil, 140–153. In "Traditional Cacao Use," McNeil discusses fermented and fresh cacao-containing beverages.

8. Archaeological and ethnographic evidence points to the deep significance of cacao for the Olmec, Maya, Zapotec, Mixtec, and Aztecs (Mexica).

9. Dreiss and Greenhill, *Chocolate*, 9–40, in which the authors discuss cacao as a World Tree, "[conduit] for departed souls and gods to travel between all three realms, [Sky, Earth, and Underworld]" (22).

10. Contemporary cacao beverage recipes among indigenous Mesoamericans are outlined in McNeil, "Traditional Cacao Use," 341–366.

11. In contemporary folklore, "chocolate is for the body, but the foam is for the soul" (Dreiss and Greenhill, *Chocolate*, 108).

12. Young, *Chocolate Tree*, 19; McNeil, "Traditional Cacao Use," 15. Dreiss and Greenhill describe the ancient symbolic connection between the heart and the cacao pod, blood and the chocolate drink, in *Chocolate*, 49. In addition to discussing *achiote*'s role in turning cacao beverages blood red, Dreiss and Greenhill relate that Classic period Maya used it to stain pottery for ritual offerings, "ensouling them with life force" (57).

13. The origins of the English words "cacao" and "chocolate" are the subject of much scholarly inquiry, according to Coe and Coe, *True History of Chocolate*, 116–120. Many authoritative sources, including the *Merriam-Webster's Dictionary* and the Royal Spanish Academy, indicate that *chocolatl* is the Aztec Nahuatl word for the liquid cacao drink. However, Coe and Coe point to the earliest texts that refer instead to the Nahuatl word *cacahuatl* (cacao water). Another possibility is that the modern word "chocolate" derives from the Mayan *chacau haa* or *chocol haa* (hot water). As for the Maya antecedents, the linguists Terrence Kaufman and John Justeson demonstrate that the term *kakawa* (cacao) emerged from the Gulf Coast of southern Mexico in antiquity and gave rise to diffused words throughout Mesoamerica, including *kakaw* in Mayan, in "The History of the Word for 'Cacao' and Related Terms in Ancient Meso-America," in *Chocolate in Mesoamerica*, ed. McNeil, 117–139. The term was taken into early Nahuatl as *kakawatl* and referred to both the seed and its processed form. Regarding the word "chocolate," the linguists' evidence rejects the notion that a word in some form of ancestral Nahuatl, such as *chokola:tl* or *chikola:tl*, gave rise to the Spanish *chocolate* or *chicolate*. The word *chocolatl* was apparently invented by the Spanish. Indeed, the earliest writing by the Spanish on the subject reference both the seed and a drink made from it as *cacao*. The first Spanish source that claims *chocolatl* as the indigenous word for the cacao beverage is from 1580, sixty years after the conquest.

14. Henderson and Joyce, "Brewing Distinction," 150–151. Lynn M. Meskell and Rosemary A. Joyce evoke the blood analogy in marriage by relaying that ancient Mesoamericans considered chocolate to represent the blood that flows between, and bonds, two families, in *Embodied Lives: Figuring Ancient Maya and Egyptian Experience* (London: Routledge, 2003), 139–140.

15. McNeil, "Introduction," 12.

16. Dreiss and Greenhill, *Chocolate*, 69–76. The authors also describe contemporary Mexican rituals for the deceased involving cacao.

17. Dreiss and Greenhill, *Chocolate*, 97–100.

18. Francisco Oviedo y Valdés, cited in Young, *Chocolate Tree*, 29–30.

19. Eight to ten, "according to how they agree," as conveyed by Oviedo, in Coe and Coe, *True History of Chocolate*, 59.

20. Dreiss and Greenhill, *Chocolate*, 101. The significance of the clay cacao bean forgeries is unresolved: they may have been counterfeit to substitute for more valuable real beans or, alternatively, may have been specially manufactured to serve a symbolic or ritual purpose.

21. Marcy Norton, *Sacred Gifts, Profane Pleasures: A History of Tobacco and Chocolate in the Atlantic World* (Ithaca, N.Y.: Cornell University Press, 2008), 35.

22. Bernard Ortiz de Montellano, *Aztec Medicine, Health, and Nutrition* (New Brunswick, N.J.: Rutgers University Press, 1990), 141.

23. Norton, *Sacred Gifts, Profane Pleasures*, 35.

24. William Gates, *An Aztec Herbal: The Classic Codex of 1552* (Toronto: Dover, 1939), 70.

25. Dreiss and Greenhill refer to eighteenth-century Yucatán Maya manuscripts, the books of Chilam Balam, and *The Ritual of the Bacabs*, in *Chocolate*, 136.

26. McNeil, "Traditional Cacao Use," 357, 360.

27. Bletter and Daly, "Cacao and Its Relatives in South America," 49–58.

28. Quoted in Young, *Chocolate Tree*, 30. A lengthier translation of this passage, attributed to an anonymous "gentleman of Hernán Cortés" (ca. 1556), appears in Coe and Coe, *True History of Chocolate*, 84.

29. Young, *Chocolate Tree*, 30–31.

30. Coe and Coe, *True History of Chocolate*, 129–131.

31. Pre-Columbian Mesoamericans prepared cacao with numerous additives, according to extant recipes. Some of the cacao mixtures included honey as a sweetener. See Dreiss and Greenhill, *Chocolate*, 136; and David Stuart, "The Language of Chocolate: References to Cacao on Classic Maya Drinking Vessels," in *Chocolate in Mesoamerica*, ed. Cameron L. McNeil, 195–196. Bernardino de Sahagún describes the Aztec practice of drinking honeyed chocolate flavored with fragrant flowers and vanilla, in *Florentine Codex: General History of the Things of New Spain*, book 8, *Kings and Lords*, trans. Charles E. Dibble and Arthur J. O. Anderson (Santa Fe, N.Mex.: School of American Research, 1954). Coe and Coe discuss the spices and herbs used as Aztec chocolate flavorings, in *True History of Chocolate*, 86–94.

32. Coe and Coe, *True History of Chocolate*, 131–134.

33. Coe and Coe, *True History of Chocolate*, 188–189.

34. Cacao reached the Philippines no later than the early eighteenth century and West Africa by the mid-nineteenth century. See Coe and Coe, *True History of Chocolate*, 176, 199, 243.

35. Young, *Chocolate Tree*, 34–36, 39–44.

36. Quoted in Young, *Chocolate Tree*, 37.

37. Young, *Chocolate Tree*, 37; Coe and Coe, *True History of Chocolate*, 241–243, 247.

38. Quoted in Young, *Chocolate Tree*, 38.

39. Young, *Chocolate Tree*, 38.

40. Coe and Coe, *True History of Chocolate*, 251–253; Mars, "History Timeline," http://www.mars.com/global/about-mars/history.aspx.

41. Consistent with humoral theory, these qualities are not to be literally read as physically hot, cold, wet, and dry.

42. Coe and Coe, *True History of Chocolate*, 122.

43. Coe and Coe, *True History of Chocolate*, 123. There was incomplete consensus among Spanish physicians whether cacao was better described as temperate rather than cool and dry rather than wet. The indigenous additives to chocolate were generally thought to be hot, except for *achiote*, whose temperature was disputed. See Norton, *Sacred Gifts, Profane Pleasures*, 237.

44. Driess and Greenhill, *Chocolate*, 141.

45. Antonio Colmonero de Ledesma, cited in Louis Evan Grivetti, "Medicinal Chocolate in New Spain, Western Europe, and North America," in *Chocolate: History, Culture, and Heritage*, ed. Louis Evan Grivetti and Howard-Yana Shapiro (Hoboken, N.J.: Wiley, 2009), 70.

46. *Needham's Mercurius Politicus* (June 12–23, 1657), quoted in Coe and Coe, *True History of Chocolate*, 167.

47. Henry Stubbe, *The Indian Nectar, or a Discourse Concerning Chocolata* (1662), quoted in Norton, *Sacred Gifts, Profane Pleasures*, 251.

48. Marie de Villars, quoted in Coe and Coe, *True History of Chocolate*, 136–137.

49. Santiago de Valverdes Turices, cited in Norton, *Sacred Gifts, Profane Pleasures*, 238.

50. Mijail Rimache Artica, *Cultivo del Cacao* (Miraflores, Peru: Empresa Editora Macro, 2008), 104. The length of fermentation depends on environmental conditions and the experience of the producer, who must turn the beans periodically to maintain temperature and dissipate byproducts of the decomposition.

51. In some locations, fermented beans are washed before drying. In modern operations, sun drying is still the most economical option, although many producers choose to use a heated platform or chamber or a drying machine. See Mijail Rimache Artica, *Cultivo del Cacao*, 105–107.

52. Driess and Greenhill, *Chocolate*, 105–106.

53. Rodney Snyder, Bradley Foliart Olsen, and Laura Pallas Brindle, "From Stone Metates to Steel Mills: The Evolution of Chocolate Manufacturing," in *Chocolate*, ed. Grivetti and Shapiro, 611–623.

54. The patent was issued in 1828. See Snyder, Olsen, and Brindle, "From Stone Metates to Steel Mills," 614; and Coe and Coe, *True History of Chocolate*, 235–236.

55. Coe and Coe, *True History of Chocolate*, 241.

56. Snyder, Olsen, and Brindle, "From Stone Metates to Steel Mills," 616; Coe and Coe, *True History of Chocolate*, 248–249.

57. Snyder, Olsen, and Brindle, "From Stone Metates to Steel Mills," 615; Coe and Coe, *True History of Chocolate*, 236.

58. Snyder, Olsen, and Brindle, "From Stone Metates to Steel Mills," 616. The conching process has been likened to kneading.

59. An outline of chocolate-manufacturing techniques and terminology is in Coe and Coe, *True History of Chocolate*, 255–259. A slightly different use of terminology is in Bennett Alan Weinberg and Bonnie K. Bealer, *The World of Caffeine: The Science and Culture of the World's Most Popular Drug* (New York: Routledge, 2001), 59–60.

60. Coe and Coe, *True History of Chocolate*, 28. Cacao solids are alternatively called cocoa solids.

61. Robert Rucker, "Nutritional Properties of Cocoa," in *Chocolate*, ed. Grivetti and Shapiro, 943–946.

62. For a review of epidemiological evidence and small-scale clinical trials on cacao for cardiovascular health, see Roberto Corti et al., "Cocoa and Cardiovascular Health," *Circulation* 119 (2009): 1433–1441; for cancer, see M. A. Martin, L. Goya, and S. Ramos, "Potential for Preventive Effects of Cocoa and Cocoa Polyphenols in Cancer," *Food and Chemical Toxicology* 56 (2013): 336–351.

63. F. Hermann et al., "Dark Chocolate Improves Endothelial and Platelet Function," *Heart* 92 (2006): 119–120.

64. E. L. Ding et al., "Chocolate and Prevention of Cardiovascular Disease: A Systematic Review," *Nutrition and Metabolism* 3 (2006): 2. Studies of cocoa or cocoa constituents have employed diverse experimental conditions that makes comparing them more challenging. Nevertheless, accumulating evidence indicates that cacao probably has blood pressure–lowering properties, among other effects, awaiting large-scale and long-term experiments to demonstrate these effects more robustly. See Karin Ried et al., "Effect of Cocoa on Blood Pressure," *Cochrane Database of Systematic Reviews* (2012): CD008893.

65. Christian Rätsch provides the concentrations of 1.45 percent theobromine and 0.02 percent caffeine in cacao beans, in *The Encyclopedia of Psychoactive Plants: Ethnopharmacology and Its Applications* (Rochester, Vt.: Park Street Press, 2005), 502. Michael Wink and Ben-Erik van Wyk give 3 percent total alkaloids in cacao beans, in *Mind-Altering and Poisonous Plants of the World* (Portland, Ore.: Timber Pres, 2008), 226. Walter Lewis and Memory P. F. Elvin-Lewis report up to 3 percent theobromine and a minute level of caffeine in beans, in *Medical Botany: Plants Affecting Human Health* (Hoboken, N.J.: Wiley, 2003), 681.

66. Lewis and Elvin-Lewis, *Medical Botany*, 681.

67. Weinberg and Bealer, *World of Caffeine*, 275; Lewis and Elvin-Lewis, *Medical Botany*, 291–292, 514–515.

68. Hendrik Jan Smit, "Theobromine and the Pharmacology of Cocoa," in *Methylxanthines: Handbook of Experimental Pharmacology*, ed. Bertil B. Fredholm (Berlin: Springer, 2011), 201–234.

13. TOBACCO

1. *Nicotiana rustica*, 1 to 1.5 meters; *N. tabacum*, 3 meters; *N. quadrivalvis*, 2 meters; *N. attenuata*, 1.5 meters; *N. trigonophylla*, 1 meter; *N. glauca*, 10 meters; *N. clevelandii*, 0.6 meter. The species used most frequently are annuals. See Joseph Winter, "Introduction to the North American Tobacco Species," in *Tobacco Use by Native North Americans;*

Sacred Smoke and Silent Killer, ed. Joseph Winter (Norman: University of Oklahoma Press, 2000), 4–8.

2. Johannes Wilbert, *Tobacco and Shamanism in South America* (New Haven, Conn.: Yale University Press, 1987), 2; Arthur W. Musk and Nicholas H. De Klerk, "History of Tobacco and Health," *Respirology* 8 (2003): 286–290.

3. Wilbert, *Tobacco and Shamanism*, 135–136.

4. A thorough study of indigenous South American tobacco use is Wilbert, *Tobacco and Shamanism*. For North American tobacco use, see Joseph Winter, "Traditional Uses of Tobacco by Native Americans," in *Tobacco Use by Native North Americans*, ed. Winter, 9–58.

5. Wilbert, *Tobacco and Shamanism*, 9–132; Winter, "Traditional Uses of Tobacco."

6. Wilbert, *Tobacco and Shamanism*, 9–132.

7. Jordan Goodman, *Tobacco in History: The Cultures of Dependence* (London: Routledge, 1993), 25.

8. Goodman, *Tobacco in History*, 25–26.

9. Winter, "Traditional Uses of Tobacco," 17.

10. Wilbert, *Tobacco and Shamanism*, 100.

11. Winter, "Traditional Uses of Tobacco," 31.

12. Wilbert, *Tobacco and Shamanism*, 34–35. Taurepan is spelled Taurepang in other sources.

13. Wilbert, *Tobacco and Shamanism*, 103.

14. Goodman, *Tobacco in History*, 27–28.

15. Marcy Norton, *Sacred Gifts, Profane Pleasures: A History of Tobacco and Chocolate in the Atlantic World* (Ithaca, N.Y.: Cornell University Press, 2008), 40–44.

16. William Gates, *An Aztec Herbal: The Classic Codex of 1552* (Toronto: Dover, 1939), 82. The ailment *morbus iterum rediens* in this passage in the Badianus manuscript is also translated as "recurrent fever" by Sarah A. Dickson, *Panacea or Precious Bane: Tobacco in Sixteenth Century Literature* (New York: New York Public Library, 1954), 31–32.

17. Norton, *Sacred Gifts, Profane Pleasures*, 41; Bernard Ortiz de Montellano, *Aztec Medicine, Health, and Nutrition* (New Brunswick, N.J.: Rutgers University Press, 1990), 155.

18. Daniel E. Moerman, *Native American Ethnobotany* (Portland, Ore.: Timber Press, 1998), 354–357.

19. Moerman, *Native American Ethnobotany*, 354–357.

20. Goodman, *Tobacco in History*, 33.

21. Alexander von Gernet, "Nicotian Dreams: The Prehistory and Early History of Tobacco in Eastern North America," in *Consuming Habits: Drugs in History and Anthropology*, ed. Jordan Goodman, Paul E. Lovejoy, and Andrew Sherratt (London: Routledge, 1995), 71; Mary Adair, "Tobacco on the Plains: Historical Use, Ethnographic Accounts, and Archaeological Evidence," in *Tobacco Use by Native North Americans*, ed. Winter, 175.

22. Wilbert, *Tobacco and Shamanism*, 29.

23. Wilbert, *Tobacco and Shamanism*, 64–123; Winter, "Traditional Uses of Tobacco," 9–58; Alexander von Gernet, "North American Indigenous *Nicotiana* Use and Tobacco Shamanism: The Early Documentary Record, 1520–1660," in *Tobacco Use by Native North Americans*, ed. Winter, 73. Pipes or tubes were also used to suck smoke from a tobacco fire for the purpose of blowing, such as onto a patient for healing.

24. Wilbert, *Tobacco and Shamanism*, 142.

25. Wilbert, *Tobacco and Shamanism*, 48–64.

26. Wilbert, *Tobacco and Shamanism*, 46–48.

27. Goodman, *Tobacco in History*, 37; *The Diario of Christopher Columbus's First Voyage to America, 1492–1493*, trans. Oliver Dunn and James E. Kelley Jr. (Norman: University of Oklahoma Press, 1989), 139.

28. Winter, "Introduction," 2; Norton, *Sacred Gifts, Profane Pleasures*, 57.

29. Winter, "Introduction," 2. The first seeds transported to Europe were probably *N. rustica* from the Caribbean, followed by *N. tabacum* from the Yucatán.

30. Bartolomé de las Casas (ca. 1527), quoted in Wilbert, *Tobacco and Shamanism*, 10.

31. André Thevet (1557), cited in Wilbert, *Tobacco and Shamanism*, 11–2

32. Francisco Oviedo y Valdés, cited in Wilbert, *Tobacco and Shamanism*, 10; and Norton, *Sacred Gifts, Profane Pleasures*, 86.

33. Oviedo, quoted in Norton, *Sacred Gifts, Profane Pleasures*, 86.

34. Winter, "Introduction," 2.

35. In North America, the best-quality tobacco is said to grow on marginal, less fertile land. See Goodman, *Tobacco in History*, 206.

36. Goodman discusses the European innovations in tobacco consumption, in *Tobacco in History*, 59–89.

37. Goodman describes the development of curing methods in the United States, in *Tobacco in History*, 207.

38. Goodman, *Tobacco in History*, 97.

39. Bernardino de Sahagún, quoted in Dickson, *Panacea or Precious Bane*, 29. *Piciyetl, picietl*, and *piciete* are alternative spellings.

40. Peter Martyr d'Anghiera, *The decades of the newe worlde or west India conteynyng the nauigations and conquestes of the Spanyardes* (1511), trans. Richard Eden (1555), quoted in Dickson, *Panacea or Precious Bane*, 24–25. The account was influenced strongly by Friar Ramón Pané's treatise of 1496/1497 on "all the ceremonies and antiquities" of the people, informed by his two years on Hispaniola after being left as missionary to the Indians by Columbus's second voyage. The full quote is: "Such is the strength of the powder of cohobba that it takes away the senses of the one using it." The identity of *cohoba* snuff is controversial, as some scholars argue that it may have been produced from an unrelated psychoactive plant. See Wilbert, *Tobacco and Shamanism*, 16–18, 54.

41. Girolamo Benzoni, quoted in Eric Burns, *Smoke of the Gods: A Social History of Tobacco* (Philadelphia: Temple University Press, 2007), 38; Dickson, *Panacea or Precious Bane*, 122. A translation of Benzoni's account is *History of the New World*, trans. W. H. Smyth (London: Haklyut Society, 1857).

42. In *Sacred Gifts, Profane Pleasures*, Norton suggests that the first Old World settlers to adapt tobacco practices were sailors and African slaves. The diffusion of tobacco, then, proceeded from the lower class to higher classes of society. A brief review of the early debate on tobacco's

medicinal properties is Musk and De Klerk, "History of Tobacco and Health"; a lengthy review is Grace Stewart, "A History of the Medicinal Use of Tobacco: 1492–1860," *Medical History* 11 (1967): 228–268.

43. Jean Liébault, *L'agriculture et maison rustique*, cited in Dickson, *Panacea or Precious Bane*, 72–75.

44. Nicolás Monardes, *Segunda parte*, cited in Dickson, *Panacea or Precious Bane*, 84–88. See also Nicolás Monardes, *Joyfull Newes out of the New-Found World*, trans. John Frampton (London, 1596), 33–45, which includes additional material from the French botanist Jean Nicot.

45. Mattias de l'Obel, *Plantarum seu stirpium historia*, quoted in Dickson, *Panacea or Precious Bane*, 44–45. In l'Obel's book, tobacco is illustrated with a woodcut of *N. tabacum* and identified as *sana sancta* (holy healing [herb]).

46. Anthony Chute (A. C.), *Tabaco: The distinct and seuerall opinions of the late and best Phisitions that have written of the diuers natures and qualities thereof*, cited in Dickson, *Panacea or Precious Bane*, 98.

47. "After meales it doth much hurt, for it infecteth the braine and the liuer" (William Vaughan, *Naturall and artificial directions for health*, quoted in Dickson, *Panacea or Precious Bane*, 98–99). In "History of the Medicinal Use of Tobacco," Stewart attributes this quote to John Gerard, *The Herball, or Generall Historie of Plants* (London, 1597).

48. Dickson, *Panacea or Precious Bane*, 156.

49. The Jamestown colonists planted tobacco beginning in 1612. See Goodman, *Tobacco in History*, 134; and Winter, "Introduction," 2.

50. Initially *N. rustica*, then *N. tabacum*, thought to produce a less harsh smoke.

51. Goodman gives an account of the relationship between colonial output and tobacco price at wholesale and retail, in *Tobacco in History*, 64.

52. Dickson, *Panacea or Precious Bane*; Anne Charlton, "Medicinal Uses of Tobacco in History," *Journal of the Royal Society of Medicine* 97 (2004): 292–296.

53. Goodman, *Tobacco in History*, 117.

54. Goodman, *Tobacco in History*, 98–99; Musk and De Klerk, "History of Tobacco and Health," 287.

55. Department of Health and Human Services, *How Tobacco Smoke Causes Disease: The Biology and Behavioral Basis for Smoking-Attributable Disease, A Report of the Surgeon General* (Atlanta: Department of Health and Human Services, Centers for Disease Control and Prevention, National Center for Chronic Disease Prevention and Health Promotion, Office on Smoking and Health, 2010), 15, and its citations.

56. Musk and De Klerk, "History of Tobacco and Health," 287.

57. For a review of the pictorial and textual content of tobacco labels, which rotate periodically and can include health warnings and cessation assistance, see David Hammond and Jessica Reid, "Health Warnings on Tobacco Products: International Practices," *Salud pública de México* 54 (2012): 270–280. The FDA has required graphic warnings on cigarette packages since 2012, under the Code of Federal Regulations, Title 21, Part 1141.

58. For a review of the chemical studies on tobacco smoke from cigars, see Dietrich Hoffmann and Ilse Hoffmann, "Chemistry and Toxicology," in *Cigars: Health Effects and Trends*, Smoking and Tobacco Control Monograph 9 (Bethesda, Md.: Department of Health and Human Services, Public Health Service, National Institutes of Health, 1998), 55–104.

59. Robert M. Julien, Claire Advokat, and Joseph E. Comaty, *A Primer of Drug Action*, 12th ed. (New York: Worth, 2011), 369.

60. Wilbert, *Tobacco and Shamanism*, 133–134.

61. Wilbert contends that alkaline additives are not required to release the nicotine but instead enhance salivation, which promotes its uptake across the mucous membrane, in *Tobacco and Shamanism*, 138. In contrast, Hoffman and Hoffman explain that in a pH gradient between pH 6.0, where unprotonated (base form) nicotine is low, and pH 8.0, where unprotonated and protonated nicotine are equally abundant, increasing alkalinity of tobacco results in increased nicotine yield, in "Chemistry and Toxicology," 67–69. Therefore, in cigars and chewing tobacco prepared for nicotine administration via the buccal route, alkalinity increases the fraction of free nicotine that crosses into the bloodstream. In cigarettes, which are prepared for smoking into the lungs, the smaller proportion of free nicotine is vaporized and rapidly taken up by the lungs while the protonated nicotine remains in the particulate (solid) phase of smoke and is taken up less efficiently.

62. Goodman, *Tobacco in History*, 5.

63. Goodman, *Tobacco in History*, 246–247, and in the text.

64. Dietrich Hoffmann and Ilse Hoffmann, "The Changing Cigarette: Chemical Studies and Bioassays," in *Risks Associated with Smoking Cigarettes with Low Machine-Measured Yields of Tar and Nicotine*, Smoking and Tobacco Control Monograph 13 (Bethesda, Md.: Department of Health and Human Services, Public Health Service, National Institutes of Health, 2001), 171–174.

65. Christian Rätsch reports a nicotine content in the plant between 0.05 and 4 percent, which may not be restricted to the leaves, in *The Encyclopedia of Psychoactive Plants: Ethnopharmacology and Its Applications* (Rochester, Vt.: Park Street Press, 2005), 387. Michael Wink and Ben-Erik van Wyk give a total leaf alkaloid content of 9 percent in *N. tabacum* and 18 percent in *N. rustica*, in *Mind-Altering and Poisonous Plants of the World* (Portland, Ore.: Timber Press, 2008), 177. Wilbert notes that nicotine content varies from 0.6 to 9 percent in *N. tabacum* leaves, in *Tobacco and Shamanism*, 134.

66. For a review of data on nicotine yield and "tobacco column" nicotine content, see Lynn Kozlowski, Richard O'Connor, and Christine Sweeney, "Cigarette Design," in *Risks Associated with Smoking Cigarettes*, 17. Julien, Advokat, and Comaty provide the nicotine content per cigarette at 0.5 to 2 milligrams and 0.1 to 0.4 milligram the amount entering the bloodstream, in *Primer of Drug Action*, 371.

67. Julien, Advokat, and Comaty, *Primer of Drug Action*, 376–378.

68. Hoffmann and Hoffmann, "Changing Cigarette," 161–165.

14. POPULAR HERBS

1. The anthropologist Nina Etkin referenced this well-known dietary maxim to set the agenda at the outset of her fascinating study *Edible Medicines: An Ethnopharmacology of Food* (Tucson: University of Arizona Press, 2006), 3. Despite its common attribution, the quote does not appear in the Hippocratic Corpus.

2. "Natural History of the American Cranberry, *Vaccinium macroparpon* Ait," http://www.umass.edu/cranberry/downloads/nathist.pdf; Paul Eck, *The American Cranberry* (New Brunswick, N.J.: Rutgers University Press, 1990), 43–46.

3. Robert S. Cox and Jacob Walker, *Massachusetts Cranberry Culture: A History from Bog to Table* (Charleston, S.C.: History Press, 2012), 12–13.

4. Eck, *American Cranberry*, 2; Cox and Walker, *Massachusetts Cranberry*, 23. Cranberries were added to pemmican in those areas where they were in commerce.

5. Cox and Walker, *Massachusetts Cranberry*, 23–24.

6. Roger Williams, *A Key into the Language of America* (London: Gregory Dexter, 1643), 99.

7. John Josselyn, *New England's Rarities Discovered in Birds, Beasts, Fishes, Serpents, and Plants of That Country* (1672; Boston: William Veazie, 1865), 120–121.

8. Eck, *American Cranberry*, 2; Gayle Engels, "Cranberry," *HerbalGram* 76 (2007): 1–2.

9. M. H. Grace et al., "Comparison of Health-Relevant Flavonoids in Commonly Consumed Cranberry Products," *Journal of Food Science* 77 (2012): H176–183; J. Côté et al., "Bioactive Compounds in Cranberries and Their Biological Properties," *Critical Reviews in Food Science and Nutrition* 50 (2010): 666–679.

10. David R. P. Guay cites a concentration of 200 milligrams of vitamin C per kilogram of fresh cranberries, in "Cranberry and Urinary Tract Infections," *Drugs* 69 (2009): 775–807. The Department of Agriculture gives 2.9 milligrams of vitamin C per ounce of unsweetened cranberry juice and 5.5 milligrams per cup of cranberry sauce, in "USDA National Nutrient Database for Standard Reference, Release 26" (2013).

11. Côté et al., "Bioactive Compounds in Cranberries."

12. Ruggero Rossi, Silvia Porta, and Brenno Canovi, "Overview on Cranberry and Urinary Tract Infections in Females," *Journal of Clinical Gastroenterology* 44 (2010): S61–62.

13. L. Y. Foo et al., "A-Type Proanthocyanidin Trimers from Cranberry That Inhibit Adherence of Uropathogenic P-fimbriated *Escherichia coli*," *Journal of Natural Products* 63 (2000): 1225–1228; Amy B. Howell et al., "Dosage Effect on Uropathogenic *Escherichia coli* Antiadhesion Activity in Urine Following Consumption of Cranberry Powder Standardized for Proanthocyanidin Content: A Multicentric Randomized Double Blind Study," *BMC Infectious Diseases*

10 (2010), 94; Tyler Smith, "Cochrane Collaboration Revises 2008 Conclusions on Cranberry for UTI Prevention: Experts, Researchers Clarify Results from Recent Meta-Analysis," *HerbalGram* 97 (2013): 28–31.

14. Guay, "Cranberry and Urinary Tract Infections," 779–783.

15. Guay, "Cranberry and Urinary Tract Infections," 788–789.

16. Ruth G. Jepson, Gabrielle Williams, and Jonathan C. Craig, "Cranberries for Preventing Urinary Tract Infections," *Cochrane Database of Systematic Reviews* (2012) CD001321; Smith, "Cochrane Collaboration."

17. For a meta-analysis of two dozen human trials on cranberry for the prevention of urinary tract infections, see Jepson, Williams, and Craig, "Cranberries for Preventing Urinary Tract Infections."

18. Theodore Hymowitz, "On the Domestication of the Soybean," *Economic Botany* 24, no. 4 (1970): 408–421.

19. A height of 75 to 125 centimeters is given in Nels R. Lersten and John B. Carlson, "Vegetative Morphology," in *Soybeans: Improvement, Production, and Uses*, 3rd ed., ed. H. Roger Boerma and James E. Specht (Madison, Wis.: American Society of Agronomy, Crop Science Society of America, Soil Science Society, 2004), 15. Pods usually contain one to three seeds, according to Theodore Hymowitz, "Speciation and Genetics," in *Soybeans*, ed. Boerma and Specht, 106.

20. James R. Wilcox, "World Distribution and Trade of Soybean," in *Soybeans*, ed. Boerma and Specht, 3–13.

21. Theodore Hymowitz and W. R. Shurtleff, "Debunking Soybean Myths and Legends in the Historical and Popular Literature," *Crop Science* 45 (2005): 473–476; H. T. Huang, "Early Uses of Soybean in Chinese History," in *The World of Soy*, ed. Christine M. DuBois, Chee-Beng Tan, and Sidney W. Mintz (Urbana: University of Illinois Press, 2008), 45–55. Huang discusses the historical development of soybean-processing techniques to render the difficult-to-digest soybean seed protein more appealing, facilitating its rise as one of China's most important food sources.

22. Dan Bensky, Steven Clavey, and Erich Stöger, *Chinese Herbal Medicine: Materia Medica*, 3rd ed. (Seattle: Eastland, 2004), 63–7, 977–978. Soybean sprouts first appeared in the *Divine Husbandman's Classic of the Materia Medica*, written nearly 2000 years ago. Note that in "Debunking Soybean Myths," Hymowitz and Shurtleff doubt the date often attributed to this work, perhaps 2700 or more years ago. In *Chinese Herbal Medicine*, Bensky, Clavey, and Stöger refer to a version of the text recorded by one or more early herbalists long after the mythical author is said to have lived. Prepared soybean was listed in the Ming-era *Treasury of Words on the Materia Medica*, and soybean seed coats in the *Grand Materia Medica* (1596).

23. K.-S. Chia et al., "Profound Changes in Breast Cancer Incidence May Reflect Changes into a Westernized Lifestyle: A Comparative Population-Based Study in Singapore and Sweden," *International Journal of Cancer* 113 (2005): 302–306; Peggy Porter, "'Westernizing' Women's Risks? Breast Cancer in Lower-Income Countries," *New England Journal of Medicine* 358 (2008): 213–216.

24. Stephen Barnes cites a concentration of 1 to 2 milligrams per gram, 100 times higher than in the chickpea, in "The Biochemistry, Chemistry, and Physiology of the Isoflavones in Soybeans and Their Food Products," *Lymphatic Research and Biology* 8 (2010): 89–98.

25. Barnes, "Biochemistry, Chemistry, and Physiology of the Isoflavones."

26. Bahram H. Arjmahdi et al., "Dietary Soybean Protein Prevents Bone Loss in an Ovariectomized Rat Model of Osteoporosis," *Journal of Nutrition* 126 (1996): 161–167; Silvina Levis and Marcio L. Griebeler, "The Role of Soy Foods in the Treatment of Menopausal Symptoms," *Journal of Nutrition* 140 (2010): 2318S–2321S.

27. Levis and Griebeler, "Role of Soy Foods."

28. K. Taku et al., "Soy Isoflavones for Osteoporosis: An Evidence-Based Approach," *Maturitas* 70 (2011): 333–338; P. Wei et al., "Systematic Review of Soy Isoflavone Supplements on Osteoporosis in Women," *Asian Pacific Journal of Tropical Medicine* (2012): 243–248.

29. Mark J. Messina and Charles E. Wood, "Soy Isoflavones, Estrogen Therapy, and Breast Cancer Risk: Analysis and Commentary," *Nutrition Journal* 7 (2008): 17.

30. Ivonne M. C. M. Rietjens et al., "Mechanisms Underlying the Dualistic Mode of Action of Major Soy Isoflavones in Relation to Cell Proliferation and Cancer Risks," *Molecular Nutrition and Food Research* 57 (2013): 100–113.

31. S. C. Ho et al., "Intake of Soy Products Is Associated with Better Plasma Lipid Profiles in the Hong Kong Chinese Population," *Journal of Nutrition* 130 (2000): 2590–2593; Baohua Liu et al., "Prevalence of the Equol-Producer Phenotype and Its Relationship with Dietary Isoflavone and Serum Lipids in Healthy Chinese Adults," *Journal of Epidemiology* 20 (2010): 377–384; X. X. Liu et al., "Effect of Soy Isoflavones on Blood Pressure: A Meta-Analysis of Randomized Controlled Trials," *Nutrition, Metabolism, and Cardiovascular Diseases* 22 (2012): 463–470.

32. Yu Qin et al., "Isoflavones for Hypercholesterolaemia in Adults," *Cochrane Database of Systematic Reviews* (2013): CD009518.

33. Bradley C. Bennett and Judith R. Hicklin, "Uses of Saw Palmetto (*Serenoa repens*, Arecaceae) in Florida," *Economic Botany* 52, no. 4 (1998): 381–393.

34. Bennett and Hicklin, "Uses of Saw Palmetto," 382. A height of 5 to10 feet (1.5–3 m) is given in Edward F. Gilman, "*Serenoa repens* Saw Palmetto," University of Florida Extension Publication FPS-547 (1999), http://edis.ifas.ufl.edu/fp547. Upright stems can reach a height of 15 to 20 feet (4.5–6 m), according to George W. Tanner, J. Jeffrey Mullahey, and David Maehr, "Saw Palmetto: An Ecologically and Economically Important Native Palm," University of Florida Extension Publication WEC-109 (1999), http://www.plantapalm.com/vpe/misc/saw-palmetto.pdf. Botanists consider the single-seeded, fleshy saw palmetto fruit to be a drupe.

35. Bennett and Hicklin, "Uses of Saw Palmetto," 383–386.

36. *Jonathan Dickinson's Journal, or God's Protecting Providence, Being the Narrative of a Journey from Port Royal in Jamaica to Philadelphia Between August 23, 1696 and April 1, 1697*, ed. Evangeline Walker Andrews and Charles McLean Andrews (New Haven, Conn.: Yale University Press, 1945), 48–49. Dickinson, a Quaker merchant, was shipwrecked in 1696 on the coast of eastern Florida and held captive by local tribes for several weeks before being released.

37. In "Uses of Saw Palmetto," Bennett and Hicklin compile dozens of reported medicinal uses of saw palmetto berries from traditional indigenous, folk, and early biomedical sources.

38. Edwin M. Hale, *Saw Palmetto (Sabal Serrulata, Serenoa Serrulata): Its History, Botany, Chemistry, Pharmacology, Provings, Clinical Experience, and Therapeutic Applications* (Philadelphia: Boericke and Tafel, 1898), 20–24.

39. Hale, *Saw Palmetto*.

40. *Pacific Record of Medicine and Surgery*, quoted in Hale, *Saw Palmetto*, 57. See also Herbert T. Webster, *Dynamical Therapeutics*, 2nd ed. (San Francisco: Webster Medical, 1898), 498.

41. The test of saw palmetto extract on a twenty-three-year-old woman reportedly increased her bust size from 81 to 85 centimeters in a period of three months, accompanied by "sharp pain" in the ovary, "headache," and "tenderness of mammary glands," according to her (excruciatingly) detailed diary, as reported in Hale, *Saw Palmetto*, 30–37. Hale also reprints the self-experimentation of Dr. Boocock, who elaborated on how saw palmetto "aroused up some amorous feelings" and "good, firm erections" (24–26). With regard to the prostate, Dr. Goss writes of saw palmetto: "As a remedy for chronic enlargement of the prostate gland, I have found this a very positive remedy" (62).

42. World Health Organization, *WHO Monographs on Selected Medicinal Plants* (Geneva: World Health Organization, 2004), 2:289–291.

43. Among many others, see Culley Carson and Roger Rittmaster, "The Role of Dihydrotestosterone in Benign Prostatic Hyperplasia," *Urology* 61, suppl. 4A (2003): 2–7; and Heike Weisser et al., "Effects of the *Sabal serrulata* Extract IDS 89 and Its Subfractions on 5α-Reductase Activity in Human Benign Prostatic Hyperplasia," *Prostate* 28 (1996): 300–306.

44. Summarized in World Health Organization, *Monographs*, 2:291.

45. Several clinical trials with positive results are reviewed in World Health Organization, *Monographs*, 2:292.

46. J. Tacklind et al., "*Serenoa repens* for Benign Prostatic Hyperplasia," *Cochrane Database of Systematic Reviews* (2012): CD001423.

47. Bennett and Hicklin, "Saw Palmetto," 387; Aisling Swift, "Palmetto Berries: Industry Has Health Benefits," *Naples (Fla.) Daily News*, September 28, 2013.

48. F. K. Habib and M. G. Wyllie, "Not All Brands Are Created Equal: A Comparison of Selected Components of Different Brands of *Serenoa repens* Extract," *Prostate Cancer and Prostatic Diseases* 7 (2004): 195–200.

49. Eric Block, *Garlic and Other Alliums: The Lore and the Science* (London: Royal Society of Chemistry, 2010),

1–10; Biljana Bauer Petrovska and Svetlana Cekovska, "Extracts from the History and Medical Properties of Garlic," *Pharmacognosy Review* 4 (2010): 106–110.

50. World Health Organization, *WHO Monographs on Selected Medicinal Plants* (Geneva: World Health Organization, 1999), 1:17. Leaf length depends on cultivar and growing conditions.

51. Block, *Garlic and Other Alliums*, 10–11.

52. Richard S. Rivlin, "Historical Perspective on the Use of Garlic," *Journal of Nutrition* 131 (2001): 951S–954S; Block, *Garlic and Other Alliums*, 19–26.

53. The Ebers papyrus, a New Kingdom medical text, documents the use of garlic in treatments for trembling limbs and for irregular menstruation, among others, and for strengthening the body. See *Ancient Egyptian Medicine: The Papyrus Ebers*, trans. Cyril P. Bryan (London: Bles, 1930), 63, 82, 114, etc. However, other sources caution that the translation of the word "garlic" in Egyptian sources has not been settled. While numerous archaeological specimens of garlic exist from ancient Egypt, its use as food and medicine can only be assumed. See John F. Nunn, *Ancient Egyptian Medicine* (Norman: University of Oklahoma Press, 1996), 14; and J. R. Campbell and A. R. David, "The Application of Archaeobotany, Phytogeography, and Phamacognosy to Confirm the Pharmacopeia of Ancient Egypt 1850–1200 B.C.," in *Pharmacy and Medicine in Ancient Egypt: Proceedings of the Conferences Held in Cairo (2007) and Manchester (2008)*, ed. Jenefer Cockitt and Rosalie David, BAR International Series 2141 (Oxford: Archaeopress, 2010), 20–29.

54. Pedanius Dioscorides of Anazarbus, *De materia medica*, trans. Lily Y. Beck (Hildesheim: Olms-Weidmann, 2005), 155–156. For the historical milieu and modern evidence related to some of Dioscorides's garlic prescriptions, see John M. Riddle, "The Medicines of Greco-Roman Antiquity as a Source of Medicine for Today," in *Prospecting for Drugs in Ancient and Medieval European Texts: A Scientific Approach*, ed. Bart K. Holland (Amsterdam: Harwood, 1996), 7–9.

55. John Parkinson, *Theatrum Botanicum* (London: Cotes, 1640), 874.

56. The World Health Organization reports 0.25 to 1.7 percent of alliin in whole garlic clove, the higher end characteristic of carefully dried material, in *Monographs*, 1:20.

57. The sharpness of freshly crushed garlic demonstrates the action of allicin on nociceptors (pain receptors) in the oral and nasal mucosa. See Block, *Garlic and Other Alliums*, 74–75. Garlic's pungent taste also belies the highly oxidant nature of allicin, which, when swallowed, reacts with food particles and proteins on the surface of cells lining the digestive tract, ultimately breaking down into other compounds. Very little allicin is likely absorbed by the intestine, and allicin does not persist in blood circulation. Therefore, the allicin in fresh garlic is not considered a healthful constituent; rather, the products of its degradation are thought to be responsible for possible health effects. See Harunobu Amagase, "Clarifying the Real Bioactive Constituents of Garlic," *Journal of Nutrition* 136 (2006): 716S–725S.

58. World Health Organization, *Monographs*, 1:19; Harunobu Amagase et al., "Intake of Garlic and Its Bioactive Components," *Journal of Nutrition* 131 (2001): 955S–962S. Block reports that allicin's stability depends on temperature, noting that allicin in water can rapidly break down within days or weeks at body temperature or room temperature but persist indefinitely if frozen, in *Garlic and Other Alliums*, 69–70.

59. Amagase et al., "Intake of Garlic." For a summary of the antioxidant properties of one of the commercial forms of medicinal garlic, an odorless, allicin-free alcohol/water extract of garlic aged up to twenty months, see Carmia Borek, "Antioxidant Health Effects of Aged Garlic Extract," *Journal of Nutrition* 131 (2001): 1010S–1015S. Other preparations include powdered garlic, essential oil of garlic, and fresh garlic. Recent research on garlic in health has employed a wide diversity of preparations, some taking care to preserve allicin, despite its poor ability to enter systemic circulation. See Sanjay K. Banerjee and Subir K. Maulik, "Effect of Garlic on Cardiovascular Disorders: A Review," *Nutrition Journal* 1 (2002): 4.

60. Anna Capasso, "Antioxidant Action and Therapeutic Efficacy of *Allium sativum* L.," *Molecules* 18 (2013): 690–700; Borek, "Antioxidant Health Effects."

61. Capasso, "Antioxidant Action"; Ana L. Colín-González et al., "The Antioxidant Mechanisms Underlying the Aged Garlic Extract- and S-Allylcysteine-Induced Protection," *Oxidative Medicine and Cellular Longevity* (2012): ID 907162; Borek, "Antioxidant Health Effects."

62. Banerjee and Maulik, "Effect of Garlic on Cardiovascular Disorders."

63. Benjamin H. S. Lau, "Suppression of LDL Oxidation by Garlic Compounds Is a Possible Mechanism of Cardiovascular Health Benefit," *Journal of Nutrition* 136 (2006): 765S–768S.

64. Khalid Rahman and Gordon M. Lowe, "Garlic and Cardiovascular Disease: A Critical Review," *Journal of Nutrition* 136 (2006): 736S–740S; C. D. Gardner et al., "Effect of Raw Garlic Versus Commercial Garlic Supplements on Plasma Lipid Concentrations in Adults with Moderate Hypercholesterolemia: A Randomized Clinical Trial," *Archives of Internal Medicine* 167 (2007): 346–353. A recent meta-analysis including the results of thirty-nine primary trials demonstrates a significant reduction in serum cholesterol and low-density lipoprotein cholesterol attributable to garlic treatment. See K. Ried, C. Toben, and P. Fakler, "Effect of Garlic on Serum Lipids: An Updated Meta-Analysis," *Nutrition Reviews* 71 (2013): 282–299.

65. K. Ried et al., "Effect of Garlic on Blood Pressure: A Systematic Review and Meta-Analysis," *BMC Cardiovascular Disorders* 8 (2008): 13; K. Ried, O. R. Frank, and N. P. Stocks, "Aged Garlic Extract Reduces Blood Pressure in Hypertensives: A Dose-Response Trial," *European Journal of Clinical Nutrition* 67 (2013): 64–70. Recent placebo-controlled studies demonstrate a reduction in blood pressure attributable to garlic, but the base of experimental data from previous trials is small and incomplete. See S. N. Stabler et al., "Garlic for the Prevention of Cardiovascular Morbidity

and Mortality in Hypertensive Patients," *Cochrane Database of Systematic Reviews* (2012): CD007653. "This makes it difficult to determine the true impact of garlic on lowering blood pressure," they write.

66. Gloria A. Benavides et al., "Hydrogen Sulfide Mediates the Vasoactivity of Garlic," *Proceedings of the National Academy of Sciences USA* 104 (2007): 17977–17982.

67. Gillian L. Allison, Gordon M. Lowe, and Khalid Rahman, "Aged Garlic Extract and Its Constituents Inhibit Platelet Aggregation Through Multiple Mechanisms," *Journal of Nutrition* 136 (2006): 782S–788S.

68. Petrovska and Cekovska, "Extracts from the History and Medical Properties of Garlic"; Amagase, "Real Bioactive Constituents"; Eikai Kyo et al., "Immunomodulatory Effects of Aged Garlic Extract," *Journal of Nutrition* 131 (2001): 1075S–1079S.

69. See, for example, Patrick S. Ruddock et al., "Garlic Natural Health Products Exhibit Variable Constituent Levels and Antimicrobial Activity Against *Neisseria gonorrhoeae, Staphylococcus aureus*, and *Enterococcus faecalis*," *Phytotherapy Research* 19 (2005): 327–34; Serge Ankri and David Mirelman, "Antimicrobial Properties of Allicin from Garlic," *Microbes and Infection* 2 (1999): 125–129; and J. C. Harris et al., "Antimicrobial Properties of *Allium sativum* (garlic)," *Applied Microbiology and Biotechnology* 57 (2001): 282–286.

70. Elizabeth Lissiman, Alice L. Bhasale, and Marc Cohen, "Garlic for the Common Cold," *Cochrane Database of Systematic Reviews* (2012): CD006206.

71. World Health Organization, *Monographs*, 1:154; Z. Pang, F. Pan, and S. He, "*Ginkgo biloba* L.: History, Current Status, and Future Prospects," *Journal of Alternative and Complementary Medicine* 2 (1996): 359–363.

72. Peter Crane, *Ginkgo: The Tree That Time Forgot* (New Haven, Conn.: Yale University Press, 2013), 53–59.

73. Crane, *Ginkgo*, 53–59, 72.

74. Bensky, Clavey, and Stöger, *Chinese Herbal Medicine*, 893–896.

75. Crane, *Ginkgo*, 230–231.

76. One of the most widely used leaf extracts, produced by a German firm, is labeled EGb 761, standardized to a particular concentration of flavone glycosides and terpene lactones.

77. The protocols particularly of some of the earlier, small-scale trials are criticized in Jacqueline Birks and John Grimley Evans, "*Ginkgo biloba* for Cognitive Impairment and Dementia," *Cochrane Database of Systematic Reviews* (2009): CD003120; and Beth E. Snitz et al., "*Ginkgo biloba* for Preventing Cognitive Decline in Older Adults: A Randomized Trial," *Journal of the American Medical Association* 302 (2009): 2663–2670. The discrepancy between the labeled chemical content and tested profiles of commercially available ginkgo products is highlighted in H. P. Fransen et al., "Assessment of Health Claims, Content, and Safety of Herbal Supplements Containing *Ginkgo biloba*," *Food & Nutrition Research* 54 (2010): 5221.

78. Mark Blumenthal, Alicia Goldberg, and Josef Brinckmann, eds., *Herbal Medicine: Expanded Commission E Monographs* (Austin, Tex.: American Botanical Council, 2000).

79. The ginkgolides (most strongly ginkgolide B) interfere with a component of the blood-clotting pathway. Other possible activities are less well understood. Some physiological effects are summarized in World Health Organization, *Monographs*, 1:158–162; and Lon S. Schneider, "*Ginkgo biloba* Extract and Preventing Alzheimer's Disease," *Journal of the American Medical Association* 300 (2008): 2306.

80. Birks and Grimley Evans, "*Ginkgo biloba*," 10.

81. Snitz et al., "*Ginkgo biloba*"; S. T. DeKosky et al., "*Ginkgo biloba* for Prevention of Dementia: A Randomized Controlled Trial," *Journal of the American Medical Association* 300 (2008): 2253–2262.

82. Jennifer R. Evans reviews two earlier trials that seem to show that ginkgo extract has some effect on macular degeneration but recommends larger studies of proper design, in "*Ginkgo biloba* Extract for Age-Related Macular Degeneration," *Cochrane Database of Systematic Reviews* (2013): CD001775.

83. See the meta-analysis conducted by Malcolm P. Hilton, Eleanor F. Zimmerman, and William T. Hunt, which found a statistically significant improvement in tinnitus in the subset of patients also experiencing dementia, an observation they concluded "is unlikely to be clinically significant" ("*Ginkgo biloba* for Tinnitus," *Cochrane Database of Systematic Reviews* [2013]: CD003852, 9).

84. S. P. A. Nicolaï et al., "*Ginkgo biloba* for Intermittent Claudication," *Cochrane Database of Systematic Reviews* (2013): CD006888, 15. The authors noted that several of the studies included in their analysis were of poor design and recommended larger and appropriately designed trials, although, they state that "it is unlikely that the effect of *Ginkgo biloba* will ever be clinically relevant."

85. World Health Organization, *Monographs*, 2:300–301; Rhonda Janke and Jeanie DeArmond, *A Grower's Guide: Milk Thistle* Silybum marianum, Kansas State University Research and Extension Publication MF-2618 (2004); P. Corchete, "Silybum marianum (L.) Gaertn: The Source of Silymarin," in *Bioactive Molecules and Medicinal Plants*, ed. K. G. Ramawat and J.-M. Mérillon (Berlin: Springer, 2008), 126. The seeds comprise the majority of the dried fruits and are not readily separable. Therefore, it makes sense that historical and modern sources consider the "seeds" to be the medicinal product.

86. Corchete, "*Silybum marianum*," 124.

87. William Coles, *Adam in Eden, or, Nature's Paradise. The History of Plants, Herbs and Flowers* (London: Nathaniel Brooke, 1657), 211.

88. Webster, *Dynamical Therapeutics*, 238. The author of this text also commented generally about milk thistle's utility in medicine: "It is a powerful and valuable remedy, and one which will repay any one for a careful clinical study. Its virtues have not half been determined" (339–340).

89. World Health Organization, *Monographs*, 2:302–303.

90. Corchete explains that the silymarin standardized extract contains 70 to 80 percent silymarin (30–50 percent silybin A and B, the remainder varying amounts

of isosilybin, dehydrosilybin, silychristin, silydianin, and taxifolin) and 20 to 30 percent polyphenolics, in "*Silybum marianum*," 130. For a summary of various extraction techniques, see Rajesh Agarwal et al., "Anticancer Potential of Silymarin: From Bench to Bed Side," *Anticancer Research* 26 (2006): 4457–4498. Note that the various techniques yield disparate concentrations of these active constituents.

91. Agarwal et al., "Anticancer Potential of Silymarin," 4461–4467.

92. Agarwal et al., "Anticancer Potential of Silymarin," 4467–4480.

93. Carmela Loguercio and Davide Festi, "Silybin and the Liver: From Basic Research to Clinical Practice," *World Journal of Gastroenterology* 17 (2011): 2288–2301.

94. The hepatoprotective effect is attributable to the isosilybin A, taxifolin, and silibinin constituents of silymarin, according to S. J. Polyak et al., "Identification of Hepatoprotective Flavonolignans from Silymarin," *Proceedings of the National Academy of Sciences USA* 107 (2010): 5995–5999.

95. Loguercio and Festi, "Silybin and the Liver," 2289.

96. In "Silybin and the Liver," Loguercio and Festi describe an analysis of ten clinical trials on silymarin, all of which failed to meet the standards of experimental design that would make their outcomes meaningful. For a meta-analysis of previous trials using milk thistle extract for alcoholic, hepatitis B, and hepatitis C viral liver diseases, see A. Rambaldi, B. P. Jacobs, and C. Gluud, "Milk Thistle for Alcoholic and/or Hepatitis B or C Virus Liver Diseases," *Cochrane Database of Systematic Reviews* (2007): CD003620. This meta-analysis found that fewer than 30 percent of them met high standards for experimental design.

97. Carmen Tamayo and Suzanne Diamond, "Review of Clinical Trials Evaluating Safety and Efficacy of Milk Thistle (*Silybum marianum* [L.] Gaertn.)," *Integrative Cancer Therapies* 6 (2007): 146–157.

98. The rationale for a new trial of silymarin for hepatitis C symptoms that accounts for several of the shortcomings in previous experiments is addressed in K. R. Reddy et al., "Rationale, Challenges, and Participants in a Phase II Trial of a Botanical Product for Chronic Hepatitis C," *Clinical Trials* 9 (2012): 102–112. For the results of a relatively well-designed trial in patients with hepatitis C unresponsive to standard interferon treatment, see M. W. Fried et al., "Effect of Silymarin (Milk Thistle) on Liver Disease in Patients with Chronic Hepatitis C Unsuccessfully Treated with Interferon Therapy: A Randomized Controlled Trial," *Journal of the American Medical Association* 308 (2012): 274–282. The study failed to show any difference in outcome parameters between those administered silymarin at a common dose, high dose, or placebo. Such data, as they accumulate, will ultimately answer long-standing questions about milk thistle's efficacy.

99. For a description of the mechanism of action and summary of clinical experiences with an injectable derivative of milk thistle extract developed to counteract *Amanita* poisoning, see, for example, F. Enjalbert et al., "Treatment of Amatoxin Poisoning: Twenty-Year Retrospective Analysis,"

Journal of Toxicology 40 (2002): 715–757; Jeanine Ward et al., "Amatoxin Poisoning: Case Reports and Review of Current Therapies," *Journal of Emergency Medicine* 44 (2013): 116–121; and Ulrich Mengs, Ralf-Torsten Pohl, and Todd Mitchell, "Legalon® SIL: The Antidote of Choice in Patients with Acute Hepatotoxicity from Amatoxin Poisoning," *Current Pharmaceutical Biotechnology* 12 (2012): 1964–1970.

100. A randomized, controlled trial involving fifty lactating women reports to have demonstrated a significant increase in milk production with silymarin treatment. See F. Di Pierro et al., "Clinical Efficacy, Safety, and Tolerability of BIO-C® (Micronized Silymarin) as a Galactagogue," *Acta Biomedica* 79 (2008): 205–210.

101. For a discussion of the natural range of the various *Echinacea* species, see Kelly Kindscher, "Ethnobotany of Purple Coneflower (*Echinacea angustifolia*, Asteraceae) and Other Echinacea Species," *Economic Botany* 43, no. 4 (1989): 498–507; Christopher Hobbs, "Echinacea: A Literature Review. Botany, History, Chemistry, Pharmacology, Toxicology, and Clinical Uses," *HerbalGram* 30 (1994): 33; and Kathleen A. McKeown, "A Review of the Taxonomy of the Genus *Echinacea*," in *Perspectives on New Crops and New Uses*, ed. J. Janick (Alexandria, Va.: ASHS, 1999), 482–489. The English common names are those accepted under the Integrated Taxonomic Information System (http://www.itis.gov).

102. Kindscher, "Ethnobotany of Purple Coneflower," 498–499.

103. In "Ethnobotany of Purple Coneflower," Kindscher reviews the ethnobotanic scholarship on *Echinacea*.

104. Kindscher cites reports of the following uses: "the Crow chewed the root for colds and drank a tea prepared from the root for colic," "Hidatsa warriors were known to chew small pieces of the root as a stimulant when traveling all night," and the "Comanche used the root for treating sore throat and toothache" ("Ethnobotany of Purple Coneflower," 500).

105. Kindscher, "Ethnobotany of Purple Coneflower," 500–502.

106. Eclectic medicine grew out of the early nineteenth century and was at its high point in the 1880s and 1890s, with its own constellation of medical schools around the country and a substantial literature. By the early twentieth century, it began to lose favor, and its last medical college closed in 1939. Eclectic medicine shares some philosophical underpinnings with similar movements antagonistic to mainstream medicine in the nineteenth century. See John S. Haller, *Medical Protestants: The Eclectics in American Medicine, 1825–1939* (Carbondale: Southern Illinois University Press, 1994).

107. The earliest lengthy account of coneflower's properties is H. C. F. Meyer, "*Echinacea Angustifolia*," *Eclectic Medical Journal* 47 (1887): 209–210. He used a preparation of its root to treat many types of infectious disease, sores, hemorrhoids, rattlesnake bites, and bee and wasp stings.

108. Finley Ellingwood and John Uri Lloyd, *American Materia Medica, Therapeutics, and Pharmacognosy* (Chicago: Ellingwood's Therapeutist, 1915), 358–367. For an

account of how physicians employed purple coneflower during this era, and of the debate within the larger medical community, see Francis Brinker, "*Echinacea* Differences Matter: Traditional Uses of *Echinacea angustifolia* Root Extracts Versus Modern Clinical Trials with *Echinacea purpurea* Fresh Plant Extracts," *HerbalGram* 97 (2013): 46–57.

109. According to Brinker in "*Echinacea* Differences Matter," the German product Echinacin, the result of a 1930s-era mistaken collection of *E. purpurea* instead of *E. angustifolia*, is the juice of the above-ground herb preserved in 22 percent ethanol. The Swiss product Echinaforce resulted from the intentional collection of *E. purpurea*, which was thought to grow better in Switzerland than *E. angustifolia*. This formulation is composed of a mixture of aerial plant extract (95 percent) and root extract (5 percent) in 65 percent ethanol. The water–alcohol extract of *E. pallida* roots is also recognized for clinical use in Germany. See Blumenthal, Goldberg, and Brinckmann, *Herbal Medicine*.

110. Brinker, "*Echinacea* Differences Matter"; Cathi Dennehy, "Need for Additional, Specific Information in Studies with *Echinacea*," *Antimicrobial Agents and Chemotherapy* 45 (2001): 369.

111. World Health Organization, *Monographs*, 1:129–130, 138; Brinker, "*Echinacea* Differences Matter." Chicoric acid is also spelled "cicoric acid."

112. Brinker, "*Echinacea* Differences Matter."

113. M. Sharma, R. Schoop, and J. B. Hudson, "*Echinacea* as an Anti-Inflammatory Agent: The Influence of Physiologically Relevant Parameters," *Phytotherapy Research* 23 (2009): 863–867; S. Pleschka et al., "Antiviral Properties and Mode of Action of Standardized *Echinacea purpurea* Extract Against Highly Pathogenic Avian Influenza Virus (H5N1, H7N7) and Swine-Origin H1N1 (S-OIV)," *Virology Journal* 6 (2009): 197. Alkamides are considered especially strong candidates for purple coneflower's antiviral activity.

114. For example, B. Barrett et al. used both *E. angustifolia* and *E. purpurea* root extract in pill form on patients with early-stage colds and found a statistically insignificant trend toward lower duration and severity in those receiving purple coneflower treatment over placebo, as reported in "*Echinacea* for Treating the Common Cold: A Randomized Controlled Trial," *Annals of Internal Medicine* 153 (2010): 769–777. Ronald B. Turner et al. used two formulations of an *E. angustifolia* root extract in liquid form on healthy patients administered as a nasal spray containing a dose of the cold virus and found no difference between groups in cold frequency or severity, as reported in "An Evaluation of *Echinacea angustifolia* in Experimental Rhinovirus Infections," *New England Journal of Medicine* 353 (2005): 341–348. A relatively large study enrolling over 700 participants gave healthy volunteers either a commercial preparation of *E. purpurea* leaves and roots (patients were treated three to five times a day and were asked to retain the liquid extract in their mouths for ten seconds to extend its possible contact with environmental viruses) or a placebo delivered similarly. See M. Jawad et al., "Safety and Efficacy Profile of *Echinacea purpurea* to Prevent Common Cold Episodes: A Randomized, Double-Blind, Placebo-Controlled Trial,"

Evidence-Based Complementary and Alternative Medicine (2012): ID 841315. The researchers' data indicate that the *Echinacea* treatment lowered the incidence and severity of colds. Each trial used a different source of purple coneflower extract, standardized to different parameters, and administered differently.

115. Black cohosh was designated *Cimicifuga racemosa* through much of the twentieth century, and some sources still refer to it by that binomial. Its taxonomy was revised in the 1990s. See James A. Compton, Alastair Culham, and Stephen L. Jury, "Reclassification of *Actaea* to Include *Cimicifuga* and *Souliea* (Ranunculaceae): Phylogeny Inferred from Morphology, nrDNA ITS, and cpDNA trnL-F Sequence Variation," *Taxon* 47 (1998): 593–634.

116. Steven Foster, "Black Cohosh: A Literature Review," *HerbalGram* 45 (1999): 35–50; World Health Organization, *Monographs*, 2:55.

117. Foster, "Black Cohosh."

118. James W. Herrick, *Iroquois Medical Botany*, ed. Dean R. Snow (Syracuse, N.Y.: Syracuse University Press, 1997), 119.

119. Paul B. Hamel and Mary U. Chiltoskey, *Cherokee Plants: Their Uses—A Four-Hundred-Year History* (Sylva, N.C.: Herald, 1975), 30.

120. John King and Robert S. Newton, *The Eclectic Dispensatory of the United States of America* (Cincinnati: Derby, 1852), 250.

121. King and Newton, *Eclectic Dispensatory*, 251–252.

122. King and Newton, *Eclectic Dispensatory*, 251–253.

123. Foster, "Black Cohosh."

124. Steven Foster, "Exploring the Peripatetic Maze of Black Cohosh Adulteration: A Review of the Nomenclature, Distribution, Chemistry, Market Status, Analytical Methods, and Safety," *HerbalGram* 98 (2013): 32–51.

125. World Health Organization, *Monographs*, 2:58; Chunhui Ma et al., "Metabolic Profiling of *Actaea* (*Cimicifuga*) Species Extracts Using High Performance Liquid Chromatography Coupled with Electrospray Ionization Time-of-Flight Mass Spectrometry," *Journal of Chromatography A* 1218, no. 11 (2011): 1461–1476. In "Exploring the Peripatetic Maze of Black Cohosh Adulteration," Foster reviews some of the recent investigations that have further contributed to the list of possible bioactive compounds in *Actaea*.

126. Minerva Mercado-Feliciano et al., "An Ethanolic Extract of Black Cohosh Causes Hematological Changes but Not Estrogenic Effects in Female Rodents," *Toxicology and Applied Pharmacology* 263 (2012): 138–147; W. Wuttke et al., "Chaste Tree (*Vitex agnus-castus*): Pharmacology and Clinical Indications," *Phytomedicine* 10 (2003): 348–357.

127. Teresa L. Johnson and Jed W. Fahey, "Black Cohosh: Coming Full Circle?" *Journal of Ethnopharmacology* 141 (2012): 775–779.

128. Mee-Ra Rhyu et al., "Black Cohosh (*Actaea racemosa*, *Cimicifuga racemosa*) Behaves as a Mixed Competitive Ligand and Partial Agonist at the Human Mu Opiate Receptor," *Journal of Agricultural and Food Chemistry* 54 (2006): 9852–9857; Nancy E. Keame et al., "Black Cohosh

Has Central Opioid Activity in Postmenopausal Women: Evidence from Naloxone Blockade and PET Neuroimaging Studies," *Menopause* 15 (2008): 832–840.

129. Sharla L. Powell et al., "*In vitro* Serotonergic Activity of Black Cohosh and Identification of N$_\omega$-Methylserotonin as a Potential Active Constituent," *Journal of Agricultural and Food Chemistry* 56 (2008): 11718–11726. There may also be evidence for black cohosh effects on the GABAergic system. See Serhat S. Cicek et al., "Bioactivity-Guided Isolation of GABA$_A$ Receptor Modulating Constituents from the Rhizomes of *Actaea racemosa*," *Journal of Natural Products* 73 (2010): 2024–2028. Black cohosh may also have effects on the immune system. See Rachel L. Ruhlen, Grace Y. Sun, and Edward R. Sauter, "Black Cohosh: Insights into Its Mechanism(s) of Action," *Integrative Medicine Insights* 3 (2008): 21–32.

130. Johnson and Fahey, "Black Cohosh."

131. Ruhlen et al., "Black Cohosh"; Matthew J. Leach and Vivienne Moore, "Black Cohosh (*Cimicifuga* spp.) for Menopausal Symptoms," *Cochrane Database of Systematic Reviews* (2012): CD007244.

132. Leach and Moore, "Black Cohosh."

133. World Health Organization, *Monographs*, 2:149; Steven Foster Group, "St. John's Wort," http://www.stevenfoster.com/education/monograph/hypericum2.html.

134. World Health Organization, *Monographs*, 2:150.

135. Dioscorides, *De materia medica*, 249. The author refers to the fruit's medicinal properties.

136. Karen Reeds, "Saint John's Wort (*Hypericum perforatum* L.) in the Age of Paracelsus and the Great Herbals: Assessing the Historical Claims for a Traditional Remedy," in *Herbs and Healers from the Ancient Mediterranean Through the Medieval West: Essays in Honor of John M. Riddle*, ed. Anne Van Arsdall and Timothy Graham (Farnham: Ashgate, 2012), 270.

137. Christopher Hobbs, "St. John's Wort: Ancient Herbal Protector," *Pharmacy History* 32 (1990): 166. Reeds cites the medieval Latin name *fuga demonum* ("flee-devil" or "devil's scourge") as evidence of its perceived protective force, in "Saint John's Wort," 266–267, 280–281. Proximity to the plant was deemed effective, whether carried on the body, placed among the bedsheets, hung above a doorway, or used as a fumigant.

138. Quoted in Reeds, "Saint John's Wort," 280.

139. Reeds, "Saint John's Wort," 287–288.

140. Paracelsus, *Von den natürlichen Dingen*, ed. Karl Sudhoff, 2:111–121, quoted in Reeds, "Saint John's Wort," 288.

141. Angelo Sala, *Processus Angeli Salae* (1630), quoted in Reeds, "Saint John's Wort," 292.

142. John Hill, *The Family Herbal* (Bungay, Eng.: Brightly, 1812), 183; A. Gautier, *Manuel des plantes médicinales* (Paris: Audot, 1822), 710.

143. Gautier, *Manuel des plantes médicinales*, 711.

144. World Health Organization, *Monographs*, 2:152. The levels of 0.05 to 0.15 percent for hypericin and pseudohypericin are given in Anna Rita Bilia, Sandra Gallori, and Franco F. Vincieri, "St. John's Wort and Depression:

Efficacy, Safety and Tolerability—An Update," *Life Sciences* 70 (2002): 3077–0396. Levels of 2 to 4.5 percent for hyperforin are given in Joanne Barnes, Linda A. Anderson, and J. David Phillipson, "St. John's Wort (*Hypericum perforatum* L.): A Review of Its Chemistry, Pharmacology, and Clinical Properties," *Journal of Pharmacy and Pharmacology* 53 (2001): 583–600.

145. In "St. John's Wort and Depression," Bilia, Gallori, and Vincieri report that manufacturers standardize St. John's wort extract using various methods that make the resulting products less comparable in terms of active chemicals. Furthermore, St. John's wort extract is susceptible to degradation over time. See also Klaus Linde, "St. John's Wort—An Overview," *Forschende Komplementärmedizin* 16 (2009): 146 155. Modern-day herbalists recommend preparing St. John's wort extract as an herbal tea or by soaking it in oil for home use. See Rebecca L. Johnson et al., *National Geographic Guide to Medicinal Herbs: The World's Most Effective Healing Plants* (Washington, D.C.: National Geographic, 2010), 51–53.

146. Barnes, Anderson, and Phillipson, "St. John's Wort." In this context, it is also worth noting that many laboratory tests use concentrations of presumed active principles not likely to be achieved in the human body.

147. Veronika Butterweck and Mathias Schmidt, "St. John's Wort: Role of Active Compounds for Its Mechanism of Action and Efficacy," *Wiener Medizinische Wochenschrift* 157 (2007): 356–361; Bilia, Gallori, and Vincieri, "St. John's Wort and Depression"; World Health Organization, *Monographs*, 2:154–157.

148. Klaus Linde, Michael M. Berner, and Levente Kriston, "St John's Wort for Major Depression," *Cochrane Database of Systematic Reviews* (2008): CD000448. The authors noted that their meta-analysis was complicated by a high degree of heterogeneity of published findings. Part of this might be the result of disparate methodologies employed in the dozens of studies conducted on St. John's wort. Since the authors chose to review only studies aiming to treat "major depression," the findings cannot be simply extended to mild depression or other psychosocial conditions. Interestingly, the analysis revealed a stronger effect of St. John's wort on patients in German-speaking countries than in other countries. It is important to recognize the role of patients' cultural setting in shaping diagnoses and therapeutic outcomes.

149. Zeb Saddiqe, Ismat Naeem, and Alya Maimoona, "A Review of the Antibacterial Activity of *Hypericum perforatum* L.," *Journal of Ethnopharmacology* 131 (2010): 511–521; World Health Organization, *Monographs*, 2:157–158.

150. Blumenthal, Goldberg, and Brinckmann, *Herbal Medicine*.

151. George V. Nash and Maurice G. Kains, *American Ginseng: Its Commercial History, Protection, and Cultivation* (Washington, D.C.: Government Printing Office, 1898), 9.

152. Nash and Kains, *American Ginseng*, 6–8. A height of 60 to 70 centimeters and two to three seeds per fruit in *P. ginseng* is given in William E. Court, *Ginseng: The Genus Panax* (Amsterdam: Harwood, 2000), 16–19.

153. David Taylor, *Ginseng, the Divine Root* (Chapel Hill, N.C.: Algonquin, 2006), 9–10.

154. Ginseng's most common Chinese name, *renshen* 人参, contains elements referring to "person" and the celestial influence of the "constellation Orion." Other names include "essence of the earth" (*tujing* 土精) and "divine herb" (*shencao* 神草). For commentary on the significance of ginseng's Chinese names, see Van Jay Symons, *Ch'ing Ginseng Management: Ch'ing Monopolies in Microcosm* (Tempe: Center for Asian Research, Arizona State University, 1981), 1–2, 85. The Chinese believed "that the root possessed divine qualities and abounded with spirits" (5).

155. Bensky, Clavey, and Stöger, *Chinese Herbal Medicine*, 711.

156. Symons describes ginseng's role in diplomacy between the Chinese Ming Empire and the rival Jurchens during the early seventeenth century, in *Ch'ing Ginseng Management*, 10. In modern-day Chinese communities, ginseng is sold as a luxury good alongside tobacco, alcohol, and fine tea, ready for gift giving.

157. Daniel E. Moerman, *Native American Ethnobotany* (Portland, Ore.: Timber Press, 1998), 376. The Jesuit missionary Joseph François Lafitau, in a report to the French Crown on the medicinal and economic potential of Canadian ginseng, relates that the Iroquois word for the root denotes its humanlike shape: *Mémoire présentée à son altesse royale monseigneur le duc d'Orléans, régent du royaume de France, concernant la précieuse plante du gin seng de Tartarie, découverte en Canada* (Paris: Joseph Mongé, 1718), 17.

158. Taylor, *Ginseng*, 74–75.

159. In the East Asian context, Symons summarizes ginseng's associations with virility, in *Ch'ing Ginseng Management*, 1. For North America, Moerman lists several applications of ginseng as "love medicine" by indigenous groups, such as "used by men as a love charm" and "to get back a divorced wife" (*Native American Ethnobotany*, 376).

160. Pierre Jartoux, "The Description of a Tartarian Plant, Called Ginseng; with an Account of Its Virtues . . . at Pekin, April 12, 1711," *Philosophical Transactions of the Royal Society of London* 28 (1713): 56–61.

161. Nash and Kains, *American Ginseng*, 5–6. While the commerce in North American ginseng was profitable, imported roots in China often suffered from improper preparation and storage at the hands of the European merchants, leading to depressed market prices and a poorer reputation relative to the Asian ginseng, as documented in Symons, *Ch'ing Ginseng Management*, 3–5.

162. Harvested ginseng is processed in one of two ways: either cleaned and dried (white ginseng) or cleaned, steamed, and dried (red ginseng). The latter form is red-orange in color rather than pale yellow and differs in the spectrum of active chemicals. See Lee Jia and Yuqing Zhao, "Current Evaluation of the Millennium Phytomedicine—Ginseng (I): Etymology, Pharmacognosy, Phytochemistry, Market, and Regulations," *Current Medicinal Chemistry* 16 (2009): 2475–2484.

163. For some of the voices in the debate, see Bensky, Clavey, and Stöger, *Chinese Herbal Medicine*, 822–823. For differences between American and Chinese ginseng, see John T. Arnason et al., "An Introduction to the Ginseng Evaluation Program," *HerbalGram* 52 (2001): 27–30.

164. World Health Organization, *Monographs*, 1:171. In "Current Evaluation of the Millennium Phytomedicine," Jia and Zhao give a concentration of 0.2 to 2 percent ginsenosides in the main root and 4 to 9 percent in the root hairs for *P. ginseng* and 4 to 10 percent in the root for *P. quinquefolius*. Ginseng also contains polysaccharides, flavonoids, and volatile oils that might be medicinal.

165. The term "adaptogen" originated in the 1950s. See Alexander Panossian and Hildebert Wagner, "Adaptogens: A Review of Their History, Biological Activity, and Clinical Benefits," *HerbalGram* 90 (2011): 52–63.

166. Lee Jia, Yuqing Zhao, and Xing-Jie Liang, "Current Evaluation of the Millennium Phytomedicine—Ginseng (II): Collected Chemical Entities, Modern Pharmacology, and Clinical Applications Emanated from Traditional Chinese Medicine," *Current Medicinal Chemistry* 16 (2009): 2924–2942.

167. In a systematic survey of all available trials of ginseng (*P. ginseng*) for human health published in English, only 65 of 475 potentially relevant studies met the inclusion criteria for comparison based on reported methodology. See J. L. Shergis et al., "*Panax ginseng* in Randomised Controlled Trials: A Systematic Review," *Phytotherapy Research* 27 (2013): 949–965. A similar task was taken on for trials on *P. ginseng* or *P. quinquefolius* published in the Korean literature and all but 30 of 227 reports were excluded for reasons of methodology. See J. Choi et al., "Ginseng for Health Care: A Systematic Review of Randomized Controlled Trials in Korean Literature," *PLoS ONE* 8 (2013): e59978. For a meta-analysis of ginseng and cognition, in which no effect over placebo was detected, see J. Geng et al., "Ginseng for Cognition," *Cochrane Database of Systematic Reviews* (2010): CD007769. Academic and industry groups seek to set standards and validation procedures for the chemical constituents reported in commercial ginseng products, an effort outlined in Arnason et al., "Ginseng Evaluation Program."

168. Blumenthal, Goldberg, and Brinckmann, *Herbal Medicine*; J. Weiss, N. Ainsworth, and I. Faithfull, *Horehound, Marrubium vulgare* (Publication of the Cooperative Research Centre for Weed Management Systems, University of Adelaide, Australia, September 2000); D. Wilken and L. Hannah, Marrubium vulgare (Lamiaceae) *Common Horehound, White Horehound* (Publication of the Santa Barbara Botanic Garden, for Channel Islands National Park, December 15, 1998). Common horehound is also known as white horehound, not to be confused with black horehound (*Ballota nigra*) and Greek horehound (*Ballota acetabulosa*).

169. Wilken and Hannah, Marrubium vulgare, 1.

170. Pliny the Elder, *Natural History* 20.89; *The Natural History of Pliny*, trans. John Bostock and H. T. Riley (London: Bohn, 1856), 4:290–292.

171. Parkinson, *Theatrum Botanicum*, 46.

172. "Ricola Natural Herb Cough Drops," http://www.ricola.com/en-us/Products/Herb-Cough-and-Throat-Drops/Original.

173. Blumenthal, Goldberg, and Brinckmann, *Herbal Medicine*.

174. Olugbenga K. Popoola et al., "Marrubiin," *Molecules* 18 (2013): 9049–9060.

175. For example, phenylpropanoid esters, polyphenolics, and volatile oils. See Sevser Sahpaz et al., "Isolation and Pharmacological Activity of Phenylpropanoid Esters from *Marrubium vulgare*," *Journal of Ethnopharmacology* 79 (2002): 389–392; and Daniela Rigano et al., "Antibacterial Activity of Flavonoids and Phenylpropanoids from *Marrubium globosum* ssp. *libanoticum*," *Phytotherapy Research* 21 (2007): 395–397, working with a related *Marrubium* species.

176. Gayle Engels, "Valerian," *HerbalGram* 79 (2008): 1–2.

177. World Health Organization, *Monographs*, 1:267–268.

178. Engels, "Valerian."

179. Dioscorides noted the root's "somewhat foulsmelling oppressiveness" (*De materia medica*, 12).

180. John Gerard, *The Herball, or Generall Historie of Plants* (London, 1597), 919. He also passed along the advice of other authors that wild valerian (root, perhaps) was useful for "the crampe and other convulsions" and bruises resulting from falls.

181. Nicholas Culpeper, *The English Physitian Enlarged* (London: Streater, 1666), 254.

182. For example, Martyn Paine, *A Therapeutical Arrangement of the Materia Medica* (New York: Langley, 1842), 205. According to Joseph Carson, "Valerian is a stimulant and antispasmodic, employed in nervous affections and in combination with tonics in cases of neuralgia, or diseases with nervous complications" (*Illustrations of Medical Botany* [Philadelphia: Smith, 1847], 2:57).

183. William Fox and Joseph Nadin, *The Working Man's Family Botanic Guide; or, Every Man His Own Doctor* (Sheffield: Dawsons, 1852), 77.

184. John K. Crellin and Jane Philpott, *Herbal Medicine Past and Present*, vol. 2, *Monographs on Select Medicinal Plants* (Durham, N.C.: Duke University Press, 1990), 434.

185. World Health Organization, *Monographs*, 1:270–271.

186. See, among others, Dietmar Benke et al., "GABA$_A$ Receptors as *in vivo* Substrate for the Anxiolytic Action of Valerenic Acid, a Major Constituent of Valerian Root Extracts," *Neuropharmacology* 56 (2009): 174–181. There is also evidence that valerian interacts with the serotonin receptor.

187. K. Murphy et al., "*Valeriana officinalis* Root Extracts Have Potent Anxiolytic Effects in Laboratory Rats," *Phytomedicine* 17 (2010): 674–678; K. P. Prabhakaran Nair, "The Neutraceutical [*sic*] Properties of Turmeric," in *The Agronomy and Economy of Turmeric and Ginger*, ed. K. P. Prabhakaran Nair (London: Elsevier, 2013), 179–204.

188. For example, Lincoln Sakiara Miyasaka, Álvaro N. Atallah, and Bernardo Soares, "Valerian for Anxiety Disorders," *Cochrane Database of Systematic Reviews* (2006): CD004515; Isabel Fernández-San-Martín et al., "Effectiveness of Valerian on Insomnia: A Meta-Analysis of Randomized Placebo-Controlled Trials," *Sleep Medicine* 11 (2010):

505–511; and Jerome Sarris and Gerard J. Byrne, "A Systematic Review of Insomnia and Complementary Medicine," *Sleep Medicine Reviews* 15 (2011): 99–106.

189. World Health Organization, *Monographs*, 1:115; Gayle Engels, "Turmeric," *HerbalGram* 84 (2009): 1–3.

190. The *Atharva Veda*, one of the ancient religious texts that gave rise to Ayurvedic medical practice, is cited in R. Remadevi, E. Surendran, and Takeatsu Kimura, "Turmeric in Traditional Medicine," in *Turmeric: The Genus Curcuma*, ed. P. N. Ravindran, K. Nirmal Babu, and K. Sivaraman (Boca Raton, Fla.: CRC Press, 2007), 410.

191. Bensky, Clavey, and Stöger cite the *Tang Materia Medica* and the *Grand Materia Medica*, in *Chinese Herbal Medicine*, 612.

192. Andrew Dalby, *Dangerous Tastes: The Story of Spices* (Berkeley: University of California Press, 2000), 95–96; Engels, "Turmeric."

193. H. Milne Edwards and P. Vavasseur, *Manuel de matière médicale* (Paris: Compère Jeune, 1826), 135.

194. Powdered turmeric loses its flavor and aroma more quickly than the intact rhizome, which retains its complement of characteristic compounds if stored in a cool and dry place.

195. World Health Organization, *Monographs*, 1:118. Venugopal P. Menon and Adluri Ram Sudheer give 2 to 8 percent for curcumin in prepared turmeric, in "Antioxidant and Anti-Inflammatory Properties of Curcumin," in *The Molecular Targets and Therapeutic Uses of Curcumin in Health and Disease*, ed. Bharat B. Aggarwal, Young-Joon Surh, and Shishir Shishodia (New York: Springer, 2007), 106.

196. Purusotam Basnet and Natasa Skalko-Basnet, "Curcumin: An Anti-Inflammatory Molecule from a Curry Spice on the Path to Cancer Treatment," *Molecules* 16 (2011): 4567–4598. Turmeric polyphenols formally have both prooxidant and antioxidant properties, depending on the cellular setting, an observation that underlines the nuance with which one must consider the possible effects of drugs such as curcumin. The importance of inflammation (a localized response to cellular damage) in the pathogenesis of a wide variety of illnesses is noted by Bharat B. Aggarwal et al.: "It is now well recognized that most chronic diseases are the result of dysregulated inflammation" ("Curcumin: The Indian Solid Gold," in *Molecular Targets and Therapeutic Uses of Curcumin in Health and Disease*, ed. Aggarwal, Surh, and Shishodia, 4).

197. Menon and Sudheer, "Antioxidant and Anti-Inflammatory Properties of Curcumin," 109–112; Basnet and Skalko-Basnet, "Curcumin"; H. Zhou, C. S. Beevers, and S. Huang, "Targets of Curcumin," *Current Drug Targets* 12 (2011): 332–347.

198. Rajesh L. Thangapazham, Anuj Sharma, and Radha K. Maheshwari, "Beneficial Role of Curcumin in Skin Diseases," in *Molecular Targets and Therapeutic Uses of Curcumin in Health and Disease*, ed. Aggarwal, Surh, and Shishodia, 343–357.

199. Devarajan Karunagaran, Jeena Joseph, and R. Kumar Santhosh Thankayyan, "Cell Growth Regulation"; and Young-Joon Surh and Kyung-Soo Chun, "Cancer

Chemopreventive Effects of Curcumin," both in *Molecular Targets and Therapeutic Uses of Curcumin in Health and Disease*, ed. Aggarwal, Surh, and Shishodia, 245–268, 149–172. For a review of the many possible cellular targets of curcumin, the current state of preclinical research, and future prospects, see Subash C. Gupta, Gorkem Kismali, and Bharat B. Aggarwal, "Curcumin, a Component of Turmeric: From Farm to Pharmacy," *BioFactors* 39 (2013): 2–13.

200. In "Curcumin," Gupta, Kismali, and Aggarwal report sixty-five completed clinical trials and more than thirty-five under way, employing a variety of designs and testing many different outcomes.

201. Preetha Anand et al., "Bioavailability of Curcumin: Problems and Promises," *Molecular Pharmaceutics* 4 (2007): 807–818; Seema Singh, "From Exotic Spice to Modern Drug?" *Cell* 130 (2007): 765–768; Basnet and Skalko-Basnet, "Curcumin," 4583–4586.

202. A. Vyas et al., "Perspectives on New Synthetic Curcumin Analogs and Their Potential Anticancer Properties," *Current Pharmaceutical Design* 19 (2013): 2047–2069.

203. Leonard E. Newton, "Aloes in Habitat," in *Aloes: The Genus Aloe*, ed. Tom Reynolds (Boca Raton, Fla.: CRC Press, 2004), 3–14; Gayle Engels, "Aloe," *HerbalGram* 87 (2010): 1–5.

204. World Health Organization, *Monographs*, 1:34; Engels, "Aloe," 1. *A. barbadensis* is a synonym of *A. vera* not accepted under the conventions of botanical nomenclature but occasionally found in technical literature and product labels.

205. In "Aloe," Engels refers to a Mesopotamian clay tablet and the Egyptian Ebers papyrus. Aloe is cited numerous times in the Ebers papyrus. See *Ancient Egyptian Medicine*; and Campbell and David, "Application of Archaeobotany, Phytogeography and Phamacognosy."

206. Pliny the Elder, *Natural History* 27.5; *The Natural History of Pliny*, trans. John Bostock and H. T. Riley (London: Bohn, 1856), 5:222–224.

207. Dioscorides, *De materia medica*, 187–188.

208. Gerard, *Herball*, 410.

209. World Health Organization, *Monographs*, 1:37; Tom Reynolds, "Aloe Chemistry," in *Aloes*, ed. Reynolds, 39–74. These are hydroxyanthrone derivatives.

210. E. Harlev et al., "Anticancer Potential of Aloes: Antioxidant, Antiproliferative, and Immunostimulatory Attributes," *Planta Medica* 78 (2012): 843–852.

211. Holly J. Bayne, "FDA Issues Final Rule Banning Use of Aloe and Cascara Sagrada in OTC Drug Products," *HerbalGram* 56 (2002): 56. An FDA review of stimulant laxatives including aloe raised questions about possible carcinogenicity, and aloe laxative manufacturers failed to supply data to address the safety concerns.

212. Josias H. Hamman, "Composition and Applications of *Aloe vera* Leaf Gel," *Molecules* 13 (2008): 1599–1616.

213. T. Reynolds and A. C. Dweck, "*Aloe vera* Leaf Gel: A Review Update," *Journal of Ethnopharmacology* 68 (1999): 3–37.

214. Nicola Mascolo et al., "Healing Powers of Aloes," in *Aloes*, ed. Reynolds, 209–238.

215. Mascolo et al., "Healing Powers of Aloes"; R. Maenthaisong et al., "The Efficacy of *Aloe vera* Used for Burn Wound Healing: A Systematic Review," *Burns* 33 (2007): 713–718; A. D. Dat et al., "*Aloe vera* for Treating Acute and Chronic Wounds," *Cochrane Database of Systematic Reviews* (2012): CD008762.

216. World Health Organization, *Monographs*, 1:277.

217. P. N. Ravindran and K. Nirmal Babu, "Introduction," in *Ginger: The Genus Zingiber*, ed. P. N. Ravindran and K. Nirmal Babu (Boca Raton, Fla.: CRC Press, 2005), 1–14.

218. Dalby, *Dangerous Tastes*, 21–26.

219. *Divine Husbandman's Classic of the Materia Medica*, in Bensky, Clavey, and Stöger, *Chinese Herbal Medicine*, 681.

220. Bensky, Clavey, and Stöger, *Chinese Herbal Medicine*, 682.

221. Dioscorides, *De materia medica*, 160. Dalby suggests that ginger was in cultivation in Eritrea and East Africa during Dioscorides's time but not in Arabia, in *Dangerous Tastes*, 22, 25.

222. Gerard, *Herball*, 55.

223. Pierre Pomet, *Histoire générale des drogues* (Paris: Jean-Baptiste Loyson and Augustin Pillon, 1694), 61–62.

224. Blumenthal, Goldberg, and Brinckmann, *Herbal Medicine*; World Health Organization, *Monographs*, 1:280.

225. Brett White, "Ginger: An Overview," *American Family Physician* 75, no. 11 (2007): 1689–1691.

226. White, "Ginger," 1689–1690.

227. R. Terry et al., "The Use of Ginger (*Zingiber officinale*) for the Treatment of Pain: A Systematic Review of Clinical Trials," *Pain Medicine* 12 (2011): 1808–1818; Matthew J. Leach and S. Kumar, "The Clinical Effectiveness of Ginger (*Zingiber officinale*) in Adults with Osteoarthritis," *International Journal of Evidence Based Healthcare* 6 (2008): 311–320.

228. Lamont Lindstrom, "History, Folklore, Traditional and Current Uses of Kava," in *Kava: From Ethnology to Pharmacology*, ed. Yadhu H. Singh (Boca Raton, Fla.: CRC Press, 2004), 10–28. Evidence is given for an origin in the southern Pacific archipelago of Vanuatu in Vincent Lebot, Mark Merlin, and Lamont Lindstrom, *Kava: The Pacific Elixir* (Rochester, Vt.: Healing Arts Press, 1997), 51–56. An origin in the vicinity of New Guinea is argued for in Ron Brunton, *The Abandoned Narcotic: Kava and Cultural Instability in Melanesia* (Cambridge: Cambridge University Press, 1989). The plant and a drink produced from it go by various names in local languages, including *kava*, *yaqona*, *kava kava*, *'awa*, *seka*, and *sakau*. See Mark Merlin and William Raynor, "Modern Use and Environmental Impact of the Kava Plant in Remote Oceania," in *Dangerous Harvest: Drug Plants and the Transformation of Indigenous Landscapes*, ed. Michael K. Steinberg, Joseph J. Hobbs, and Kent Mathewson (New York: Oxford University Press, 2004), 274–293.

229. These characteristics are for kava on the island of Pohnpei. See Michael J. Balick and Roberta A. Lee, "The Sacred Root: *Sakau* en Pohnpei," in *Ethnobotany of Pohnpei: Plants, People, and Island Culture*, ed. Michael J. Balick

(Honolulu: University of Hawai'i Press, 2009), 166. Older, larger root masses are reserved for special occasions.

230. Lebot, Merlin, and Lindstrom, *Kava*, 13.

231. Some kava plants have escaped cultivation and grow (but do not reproduce sexually) in semiwild settings. Domesticated kava's probable ancestor, the fertile *Piper wichmannii*, can also be called kava, is sometimes employed medicinally, and visually resembles *P. methysticum*. See Lebot, Merlin, and Lindstrom, *Kava*, 14–23.

232. Lebot, Merlin, and Lindstrom, *Kava*, 90–103.

233. Lebot, Merlin, and Lindstrom, *Kava*, 119–129; Balick and Lee, "Sacred Root," 168–174.

234. Lindstrom, "History, Folklore"; Lebot, Merlin, and Lindstrom, *Kava*, 141–152.

235. Chris Kilham, *Kava: Medicine Hunting in Paradise* (Rochester, Vt.: Park Street Press, 1996), 5, 13, 91–92; Lebot, Merlin, and Lindstrom, *Kava*, 120–121.

236. Lebot, Merlin, and Lindstrom, *Kava*, 186–189. Similarly, kava is sold bottled at markets and in *sakau* (kava) bars on modern-day Pohnpei. See Balick and Lee, "Sacred Root," 189–192.

237. Lebot, Merlin, and Lindstrom, *Kava*, 112–117.

238. Kilham, *Kava*, 80–85.

239. Lebot, Merlin, and Lindstrom, *Kava*, 195.

240. Parke, Davis, and Company, *Manual of Therapeutics* (Detroit, Mich.: Parke, Davis, 1909), 394.

241. Merlin and Raynor, "Modern Use and Environmental Impact of the Kava Plant."

242. Kilham, *Kava*, 153–156; Lebot, Merlin, and Lindstrom, *Kava*, 196–197; Yadhu N. Singh, "Kava: Production, Marketing, and Quality Assurance," in *Kava*, ed. Singh, 29–49.

243. Lebot, Merlin, and Lindstrom, *Kava*, 60–71. Kavalactone concentration "can vary between 3% and 20% of rootstock dry weight depending on the age of the plant and the cultivar."

244. Yadhu N. Singh, "Pharmacology and Toxicology of Kava and Kavalactones," in *Kava*, ed. Singh, 104–139. Kavalactones appear to cause anesthesia as well as muscle relaxation by binding to particular neuron channels and interfering with the electrical signaling of the peripheral nervous system.

245. Edzard Ernst, "Herbal Remedies for Anxiety—A Systematic Review of Controlled Clinical Trials," *Phytomedicine* 13 (2006): 205–208; J. Sarris et al., "Kava in the Treatment of Generalized Anxiety Disorder," *Journal of Clinical Psychopharmacology* 33 (2013): 643–648. Many of the early human studies on kava's mood-altering properties suffered weaknesses in design, such as poor controls, inappropriate anxiety rating scales, nonstandardized extracts, and so forth. See Singh, "Pharmacology and Toxicology."

246. J. Sarris et al., "Herbal Medicine for Depression, Anxiety, and Insomnia: A Review of Psychopharmacology and Clinical Evidence," *European Neuropsychopharmacology* (2011): 841–860.

247. Merlin and Raynor, "Modern Use and Environmental Impact of the Kava Plant," 283–292.

248. The reporting of kava-related events was problematic because of errors in bookkeeping and other inaccuracies, according to C. Ulbricht et al., "Safety Review of Kava (*Piper methysticum*) by the Natural Standard Research Collaboration," *Expert Opinion on Drug Safety* 4 (2005): 779–794.

249. Ulbricht et al., "Safety Review of Kava"; R. Teschke et al., "Kava, the Anxiolytic Herb: Back to Basics to Prevent Liver Injury?" *British Journal of Clinical Pharmacology* 71 (2011): 445–448. The matter of kava's legal regulatory status is not fully settled. For example, the German court system is considering whether the original decision to restrict kava sales in that nation was justified. See Mathias Schmidt, "German Court Ruling Reverses Kava Ban; German Regulatory Authority Appeals Decision," *HerbalEGram* 11, no. 7 (2014).

250. Teschke et al., "Kava." One analysis failed to uncover evidence that might explain the basis of kava's suspected harm to the liver. See Line R. Olsen, Mark P. Grillo, and Christian Skonberg, "Constituents in Kava Extracts Potentially Involved in Hepatotoxicity: A Review," *Chemical Research in Toxicology* 24 (2011): 992–1002.

15. THE FUTURE OF MEDICINAL PLANTS

1. *Ancient Egyptian Medicine: The Papyrus Ebers*, trans. Cyril P. Bryan (London: Bles, 1930); Reginald Campbell Thompson, *A Dictionary of Assyrian Botany* (London: British Academy, 1949); Paul U. Unschuld, *Medicine in China: A History of Pharmaceutics* (Berkeley: University of California Press, 1986), 15.

2. For an argument for the value of ancient texts as catalogs of prospective new drugs and a criticism of the wholesale rejection of classical medicine, see Bart K. Holland, "Prospecting for Drugs in Ancient Texts," *Nature* 369 (1994): 702. "This attitude has persisted," he laments, "causing us to regard ancient and mediaeval medicine as an historical curiosity rather than a source of potential therapies." See also John Riddle, "Historical Data as an Aid in Pharmaceutical Prospecting and Drug Safety Determination," *Journal of Alternative of Complementary Medicine* 5 (1999): 195–201.

3. Anne Van Arsdall, "The Medicines of Medieval and Renaissance Europe as a Source of Medicines for Today," in *Prospecting for Drugs in Ancient and Medieval European Texts: A Scientific Approach*, ed. Bart K. Holland (Amsterdam: Harwood, 1996), 19–37.

4. John Riddle, "The Medicines of Greco-Roman Antiquity as a Source of Medicine for Today," in *Prospecting for Drugs in Ancient and Medieval European Texts*, ed. Holland, 7–18; Paula De Vos, "European *Materia Medica* in Historical Texts: Longevity of a Tradition and Implications for Future Use," *Journal of Ethnopharmacology* 132 (2010): 28–47; Valerie Thomas, "Do Modern-Day Medical Herbalists Have Anything to Learn from Anglo-Saxon Medical Writings?" *Journal of Herbal Medicine* 1 (2011): 42–52.

5. Norman R. Farnsworth, "Ethnopharmacology and Drug Development," in *Ethnobotany and the Search for New Drugs*, ed. Ghillean T. Prance, Derek J. Chadwick, and Joan Marsh, Ciba Foundation Symposium 185 (Chichester: Wiley, 1994), 42–50.

6. For a discussion of ancient Greek references to wormwood (*Artemisia* spp.) against malarial symptoms, see John Riddle, *Goddesses, Elixirs, and Witches: Plants and Sexuality Throughout Human History* (New York: Palgrave Macmillan, 2010), 89–98.

7. Elisabeth Hsu, "*Qing hao* 青蒿 (*Herba Artemisiae annuae*) in the Chinese *Materia Medica*," in *Plants, Health, and Healing: On the Interface of Ethnobotany and Medical Anthropology*, ed. Elisabeth Hsu and Stephen Harris (New York: Berghahn, 2010), 83–133; Louis Miller and Xinzhuan Su, "Artemisinin: Discovery from the Chinese Herbal Garden," *Cell* 146 (2011): 855–858.

8. Pedanius Dioscorides of Anazarbus, *De materia medica*, trans. Lily Y. Beck (Hildesheim: Olms-Weidmann, 2005), 279.

9. Van Arsdall, "Medieval and Renaissance Europe," 27–28; Frances Watkins et al., "Anglo-Saxon Pharmacopoeia Revisited: A Potential Treasure in Drug Discovery," *Drug Discovery Today* 16 (2011): 1069–1075.

10. Thompson, *Dictionary of Assyrian Botany*.

11. *Ancient Egyptian Medicine*; Lise Manniche, *An Ancient Egyptian Herbal*, rev. ed. (London: British Museum Press, 2006).

12. John Riddle, *Dioscorides on Pharmacy and Medicine* (Austin: University of Texas Press, 1985).

13. For example, Mark J. Plotkin, *Tales of a Shaman's Apprentice: An Ethnobotanist Searches for New Medicines in the Amazon Rain Forest* (New York: Penguin, 1993).

14. Richard Evans Schultes, "Amazonian Ethnobotany and the Search for New Drugs," in *Ethnobotany and the Search for New Drugs*, ed. Prance, Chadwick, and Marsh, 106–112.

15. Paul Alan Cox, "Will Tribal Knowledge Survive the Millennium?" *Science* 287 (2000): 44–45; Michael Heinrich et al., "Ethnopharmacological Field Studies: A Critical Assessment of Their Conceptual Basis and Methods," *Journal of Ethnopharmacology* 124 (2009): 1–17. It should be noted that in addition to recording plant uses, many ethnobotanists' projects seek to understand the cultural milieu and meaning of plants among people.

16. Investigators working with traditional societies ought to gather precise information on how plants are harvested and prepared for their health-related effects. While many projects aim to isolate specific chemical agents responsible for ascribed medicinal properties, simply noting the plant's traditional function is often insufficient. For example, a Brazilian ethnobotanist relates a story that highlights the shortcomings of incomplete cultural knowledge in ethnobotanic cataloging projects:

> We were studying a plant from the Euphorbiaceae family, called *Croton sonderianus*. This was brought to our laboratory with the information that it is a very effective contraceptive. . . . After we started these studies, we decided to find out why this plant is used by that local population and not by another population. Then we learned that the long sticks that this plant yields are kept by the wife next to the bed and when the husband comes at night during her fertile period, she just hits the husband three times with the stick and that is a very effective contraceptive!" (quoted in Norman R. Farnsworth, "Ethnopharmacology and Drug Development," in *Ethnobotany and the Search for New Drugs*, ed. Prance, Chadwick, and Marsh, 55)

17. In an attempt to ease cross-cultural communication in the medical sphere, the World Health Organization has developed the International Classification of Diseases (http://www.who.int/classifications/icd/en/), which gives an alphanumeric code to every dysfunction or structural abnormality that might befall a patient. While such a guide might simplify information exchange between biomedically trained investigators, it does not accommodate traditional concepts of anatomy and physiology.

18. Edward M. Croom, "Documenting and Evaluating Herbal Remedies," *Economic Botany* 37 (1983): 13–27.

19. German chamomile is *Matricaria* spp., golden chamomile is *Cota tinctoria*, Roman chamomile is *Chamaemelum nobile*, and a number of chamomiles are *Anthemis* spp. Adobe snakeroot is *Sanicula maritime*, black snakeroot is *Actaea racemosa*, button snakeroot is *Eryngium yuccifolium*, and fragrant snakeroot is *Ageratina herbacea*, among many others.

20. Michael J. Balick and Paul Alan Cox, *Plants, People, and Culture: The Science of Ethnobotany* (New York: Scientific American, 1996), 46–51; Judith Sumner, *The Natural History of Medicinal Plants* (Portland, Ore.: Timber Press, 2000), 51–54. A variation on this procedure (without taking herbarium specimens) has been employed by a group generating a collection of "authenticated" samples of Chinese traditional herbs. See D. M. Eisenberg et al., "Developing a Library of Authenticated Traditional Chinese Medicinal (TCM) Plants for Systematic Biological Evaluation—Rationale, Methods, and Preliminary Results from a Sino-American Collaboration," *Fitoterapia* 82 (2011): 17–33.

21. Bradley C. Bennett and Michael J. Balick, "Phytomedicine 101: Plant Taxonomy for Preclinical and Clinical Medicinal Plant Researchers," *Journal of the Society for Integrative Oncology* 6 (2008): 150–157.

22. William Withering, *An Account of the Foxglove and Some of Its Medical Uses* (Birmingham, 1785); Paul Hauptman and Ralph Kelly, "Digitalis," *Circulation* 99 (1999): 1265–1270.

23. Jesse W.-H. Li and John C. Vederas, "Drug Discovery and Natural Products: End of an Era or an Endless Frontier?" *Science* 325 (2009): 161–165.

24. More than 70 percent of new drugs released between 1981 and 2010 were derived from natural products, either directly or semisynthetically. See David J. Newman and

Gordon M. Cragg, "Natural Products as Sources of New Drugs over the Thirty Years from 1981 to 2010," *Journal of Natural Products* 75 (2012): 311–335.

25. Gordon M. Cragg and David J. Newman, "Natural Products: A Continuing Source of Novel Drug Leads," *Biochimica et Biophysica Acta* 1830 (2013): 3670–3695.

26. Paul M. O'Neill, Victoria E. Barton, and Stephen A. Ward, "The Molecular Mechanism of Action of Artemisinin—The Debate Continues," *Molecules* 15 (2010): 1705–1721; Cragg and Newman, "Natural Products," 3671–3672.

27. Dan Bensky, Steven Clavey, and Erich Stöger, *Chinese Herbal Medicine: Materia Medica* (Seattle: Eastland, 2004), 470.

28. John Gerard, *The Herball, or Generall Historie of Plants* (London, 1597), 440.

29. A. Bradford Hill, "The Clinical Trial," *New England Journal of Medicine* 247, no. 4 (1952): 113–119.

30. Joel J. Gagnier et al., "Randomized Controlled Trials of Herbal Interventions Underreport Important Details of the Intervention," *Journal of Clinical Epidemiology* 64 (2011): 760–769.

31. Donald P. Briskin, "Medicinal Plants and Phytomedicines: Linking Plant Biochemistry and Physiology to Human Health," *Plant Physiology* 124, no. 2 (2000): 507–514.

32. The CONSORT (Consolidated Standards of Reporting Trials, http://www.consort-statement.org) group developed a widely adopted set of guidelines on clinical trial design and reporting. For elaborations on the CONSORT statement specific to herbal medicine trials, see Joel J. Gagnier et al., "Recommendations for Reporting Randomized Controlled Trials of Herbal Interventions: Explanation and Elaboration." *Journal of Clinical Epidemiology* 59 (2006): 1134–1149.

33. Steve Hickey and Hilary Roberts, *Tarnished Gold: The Sickness of Evidence-Based Medicine* (CreateSpace, 2011).

34. The Cochrane Collaboration (http://www.thecochranelibrary.com) advocates for such systematic reviews in evidence-based medical decision making.

35. A. Timmer et al., "*Pelargonium sidoides* Extract for Treating Acute Respiratory Tract Infections," *Cochrane Database of Systematic Reviews* (2013): CD006323.

36. Jerry Cott, quoted in Tyler Smith, "Ginkgo and Alzheimer's Disease Prevention: Researchers, Herbal Experts Interpret Results of the Five-Year GuidAge Clinical Trial," *HerbalGram* 97 (2013): 32–37.

37. Gordon M. Cragg and David J. Newman, "Plants as a Source of Anticancer Agents," *Journal of Ethnopharmacology* 100 (2005): 72–79. Paclitaxel and its chemical precursors are present in the bark and leaves of several *Taxus* species, allowing for ready harvest and semisynthetic preparation.

38. Mary Ann Jordan et al., "Mechanism of Mitotic Block and Inhibition of Cell Proliferation by Taxol at Low Concentrations," *Proceedings of the National Academy of Sciences USA* 90 (1993): 9552–9556.

39. Jürg Gertsch, "Botanical Drugs, Synergy, and Network Pharmacology: Forth and Back to Intelligent Mixtures," *Planta Medica* 77 (2011): 1086–1098.

40. Freddie Ann Hoffman and Steven R. Kishter, "Botanical New Drug Applications—The Final Frontier," *HerbalGram* 99 (2013): 66–69.

41. T. G. Tzellos et al., "Efficacy, Safety, and Tolerability of Green Tea Catechins in the Treatment of External Anogenital Warts: A Systematic Review and Meta-Analysis," *Journal of the European Academy of Dermatology and Venereology* 25 (2011): 345–353. While mixtures of plant-derived compounds have been marketed as pharmaceuticals for many years in Europe, China, and elsewhere, the green-tea extract was the first approved under the FDA's category of botanical drug.

42. Gayle Engels, "Dragon's Blood," *HerbalGram* 92 (2011): 1–4.

43. Lukmanee Tradtrantip, Wan Namkung, and A. S. Verkman, "Crofelemer, an Antisecretory Antidiarrheal Proanthocyanidin Oligomer Extracted from *Croton lechleri*, Targets Two Distinct Intestinal Chloride Channels," *Molecular Pharmacology* 77 (2010): 69–78.

44. Rodger D. MacArthur et al., "Efficacy and Safety of Crofelemer for Noninfectious Diarrhea in HIV-Seropositive Individuals (ADVENT Trial): A Randomized, Double-Blind, Placebo-Controlled, Two-Stage Study," *HIV Clinical Trials* 14 (2013): 261–273.

45. M. C. Berenbaum, "What Is Synergy?" *Pharmacological Reviews* 41 (1989): 93–141; E. M. Williamson, "Synergy and Other Interactions in Phytomedicines," *Phytomedicine* 8 (2001): 401–409.

46. Ethan B. Russo, "Taming THC: Potential Cannabis Synergy and Phytocannabinoid-Terpenoid Entourage Effects," *British Journal of Pharmacology* 163 (2011): 1344–1364; Robert C. Clarke and Mark D. Merlin, *Cannabis: Evolution and Ethnobotany* (Berkeley: University of California Press, 2013), 252–256.

47. Richard Evans Schultes, Albert Hofmann, and Christian Rätsch, *Plants of the Gods: Their Sacred, Healing, and Hallucinogenic Powers* (Rochester, Vt.: Healing Arts, 2001), 124–127.

48. "At present, most prescriptions contain between six and twelve substances" (Bensky, Clavey, and Stöger, *Chinese Herbal Medicine*, xix).

49. Bensky, Clavey, and Stöger, *Chinese Herbal Medicine*, xx.

50. Some examples of experimental approaches to studying synergy in traditional Chinese formulas include Aihua Zhang et al., "An *in vivo* Analysis of the Therapeutic and Synergistic Properties of Chinese Medicinal Formula Yin-Chen-Hao-Tang Based on Its Active Constituents," *Fitoterapia* 82 (2011): 1160–1168; and Kit-Man Lau et al., "Synergistic Interaction Between Astragali Radix and Rehmanniae Radix in a Chinese Herbal Formula to Promote Diabetic Wound Healing," *Journal of Ethnopharmacology* 141 (2012): 250–256. The former study addresses synergy at the chemical level; the latter, at the whole-herb level.

51. The peppers (*Piper* spp.) were widely used for their pungency and "heat" in the Old World before the introduction of the chili pepper (*Capsicum* spp.) in the sixteenth century.

52. Preetha Anand, Ajaikumar B. Kunnumakkara, et al., "Bioavailability of Curcumin: Problems and Promises," *Molecular Pharmaceutics* 4 (2007): 807–818; Preetha Anand, Sherin G. Thomas, et al., "Biological Activities of Curcumin and Its Analogues (Congeners) Made by Man and Mother Nature," *Biochemical Pharmacology* 76 (2008): 1590–1611. It is also speculated that cooking turmeric with oil enhances its ability to cross into the bloodstream when eaten in food, a suggestion supported by laboratory findings. See I. N. Dahmke et al., "Cooking Enhances Curcumin Anticancerogenic Activity Through Pyrolytic Formation of 'Deketene Curcumin,'" *Food Chemistry* 151 (2014): 514–519; and Dong-Jin Jang et al., "Enhanced Oral Bioavailability and Antiasthmatic Efficacy of Curcumin Using Redispersible Dry Emulsion," *Bio-Medical Materials and Engineering* 24 (2014): 917–930.

53. C. K. Atal, Raghvendra K. Dubey, and Jaswant Singh, "Biochemical Basis of Enhanced Drug Bioavailability by Piperine: Evidence That Piperine Is a Potent Inhibitor of Drug Metabolism," *Journal of Pharmacology and Experimental Therapeutics* 232 (1985): 258–262; Anand, Kunnumakkara, et al., "Bioavailability of Curcumin," 814; Ricky A. Sharma, William P. Steward, and Andreas J. Gescher, "Pharmacokinetics and Pharmacodynamics of Curcumin," in *The Molecular Targets and Therapeutic Uses of Curcumin in Health and Disease*, ed. Bharat B. Aggarwal, Young-Joon Surh, and Shishir Shishodia (New York: Springer, 2007), 455–456.

54. Justo Renquifo, personal communication.

55. Dioscorides, *De materia medica*, 177–178.

56. Bensky, Clavey, and Stöger, *Chinese Herbal Medicine*, 1056.

57. Frédéric D. Debelle, Jean-Louis Vanherweghem, and Joëlle Nortier, "Aristolochic Acid Nephropathy: A Worldwide Problem," *Kidney International* 74 (2008): 158–169; C.-H. Chen et al., "Aristolochic Acid–Associated Urothelial Cancer in Taiwan," *Proceedings of the National Academy of Sciences USA* 109 (2012): 8241–8246.

58. A. P. Grollman et al., "Aristolochic Acid and the Etiology of Endemic (Balkan) Nephropathy," *Proceedings of the National Academy of Sciences USA* 104 (2007): 12129–12134; Debelle, Vanherweghem, and Nortier, "Aristolochic Acid Nephropathy."

59. World Health Organization, International Agency for Research on Cancer, *Some Traditional Herbal Medicines, Some Mycotoxins, Naphthalene, and Styrene*, IARC Monographs on the Evaluation of Carcinogenic Risks to Humans 82 (Lyon: IARC, 2002), 82:69–86; Lois Swirsky Gold and Thomas H. Slone, "Aristolochic Acid, an Herbal Carcinogen, Sold on the Web After FDA Alert," *New England Journal of Medicine* 349 (2003): 1576–1577; Dawn Tung Au et al., "Ethnobotanical Study of Medicinal Plants Used by Hakka in Guangdong, China," *Journal of Ethnopharmacology* 117 (2008): 41–50. Contrary to the unequivocal biomedical evidence of carcinogenicity, the entry for *Aristolochia fangchi* on the Web site Natural Medicinal Herbs (http://www.naturalmedicinalherbs.net) explains that aristolochic acid is "an active antitumour agent" with "anticancer properties."

60. David A. Taylor, "Botanical Supplements: Weeding out the Health Risks," *Environmental Health Perspectives* 112 (2004): A750–A753. The National Academies of Science's Institute of Medicine released a detailed report on dietary supplement safety and regulations in 2005, and the Government Accountability Office made further recommendations in 2009.

61. Food and Drug Administration, *Draft Guidance for Industry: Dietary Supplements: New Dietary Ingredient Notifications and Related Issues* (2011, http://www.fda.gov/food/guidanceregulation/guidancedocumentsregulatoryinformation/dietarysupplements/ucm257563.htm), has also issued a draft, nonbinding set of regulations that would require a more substantial demonstration of safety for new dietary supplement ingredients.

62. Donald M. Marcus and Arthur P. Grollman, "The Consequences of Ineffective Regulation of Dietary Supplements," *Archives of Internal Medicine* 171 (2012): 1035–1036.

63. Government Accountability Office, "Herbal Dietary Supplements: FDA Should Take Further Actions to Improve Oversight and Consumer Understanding," Report to Congressional Requesters (GAO-09-250, 2009), and "Herbal Dietary Supplements: FDA May Have Opportunities to Expand Its Use of Reported Health Problems to Oversee Products," Report to Congressional Requesters (GAO-13-244, 2013). See also Government Accountability Office, "FDA Should Take Further Actions," 17.

64. Government Accountability Office, "Herbal Dietary Supplements: Examples of Deceptive or Questionable Marketing Practices and Potentially Dangerous Advice," Testimony Before the Special Committee on Aging, U.S Senate. (GAO-10-662T, 2010).

65. Scott A. Jordan, David G. Cunningham, and Robin J. Marles, "Assessment of Herbal Medicinal Products: Challenges, and Opportunities to Increase the Knowledge Base for Safety Assessment," *Toxicology and Applied Pharmacology* 243 (2010): 198–216; A. Ouarghidi et al., "Species Substitution in Medicinal Roots and Possible Implications for Toxicity of Herbal Remedies in Morocco," *Economic Botany* 66 (2012): 370–382.

66. Steven Foster, "Exploring the Peripatetic Maze of Black Cohosh Adulteration: A Review of the Nomenclature, Distribution, Chemistry, Market Status, Analytical Methods, and Safety," *HerbalGram* 98 (2013): 32–51.

67. Susan C. Smolinske, "Herbal Product Contamination and Toxicity," *Journal of Pharmacy Practice* 18 (2005): 188–208.

68. Debelle, Vanherweghem, and Nortier, "Aristolochic Acid Nephropathy." *Hanfangji* is 汉防己; *guangfangji* is 广防己. Therefore, both plants are part of the same traditional group of *fangji* herbs, but they are now known to vary in toxicity.

69. Karen M. Walker and Wendy L. Applequist, "Adulteration of Selected Unprocessed Botanicals in the U.S. Retail Herbal Trade," *Economic Botany* 66 (2012): 321–327.

70. For a history of intentional herbal adulteration, see Steven Foster, "A Brief History of Adulteration of Herbs, Spices, and Botanical Drugs," *HerbalGram* 92 (2011): 42–57.

The misrepresentation of drugs and spices for convenience or profit is a very old practice. A consortium of industry, academic researchers, media, and professional societies is engaged in a program of education, testing, and self-monitoring related to unintentional and deliberate adulteration of commercial herbal products. See American Botanical Council, "Botanical Adulterants Program," http://abc .herbalgram.org/site/PageServer?pagename=Adulterants.

71. Steven G. Newmaster et al., "DNA Barcoding Detects Contamination and Substitution in North American Herbal Products," *BMC Medicine* 11 (2013): 222; Anahad O'Connor, "New York Attorney General Targets Supplements at Major Retailers," *New York Times*, February 3, 2015.

72. "ABC Says New York Attorney General Misused DNA Testing for Herbal Supplements, Should Also Have Used Other Test Methods as Controls" (press release), February 3, 2015, American Botanical Council, http://cms .herbalgram.org/press/2015/ABCSaysNYAttyMisusedDNA .html; Charlotte Simmler et al., "Botanical Integrity: The Importance of the Integration of Chemical, Biological, and Botanical Analyses, and the Role of DNA Barcoding," *HerbalEGram* 12 (2015).

73. Jian Xue, Lili Hao, and Fei Peng, "Residues of Eighteen Organochlorine Pesticides in Thirty Traditional Chinese Medicines," *Chemosphere* 71 (2008): 1051–1055; Government Accountability Office, "Examples of Deceptive or Questionable Marketing Practices"; E. S. J. Harris et al., "Heavy Metal and Pesticide Content in Commonly Prescribed Individual Raw Chinese Herbal Medicines," *Science of the Total Environment* 409 (2011): 4297–4305.

74. R. B. Saper et al., "Heavy Metal Content of Ayurvedic Herbal Medicine Products," *Journal of the American Medical Association* 292 (2004): 2868–2873; H. Sarma et al., "Accumulation of Heavy Metals in Selected Medicinal Plants," *Reviews of Environmental Contamination and Toxicology* 214 (2011): 63–86; Harris et al., "Heavy Metal and Pesticide Content"; Paul Posadzki, Leala Watson, and Edzard Ernst, "Contamination and Adulteration of Herbal Medicinal Products (HMPs): An Overview of Systematic Reviews," *European Journal of Clinical Pharmacology* 69 (2013): 295–307.

75. Emma Lynch and Robin Braithwaite, "A Review of the Clinical and Toxicological Aspects of 'Traditional' (Herbal) Medicines Adulterated with Heavy Metals," *Expert Opinion on Drug Safety* 4 (2005): 769–778.

76. Wolfgang Kneifel, Erich Czech, and Brigitte Kopp, "Microbial Contamination of Medicinal Plants—A Review," *Planta Medica* 68 (2002): 5–15.

77. Junhua Zhang et al., "Quality of Herbal Medicines: Challenges and Solutions," *Complementary Therapies in Medicine* 20 (2012): 100–106.

78. Pharmacological scenarios in which combined drug effects are less than would be expected from the individual components of the mixture are called antagonistic, as elaborated in Berenbaum, "What Is Synergy?" In the context of human health, antagonism may be a positive outcome, such as for antidotes. For example, the specific antagonism of the cholinesterase inhibitor physostigmine, from the calabar bean (*Physostigma venenosum*), by the acetylcholine receptor antagonist atropine, from the nightshade plants (some members of Solanaceae), was among the first negative pharmacological relationships to be described in detail. See Thomas R. Fraser, "Lecture on the Antagonism Between the Actions of Active Substances," *British Medical Journal* 2, no. 617 (1872): 457–459; no. 618 (1872): 485–487. Because one is an antidote for the other, the counteracting relationship between the herbs is rather positive for the poisoning victim.

79. Phyllissa Schmiedlin-Ren et al., "Mechanisms of Enhanced Oral Availability of CYP3A4 Substrates by Grapefruit Constituents," *Drug Metabolism and Disposition* 25 (1997): 1228–1233. Prescription pharmaceuticals affected include the immune suppressant cyclosporin, blood pressure drug felodipine, anxiolytic midazolam, cardiac drug verapamil, protease inhibitor (in HIV treatment) saquinavir, and synthetic hormone (and birth control drug) ethinyl estradiol, among others.

80. Francesca Borrelli and Angelo A. Izzo, "Herb–Drug Interactions with St John's Wort (*Hypericum perforatum*): An Update on Clinical Observations," *American Association of Pharmaceutical Scientists Journal* 11 (2009): 710–727.

81. S. F. Zhou, "Structure, Function, and Regulation of P-glycoprotein and Its Clinical Relevance in Drug Disposition," *Xenobiotica* 38 (2008): 802–832; Borelli and Izzo, "Herb–Drug Interactions with St John's Wort," 710–712.

82. B. J. Gurley, E. K. Fifer, and Z. Gardner, "Pharmacokinetic Herb–Drug Interactions (Part 2): Drug Interactions Involving Popular Botanical Dietary Supplements and Their Clinical Relevance," *Planta Medica* 78 (2012): 1490–1514.

83. Robert Hermann and Oliver von Richter, "Clinical Evidence of Herbal Drugs as Perpetrators of Pharmacokinetic Drug Interactions," *Planta Medica* 78 (2012): 1458–1477; Gurley, Fifer, and Gardner, "Pharmacokinetic Herb–Drug Interactions."

84. Pius S. Fasinu, Patrick J. Bouic, and Bernd Rosenkranz, "An Overview of the Evidence and Mechanisms of Herb–Drug Interactions," *Frontiers in Pharmacology* 3 (2012): 69.

85. John M. Riddle and J. Worth Estes, "Oral Contraceptives in Ancient and Medieval Times," *American Scientist* 80, no. 3 (1992): 226–233; John Riddle, *Eve's Herbs: A History of Contraception and Abortion in the West* (Cambridge, Mass.: Harvard University Press, 1997), 44–46; Monika Kiehn, "Silphion Revisited," *Medicinal Plant Conservation* 13 (2007): 4–8.

86. Henry Koerper and A. L. Kolls, "The Silphium Motif Adorning Ancient Libyan Coinage: Marketing a Medicinal Plant," *Economic Botany* 53, no. 2 (1999): 133–143.

87. Koerper and Kolls, "Silphium Motif," 137–139.

88. Pliny the Elder, *Natural History* 19.15; *The Natural History of Pliny*, trans. John Bostock and H. T. Riley (London: Bohn, 1856), 4:145.

89. Van Jay Symons, *Ch'ing Ginseng Management: Ch'ing Monopolies in Microcosm* (Tempe: Center for Asian Research, Arizona State University, 1981), 3, 33; David A. Taylor,

Ginseng, the Divine Root (Chapel Hill, N.C.: Algonquin, 2006), 71.

90. Robert Beyfuss, letter to CBS News on Ginseng Poaching, September 8, 2013, Medicinal Plant Working Group Listserv. See also Robert L. Beyfuss, "How Do We Save Wild American Ginseng?" *HerbalGram* 102 (2014): 67–69.

91. The organization United Plant Savers (http://www .unitedplantsavers.org) advocates the conservation of endangered North American medicinal plants.

92. Gordon M. Cragg and David J. Newman, "Nature: A Vital Source of Leads for Anticancer Drug Development," *Phytochemistry Reviews* 8 (2009): 313–331.

93. Susan Mayor, "Tree That Provides Paclitaxel Is Put on List of Endangered Species," *BMJ* 343 (2011): d7411.

94. A. C. Hamilton, ed. *Medicinal Plants in Conservation and Development: Case Studies and Lessons Learnt* (Salisbury: Plantlife International, 2008), 8.

95. Royal Botanic Gardens, "Science and Conservation," http://www.kew.org/science/plants-at-risk/plant-groups .htm. There are more than 300,000 species of plants.

96. Mohamed Ali Ibrahim et al., "Significance of Endangered and Threatened Plant Natural Products in the Control of Human Disease," *Proceedings of the National Academy of Sciences USA* 110 (2013): 16832–16837.

97. James S. Miller provides an estimate of 35,000 species in the National Cancer Institute large-scale screens and 60,000 species in all the various screening programs combined, in "The Discovery of Medicines from Plants: A Current Biological Perspective," *Economic Botany* 65 (2011): 396–407.

98. Using conservative assumptions in "Discovery of Medicines from Plants," Miller calculates that approximately 600 new drugs remain to be discovered in plants. A more generous formula pushes the number above 20,000. In any case, the point is made that there is much untapped potential in plant diversity.

99. David Rapport and Luisa Maffi, "The Dual Erosion of Biological and Cultural Diversity: Implications for the Health of Ecocultural Systems," in *Nature and Culture: Rebuilding Lost Connections*, ed. Sarah Pilgrim and Jules Pretty (London: Earthscan, 2010), 103.

100. Daniel Nettle and Suzanne Romaine, *Vanishing Voices: The Extinction of the World's Languages* (New York: Oxford University Press, 2000), 8.

101. Richard Evans Schultes, "The Importance of Ethnobotany in Environmental Conservation," *American Journal of Economics and Sociology* 53, no. 2 (1994): 202–206; Jonathan Loh and David Harmon, "A Global Index of Biocultural Diversity," *Ecological Indicators* 5 (2005): 231–241;

Luisa Maffi, "Biocultural Diversity and Sustainability," in *The SAGE Handbook of Environment and Society*, ed. Jules Pretty et al. (London: SAGE, 2007), 267–277.

102. Schultes, "Importance of Ethnobotany in Environmental Conservation," 205.

103. Cox, "Tribal Knowledge"; Stanford Zent and Egleé L. Zent, "On Biocultural Diversity from a Venezuelan Perspective: Tracing the Interrelationships Among Biodiversity, Culture Change, and Legal Reforms"; and Michael J. Balick, "Traditional Knowledge: Lessons from the Past, Lessons for the Future," both in *Biodiversity and the Law: Intellectual Property, Biotechnology, and Traditional Knowledge*, ed. Charles R. McManis (London: Earthscan, 2007), 91–114, 280–296.

104. James S. Miller et al., "History of a Landmark Collecting Agreement: The Origin of the National Cancer Institute's Letter of Intent, a Precursor to Modern Bioprospecting Agreements," in *Biodiversity and the Law*, ed. McManis, 68–70; Cragg and Newman, "Natural Products," 3688. See also D. D. Soejarto et al., "Ethnobotany/Ethnopharmacology and Mass Bioprospecting: Issues on Intellectual Property and Benefit Sharing," *Journal of Ethnopharmacology* 100 (2005): 15–22.

105. Paul Alan Cox, "Ensuring Equitable Benefits: The Falealupo Covenant and the Isolation of Antiviral Drug Prostratin from a Samoan Medicinal Plant," *Pharmaceutical Biology* 39, supp. 1 (2001): 33–40.

106. The Convention on Biological Diversity (http:// www.cbd.int) entered into force in December 1993. The United States signed but has not ratified the convention. Another international agreement relevant to indigenous medical knowledge is the World Intellectual Property Organization, which has established the Intergovernmental Committee on Intellectual Property and Genetic Resources, Traditional Knowledge, and Folklore.

107. James S. Miller, "Impact of the Convention on Biological Diversity: The Lessons of Ten Years of Experience with Models for Equitable Sharing of Benefits," in *Biodiversity and the Law*, ed. McManis, 58–67.

108. Memory Elvin-Lewis, "Evolving Concepts Related to Achieving Benefit Sharing for Custodians of Traditional Knowledge," *Ethnobotany Research and Applications* 4 (2006): 75–96.

109. Will McClatchey argues that few new drugs are likely to emerge from ethnobotanic prospecting and that, regardless of new protocols governing international research, useful plant knowledge will be freely shared because it is human nature to do so, in "Medicinal Bioprospecting and Ethnobotany Research," *Ethnobotany Research & Applications* 3 (2005): 189–190.

Bibliography

Abel, Ernest L. *Marihuana: The First Twelve Thousand Years*. New York: Plenum, 1980.

Abourashed, Ehab A., Abir T. El-Alfy, Ikhlas A. Khan, et al. "*Ephedra* in Perspective—A Current Review." *Phytotherapy Research* 17 (2003): 703–712.

Adair, Mary. "Tobacco on the Plains: Historical Use, Ethnographic Accounts, and Archaeological Evidence." In *Tobacco Use by Native North Americans: Sacred Smoke and Silent Killer*, edited by Joseph Winter, 171–184. Norman: University of Oklahoma Press, 2000.

Adams, Jad. *Hideous Absinthe: A History of the Devil in a Bottle*. Madison: University of Wisconsin Press, 2004.

Ådén, Ulrika. "Methylxanthines During Pregnancy and Early Postnatal Life." In *Methylxanthines: Handbook of Experimental Pharmacology*, edited by Bertil B. Fredholm, 373–389. Berlin: Springer, 2011.

Agarwal, Rajesh, Charu Agarwal, Haruyo Ichikawa, et al. "Anticancer Potential of Silymarin: From Bench to Bed Side." *Anticancer Research* 26 (2006): 4457–4498.

Aggarwal, Bharat B., Chitra Sundaram, Nikita Malani, et al. "Curcumin: The Indian Solid Gold." In *The Molecular Targets and Therapeutic Uses of Curcumin in Health and Disease*, edited by Bharat B. Aggarwal, Young-Joon Surh, and Shishir Shishodia, 1–76. New York: Springer, 2007.

Allison, Gillian L., Gordon M. Lowe, and Khalid Rahman. "Aged Garlic Extract and Its Constituents Inhibit Platelet Aggregation Through Multiple Mechanisms." *Journal of Nutrition* 136 (2006): 782S–788S.

Amagase, Harunobu. "Clarifying the Real Bioactive Constituents of Garlic." *Journal of Nutrition* 136 (2006): 716S–725S.

Amagase, Harunobu, Brenda L. Petesch, Hiromichi Matsuura, et al. "Intake of Garlic and Its Bioactive Components." *Journal of Nutrition* 131 (2001): 955S–962S.

American Botanical Council. "ABC Says New York Attorney General Misused DNA Testing for Herbal Supplements, Should Also Have Used Other Test Methods as Controls." February 3, 2015. American Botanical Council. http://cms.herbalgram.org/press/2015/ABCSaysNYAtty MisusedDNA.html

Amin, Rakesh. "FDA Issues Final Rule Banning *Ephedra*." *HerbalGram* 62 (2004): 63–67.

Anand, Preetha, and K. Bley. "Topical Capsaicin for Pain Management: Therapeutic Potential and Mechanisms of Action of the New High-Concentration Capsaicin 8% Patch." *British Journal of Anaesthesia* 107 (2011): 490–502.

Anand, Preetha, Ajaikumar B. Kunnumakkara, Robert A. Newman, et al. "Bioavailability of Curcumin: Problems and Promises." *Molecular Pharmaceutics* 4 (2007): 807–818.

Anand, Preetha, Sherin G. Thomas, Ajaikumar B. Kunnumakkara, et al. "Biological Activities of Curcumin and Its Analogues (Congeners) Made by Man and Mother Nature." *Biochemical Pharmacology* 76 (2008): 1590–1611.

Ancient Egyptian Medicine: The Papyrus Ebers. Translated by Cyril P. Bryan. London: Bles, 1930.

Anderson, Edward F. "The Biogeography, Ecology, and Taxonomy of *Lophophora* (Cactaceae)." *Brittonia* 21 (1969): 299–310.

——. *Peyote: The Divine Cactus*. 2nd ed. Tucson: University of Arizona Press, 1996.

Anderson, Stuart. "Researching and Writing the History of Pharmacy." In *Making Medicines: A Brief History of Pharmacy and Pharmaceuticals*, edited by Stuart Anderson, 3–18. London: Pharmaceutical Press, 2005.

Ankri, Serge, and David Mirelman. "Antimicrobial Properties of Allicin from Garlic." *Microbes and Infection* 2 (1999): 125–129.

Anthony, F., B. Bertrand, O. Quiros, et al. "Genetic Diversity of Wild Coffee (*Coffea arabica* L.) Using Molecular Markers." *Euphytica* 118 (2001): 53–65.

Aragane, Masako, Yohei Sasaki, Jun'ichi Nakajima, et al. "Peyote Identification on the Basis of Differences in Morphology, Mescaline Content, and trnL/trnF Sequence Between *Lophophora williamsii* and *L. diffusa*." *Journal of Natural Medicine* 65 (2011): 103–110.

Arikha, Noga. *Passions and Tempers: A History of the Humours*. New York: HarperCollins, 2007.

Arjmahdi, Bahram H., Lee Alekel, Bruce W. Mollis, et al. "Dietary Soybean Protein Prevents Bone Loss in an Ovariectomized Rat Model of Osteoporosis." *Journal of Nutrition* 126 (1996): 161–167.

Arnason, John T., Dennis V. C. Awang, Mark Blumenthal, et al. "An Introduction to the Ginseng Evaluation Program." *HerbalGram* 52 (2001): 27–30.

Artica, Mijail Rimache. *Cultivo del Cacao*. Miraflores, Peru: Empresa Editora Macro, 2008.

Asea, Alexzander, Punit Kaur, Alexander Panossian, et al. "Evaluation of Molecular Chaperons Hsp72 and Neuropeptide Y as Characteristic Markers of Adaptogenic Activity of Plant Extracts." *Phytomedicine* 20 (2013): 1323–1329.

Atal, C. K., Raghvendra K. Dubey, and Jaswant Singh. "Biochemical Basis of Enhanced Drug Bioavailability by Piperine: Evidence That Piperine Is a Potent Inhibitor of Drug Metabolism." *Journal of Pharmacology and Experimental Therapeutics* 232 (1985): 258–262.

Atal, C. K., O. P. Gupta, and S. H. Afaq. "*Commiphora mukul*: Source of Guggal in Indian Systems of Medicine." *Economic Botany* 29 (1975): 208–218.

Atri, Alireza, Michael S. Chang, and Gary R. Strichartz. "Cholinergic Pharmacology." In *Principles of Pharmacology: The Pathophysiologic Basis of Drug Therapy*, 3rd ed., edited by David E. Golan, Armen H. Tahsijian, Ehrin J. Armstrong, et al., 110–131. Philadelphia: Lippincott Williams & Wilkins, 2012.

Au, Dawn Tung, Jialin Wu, Zhihong Jiang, et al. "Ethnobotanical Study of Medicinal Plants Used by Hakka in Guangdong, China." *Journal of Ethnopharmacology* 117 (2008): 41–50.

Baker, Phil. *The Book of Absinthe: A Cultural History*. New York: Grove Press, 2001.

Balick, Michael J. *Rodale's Twenty-First-Century Herbal: A Practical Guide for Healthy Living Using Nature's Most Powerful Plants*. New York: Rodale, 2014.

——. "Traditional Knowledge: Lessons from the Past, Lessons for the Future." In *Biodiversity and the Law: Intellectual Property, Biotechnology, and Traditional Knowledge*, edited by Charles R. McManis, 280–296. London: Earthscan, 2007.

Balick, Michael J., and Paul Alan Cox. *Plants, People, and Culture: The Science of Ethnobotany*. New York: Scientific American, 1996.

Balick, Michael J., and Roberta A. Lee. "The Sacred Root: *Sakau* en Pohnpei." In *Ethnobotany of Pohnpei: Plants, People, and Island Culture*, edited by Michael J. Balick, 165–203. Honolulu: University of Hawai'i Press, 2009.

Ball, Philip. *The Devil's Doctor: Paracelsus and the World of Renaissance Magic and Science*. New York: Farrar, Straus and Giroux, 2006.

Banerjee, Sanjay K., and Subir K. Maulik. "Effect of Garlic on Cardiovascular Disorders: A Review." *Nutrition Journal* 1 (2002): 4.

Barnes, Joanne, Linda A. Anderson, and J. David Phillipson. "St. John's Wort (*Hypericum perforatum* L.): A Review of Its Chemistry, Pharmacology, and Clinical Properties." *Journal of Pharmacy and Pharmacology* 53 (2001): 583–600.

Barnes, Peter J. "Pulmonary Pharmacology." In *Goodman and Gilman's The Pharmacological Basis of Therapeutics*, 12th ed., edited by Laurence L. Brunton, Bruce A. Chabner, and Björn C. Knollmann, 1031–1066. New York: McGraw-Hill, 2011.

——. "Theophylline: New Perspectives for an Old Drug." *American Journal of Respiratory and Critical Care Medicine* 167 (2003): 813–818.

Barnes, Peter J., and R. A. Stockley. "COPD: Current Therapeutic Interventions and Future Approaches." *European Respiratory Journal* 25 (2005): 1084–1106.

Barnes, Stephen. "The Biochemistry, Chemistry, and Physiology of the Isoflavones in Soybeans and Their Food Products." *Lymphatic Research and Biology* 8 (2010): 89–98.

Barrett, B. P., R. L. Brown, K. Locken, et al. "Treatment of the Common Cold with Unrefined *Echinacea*. A Randomized, Double-Blind, Placebo-Controlled Trial." *Annals of Internal Medicine* 137 (2002): 939–946.

Barrett, Bruce, Roger Brown, Dave Rakel, et al. "*Echinacea* for Treating the Common Cold: A Randomized Controlled Trial." *Annals of Internal Medicine* 153 (2010): 769–777.

Bartholow, Roberts. *A Manual of Hypodermic Medication*. Philadelphia: Lippincott, 1869.

Bartsch, G., R. S. Rittmaster, and H. Klocker. "Dihydrotestosterone and Concept of 5α-Reductase Inhibition in Benign Prostatic Hyperplasia." *European Urology* 37 (2000): 367–380.

Basavarajappa, Balapal S. "Neuropharmacology of the Endocannabinoid Signaling System—Molecular Mechanisms, Biological Actions, and Synaptic Plasticity." *Current Neuropharmacology* 5 (2007): 81–97.

Basnet, Purusotam, and Natasa Skalko-Basnet. "Curcumin: An Anti-Inflammatory Molecule from a Curry Spice on the Path to Cancer Treatment." *Molecules* 16 (2011): 4567–4598.

Bause, George S. "From Fish Poison to Merck Picrotoxin." *Anesthesiology* 118 (2013): 1261–1263.

Bausell, R. Barker. *Snake Oil Science: The Truth About Complementary and Alternative Medicine*. New York: Oxford University Press, 2007.

Bayne, Holly J. "FDA Issues Final Rule Banning Use of Aloe and Cascara Sagrada in OTC Drug Products." *HerbalGram* 56 (2002): 56.

Becerra, Judith X., Koji Noge, and D. Lawrence Venable. "Macroevolutionary Chemical Escalation in an Ancient Plant–Herbivore Arms Race." *Proceedings of the National Academy of Sciences USA* 106 (2009): 18062–18066.

Becker, Daniel, and Kenneth Reed. "Essentials of Local Anesthetic Pharmacology." *Anesthesia Progress* 53 (2006): 98–109.

Bee, Robert L. "Peyotism in North American Indian Groups." *Transactions of the Kansas Academy of Science* 68 (1965): 13–61.

Beecher, Henry K. "The Powerful Placebo." *Journal of the American Medical Association* 159 (1955): 1602–1606.

Benavides, Gloria A., Guiseppe L. Squadrito, Robert W. Mills, et al. "Hydrogen Sulfide Mediates the Vasoactivity of Garlic." *Proceedings of the National Academy of Sciences USA* 104 (2007): 17977–17982.

Benedetti, F. "Placebo Analgesia." *Neurological Science* 27 (2006): S100–S102.

Benedict, Carol. *Golden-Silk Smoke: A History of Tobacco in China, 1550–2000.* Berkeley: University of California Press, 2011.

Benet, Sula. "Early Diffusion and Folk Uses of Hemp." In *Cannabis and Culture*, edited by Vera Rubin, 39–50. The Hague: Mouton, 1975.

Benke, Dietmar, Andrea Barberis, Sascha Kopp, et al. "GABA$_A$ Receptors as *in vivo* Substrate for the Anxiolytic Action of Valerenic Acid, a Major Constituent of Valerian Root Extracts." *Neuropharmacology* 56 (2009): 174–181.

Benner, Dagmar. "Traditional Indian Systems of Healing and Medicine: Ayurveda." In *Encyclopedia of Religion*, 2nd ed., edited by Lindsay Jones, 3852–3858. New York: Macmillan, 2005.

Bennett, Bradley C. "Doctrine of Signatures: An Explanation of Medicinal Plant Discovery or Dissemination of Knowledge?" *Economic Botany* 61, no. 3 (2007): 246–255.

Bennett, Bradley C., and Michael J. Balick. "Phytomedicine 101: Plant Taxonomy for Preclinical and Clinical Medicinal Plant Researchers." *Journal of the Society for Integrative Oncology* 6 (2008): 150–157.

Bennett, Bradley C., and Judith R. Hicklin. "Uses of Saw Palmetto (*Serenoa repens*, Arecaceae) in Florida." *Economic Botany* 52, no. 4 (1998): 381–393.

Bensky, Dan, Steven Clavey, and Erich Stöger. *Chinese Herbal Medicine: Materia Medica.* 3rd ed. Seattle: Eastland, 2004.

Benzoni, Girolamo. *History of the New World.* Translated by W. H. Smyth. London: Haklyut Society, 1857.

Berenbaum, M. C. "What Is Synergy?" *Pharmacological Reviews* 41 (1989): 93–141.

Bernáth, Jenő. "Introduction." In *Poppy: The Genus Papaver*, edited by Jenő Bernáth, 1–7. Amsterdam: Harwood, 1998.

Berridge, Virginia. "Our Own Opium: Cultivation of the Opium Poppy in Britain, 1740–1823." *British Journal of Addiction* 72 (1977): 90–94.

Berridge, Virginia, and Griffith Edwards. *Opium and the People: Opiate Use in Nineteenth-Century England.* London: Allen Lane/St. Martin's Press, 1981.

Beyfuss, Robert. "How Do We Save Wild American Ginseng?" *HerbalGram* 102 (2014): 67–69.

——. Letter to CBS News on Ginseng Poaching. September 8, 2013. Medicinal Plant Working Group Listserv.

Bialous, Stella, and Silvy Peeters. "A Brief Overview of the Tobacco Industry in the Last Twenty Years." *Tobacco Control* 21 (2012): 92–94.

Bilia, Anna Rita, Sandra Gallori, and Franco F. Vincieri. "St. John's Wort and Depression: Efficacy, Safety and Tolerability—An Update." *Life Sciences* 70 (2002): 3077–3096.

Birks, Jacqueline, and John Grimley Evans. "Ginkgo biloba for Cognitive Impairment and Dementia." *Cochrane Database of Systematic Reviews* (2009): CD003120.

Bisset, Norman G. "Arrow Poisons and Their Role in the Development of Medicinal Agents." In *Ethnobotany: Evolution of a Discipline*, edited by Richard Evans Schultes and Siri von Reis, 289–302. Portland, Ore.: Dioscorides, 1995.

Blackwell, Will H. *Poisonous and Medicinal Plants.* Englewood Cliffs, N.J.: Prentice-Hall, 1990.

Bletter, Nathaniel, and Douglas C. Daly. "Cacao and Its Relatives in South America: An Overview of Taxonomy, Ecology, Biogeography, Chemistry, and Ethnobotany." In *Chocolate in Mesoamerica: A Cultural History of Cacao*, edited by Cameron L. McNeil, 31–68. Gainesville: University Press of Florida, 2006.

Bloch, Enid. "Hemlock Poisoning and the Death of Socrates: Did Plato Tell the Truth?" In *The Death of Socrates*, by Emily Wilson, 255–278. London: Profile, 2007.

Block, Eric. *Garlic and Other Alliums: The Lore and the Science.* London: Royal Society of Chemistry, 2010.

Blue, Gregory. "Opium for China: The British Connection." In *Opium Regimes: China, Britain, and Japan, 1839–1952*, edited by Timothy Brook and Bob Tadashi Wakabayashi, 31–54. Berkeley: University of California Press, 2000.

Blumenthal, Mark, Alicia Goldberg, and Josef Brinckmann, eds. *Herbal Medicine: Expanded Commission E Monographs.* Austin, Tex.: American Botanical Council, 2000.

Bock, Hieronymus. *Kreüter Buoch.* Strassburg: Wendel Rihel, 1546.

Bohm, Bruce, Fred Ganders, and Timothy Plowman. "Biosystematics and Evolution of Cultivated Coca (*Erythroxylaceae*)." *Systematic Botany* 7, no. 2 (1982): 121–133.

Bonnie, Richard J., and Charles H. Whitebread. *The Marihuana Conviction: A History of Marihuana Prohibition in the United States.* Charlottesville: University Press of Virginia, 1974.

Booth, Martin. *Cannabis: A History.* New York: St. Martin's Press, 2003.

——. *Opium: A History.* New York: St. Martin's Press, 1996.

Borek, Carmia. "Antioxidant Health Effects of Aged Garlic Extract." *Journal of Nutrition* 131 (2001): 1010S–1015S.

Borrelli, Francesca, and Angelo A. Izzo. "Herb–Drug Interactions with St John's Wort (*Hypericum perforatum*): An Update on Clinical Observations." *American Association of Pharmaceutical Scientists Journal* 11 (2009): 710–727.

Bowers, M. Deane. "Iridioid Glycosides." In *Herbivores: Their Interactions with Secondary Plant Metabolites*, 2nd ed., edited by Gerald A. Rosenthal and May R. Berenbaum, 1:297–326. San Diego: Academic Press, 1991.

Bowman, W. C. "Neuromuscular Block." *British Journal of Pharmacology* 147 (2006): S277–S286.

Brant, Charles S. "Peyotism Among the Kiowa-Apache and Neighboring Tribes." *Southwestern Journal of Anthropology* 6, no. 2 (1950): 212–222.

Bray, Elizabeth A., Julia Bailey-Serres, and Elizabeth Weretilnik. "Responses to Abiotic Stresses." In *Biochemistry and Molecular Biology of Plants*, edited by Bob B. Buchanan, Wilhelm Gruissem, and Russell L. Jones,

1158–1203. Rockville, Md.: American Society of Plant Biologists, 2000.

Brekhman, I. I., and I. V. Dardymov. "New Substances of Plant Origin Which Increase Nonspecific Resistance." *Annual Review of Pharmacology* 9 (1969): 419–430.

Brenneisen, Rudolf. "Chemistry and Analysis of Phytocannabinoids and Other *Cannabis* Constituents." In *Marijuana and the Cannabinoids*, edited by Mahmoud A. ElSohly, 17–50. Totowa, N.J.: Humana, 2007.

Brinker, Francis. "*Echinacea* Differences Matter: Traditional Uses of *Echinacea angustifolia* Root Extracts Versus Modern Clinical Trials with *Echinacea purpurea* Fresh Plant Extracts." *HerbalGram* 97 (2013): 46–57.

Briskin, Donald P. "Medicinal Plants and Phytomedicines: Linking Plant Biochemistry and Physiology to Human Health." *Plant Physiology* 124, no. 2 (2000): 507–514.

Brody, Howard. "The Placebo Response: Recent Research and Implications for Family Medicine." *Journal of Family Practice* 49, no. 7 (2000): 649–654.

Brody, Jane E. "Tapping Medical Marijuana's Potential." *New York Times*, November 4, 2013.

Brook, Timothy. "Smoking in Imperial China." In *Smoke: A Global History of Smoking*, edited by Sander L. Gilman and Zhou Xun, 84–91. London: Reaktion, 2004.

Brook, Timothy, and Bob Tadashi Wakabayashi. "Opium's History in China." In *Opium Regimes: China, Britain, and Japan, 1839–1952*, edited by Timothy Brook and Bob Tadashi Wakabayashi, 1–27. Berkeley: University of California Press, 2000.

Brown, Geoffrey D. "The Biosynthesis of Artemisinin (Qinghaosu) and the Phytochemistry of *Artemisia annua* L. (Qinghao)." *Molecules* 15 (2010): 7603–7698.

Brown, Joan Heller, and Nora Laiken. "Muscarinic Receptor Agonists and Antagonists." In *Goodman and Gilman's The Pharmacological Basis of Therapeutics*, 12th ed., edited by Laurence L. Brunton, Bruce A. Chabner, and Björn C. Knollmann, 219–238. New York: McGraw-Hill, 2011.

Bruhn, J. G., J.-E. Lindgren, J. Holmstedt, et al. "Peyote Alkaloids: Identification in a Prehistoric Specimen of *Lophophora* from Coahuila, Mexico." *Science* 199 (1978): 1437–1438.

Bruhn, Jan G., and Peter A. G. M. De Smet. "Ceremonial Peyote Use and Its Antiquity in the Southern United States." *HerbalGram* 58 (2003): 30–33.

Brunschwig, Hieronymus. *The Vertuose Boke of Distyllacyon*. London, 1527.

Brunton, Ron. *The Abandoned Narcotic: Kava and Cultural Instability in Melanesia*. Cambridge: Cambridge University Press, 1989.

Burdock, George. "Dietary Supplements and Lessons to Be Learned from GRAS." *Regulatory Toxicology and Pharmacology* 31 (2000): 68–76.

Burdock, George A., Ioana G. Carabin, and James C. Griffiths. "Toxicology and Pharmacology of Sodium Ricinoleate." *Food and Chemical Toxicology* 44 (2006): 1689–1698.

Burkhart, Craig, Dean Morrell, and Lowell Goldsmith. "Dermatological Pharmacology." In *Goodman and Gilman's The Pharmacological Basis of Therapeutics*, 12th ed., edited by Laurence L. Brunton, Bruce A. Chabner, and Björn C. Knollmann, 1803–1832. New York: McGraw-Hill, 2011.

Burns, Eric. *Smoke of the Gods: A Social History of Tobacco*. Philadelphia: Temple University Press, 2007.

Butrica, James L. "The Medical Use of Cannabis Among the Greeks and Romans." *Journal of Cannabis Therapeutics* 2 (2002): 51–70.

Butterweck, Veronika, and Mathias Schmidt. "St. John's Wort: Role of Active Compounds for Its Mechanism of Action and Efficacy." *Wiener Medizinische Wochenschrift* 157 (2007): 356–361.

Cadéac, C., and A. Meunier. "Nouvelle note sur l'étude physiologique de la liqueur d'absinthe." *Comptes-rendus de la Société de biologie* (1889): 633–638.

Calkins, Alonzo. *Opium and the Opium-Appetite*. Philadelphia: Lippincott, 1871.

Callaway, J. C. "Hempseed as a Nutritional Resource: An Overview." *Euphytica* 140 (2004): 65–72.

Cameron, Melainie, and Sigrun Chrubasik. "Topical Herbal Therapies for Treating Osteoarthritis." *Cochrane Database of Systematic Reviews* (2013): CD010538.

Campbell, J. R., and A. R. David. "The Application of Archaeobotany, Phytogeography, and Phamacognosy to Confirm the Pharmacopeia of Ancient Egypt, 1850–1200 B.C." In *Pharmacy and Medicine in Ancient Egypt: Proceedings of the Conferences Held in Cairo (2007) and Manchester (2008)*, edited by Jenefer Cockitt and Rosalie David, 20–29. BAR International Series 2141. Oxford: Archaeopress, 2010.

Capasso, Anna. "Antioxidant Action and Therapeutic Efficacy of *Allium sativum* L." *Molecules* 18 (2013): 690–700.

Carolan, James C., Ingrid L. I. Hook, Mark W. Chase, et al. "Phylogenetics of *Papaver* and Related Genera Based on DNA Sequences from ITS Nuclear Ribosomal DNA and Plastid *trnL* Intron and *trnL–F* Intergenic Spacers." *Annals of Botany* 98 (2006): 141–155.

Carson, Culley, and Roger Rittmaster. "The Role of Dihydrotestosterone in Benign Prostatic Hyperplasia." *Urology* 61, suppl. 4A (2003): 2–7.

Carson, Joseph. *Illustrations of Medical Botany*. 2 vols. Philadelphia: Smith, 1847.

Catterall, William A., and Kenneth Mackie. "Local Anesthetics." In *Goodman and Gilman's The Pharmacological Basis of Therapeutics*, 12th ed., edited by Laurence L. Brunton, Bruce A. Chabner, and Björn C. Knollmann, 565–582. New York: McGraw-Hill, 2011.

Cazzola, Mario, Clive P. Page, Luigino Calzetta, et al. "Pharmacology and Therapeutics of Bronchodilators." *Pharmacological Reviews* 64 (2012): 450–504.

Charlton, Anne. "Medicinal Uses of Tobacco in History." *Journal of the Royal Society of Medicine* 97 (2004): 292–296.

Chaudhri, I. I. "Pakistani Ephedra." *Economic Botany* 11 (1957): 257–263.

Chen, C.-H., K. G. Dickman, M. Moriya, et al. "Aristolochic Acid–Associated Urothelial Cancer in Taiwan." *Proceedings of the National Academy of Sciences USA* 109 (2012): 8241–8246.

Chen, Huai-Yuan, Shoei-Yn Lin-Shiau, and Jen-Kun Lin. "Pu-crh Tea: Its Manufacturing and Health Benefits." In *Tea and Health Products: Chemistry and Health-Promoting Properties*, edited by Chi-Tang Ho, Jen-Kun Lin, and Fereidoon Shahidi, 9–16. Boca Raton, Fla.: CRC Press, 2009.

Chen, John K., and Tina T. Chen. *Chinese Medical Herbology and Pharmacology*. City of Industry, Calif.: Art of Medicine Press, 2004.

Chen, Kuang-Kuo, and Jen-Hwey Chiu. "Effect of *Epimedium brevicornum* Maxim Extract on Elicitation of Penile Erection in the Rat." *Urology* 67 (2006): 631–635.

Chen, S.-N., J. B. Friesen, D. Webster, et al. "Phytoconstituents from *Vitex agnus-castus* Fruits." *Fitoterapia* 82 (2011): 528–533.

Chia, K.-S., M. Reilly, C.-S. Tan, et al. "Profound Changes in Breast Cancer Incidence May Reflect Changes into a Westernized Lifestyle: A Comparative Population-Based Study in Singapore and Sweden." *International Journal of Cancer* 113 (2005): 302–306.

Choi, J., T.-H. Kim, T.-Y. Choi, et al. "Ginseng for Health Care: A Systematic Review of Randomized Controlled Trials in Korean Literature." *PLoS ONE* 8 (2013): e59978.

Chopra, I. C., and R. N. Chopra. "The Use of *Cannabis* Drugs in India." *Bulletin on Narcotics* 9 (1957): 4–29.

Chopra, R. N., and I. C. Chopra. *Indigenous Drugs of India*. Calcutta: Art Press, 1933.

Christenson, Allen J. *Art and Society in a Highland Maya Community: The Altarpiece of Santiago Atitlán*. Austin: University of Texas Press, 2001.

Cicek, Serhat S., Sophia Khom, Barbara Taferner, et al. "Bioactivity-Guided Isolation of GABA$_A$ Receptor Modulating Constituents from the Rhizomes of *Actaea racemosa*." *Journal of Natural Products* 73 (2010): 2024–2028.

Cilliers, L., and F. P. Retief. "Poisons, Poisoning, and the Drug Trade in Ancient Rome." *Akroterion* 45 (2000): 88–100.

Clarke, Robert C., and Mark D. Merlin. *Cannabis: Evolution and Ethnobotany*. Berkeley: University of California Press, 2013.

Clarke, Robert C., and David P. Watson. "*Cannabis* and Natural *Cannabis* Medicines." In *Marijuana and the Cannabinoids*, edited by Mahmoud A. ElSohly, 1–16. Totowa, N.J.: Humana, 2007.

Cockayne, Oswald, ed. *Leechdoms, Wortcunning, and Starcraft of Early England*. Vol. 2. London: Longman, Green, Longman, Roberts, and Green, 1863.

Coe, Sophie D., and Michael D. Coe. *The True History of Chocolate*. 3rd ed. New York: Thames & Hudson, 2013.

Cohen, Peter J. "Medical Marijuana: The Conflict Between Scientific Evidence and Political Ideology (Part One)." *Journal of Pain & Palliative Care Pharmacotherapy* 23 (2009): 4–25.

——. "Medical Marijuana: The Conflict Between Scientific Evidence and Political Ideology (Part Two)." *Journal of Pain & Palliative Care Pharmacotherapy* 23 (2009): 120–140.

Cole, James M. "Memorandum for All United States Attorneys: Guidance Regarding Marijuana Enforcement." Department of Justice, Office of the Deputy Attorney General, August 29, 2013.

Coles, William. *Adam in Eden; or, Nature's Paradise. The History of Plants, Herbs, and Flowers*. London: Nathaniel Brooke, 1657.

Colín-González, Ana L., Ricardo A. Santana, Carlos A. Silva-Islas, et al. "The Antioxidant Mechanisms Underlying the Aged Garlic Extract- and S-Allylcysteine-Induced Protection." *Oxidative Medicine and Cellular Longevity* 2012 (2012): ID 907162.

Collins, James. "A Descriptive Introduction to the Taos Peyote Ceremony." *Ethnology* 7 (1968): 427–449.

Columbus, Christopher. *The* Diario *of Christopher Columbus's First Voyage to America, 1492–1493*. Translated by Oliver Dunn and James E. Kelley Jr. Norman: University of Oklahoma Press, 1989.

Compton, James A., Alastair Culham, and Stephen L. Jury. "Reclassification of *Actaea* to Include *Cimicifuga* and *Souliea* (Ranunculaceae): Phylogeny Inferred from Morphology, nrDNA ITS, and cpDNA trnL-F Sequence Variation." *Taxon* 47 (1998): 593–634.

Conrad, Barnaby. *Absinthe: History in a Bottle*. San Francisco: Chronicle, 1988.

Cooper, Courtney Riley. *Here's to Crime*. Boston: Little, Brown, 1937.

Corchete, P. "*Silybum marianum* (L.) Gaertn: The Source of Silymarin." In *Bioactive Molecules and Medicinal Plants*, edited by K. G. Ramawat and J.-M. Mérillon, 124–148. Berlin: Springer, 2008.

Corrêa, Geone M., and Antônio F. de C. Alcântara. "Chemical Constituents and Biological Activities of Species of *Justicia*—A Review." *Brazilian Journal of Pharmacognosy* 22 (2012): 220–238.

Corti, R., A. J. Flammer, N. K. Hollenberg, et al. "Cocoa and Cardiovascular Health." *Circulation* 119 (2009): 1433–1441.

Côté, J., S. Caillet, G. Doyon, et al. "Bioactive Compounds in Cranberries and Their Biological Properties." *Critical Reviews in Food Science and Nutrition* 50 (2010): 666–679.

Court, William E. *Ginseng: The Genus* Panax. Amsterdam: Harwood, 2000.

——. "Pharmacy from the Ancient World to 1100 A.D." In *Making Medicines: A Brief History of Pharmacy and Pharmaceuticals*, edited by Stuart Anderson, 21–36. London: Pharmaceutical Press, 2005.

Courtwright, David T. *Dark Paradise: Opiate Addiction in the United States Before 1940*. Cambridge, Mass.: Harvard University Press, 1982.

——. "The Rise and Fall and Rise of Cocaine in the United States." In *Consuming Habits: Drugs in History and Anthropology*, edited by J. Goodman, P. E. Lovejoy, and A. Sherratt, 206–228. London: Routledge, 1995.

Cox, Paul Alan. "Ensuring Equitable Benefits: The Falealupo Covenant and the Isolation of Antiviral Drug Prostratin from a Samoan Medicinal Plant." *Pharmaceutical Biology* 39, suppl. 1 (2001): 33–40.

——. "Will Tribal Knowledge Survive the Millennium?" *Science* 287 (2000): 44–45.

Cox, Robert S., and Jacob Walker. *Massachusetts Cranberry Culture: A History from Bog to Table*. Charleston, S.C.: History Press, 2012.

Crafts, Wilbur, and Mary Leitch. *Intoxicating Drinks and Drugs in All Lands and Times*. Washington, D.C.: International Reform Bureau, 1909.

Cragg, Gordon M., and David J. Newman. "Natural Products: A Continuing Source of Novel Drug Leads." *Biochimica et Biophysica Acta* 1830 (2013): 3670–3695.

——. "Nature: A Vital Source of Leads for Anticancer Drug Development." *Phytochemistry Reviews* 8 (2009): 313–331.

——. "Plants as a Source of Anticancer Agents." *Journal of Ethnopharmacology* 100 (2005): 72–79.

Crane, Peter. *Ginkgo: The Tree That Time Forgot*. New Haven, Conn.: Yale University Press, 2013.

Crellin, John K. "Aboriginal/Traditional Medicine in North America: A Practical Approach for Practitioners." In *Traditional Medicine: A Global Perspective*, edited by Steven B. Kayne, 44–64. London: Pharmaceutical Press, 2010.

Crellin, John K., and Jane Philpott. *Herbal Medicine Past and Present*. Vol. 1, *Trying to Give Ease*. Durham, N.C.: Duke University Press, 1990.

——. *Herbal Medicine Past and Present*. Vol. 2, *Monographs on Select Medicinal Plants*. Durham, N.C.: Duke University Press, 1990.

Croll, Oswald. *Basilica Chymica*. Savoy: Paul Marcell, 1610.

Croom, Edward M. "Documenting and Evaluating Herbal Remedies." *Economic Botany* 37 (1983): 13–27.

Croteau, Rodney, Toni M. Kutchan, and Norman G. Lewis. "Natural Products (Secondary Metabolites)." In *Biochemistry and Molecular Biology of Plants*, edited by Bob B. Buchanan, Wilhelm Gruissem, and Russell L. Jones, 1250–1318. Rockville, Md.: American Society of Plant Biologists, 2000.

Crumpe, Samuel. *An Inquiry into the Nature and Properties of Opium*. London: Robinson, 1793.

Culpeper, Nicholas. *The English Physitian Enlarged*. London: Streater, 1666.

Cuttle, L., M. Kempf, O. Kravchuk, et al. "The Efficacy of *Aloe vera*, Tea Tree Oil, and Saliva as First-Aid Treatment for Partial Thickness Burn Injuries." *Burns* 34 (2008): 1176–1182.

Dahmke, I. N., S. P. Boettcher, M. Groh, et al. "Cooking Enhances Curcumin Anticancerogenic Activity Through Pyrolytic Formation of 'Deketene Curcumin.'" *Food Chemistry* 151 (2014): 514–519.

Dalby, Andrew. *Dangerous Tastes: The Story of Spices*. Berkeley: University of California Press, 2000.

Daly, John W., and Bertil B. Fredholm. "Caffeine—An Atypical Drug of Dependence." *Drug and Alcohol Dependence* 51, no. 1 (1998): 199–206.

Das, Kaushik, and Jharma Bhattacharttya. "Antioxidant Functions of Green and Black Tea." In *Tea in Health and Disease Prevention*, edited by Victor R. Preedy, 521–528. London: Elsevier, 2013.

Dat, A. D., F. Poon, K. B. T. Pham, et al. "*Aloe vera* for Treating Acute and Chronic Wounds." *Cochrane Database of Systematic Reviews* (2012): CD008762.

Davidson, E., B. F. Zimmermann, E. Jungfer, et al. "Prevention of Urinary Tract Infections with *Vaccinium* Products." *Phytotherapy Research* 28 (2014): 465–470.

Davidson, John. *The Cascara Tree in British Columbia*. Victoria, B.C.: Ministry of Agriculture, 1942.

Davis, George A. *The Pharmacology of the Newer Materia Medica*. Detroit: Davis, 1892.

Dawson, Ian, James Were, and Ard Lengkeek. "Conservation of *Prunus africana*, an Over-exploited African Medicinal Tree." *Forest Genetic Resources* 28 (2000): 27–33.

Debelle, Frédéric D., Jean-Louis Vanherweghem, and Joëlle Nortier. "Aristolochic Acid Nephropathy: A Worldwide Problem." *Kidney International* 74 (2008): 158–169.

DeKosky, S. T., J. D. Williamson, A. L. Fitzpatrick, et al. "*Ginkgo biloba* for Prevention of Dementia: A Randomized Controlled Trial." *Journal of the American Medical Association* 300 (2008): 2253–2262.

Delahaye, Marie-Claude. *L'absinthe: Son histoire*. Auvers-sur-Oise: Musée de l'Absinthe, 2001.

De la Vega, Garcilaso. *The Royal Commentaries of Peru*. Translated by Paul Ricaut. London: Christopher Wilkinson, 1688.

Del Rio, Daniele, Ana Rodriguez-Mateos, Jeremy P. E. Spencer, et al. "Dietary (Poly)phenolics in Human Health: Structures, Bioavailability, and Evidence of Protective Effects Against Chronic Diseases." *Antioxidants and Redox Signaling* 18 (2013): 1818–1892.

Dennehy, Cathi. "Need for Additional, Specific Information in Studies with *Echinacea*." *Antimicrobial Agents and Chemotherapy* 45 (2001): 369.

Densmore, Frances. "Uses of Plants by the Chippewa Indians." *Smithsonian Institution–Bureau of American Ethnology Annual Report* 44 (1928): 273–379.

Des Forges, Alexander. "Opium/Leisure/Shanghai: Urban Economies of Consumption." In *Opium Regimes: China, Britain, and Japan, 1839–1952*, edited by Timothy Brook and Bob Tadashi Wakabayashi, 167–185. Berkeley: University of California Press, 2000.

De Vos, Paula. "European *Materia Medica* in Historical Texts: Longevity of a Tradition and Implications for Future Use." *Journal of Ethnopharmacology* 132 (2010): 28–47.

Dewey, Lyster H. "Hemp." In *Yearbook of the United States Department of Agriculture*, 283–346. Washington, D.C.: Government Printing Office, 1914.

Dickinson, Jonathan. *Jonathan Dickinson's Journal, or God's Protecting Providence, Being the Narrative of a Journey from Port Royal in Jamaica to Philadelphia Between August 23, 1696, and April 1, 1697.* Edited by Evangeline Walker Andrews and Charles McLean Andrews. New Haven, Conn.: Yale University Press, 1945.

Dickson, Sarah A. *Panacea or Precious Bane: Tobacco in Sixteenth-Century Literature.* New York: New York Public Library, 1954.

Ding, E. L., S. M. Hutfless, X. Ding, et al. "Chocolate and Prevention of Cardiovascular Disease: A Systematic Review." *Nutrition and Metabolism* 3 (2006): 2.

Dioscorides, Pedanius. *De materia medica.* Translated by Lily Y. Beck. Hildesheim: Olms-Weidmann, 2005.

——. *The Greek Herbal of Dioscorides, Illustrated by a Byzantine, 512 C.E.* Edited by Robert T. Gunther. Translated by John Goodyer. New York: Hafner, 1959.

Di Pierro, F., A. Callegari, D. Carotenuto, et al. "Clinical Efficacy, Safety, and Tolerability of BIO-C® (Micronized Silymarin) as a Galactagogue." *Acta Biomedica* 79 (2008): 205–210.

Dodge, Charles Richard. *A Report on the Culture of Hemp and Jute in the United States.* Washington, D.C.: Government Printing Office, 1896.

Dreiss, Meredith L., and Sharon Edgar Greenhill. *Chocolate: Pathway to the Gods.* Tucson: University of Arizona Press, 2008.

Duke, James. *Handbook of Medicinal Herbs.* 2nd ed. Boca Raton, Fla.: CRC Press, 2002.

——. "Utilization of Papaver." *Economic Botany* 27 (1973): 390–400.

Duncan, Daniel. *Wholesome Advice Against the Abuse of Hot Liquors, Particularly Coffee, Tea, Chocolate, Brandy, and Strong Waters.* London, 1706.

Eccles, Ronald. "Menthol: Effects on Nasal Sensation of Airflow and the Drive to Breathe." *Current Allergy and Asthma Reports* 3 (2003): 210–214.

Eck, Paul. *The American Cranberry.* New Brunswick, N.J.: Rutgers University Press, 1990.

Edkins, J. *Opium: Historical Note on the Poppy in China.* Shanghai: Statistical Department of the Inspectorate General of Customs, 1889.

Edwards, H. Milne, and P. Vavasseur. *Manuel de matière médicale.* Paris: Compère Jeune, 1826.

Eisenberg, D. M., E. S. J. Harris, B. A. Littlefield, et al. "Developing a Library of Authenticated Traditional Chinese Medicinal (TCM) Plants for Systematic Biological Evaluation—Rationale, Methods, and Preliminary Results from a Sino-American Collaboration." *Fitoterapia* 82 (2011): 17–33.

Ellingwood, Finley, and John Uri Lloyd. *American Materia Medica, Therapeutics, and Pharmacognosy.* Chicago: Ellingwood's Therapeutist, 1915.

Ellis, John. "A Catalogue of Such Foreign Plants as Are Worthy of Being Encouraged in the American Colonies, for the Purposes of Medicine, Agriculture, and Commerce." *Transactions of the American Philosophical Society* 1 (1769–1771): 255–266.

Ellis, Markman. *The Coffee House: A Cultural History.* London: Weidenfeld and Nicholson, 2004.

Elphick, Maurice R., and Michaela Egertová. "The Neurobiology and Evolution of Cannabinoid Signalling." *Philosophical Transactions of the Royal Society of London B* 356 (2001): 381–408.

El-Seedi, Hesham R., Peter De Smet, Olef Beck, et al. "Prehistoric Peyote: Alkaloid Analysis and Radiocarbon Dating of Archaeological Specimens of *Lophophora* from Texas." *Journal of Ethnopharmacology* 101 (2005): 238–242.

Elvin-Lewis, Memory. "Evolving Concepts Related to Achieving Benefit Sharing for Custodians of Traditional Knowledge." *Ethnobotany Research & Applications* 4 (2006): 75–96.

Elworthy, Frederick Thomas. *The Evil Eye: An Account of This Ancient and Widespread Superstition.* London: Murray, 1895.

Engels, Gayle. "Aloe." *HerbalGram* 87 (2010): 1–5.

——. "Cranberry." *HerbalGram* 76 (2007): 1–2.

——. "Dragon's Blood." *HerbalGram* 92 (2011): 1–4.

——. "Licorice." *HerbalGram* 70 (2006): 1, 4–5.

——. "Turmeric." *HerbalGram* 84 (2009): 1–3.

——. "Valerian." *HerbalGram* 79 (2008): 1–2.

Enjalbert, F., S. Rapior, J. Nouguier-Soulé, et al. "Treatment of Amatoxin Poisoning: Twenty-Year Retrospective Analysis." *Journal of Toxicology* 40 (2002): 715–757.

Ernst, Edzard. "Herbal Remedies for Anxiety—A Systematic Review of Controlled Clinical Trials." *Phytomedicine* 13 (2006): 205–208.

——. "The Importance of Having a Robust Evidence Base—A Personal View." In *Homeopathic Practice*, edited by Steven B. Kayne, 33–42. London: Pharmaceutical Press, 2008.

Etkin, Nina L. *Edible Medicines: An Ethnopharmacology of Food.* Tucson: University of Arizona Press, 2006.

——. *Foods of Association: Biocultural Perspectives on Foods and Beverages That Mediate Sociability.* Tucson: University of Arizona Press, 2009.

Evans, Jennifer R. "*Ginkgo biloba* Extract for Age-Related Macular Degeneration." *Cochrane Database of Systematic Reviews* (2013): CD001775.

Fabre, François Antoine. *Dictionnaire des dictionnaires de médecine, français et étranger.* Paris: Béthune et Plon, 1840.

Fairhurst, R. M., G. M. L. Nayyar, J. G. Breman, et al. "Artemisinin-Resistant Malaria: Research Challenges, Opportunities, and Public Health Implications." *American Journal of Tropical Medicine and Hygiene* 87 (2012): 231–241.

Fantegrossi, William E., Aeneas C. Murnane, and Chad J. Reissig. "The Behavioral Pharmacology of Hallucinogens." *Biochemical Pharmacology* 75 (2008): 17–33.

Farnsworth, Norman R. "Ethnopharmacology and Drug Development." In *Ethnobotany and the Search for New Drugs*, edited by Ghillean T. Prance, Derek J. Chadwick, and Joan Marsh, 42–59. Ciba Foundation Symposium 185. Chichester: Wiley, 1994.

Fasinu, Pius S., Patrick J. Bouic, and Bernd Rosenkranz. "An Overview of the Evidence and Mechanisms of Herb–Drug Interactions." *Frontiers in Pharmacology* 3 (2012): 69.

Feeny, Paul. "The Evolution of Chemical Ecology: Contributions from the Study of Herbivorous Insects." In *Herbivores: Their Interactions with Secondary Plant Metabolites*, 2nd ed., edited by G. A. Rosenthal and M. R. Berenbaum, 2:1–44. San Diego: Academic Press, 1992.

Fernández-San-Martín, Isabel, Roser Masa-Font, Laura Palacios-Soler, et al. "Effectiveness of Valerian on Insomnia: A Metaanalysis of Randomized Placebo-Controlled Trials." *Sleep Medicine* 11 (2010): 505–511.

Finniss, Damien G., Ted J. Kaptchuk, Franklin Miller, et al. "Placebo Effects: Biological, Clinical, and Ethical Advances." *Lancet* 375 (2010): 686–695.

Fiore, C., M. Eisenhut, R. Krausse, et al. "Antiviral Effects of Glycyrrhiza Species." *Phytotherapy Research* 22 (2008): 141–148.

Fiore, C., M. Eisenhut, E. Ragazzi, et al. "A History of the Therapeutic Use of Liquorice in Europe." *Journal of Ethnopharmacology* 99 (2005): 317–324.

Flemming, Percy. "The Medical Aspects of the Medieval Monastery in England." *Proceedings of the Royal Society of Medicine* 22 (1929): 771–782.

Flückiger, Friedrich A., and Daniel Hanbury. *Pharmacographia: A History of the Principal Drugs of Vegetable Origin Met with in Great Britain and British India*. 2nd ed. London: Macmillan, 1879.

Foo, L. Y., Y. Lu, A. B. Howell, et al. "A-Type Proanthocyanidin Trimers from Cranberry That Inhibit Adherence of Uropathogenic P-fimbriated *Escherichia coli*." *Journal of Natural Products* 63 (2000): 1225–1228.

Food and Drug Administration. *Draft Guidance for Industry: Dietary Supplements: New Dietary Ingredient Notifications and Related Issues*. July 2011. http://www.fda.gov/food/guidanceregulation/guidancedocumentsregulatoryinformation/dietarysupplements/ucm257563.htm.

——. *FDA 101: Dietary Supplements*. August 4, 2008. http://www.fda.gov/ForConsumers/ConsumerUpdates/ucm050803.htm.

Foster, Steven. "Black Cohosh: A Literature Review." *HerbalGram* 45 (1999): 35–50.

——. "A Brief History of Adulteration of Herbs, Spices, and Botanical Drugs." *HerbalGram* 92 (2011): 42–57.

——. "Exploring the Peripatetic Maze of Black Cohosh Adulteration: A Review of the Nomenclature, Distribution, Chemistry, Market Status, Analytical Methods, and Safety." *HerbalGram* 98 (2013): 32–51.

Fox, William, and Joseph Nadin. *The Working Man's Family Botanic Guide; or, Every Man His Own Doctor*. Sheffield: Dawsons, 1852.

Francis, John K. *Coffea arabica*. Department of Agriculture, Forest Service, International Institute of Tropical Forestry. http://www.fs.fed.us/global/iitf/pdf/shrubs/Coffea%20arabica.pdf.

Fransen, H. P., S. M. G. J. Pelgrom, B. Stewart-Knox, et al. "Assessment of Health Claims, Content, and Safety of Herbal Supplements Containing *Ginkgo biloba*." *Food & Nutrition Research* 54 (2010): 5221.

Fraser, Thomas R. "Lecture on the Antagonism Between the Actions of Active Substances." *British Medical Journal* 2, no. 617 (1872): 457–459.

——. "Lecture on the Antagonism Between the Actions of Active Substances." *British Medical Journal* 2, no. 618 (1872): 485–487.

Fraser-Harris, D. "The Former Importance of Our Sea-Borne Trade in Drugs." *Canadian Medical Association Journal* 18 (1928): 468.

Freud, Sigmund. *Cocaine Papers*. Edited by Robert Byck. New York: Stone Hill, 1974.

Fried, M. W., V. J. Navarro, N. Afdhal, et al. "Effect of Silymarin (Milk Thistle) on Liver Disease in Patients with Chronic Hepatitis C Unsuccessfully Treated with Interferon Therapy: A Randomized Controlled Trial." *Journal of the American Medical Association* 308 (2012): 274–282.

Furst, Peter T., ed. *Flesh of the Gods: The Ritual Use of Hallucinogens*. New York: Praeger, 1972.

Fürst, Susanna, and Sándor Hosztafi. "Pharmacology of Poppy Alkaloids." In *Poppy: The Genus Papaver*, edited by Jenö Bernáth, 291–318. Amsterdam: Harwood, 1998.

Futuyma, Douglas J., and Anurag A. Agrawal. "Macroevolution and the Biological Diversity of Plants and Herbivores." *Proceedings of the National Academy of Sciences USA* 106 (2009): 18054–18061.

Gagnier, Joel J., Heather Boon, Paula Rochon, et al., for the CONSORT Group. "Recommendations for Reporting Randomized Controlled Trials of Herbal Interventions: Explanation and Elaboration." *Journal of Clinical Epidemiology* 59 (2006): 1134–1149.

Gagnier, Joel J., David Moher, Heather Boon, et al. "Randomized Controlled Trials of Herbal Interventions Underreport Important Details of the Intervention." *Journal of Clinical Epidemiology* 64 (2011): 760–769.

Gahlinger, Paul. *Illegal Drugs: A Complete Guide to Their History, Chemistry, Use, and Abuse*. New York: Plume, 2004.

Galanter, Joshua M., Susannah B. Cornes, and Daniel H. Lowenstein. "Principles of Nervous System Physiology and Pharmacology." In *Principles of Pharmacology: The Pathophysiologic Basis of Drug Therapy*, 3rd ed., edited by David E. Golan, Armen H. Tashjian, Ehrin J. Armstrong, et al., 93–108. Philadelphia: Lippincott Williams & Wilkins, 2012.

Gardner, C. D., L. D. Lawson, E. Block, et al. "Effect of Raw Garlic Versus Commercial Garlic Supplements on Plasma Lipid Concentrations in Adults with Moderate Hypercholesterolemia: A Randomized Clinical Trial." *Archives of Internal Medicine* 167 (2007): 346–353.

Garrett, J. T. *The Cherokee Herbal: Native Plant Medicine from the Four Directions*. Rochester, Vt.: Bear, 2003.

Gates, William. *An Aztec Herbal: The Classic Codex of 1552*. Toronto: Dover, 1939.

Gautier, A. *Manuel des plantes médicinales*. Paris: Audot, 1822.

Geng, J., J. Dong. H. Ni, et al. "Ginseng for Cognition." *Cochrane Database of Systematic Reviews* (2010): CD007769.

Gerard, John. *The Herball, or Generall Historie of Plants*. London, 1597 and 1633.

Gershenzon, Jonathan, and Rodney Croteau. "Terpenoids." In *Herbivores: Their Interactions with Secondary Plant Metabolites*, 2nd ed., edited by Gerald A. Rosenthal and May R. Berenbaum, 1:165–219. San Diego: Academic Press, 1991.

Gertsch, Jürg. "Botanical Drugs, Synergy, and Network Pharmacology: Forth and Back to Intelligent Mixtures." *Planta Medica* 77 (2011): 1086–1098.

Gilman, Edward F. "*Serenoa repens* Saw Palmetto." University of Florida Extension Publication FPS-547. October 1999. http://edis.ifas.ufl.edu/fp547.

Goenka, P., A. Sarawgi, V. Karun, et al. "*Camellia sinensis* (Tea): Implications and Role in Preventing Dental Decay." *Pharmacognosy Review* 7, no. 14 (2013): 152–156.

Gold, Lois Swirsky, and Thomas H. Slone. "Aristolochic Acid, an Herbal Carcinogen, Sold on the Web After FDA Alert." *New England Journal of Medicine* 349 (2003): 1576–1577.

Gold, M. S. "Cocaine (and Crack): Clinical Aspects." In *Substance Abuse: A Comprehensive Textbook*, 3rd ed., edited by J. H. Lowinson, P. Ruiz, R. B. Millman, et al., 181–198. Baltimore: Williams & Wilkins, 1997.

Gomila, Frank R., and Madeline C. Gomila Lambou. "Present Status of the Marihuana Vice in the United States." In *Marihuana: America's New Drug Problem*, by Robert P. Walton, 27–39. Philadelphia: Lippincott, 1938.

Gonzales de Mejia, Elvira, Sirima Puangpraphant, and Rachel Eckhoff. "Tea and Inflammation." In *Tea in Health and Disease Prevention*, edited by Victor R. Preedy, 563–580. London: Elsevier, 2013.

González, José Antonio, Mónica García-Barriuso, Manual Pardo-de-Santayana, et al. "Plant Remedies Against Witches and the Evil Eye in a Spanish 'Witches' Village.'" *Economic Botany* 66 (2012): 35–45.

González-Castejón, Marta, Francesco Visioli, and Arantxa Rodriguez-Casado. "Diverse Biological Activities of Dandelion." *Nutrition Reviews* 70 (2012): 534–547.

Goodman, Jordan. *Tobacco in History: The Cultures of Dependence*. London: Routledge, 1993.

Gootenberg, Paul. *Andean Cocaine: The Making of a Global Drug*. Chapel Hill: University of North Carolina Press, 2008.

Government Accountability Office. "Herbal Dietary Supplements: Examples of Deceptive or Questionable Marketing Practices and Potentially Dangerous Advice." Testimony Before the Special Committee on Aging, U.S. Senate. GAO-10-662T. 2010.

——. "Herbal Dietary Supplements: FDA May Have Opportunities to Expand Its Use of Reported Health Problems to Oversee Products." Report to Congressional Request ers. GAO-13-244. 2013.

——. "Herbal Dietary Supplements: FDA Should Take Further Actions to Improve Oversight and Consumer Understanding." Report to Congressional Requesters. GAO-09-250. 2009.

Grace, M. H., A. R. Massey, F. Mbeunkui, et al. "Comparison of Health-Relevant Flavonoids in Commonly Consumed Cranberry Products." *Journal of Food Science* 77 (2012): H176–H183.

Grant, Mark. *Galen on Food and Diet*. London: Routledge, 2000.

Greenberg, Michael. *British Trade and the Opening of China, 1800–1842*. Cambridge: Cambridge University Press, 1951.

Grivetti, Louis Evan. "Medicinal Chocolate in New Spain, Western Europe, and North America." In *Chocolate: History, Culture, and Heritage*, edited by Louis Evan Grivetti and Howard-Yana Shapiro, 67–88. Hoboken, N.J.: Wiley, 2009.

Grollman, A. P., S. Shibutani, M. Moriya, et al. "Aristolochic Acid and the Etiology of Endemic (Balkan) Nephropathy." *Proceedings of the National Academy of Sciences USA* 104 (2007): 12129–12134.

Groom, S. N., T. Johns, and P. R. Oldfield. "The Potency of Immunomodulatory Herbs May Be Primarily Dependent upon Macrophage Activation." *Journal of Medicinal Food* 10 (2007): 73–79.

Gruenwald, Joerg, Thomas Brendler, and Christof Jaenicke, eds. *PDR for Herbal Medicines*. 4th ed. Montvale, N.J.: Thomson, 2007.

Guay, A. T., R. F. Spark, J. Jacobson, et al. "Yohimbine Treatment of Organic Erectile Dysfunction in a Dose-Escalation Trial." *International Journal of Impotence Research* 14 (2004): 25–31.

Guay, David R. P. "Cranberry and Urinary Tract Infections." *Drugs* 69 (2009): 775–807.

Gupta, Subash C., Gorkem Kismali, and Bharat B. Aggarwal. "Curcumin, a Component of Turmeric: From Farm to Pharmacy." *BioFactors* 39 (2013): 2–13.

Gurley, B. J., E. K. Fifer, and Z. Gardner. "Pharmacokinetic Herb–Drug Interactions (Part 2): Drug Interactions Involving Popular Botanical Dietary Supplements and Their Clinical Relevance." *Planta Medica* 78 (2012): 1490–1514.

Habib, F. K. "*Serenoa repens*: The Scientific Basis for the Treatment of Benign Prostatic Hyperplasia." *European Urology Supplements* 8 (2009): 887–893.

Habib, F. K., and M. G. Wyllie. "Not All Brands Are Created Equal: A Comparison of Selected Components of Different Brands of *Serenoa repens* Extract." *Prostate Cancer and Prostatic Diseases* 7 (2004): 195–200.

Hale, Edwin M. *Saw Palmetto (Sabal Serrulata, Serenoa Serrulata): Its History, Botany, Chemistry, Pharmacology, Provings, Clinical Experience, and Therapeutic Applications*. Philadelphia: Boericke and Tafel, 1898.

Halioula, Bruno, and Bernard Ziskind. *Medicine in the Days of the Pharaohs*. Translated by M. B. DeBevoise. Cambridge, Mass.: Harvard University Press, 2005.

Haller, John S. *Medical Protestants: The Eclectics in American Medicine, 1825–1939*. Carbondale: Southern Illinois University Press, 1994.

Hamel, Paul B., and Mary U. Chiltoskey. *Cherokee Plants: Their Uses—A Four-Hundred-Year History.* Sylva, N.C.: Herald, 1975.

Hamilton, A. C., ed. *Medicinal Plants in Conservation and Development: Case Studies and Lessons Learnt.* Salisbury: Plantlife International, 2008.

Hamman, Josias H. "Composition and Applications of *Aloe vera* Leaf Gel." *Molecules* 13 (2008): 1599–1616.

Hammond, David, and Jessica Reid. "Health Warnings on Tobacco Products: International Practices." *Salud pública de México* 54 (2012): 270–280.

Hara, Yukihiko. "Tea Catechins and Their Applications as Supplements and Pharmaceutics." *Pharmacological Research* 64 (2011): 100–104.

Harlev, E., E. Nevo, E. Lansky, et al. "Anticancer Potential of Aloes: Antioxidant, Antiproliferative, and Immunostimulatory Attributes." *Planta Medica* 78 (2012): 843–852.

Harper, Donald J. *Early Chinese Medical Literature: The Mawangdui Medical Manuscripts.* London: Kegan Paul, 1998.

Harris, E. S. J., S. Cao, B. A. Littlefield, et al. "Heavy Metal and Pesticide Content in Commonly Prescribed Individual Raw Chinese Herbal Medicines." *Science of the Total Environment* 409 (2011): 4297–4305.

Harris, J. C., S. L. Cottrell, S. Plummer, et al. "Antimicrobial Properties of *Allium sativum* (Garlic)." *Applied Microbiology and Biotechnology* 57 (2001): 282–286.

Hartmann, Thomas. "Alkaloids." In *Herbivores: Their Interactions with Secondary Plant Metabolites*, 2nd ed., edited by Gerald A. Rosenthal and May R. Berenbaum, 1:79–121. San Diego: Academic Press, 1991.

Hattox, Ralph S. *Coffee and Coffeehouses: The Origins of a Social Beverage in the Medieval Near East.* Seattle: University of Washington Press, 1985.

Hauptman, Paul, and Ralph Kelly. "Digitalis." *Circulation* 99 (1999): 1265–1270.

Heinrich, Michael, Sarah Edwards, Daniel E. Moerman, et al. "Ethnopharmacological Field Studies: A Critical Assessment of Their Conceptual Basis and Methods." *Journal of Ethnopharmacology* 124 (2009): 1–17.

Heiss, Mary Lou, and Robert J. Heiss. *The Story of Tea: A Cultural History and Drinking Guide.* Berkeley, Calif.: Ten Speed Press, 2007.

Helfand, William H. "Mariani et le Vin de Coca." *Revue d'histoire de pharmacie* 247 (1980): 227–234.

Henderer, Jeffrey D., and Christopher J. Rapuano. "Ocular Pharmacology." In *Goodman and Gilman's The Pharmacological Basis of Therapeutics*, 12th ed., edited by Laurence L. Brunton, Bruce A. Chabner, and Björn C. Knollmann, 1773–1802. New York: McGraw-Hill, 2011.

Henderson, J. S., and R. A. Joyce. "Brewing Distinction: The Development of Cacao Beverage in Formative Mesoamerica." In *Chocolate in Mesoamerica: A Cultural History of Cacao*, edited by Cameron L. McNeil, 140–153. Gainesville: University Press of Florida, 2006.

Henderson, J. S., R. A. Joyce, G. R. Hall, et al. "Chemical and Archaeological Evidence for the Earliest Cacao Beverages." *Proceedings of the National Academy of Sciences USA* 104 (2007): 18937–18940.

Henn, Brenna M., L. L. Cavalli-Sforza, and Marcus W. Feldman. "The Great Human Expansion." *Proceedings of the National Academy of Sciences USA* 109 (2012): 17758–17764.

Hermann, F., L. E. Spieker, F. Ruschitzka, et al. "Dark Chocolate Improves Endothelial and Platelet Function." *Heart* 92 (2006): 119–120.

Hermann, Robert, and Oliver von Richter. "Clinical Evidence of Herbal Drugs as Perpetrators of Pharmacokinetic Drug Interactions." *Planta Medica* 78 (2012): 1458–1477.

Hernández, Francisco. *Historia natural de Nueva España.* Vol. 1. Mexico City: Universidad Nacional de México, 1959.

Herodotus. Translated by A. D. Godley. Vol. 1. New York: Putnam, 1920.

Herrick, James W. *Iroquois Medical Botany.* Edited by Dean R. Snow. Syracuse, N.Y.: Syracuse University Press, 1997.

Hewson, Mariana G. "Indigenous Knowledge Systems: Southern African Healing." In *Fundamentals of Complementary and Alternative Medicine*, 4th ed., edited by Marc S. Micozzi, 522–530. St. Louis: Saunders Elsevier, 2011.

Hibbs, Ryan E., and Alexander C. Zambon. "Agents Acting at the Neuromuscular Junction and Autonomic Ganglia." In *Goodman and Gilman's The Pharmacological Basis of Therapeutics*, 12th ed., edited by Laurence L. Brunton, Bruce A. Chabner, and Björn C. Knollmann, 255–276. New York: McGraw-Hill, 2011.

Hickey, Steve, and Hilary Roberts. *Tarnished Gold: The Sickness of Evidence-Based Medicine.* CreateSpace, 2011.

Hill, A. Bradford. "The Clinical Trial." *New England Journal of Medicine* 247, no. 4 (1952): 113–119.

Hill, A. J., C. M. Williams, B. J. Whalley, et al. "Phytocannabinoids as Novel Therapeutic Agents in CNS Disorders." *Pharmacology & Therapeutics* 133 (2012): 79–97.

Hill, John. *The Family Herbal.* Bungay, Eng.: Brightly, 1812.

Hillig, Karl W., and Paul G. Mahlberg. "A Chemotaxonomic Analysis of Cannabinoid Variation in *Cannabis* (Cannabaceae)." *American Journal of Botany* 91 (2004): 966–975.

Hilton, Malcolm P., Eleanor F. Zimmerman, and William T. Hunt. "*Ginkgo biloba* for Tinnitus." *Cochrane Database of Systematic Reviews* (2013): CD003852.

Hilts, Philip J. *Protecting America's Health: The FDA, Business, and One Hundred Years of Regulation.* New York: Knopf, 2003.

Ho, Christopher C. K., and Hui Meng Tan. "Rise of Herbal and Traditional Medicine in Erectile Dysfunction Management." *Current Urology Reports* 12 (2011): 470–478.

Ho, S. C., J. L. F. Woo, S. S. F. Leung, et al. "Intake of Soy Products Is Associated with Better Plasma Lipid Profiles in the Hong Kong Chinese Population." *Journal of Nutrition* 130 (2000): 2590–2593.

Hobbs, Christopher. "Echinacea: A Literature Review. Botany, History, Chemistry, Pharmacology, Toxicology, and Clinical Uses." *HerbalGram* 30 (1994): 33.

——. "St. John's Wort: Ancient Herbal Protector." *Pharmacy History* 32 (1990): 166.

Hobbs, Joseph J. "Troubling Fields: The Opium Poppy in Egypt." *Geographical Review* 88 (1998): 64–85.

Hoffman, Brian B., and Freddie M. Williams. "Adrenergic Pharmacology." In *Principles of Pharmacology: The Pathophysiologic Basis of Drug Therapy*, 3rd ed., edited by David E. Golan, Armen H. Tashjian, Ehrin J. Armstrong, et al., 132–146. Philadelphia: Lippincott Williams & Wilkins, 2012.

Hoffmann, Dietrich, and Ilse Hoffmann. "The Changing Cigarette: Chemical Studies and Bioassays." In *Risks Associated with Smoking Cigarettes with Low Machine-Measured Yields of Tar and Nicotine*, 159–191. Smoking and Tobacco Control Monograph 13. Bethesda, Md.: Department of Health and Human Services, Public Health Service, National Institutes of Health, 2001.

——. "Chemistry and Toxicology." In *Cigars: Health Effects and Trends*, 55–104. Smoking and Tobacco Control Monograph 9. Bethesda, Md.: Department of Health and Human Services, Public Health Service, National Institutes of Health, 1998.

Hoffman, Freddie Ann, and Steven R. Kishter. "Botanical New Drug Applications—The Final Frontier." *HerbalGram* 99 (2013): 66–69.

Hohenegger, Beatrice. *Liquid Jade: The Story of Tea from East to West*. New York: St. Martin's Press, 2006.

Höld, Karin M., Nilantha Sirisoma, Tomoko Ikeda, et al. "Alpha-Thujone (the Active Component of Absinthe): Gamma-Aminobutyric Acid Type A Receptor Modulation and Metabolic Detoxification." *Proceedings of the National Academy of Sciences USA* 97, no. 8 (2000): 3826–3831.

Holland, Bart K. "Prospecting for Drugs in Ancient Texts." *Nature* 369 (1994): 702.

Homan, Peter G. "The Development of Community Pharmacy." In *Making Medicines: A Brief History of Pharmacy and Pharmaceuticals*, edited by Stuart Anderson, 115–134. London: Pharmaceutical Press, 2005.

Homan, Peter G., Briony Hudson, and Raymond Rowe. *Popular Medicines: An Illustrated History*. London: Pharmaceutical Press, 2008.

Hosking, R. D., and J. P. Zajicek. "Therapeutic Potential of Cannabis in Pain Medicine." *British Journal of Anaesthesia* 101, no. 1 (2008): 59–68.

Howell, Amy B., Henry Botto, Christophe Combescure, et al. "Dosage Effect on Uropathogenic *Escherichia coli* Antiadhesion Activity in Urine Following Consumption of Cranberry Powder Standardized for Proanthocyanidin Content: A Multicentric Randomized Double-Blind Study." *BMC Infectious Diseases* 10 (2010): 94.

Hsu, Elisabeth. "*Qing hao* 青蒿 (*Herba Artemisiae annuae*) in the Chinese *Materia Medica*." In *Plants, Health, and Healing: On the Interface of Ethnobotany and Medical Anthropology*, edited by Elisabeth Hsu and Stephen Harris, 83–133. New York: Berghahn, 2010.

Huang, H. T. "Early Uses of Soybean in Chinese History." In *The World of Soy*, edited by Christine M. Du Bois, Chee-Beng Tan, and Sidney W. Mintz, 45–55. Urbana: University of Illinois Press, 2008.

Hudson, James B. "Applications of the Phytomedicine *Echinacea purpurea* (Purple Coneflower) in Infectious Diseases." *Journal of Biomedicine and Biotechnology* (2012): ID 769896.

Huestis, Marilyn A., and Michael L. Smith. "Human Cannabinoid Pharmacokinetics and Interpretation of Cannabinoid Concentrations in Biological Fluids and Tissues." In *Marijuana and the Cannabinoids*, edited by Mahmoud A. ElSohly, 205–236. Totowa, N.J.: Humana, 2007.

Hughes, George. "Amoy Trade Report for the Year 1870." In *Reports on Trade at the Treaty Ports in China for the Year 1870*, 83–104. Shanghai: Customs Press, 1871.

Hughes, Samuel C. "Intraspinal Narcotics for Obstetric Analgesia." *Western Journal of Medicine* 162, no. 1 (1995): 54–55.

Hutchison, Annie, Richard Farmer, Katia Verhamme, et al. "The Efficacy of Drugs for the Treatment of LUTS/BPH: A Study in Six European Countries." *European Urology* 51 (2007): 207–216.

Huxley, Aldous. *The Doors of Perception*. New York: Harper, 1954.

Hymowitz, Theodore. "On the Domestication of the Soybean." *Economic Botany* 24, no. 4 (1970): 408–421.

——. "Speciation and Genetics." In *Soybeans: Improvement, Production, and Uses*, 3rd ed., edited by H. Roger Boerma and James E. Specht, 97–136. Madison, Wis.: American Society of Agronomy, Crop Science Society of America, Soil Science Society, 2004.

Hymowitz, Theodore, and W. R. Shurtleff. "Debunking Soybean Myths and Legends in the Historical and Popular Literature." *Crop Science* 45 (2005): 473–476.

Ibrahim, Mohamed Ali, MinKyun Na, Joonseok Oh, et al. "Significance of Endangered and Threatened Plant Natural Products in the Control of Human Disease." *Proceedings of the National Academy of Sciences USA* 110 (2013): 16832–16837.

Institute of Medicine. *Complementary and Alternative Medicine in the United States*. Washington, D.C.: National Academies Press, 2005.

Izzo, Angelo A., Francesca Borrelli, Raffaele Capasso, et al. "Nonpsychotropic Plant Cannabinoids: New Therapeutic Opportunities from an Ancient Herb." *Trends in Pharmacological Sciences* 30 (2009): 515–527.

Jang, Dong-Jin, Sun Tae Kim, Euichaul Oh, et al. "Enhanced Oral Bioavailability and Antiasthmatic Efficacy of Curcumin Using Redispersible Dry Emulsion." *Bio-Medical Materials and Engineering* 24 (2014): 917–930.

Janke, Rhonda, and Jeanie DeArmond. *A Grower's Guide: Milk Thistle* Silybum marianum. Kansas State University Research and Extension Publication MF-2618. May 2004.

Jartoux, Pierre. "The Description of a Tartarian Plant, Called Ginseng; with an Account of Its Virtues . . . at Pekin, April 12, 1711." *Philosophical Transactions of the Royal Society of London* 28 (1713): 56–61.

Jawad, M., R. Schoop, A. Suter, et al. "Safety and Efficacy Profile of *Echinacea purpurea* to Prevent Common Cold

Episodes: A Randomized, Double-Blind, Placebo-Controlled Trial." *Evidence-Based Complementary and Alternative Medicine* (2012): ID 841315.

Jepson, Ruth G., Gabrielle Williams, and Jonathan C. Craig. "Cranberries for Preventing Urinary Tract Infections." *Cochrane Database of Systematic Reviews* (2012): CD001321.

Jia, Lee, and Yuqing Zhao. "Current Evaluation of the Millennium Phytomedicine—Ginseng (I): Etymology, Pharmacognosy, Phytochemistry, Market, and Regulations." *Current Medicinal Chemistry* 16 (2009): 2475–2484.

Jia, Lee, Yuqing Zhao, and Xing-Jie Liang. "Current Evaluation of the Millennium Phytomedicine—Ginseng (II): Collected Chemical Entities, Modern Pharmacology, and Clinical Applications Emanated from Traditional Chinese Medicine." *Current Medicinal Chemistry* 16 (2009): 2924–2942.

Johnson, Emanuel, James Saunders, Sue Mischke, et al. "Identification of *Erythroxylum* Taxa by AFLP DNA Analysis." *Phytochemistry* 64 (2003): 187–197.

Johnson, Rebecca L., Steven Foster, Tieraona Low Dog, et al. *National Geographic Guide to Medicinal Herbs: The World's Most Effective Healing Plants.* Washington, D.C.: National Geographic, 2010.

Johnson, Teresa L., and Jed W. Fahey. "Black Cohosh: Coming Full Circle?" *Journal of Ethnopharmacology* 141 (2012): 775–779.

Johnson, Tom. "A Rough Pictorial Guide to the Roast Process." 2013. Sweet Maria's. http://www.sweetmarias.com/roasted.pict-guide.php.

Jordan, Mary Ann, Robert J. Toso, Doug Thrower, et al. "Mechanism of Mitotic Block and Inhibition of Cell Proliferation by Taxol at Low Concentrations." *Proceedings of the National Academy of Sciences USA* 90 (1993): 9552–9556.

Jordan, Scott A., David G. Cunningham, and Robin J. Marles. "Assessment of Herbal Medicinal Products: Challenges and Opportunities to Increase the Knowledge Base for Safety Assessment." *Toxicology and Applied Pharmacology* 243 (2010): 198–216.

Josselyn, John. *New England's Rarities Discovered in Birds, Beasts, Fishes, Serpents, and Plants of That Country.* 1672. Boston: William Veazie, 1865.

Julien, Robert M., Claire Advokat, and Joseph E. Comaty. *A Primer of Drug Action.* 12th ed. New York: Worth, 2011.

Kaefer, Christine M., and John A. Milner. "The Role of Herbs and Spices in Cancer Prevention." *Journal of Nutritional Biochemistry* 19 (2008): 347–361.

Kandel, Eric R., J. H. Schwartz, T. M. Jessell, et al., eds. *Principles of Neural Science.* 5th ed. New York: McGraw-Hill, 2013.

Kao, Tzu-Chien, Chi-Hao Wu, and Gow-Chin Yen. "Bioactivity and Potential Health Benefits of Licorice." *Journal of Agricultural and Food Chemistry* 62 (2014): 542–553.

Karch, Steven B. *A Brief History of Cocaine.* 2nd ed. Boca Raton, Fla.: CRC Press, 2005.

Karunagaran, Devarajan, Jeena Joseph, and R. Kumar Santhosh Thankayyan. "Cell Growth Regulation." In *The Molecular Targets and Therapeutic Uses of Curcumin in Health and Disease,* edited by Bharat B. Aggarwal, Young-Joon Surh, and Shishir Shishodia, 245–268. New York: Springer, 2007.

Kaufman, Terrence, and John Justeson. "The History of the Word for 'Cacao' and Related Terms in Ancient Mesoamerica." In *Chocolate in Mesoamerica: A Cultural History of Cacao,* edited by Cameron L. McNeil, 117–139. Gainesville: University Press of Florida, 2006.

Kaul, M. K. *Medicinal Plants of Kashmir and Ladakh: Temperate and Cold Arid Himalaya.* New Delhi: Indus, 1997.

Kaye, Alan S. "The Etymology of 'Coffee': The Dark Brew." *Journal of the American Oriental Society* 106 (1986): 557–558.

Kayne, Steven B. "Homeopathy—An Overview." In *Homeopathic Practice,* edited by Steven B. Kayne, 1–32. London: Pharmaceutical Press, 2008.

——, ed. *Traditional Medicine: A Global Perspective.* London: Pharmaceutical Press, 2010.

Kennedy, David O. *Plants and the Human Brain.* Oxford: Oxford University Press, 2014.

Kennedy, David O., and Emma L. Wightman. "Herbal Extracts and Phytochemicals: Plant Secondary Metabolites and the Enhancement of Human Brain Function." *Advances in Nutrition* 2 (2011): 32–50.

Khanna, Dinesh, Gautam Sethi, Kwang Seok Ahn, et al. "Natural Products as a Gold Mine for Arthritis Treatment." *Current Opinion in Pharmacology* 7 (2007): 344–351.

Kiehn, Monika. "Silphion Revisited." *Medicinal Plant Conservation* 13 (2007): 4–8.

Kilham, Chris. *Kava: Medicine Hunting in Paradise.* Rochester, Vt.: Park Street Press, 1996.

Kindscher, Kelly. "Ethnobotany of Purple Coneflower (*Echinacea angustifolia,* Asteraceae) and Other Echinacea Species." *Economic Botany* 43, no. 4 (1989): 498–507.

King, John, and Robert S. Newton. *The Eclectic Dispensatory of the United States of America.* Cincinnati: Derby, 1852.

Kneifel, Wolfgang, Erich Czech, and Brigitte Kopp. "Microbial Contamination of Medicinal Plants—A Review." *Planta Medica* 68 (2002): 5–15.

Koerper, Henry, and A. L. Kolls. "The Silphium Motif Adorning Ancient Libyan Coinage: Marketing a Medicinal Plant." *Economic Botany* 53, no. 2 (1999): 133–143.

Konadu, Kwasi. *Indigenous Medicine and Knowledge in African Society.* New York: Routledge, 2007.

Kozlowski, Lynn, Richard O'Connor, and Christine Sweeney. "Cigarette Design." In *Risks Associated with Smoking Cigarettes with Low Machine-Measured Yields of Tar and Nicotine,* 13–37. Smoking and Tobacco Control Monograph 13. Bethesda, Md.: Department of Health and Human Services, Public Health Service, National Institutes of Health, 2001.

Kritikos, P. G., and S. P. Papadaki. "The History of the Poppy and of Opium and Their Expansion in Antiquity in the Eastern Mediterranean Area." *Bulletin on Narcotics* 3 (1967): 17–38.

——. "The History of the Poppy and of Opium and Their Expansion in Antiquity in the Eastern Mediterranean Area." *Bulletin on Narcotics* 4 (1967): 5–10.

Kuhnert, Nikolai. "Chemistry and Biology of the Black Tea Thearubigins and of Tea Fermentation." In *Tea in Health and Disease Prevention*, edited by Victor R. Preedy, 343–360. London: Elsevier, 2013.

Kyo, Eikai, Naoto Uda, Shigeo Kasuga, et al. "Immunomodulatory Effects of Aged Garlic Extract." *Journal of Nutrition* 131 (2001): 1075S–1079S.

La Barre, Weston. *The Peyote Cult.* 1938. Hamden, Conn.: Archon, 1975.

——. "Twenty Years of Peyote Studies." *Current Anthropology* 1, no. 1 (1960): 45–60.

Lachenmeier, Dirk W., David Nathan-Maister, Theodora Breaux, et al. "Chemical Composition of Vintage Preban Absinthe with Special Reference to Thujone, Fenchone, Pinocamphone, Methanol, Copper, and Antimony Concentrations." *Journal of Agricultural and Food Chemistry* 56 (2008): 3073–3081.

Lachenmeier, Dirk W., Constanze Sproll, and Frank Musshoff. "Poppy Seed Foods and Opiate Drug Testing—Where Are We Today?" *Therapeutic Drug Monitoring* 32 (2010): 11–18.

Lafitau, Joseph François. *Mémoire présentée à son altesse royale monseigneur le Duc d'Orléans, régent du royaume de France, concernant la précieuse plante du gin seng de Tartarie, découverte en Canada.* Paris: Joseph Mongé, 1718.

Laizure, S. C., T. Mandrell, N. M. Gades, et al. "Cocaethylene Metabolism and Interaction with Cocaine and Ethanol: Role of Carboxylesterases." *Drug Metabolism and Disposition* 31 (2003): 16–20.

Lau, Benjamin H. S. "Suppression of LDL Oxidation by Garlic Compounds Is a Possible Mechanism of Cardiovascular Health Benefit." *Journal of Nutrition* 136 (2006): 765S–768S.

Lau, Kit-Man, Kwok-Kin Lai, Cheuk-Lun Liu, et al. "Synergistic Interaction Between Astragali Radix and Rehmanniae Radix in a Chinese Herbal Formula to Promote Diabetic Wound Healing." *Journal of Ethnopharmacology* 141 (2012): 250–256.

Leach, Matthew J., and S. Kumar. "The Clinical Effectiveness of Ginger (*Zingiber officinale*) in Adults with Osteoarthritis." *International Journal of Evidence Based Healthcare* 6 (2008): 311–320.

Leach, Matthew J., and Vivienne Moore. "Black Cohosh (*Cimicifuga* spp.) for Menopausal Symptoms." *Cochrane Database of Systematic Reviews* (2012): CD007244.

Lebot, Vincent, Mark Merlin, and Lamont Lindstrom. *Kava: The Pacific Elixir.* Rochester, Vt.: Healing Arts Press, 1997. [Previously published as *Kava: The Pacific Drug.* New Haven, Conn.: Yale University Press, 1992]

Lee, M. R. "Curare: The South American Arrow Poison." *Journal of the Royal College of Physicians, Edinburgh* 35 (2005): 83–92.

——. "The History of Ephedra (ma-huang)." *Journal of the Royal College of Physicians, Edinburgh* 41 (2011): 78–84.

——. "Ipecacuanha: The South American Vomiting Root." *Journal of the Royal College of Physicians, Edinburgh* 38 (2008): 355–360.

Lee, Peter. *Opium Culture: The Art and Ritual of the Chinese Tradition.* Rochester, Vt.: Park Street Press, 2006.

Leigh, John. *An Experimental Investigation into the Properties of Opium and Its Effects on Living Subjects.* Edinburgh: Charles Elliot, 1786.

Leonard, Irving. "Peyote and the Mexican Inquisition, 1620." *American Anthropologist* 44, no. 2 (1942): 324–326.

Lersten, Nels R., and John B. Carlson. "Vegetative Morphology." In *Soybeans: Improvement, Production, and Uses*, 3rd ed., edited by H. Roger Boerma and James E. Specht, 15–58. Madison, Wis.: American Society of Agronomy, Crop Science Society of America, Soil Science Society, 2004.

Levis, Silvina, and Marcio L. Griebeler. "The Role of Soy Foods in the Treatment of Menopausal Symptoms." *Journal of Nutrition* 140 (2010): 2318S–2321S.

Lewin, Louis. *Phantastica: A Classic Survey on the Use and Abuse of Mind-Altering Plants.* Translated by P. H. A. Wirth. 1924. Rochester, Vt.: Park Street Press, 1998.

Lewis, Walter, and Memory P. F. Elvin-Lewis. *Medical Botany: Plants Affecting Human Health.* Hoboken, N.J.: Wiley, 2003.

Li, Hui-Lin. "The Origin and Use of Cannabis in Eastern Asia: Their Linguistic-Cultural Implications." In *Cannabis and Culture*, edited by Vera Rubin, 51–62. The Hague: Mouton, 1975.

Li, Jesse W.-H., and John C. Vederas. "Drug Discovery and Natural Products: End of an Era or an Endless Frontier?" *Science* 325 (2009): 161–165.

Linde, Klaus. "St. John's Wort—An Overview." *Forschende Komplementärmedizin* 16 (2009): 146–155.

Linde, Klaus, Michael M. Berner, and Levente Kriston. "St. John's Wort for Major Depression." *Cochrane Database of Systematic Reviews* (2008): CD000448.

Lindstrom, Lamont. "History, Folklore, Traditional and Current Uses of Kava." In *Kava: From Ethnology to Pharmacology*, edited by Yadhu H. Singh, 10–28. Boca Raton, Fla.: CRC Press, 2004.

Lissiman, Elizabeth, A. L. Bhasale, and M. Cohen. "Garlic for the Common Cold." *Cochrane Database of Systematic Reviews* (2012): CD006206.

Liu, Baohua, L. Qin, A. Liu, et al. "Prevalence of the Equol-Producer Phenotype and Its Relationship with Dietary Isoflavone and Serum Lipids in Healthy Chinese Adults." *Journal of Epidemiology* 20 (2010): 377–384.

Liu, Chongyun, Angela Tseng, and Sue Yang. *Chinese Herbal Medicine: Modern Applications of Traditional Formulas.* Boca Raton, Fla.: CRC Press, 2005.

Liu, X. X., S. H. Li, J. Z. Chen, et al. "Effect of Soy Isoflavones on Blood Pressure: A Meta-Analysis of Randomized Controlled Trials." *Nutrition, Metabolism, and Cardiovascular Diseases* 22 (2012): 463–470.

Liu, Yanchi. *The Essential Book of Traditional Chinese Medicine.* Vol. 1, *Theory.* New York: Columbia University Press, 1988.

——. *The Essential Book of Traditional Chinese Medicine.* Vol. 2, *Clinical Practice.* New York: Columbia University Press, 1988.

Lloyd, John Uri. "History of the Vegetable Drugs of the Pharmacopeia of the United States." *Bulletin of the Lloyd Library of Botany, Pharmacy, and Materia Medica* 18, no. 4 (1911).

Loguercio, Carmela, and Davide Festi. "Silybin and the Liver: From Basic Research to Clinical Practice." *World Journal of Gastroenterology* 17 (2011): 2288–2301.

Loh, Jonathan, and David Harmon. "A Global Index of Biocultural Diversity." *Ecological Indicators* 5 (2005): 231–241.

Lopez, Carlos. "Atharva Veda." *Oxford Bibliographies: Hinduism.* November 21, 2012. http://www.oxfordbiblio graphies.com/view/document/obo-9780195399318 /obo-9780195399318-0008.xml?rskey=Eago7N &result=12.

López Austin, Alfredo. *The Human Body and Ideology Concepts of the Ancient Nahuas.* Vol. 1. Translated by Thelma Ortiz de Montellano and Bernard Ortiz de Montellano. Salt Lake City: University of Utah Press, 1988.

Lorenzi, Harri, and F. J. Abreu Matos. *Plantas medicinais no Brasil.* 2nd ed. Nova Odessa: Instituto Plantarum de Estudos da Flora, 2008.

Lu Yü. *The Classic of Tea: Origins and Rituals.* Translated by Francis Ross. Boston: Little, Brown, 1974.

Lumholtz, Carl. *Unknown Mexico.* Vol. 2. London: Macmillan, 1903.

Lynch, Emma, and Robin Braithwaite. "A Review of the Clinical and Toxicological Aspects of 'Traditional' (Herbal) Medicines Adulterated with Heavy Metals." *Expert Opinion on Drug Safety* 4 (2005): 769–778.

Ma, Chunhui, Adam R. Kavalier, Bei Jiang, et al. "Metabolic Profiling of *Actaea* (*Cimicifuga*) Species Extracts Using High-Performance Liquid Chromatography Coupled with Electrospray Ionization Time-of-Flight Mass Spectrometry." *Journal of Chromatography A* 1218, no. 11 (2011): 1461–1476.

Ma, Huiping, Xirui He, Yan Yang, et al. "The Genus *Epimedium*: An Ethnopharmacological and Phytochemical Review." *Journal of Ethnopharmacology* 134 (2011): 519–541.

MacArthur, Rodger D., T. N. Hawkins, S. J. Brown, et al. "Efficacy and Safety of Crofelemer for Noninfectious Diarrhea in HIV-Seropositive Individuals (ADVENT Trial): A Randomized, Double-Blind, Placebo-Controlled, Two-Stage Study." *HIV Clinical Trials* 14 (2013): 261–273.

Mack, Richard N. "Plant Naturalizations and Invasions in the Eastern United States: 1634–1860." *Annals of the Missouri Botanical Garden* 90 (2003): 77–90.

Madge, Clare. "Therapeutic Landscapes of the Jola, the Gambia, West Africa." *Health & Place* 4 (1998): 293–311.

Maenthaisong, R., N. Chaiyakunapruk, S. Niruntraporn, et al. "The Efficacy of *Aloe vera* Used for Burn Wound Healing: A Systematic Review." *Burns* 33 (2007): 713–718.

Maffi, Luisa. "Biocultural Diversity and Sustainability." In *The SAGE Handbook of Environment and Society*, edited by Jules Pretty, Andrew W. Ball, Ted Benton, et al., 267–277. London: Sage, 2007.

Magendie, F. *Formulaire pour la préparation et l'emploi de plusieurs nouveaux médicamens.* Paris: Méquignon-Marvis, 1821.

Magnan, V. *On Alcoholism, the Various Forms of Alcoholic Delirium, and Their Treatment.* Translated by W. S. Greenfield. London: Lewis, 1876.

Mair, Victor H., and Erling Hoh. *The True History of Tea.* New York: Thames & Hudson, 2009.

Malcolm, Stephen B. "Cardenolide-Mediated Interactions Between Plants and Herbivores." In *Herbivores: Their Interactions with Secondary Plant Metabolites*, 2nd ed., edited by Gerald A. Rosenthal and May R. Berenbaum, 1:251–296. San Diego: Academic Press, 1991.

Manniche, Lise. *An Ancient Egyptian Herbal.* Rev. ed. London: British Museum Press, 2006.

Marcus, Donald M., and Arthur P. Grollman. "The Consequences of Ineffective Regulation of Dietary Supplements." *Archives of Internal Medicine* 171 (2012): 1035–1036.

Mariani and Company. *Vin Mariani, Erythroxylon Coca: Its Uses in the Treatment of Disease.* 2nd ed. Paris: Mariani, 1884.

Marieb, Elaine M., and Katja Hoehn. *Human Anatomy and Physiology.* 7th ed. San Francisco: Cummings, 2007.

Maron, Bradley A., and Thomas P. Rocco. "Pharmacotherapy of Congestive Heart Failure." In *Goodman and Gilman's The Pharmacological Basis of Therapeutics*, 12th ed., edited by Laurence L. Brunton, Bruce A. Chabner, and Björn C. Knollmann, 789–814. New York: McGraw-Hill, 2011.

Maroukis, Thomas Constantine. *The Peyote Road: Religious Freedom and the Native American Church.* Norman: University of Oklahoma Press, 2010.

Martin, Billy R. "The Endocannabinoid System and the Therapeutic Potential of Cannabinoids." In *Marijuana and the Cannabinoids*, edited by Mahmoud A. ElSohly, 125–144. Totowa, N.J.: Humana, 2007.

Martin, M. A., L. Goya, and S. Ramos. "Potential for Preventive Effects of Cocoa and Cocoa Polyphenols in Cancer." *Food and Chemical Toxicology* 56 (2013): 336–351.

Martin, Peter R., Sachin Patel, and Robert M. Swift. "Pharmacology of Drugs of Abuse." In *Principles of Pharmacology: The Pathophysiologic Basis of Drug Therapy*, 3rd ed., edited by David E. Golan, Armen H. Tashjian, Ehrin J. Armstrong, et al., 284–309. Philadelphia: Lippincott Williams & Wilkins, 2012.

Martin, Richard. "The Role of Coca in the History, Religion, and Medicine of South American Indians." *Economic Botany* 24, no. 4 (1970): 422–438.

Mascolo, Nicola, A. A. Izzo, F. Borrelli, et al. "Healing Powers of Aloes." In *Aloes: The Genus Aloe*, edited by Tom Reynolds, 209–238. Boca Raton, Fla.: CRC Press, 2004.

Mattern, Susan P. *The Prince of Medicine: Galen in the Roman Empire.* Oxford: Oxford University Press, 2013.

Matthee, Rudi. "Tobacco in Iran." In *Smoke: A Global History of Smoking*, edited by Sander L. Gilman and Zhou Xun, 58–67. London: Reaktion, 2004.

Mayor, Susan. "Tree That Provides Paclitaxel Is Put on List of Endangered Species." *BMJ* 343 (2011): d7411.

McClatchey, Will. "Medicinal Bioprospecting and Ethnobotany Research." *Ethnobotany Research and Applications* 3 (2005): 189–190.

McKeown, Kathleen A. "A Review of the Taxonomy of the Genus *Echinacea*." In *Perspectives on New Crops and New Uses*, edited by J. Janick, 482–489. Alexandria, Va.: ASHS, 1999.

McNeil, Cameron L. "Introduction: The Biology, Antiquity, and Modern Uses of the Chocolate Tree (*Theobroma cacao* L.)." In *Chocolate in Mesoamerica: A Cultural History of Cacao*, edited by Cameron L. McNeil, 1–28. Gainesville: University Press of Florida, 2006.

——. "Traditional Cacao Use in Modern Mesoamerica." In *Chocolate in Mesoamerica: A Cultural History of Cacao*, edited by Cameron L. McNeil, 341–366. Gainesville: University Press of Florida, 2006.

Meaney, Audrey. "The Practice of Medicine in England About the Year 1000." *Social History of Medicine* 13, no. 2 (2000): 221–237.

Mechoulam, Raphael. "The Pharmacohistory of *Cannabis sativa*." In *Cannabinoids as Therapeutic Agents*, edited by Raphael Mechoulam, 1–19. Boca Raton, Fla.: CRC Press, 1986.

Mechoulam, Raphael, and Lumir Hanus. "A Historical Overview of Chemical Research on Cannabinoids." *Chemistry and Physics of Lipids* 108 (2000): 1–13.

Mengs, Ulrich, Ralf-Torsten Pohl, and Todd Mitchell. "Legalon® SIL: The Antidote of Choice in Patients with Acute Hepatotoxicity from Amatoxin Poisoning." *Current Pharmaceutical Biotechnology* 12 (2012): 1964–1970.

Menon, Venugopal P., and Adluri Ram Sudheer. "Antioxidant and Anti-inflammatory Properties of Curcumin." In *The Molecular Targets and Therapeutic Uses of Curcumin in Health and Disease*, edited by Bharat B. Aggarwal, Young-Joon Surh, and Shishir Shishodia, 105–126. New York: Springer, 2007.

Mercado-Feliciano, Minerva, M. C. Cora, K. L. Witt, et al. "An Ethanolic Extract of Black Cohosh Causes Hematological Changes but Not Estrogenic Effects in Female Rodents." *Toxicology and Applied Pharmacology* 263 (2012): 138–147.

Merlin, Mark D. "Archaeological Evidence for the Tradition of Psychoactive Plant Use in the Old World." *Economic Botany* 57 (2003): 295–323.

——. *On the Trail of the Ancient Opium Poppy*. Rutherford, N.J.: Farleigh Dickinson University Press, 1984.

Merlin, Mark, and William Raynor. "Modern Use and Environmental Impact of the Kava Plant in Remote Oceania." In *Dangerous Harvest: Drug Plants and the Transformation of Indigenous Landscapes*, edited by Michael K. Steinberg, Joseph J. Hobbs, and Kent Mathewson, 274–293. New York: Oxford University Press, 2004.

Meskell, Lynn M., and Rosemary A. Joyce. *Embodied Lives: Figuring Ancient Maya and Egyptian Experience*. London: Routledge, 2003.

Messina, Mark J., and Charles E. Wood. "Soy Isoflavones, Estrogen Therapy, and Breast Cancer Risk: Analysis and Commentary." *Nutrition Journal* 7 (2008): 17.

Meyer, H. C. F. "*Echinacea Angustifolia*." *Eclectic Medical Journal* 47 (1887): 209–210.

Meyer, Jerrold, and Linda Quenzer. *Psychopharmacology: Drugs, the Brain, and Behavior*. 3rd ed. Sunderland, Mass.: Sinauer, 2013.

Michel, Thomas, and Brian B. Hoffman. "Treatment of Myocardial Ischemia and Hypertension." In *Goodman and Gilman's The Pharmacological Basis of Therapeutics*, 12th ed., edited by Laurence L. Brunton, Bruce A. Chabner, and Björn C. Knollmann, 745–788. New York: McGraw-Hill, 2011.

Mikuriya, Tod H. "Marijuana in Medicine: Past, Present, and Future." *California Medicine* 110 (1969): 34–40.

Miller, James S. "The Discovery of Medicines from Plants: A Current Biological Perspective." *Economic Botany* 65 (2011): 396–407.

——. "Impact of the Convention on Biological Diversity: The Lessons of Ten Years of Experience with Models for Equitable Sharing of Benefits." In *Biodiversity and the Law: Intellectual Property, Biotechnology, and Traditional Knowledge*, edited by Charles R. McManis, 58–67. London: Earthscan, 2007.

Miller, James S., R. Andriantsiferana, G. M. Cragg, et al. "History of a Landmark Collecting Agreement: The Origin of the National Cancer Institute's Letter of Intent, a Precursor to Modern Bioprospecting Agreements." In *Biodiversity and the Law: Intellectual Property, Biotechnology, and Traditional Knowledge*, edited by Charles R. McManis, 68–70. London: Earthscan, 2007.

Miller, Louis, and Xinzhuan Su. "Artemisinin: Discovery from the Chinese Herbal Garden." *Cell* 146 (2011): 855–858.

Milligan, Barry. "The Opium Den in Victorian London." In *Smoke: A Global History of Smoking*, edited by Sander L. Gilman and Zhou Xun, 118–125. London: Reaktion, 2004.

Minelli, Alba, and Ilaria Bellezza. "Methylxanthines and Reproduction." In *Methylxanthines: Handbook of Experimental Pharmacology*, edited by Bertil B. Fredholm, 349–372. Berlin: Springer, 2011.

Miyasaka, Lincoln Sakiara, Álvaro N. Atallah, and Bernardo Soares. "Valerian for Anxiety Disorders." *Cochrane Database of Systematic Reviews* (2006): CD004515.

Moerman, Daniel E. *Meaning, Medicine, and the "Placebo Effect."* Cambridge: Cambridge University Press, 2002.

——. *Native American Ethnobotany*. Portland, Ore.: Timber Press, 1998.

Monachino, Joseph. "*Rauvolfia serpentina*: Its History, Botany, and Medical Use." *Economic Botany* 8 (1954): 349–365.

Monardes, Nicolás. *Joyfull Newes Out of the New-Found World*. Translated by John Frampton. London, 1596. [Spanish editions: 1565, 1571, 1574]

Morales, A. "Yohimbine in Erectile Dysfunction: The Facts." *International Journal of Impotence Research* 12 (2000): S70–S74.

Moreau, J. *Du hachisch et de l'aliénation mentale: Études psychologiques*. Paris: Fortin, Masson, 1845.

Morelli, Micaela, and Nicola Simola. "Methylxanthines and Drug Dependence: A Focus on Interactions with Substances of Abuse." In *Methylxanthines: Handbook of Experimental Pharmacology*, edited by Bertil B. Fredholm, 487–507. Berlin: Springer, 2011.

Mortimer, W. Golden. *Peru: History of Coca: "The Divine Plant" of the Incas*. New York: Vail, 1901.

Moteetee, A., and B.-E. Van Wyk. "The Medical Ethnobotany of Lesotho: A Review." *Bothalia* 41 (2011): 209–228.

Moxham, Roy. *Tea: Addiction, Empire, and Exploitation*. New York: Carroll and Graf, 2003.

Murphy, K., Z. J. Kubin, J. N. Shepherd, et al. "*Valeriana officinalis* Root Extracts Have Potent Anxiolytic Effects in Laboratory Rats." *Phytomedicine* 17 (2010): 674–678.

Musk, Arthur W., and Nicholas H. De Klerk. "History of Tobacco and Health." *Respirology* 8 (2003): 286–290.

Musto, David F. *The American Disease: Origins of Narcotic Control*. 3rd ed. Oxford: Oxford University Press, 1999.

——. "America's First Cocaine Epidemic." *Wilson Quarterly* 13, no. 3 (1989): 59–64.

——. "Iatrogenic Addiction: The Problem, Its Definition and History." *Bulletin of the New York Academy of Medicine* 61, no. 8 (1985): 694–705.

Myerhoff, Barbara G. "The Deer-Maize-Peyote Symbol Complex Among the Huichol Indians of Mexico." *Anthropological Quarterly* 43 (1970): 64–78.

——. "Peyote and Huichol Worldview: The Structure of a Mystic Vision." In *Cannabis and Culture*, edited by Vera Rubin, 417–438. The Hague: Mouton, 1975.

——. *Peyote Hunt: The Sacred Journey of the Huichol Indians*. Ithaca, N.Y.: Cornell University Press, 1974.

Nair, K. P. Prabhakaran. "The Neutraceutical [*sic*] Properties of Turmeric." In *The Agronomy and Economy of Turmeric and Ginger*, edited by K. P. Prabhakaran Nair, 179–204. London: Elsevier, 2013.

Narotzki, B., A. Z. Reznick, D. Aizenbud, et al. "Green Tea: A Promising Natural Product in Oral Health." *Archives of Oral Biology* 57 (2012): 429–435.

Nash, George V., and Maurice G. Kains. *American Ginseng: Its Commercial History, Protection, and Cultivation*. Washington, D.C.: Government Printing Office, 1898.

Nassiri Asl, Marjan, and Hossein Hosseinzadeh. "Review of Pharmacological Effects of *Glycyrrhiza* sp. and Its Bioactive Compounds." *Phytotherapy Research* 22 (2008): 709–724.

Nayyar, Gaurvika M. L., J. G. Breman, P. N. Newton, et al. "Poor-Quality Antimalarial Drugs in Southeast Asia and Sub-Saharan Africa." *Lancet Infectious Disease* 12 (2012): 488–496.

Nencini, Paolo. "The Rules of Drug Taking: Wine and Poppy Derivatives in the Ancient World. VI. Poppies as a Source of Food and Drug." *Substance Use & Misuse* 32, no. 6 (1997): 757–766.

Nestler, Eric J., Steven E. Hyman, and Robert C. Malenka. *Molecular Neuropharmacology: A Foundation for Clinical Neuroscience*. 2nd ed. New York: McGraw-Hill, 2009.

Nettle, Daniel, and Suzanne Romaine. *Vanishing Voices: The Extinction of the World's Languages*. New York: Oxford University Press, 2000.

Newman, David J., and Gordon M. Cragg. "Natural Products as Sources of New Drugs over the Thirty Years from 1981 to 2010." *Journal of Natural Products* 75 (2012): 311–335.

Newmaster, Steven G., Meghan Grguric, Dhivya Shanmughanandhan, et al. "DNA Barcoding Detects Contamination and Substitution in North American Herbal Products." *BMC Medicine* 11 (2013): 222.

Newton, Leonard E. "Aloes in Habitat." In *Aloes: The Genus Aloe*, edited by Tom Reynolds, 3–14. Boca Raton, Fla.: CRC Press, 2004.

Nichols, David E. "Hallucinogens." In *Encyclopedia of Psychopharmacology*, edited by Ian P. Stolerman, 572–577. Berlin: Springer, 2010.

——. "Hallucinogens." *Pharmacology & Therapeutics* 101 (2004): 131–181.

Nicholson, Tristan M., and William A. Ricke. "Androgens and Estrogens in Benign Prostatic Hyperplasia: Past, Present, and Future." *Differentiation* 82 (2011): 184–199.

Nicolaï, S. P. A., L. M. Kruidenier, B. L. Bendermacher, et al. "*Ginkgo biloba* for Intermittent Claudication." *Cochrane Database of Systematic Reviews* (2013): CD006888.

Nicolai, Ton. "Important Concepts and the Approach to Prescribing." In *Homeopathic Practice*, edited by Steven B. Kayne, 43–61. London: Pharmaceutical Press, 2008.

Njamen, D., M. A. Mvondo, S. Djiogue, et al. "Phytotherapy and Women's Reproductive Health: The Cameroonian Perspective." *Planta Medica* 79 (2013): 600–611.

Norton, Marcy. *Sacred Gifts, Profane Pleasures: A History of Tobacco and Chocolate in the Atlantic World*. Ithaca, N.Y.: Cornell University Press, 2008.

Nunn, John F. *Ancient Egyptian Medicine*. Norman: University of Oklahoma Press, 1996.

Nutton, Vivian. "Medicine in the Greek World, 800–50 B.C." In *The Western Medical Tradition, 800 B.C. to 1800*, by Lawrence I. Conrad, Michael Neve, Vivian Nutton, et al., 11–38. Cambridge: Cambridge University Press, 1995.

O'Brien, Charles P. "Drug Addiction." In *Goodman and Gilman's The Pharmacological Basis of Therapeutics*, 12th ed., edited by Laurence L. Brunton, Bruce A. Chabner, and Björn C. Knollmann, 649–668. New York: McGraw-Hill, 2011.

O'Brien, Charles P., Nora Volkow, and T.-K. Li. "What's in a Word? Addiction Versus Dependence in DSM-V." *American Journal of Psychiatry* 163 (2006): 764–765.

O'Connor, Anahad. "New York Attorney General Targets Supplements at Major Retailers." *New York Times*, February 3, 2015.

Ody, Penelope. *The Complete Medicinal Herbal: A Practical Guide to the Healing Properties of Herbs, with More Than 250 Remedies for Common Ailments*. New York: Dorling Kindersley, 1993.

Oliver, James. "On the Action of *Cannabis indica*." *British Medical Journal* 1 (1883): 905–906.

Olsen, Line R., Mark P. Grillo, and Christian Skonberg. "Constituents in Kava Extracts Potentially Involved in Hepatotoxicity: A Review." *Chemical Research in Toxicology* 24 (2011): 992–1002.

Olsen, Richard W. "Absinthe and γ-aminobutyric Acid Receptors." *Proceedings of the National Academy of Sciences USA* 97, no. 9 (2000): 4417–4418.

O'Neill, Paul M., Victoria E. Barton, and Stephen A. Ward. "The Molecular Mechanism of Action of Artemisinin—The Debate Continues." *Molecules* 15 (2010): 1705–1721.

Ortiz de Montellano, Bernard. *Aztec Medicine, Health, and Nutrition*. New Brunswick, N.J.: Rutgers University Press, 1990.

——. "Empirical Aztec Medicine." *Science* 188, no. 4185 (1975): 215–220.

O'Shaughnessy, W. B. "On the Preparations of the Indian Hemp, or Gunjah." *Provincial Medical Journal*, January 28, 1843; February 4, 1843.

Osswald, Harmut, and Jürgen Schnermann. "Methylxanthines and the Kidney." In *Methylxanthines: Handbook of Experimental Pharmacology*, edited by Bertil B. Fredholm, 391–412. Berlin: Springer, 2011.

Ouarghidi, A., B. Powell, G. J. Martin, et al. "Species Substitution in Medicinal Roots and Possible Implications for Toxicity of Herbal Remedies in Morocco." *Economic Botany* 66 (2012): 370–382.

Padosch, S. A., D. W. Lachenmeier, and L. U. Kröner. "Absinthism: A Fictitious Nineteenth-Century Syndrome with Present Impact." *Substance Abuse Treatment, Prevention, and Policy* 1 (2006): 1–14.

Paine, Martyn. *A Therapeutical Arrangement of the Materia Medica*. New York: Langley, 1842.

Pang, Z., F. Pan, and S. He. "*Ginkgo biloba* L.: History, Current Status, and Future Prospects." *Journal of Alternative and Complementary Medicine* 2 (1996): 359–363.

Panossian, A., G. Wikman, P. Kaur, et al. "Adaptogens Exert a Stress-Protective Effect by Modulation of Expression of Molecular Chaperones." *Phytomedicine* 16 (2009): 617–622.

Panossian, Alexander, and Hildebert Wagner. "Adaptogens: A Review of Their History, Biological Activity, and Clinical Benefits." *HerbalGram* 90 (2011): 52–63.

Pappas, Diane, and J. Owen Hendley. "The Common Cold and Decongestant Therapy." *Pediatrics in Review* 32 (2011): 47–54.

Park, Jongbae, and Edzard Ernst. "Ayurvedic Medicine for Rheumatoid Arthritis: A Systematic Review." *Seminars in Arthritis and Rheumatism* 34 (2005): 705–713.

Parke, Davis, and Company. *Manual of Therapeutics*. Detroit: Parke, Davis, 1909.

Parkinson, John. *Theatrum Botanicum*. London: Cotes, 1640.

Pendergrast, Mark. *For God, Country, and Coca-Cola: The Definitive History of the Great American Soft Drink and the Company That Makes It*. 3rd ed. New York: Basic Books, 2013.

——. *Uncommon Grounds: The History of Coffee and How It Transformed Our World*. 2nd ed. New York: Basic Books, 2010.

Persson, Ingrid A.-L. "Tea Flavanols: An Overview." In *Tea in Health and Disease Prevention*, edited by Victor R. Preedy, 73–78. London: Elsevier, 2013.

Petrovska, Biljana Bauer, and Svetlana Cekovska. "Extracts from the History and Medical Properties of Garlic." *Pharmacognosy Review* 4 (2010): 106–110.

Petrullo, Vincenzo. *The Diabolic Root: A Study of Peyotism, the New Indian Religion, Among the Delawares*. Philadelphia: University of Pennsylvania Press, 1934.

Philippe, G., L. Angenot, M. Tits, et al. "About the Toxicity of Some *Strychnos* Species and Their Alkaloids." *Toxicon* 44 (2004): 405–416.

Pleschka, S., M. Stein, R. Schoop, et al. "Antiviral Properties and Mode of Action of Standardized *Echinacea purpurea* Extract Against Highly Pathogenic Avian Influenza Virus (H5N1, H7N7) and Swine-Origin H1N1 (S-OIV)." *Virology Journal* 6 (2009): 197.

Pliny. *The Natural History of Pliny*. Translated by John Bostock and H. T. Riley. 6 vols. London: Bohn, 1855–1857.

Plotkin, Mark J. *Tales of a Shaman's Apprentice: An Ethnobotanist Searches for New Medicines in the Amazon Rain Forest*. New York: Penguin, 1993.

Plowman, Timothy. "The Ethnobotany of Coca (*Erythroxylum* spp., *Erythroxylaceae*)." *Advances in Economic Botany* 1 (1984): 62–111.

Polyak, S. J., C. Morishima, V. Lohmann, et al. "Identification of Hepatoprotective Flavonolignans from Silymarin." *Proceedings of the National Academy of Sciences USA* 107 (2010): 5995–5999.

Pomet, Pierre. *Histoire générale des drogues*. Paris: Jean-Baptiste Loyson and Augustin Pillon, 1694.

Popoola, O. K., A. M. Elbagory, M. Abdulrahman, et al. "Marrubiin." *Molecules* 18 (2013): 9049–9060.

Porkka-Heiskanen, Tarja, and Anna V. Kalinchuk. "Adenosine, Energy Metabolism, and Sleep Homeostasis." *Sleep Medicine Reviews* 15 (2011): 123–135.

Porter, Peggy. "'Westernizing' Women's Risks? Breast Cancer in Lower-Income Countries." *New England Journal of Medicine* 358 (2008): 213–216.

Porter, Roy. *The Greatest Benefit to Mankind: A Medical History of Humanity*. New York: Norton, 1997.

——. "What Is Disease?" In *The Cambridge History of Medicine*, edited by Roy Porter, 71–102. New York: Cambridge University Press, 2006.

Posadzki, Paul, Leala Watson, and Edzard Ernst. "Contamination and Adulteration of Herbal Medicinal Products (HMPs): An Overview of Systematic Reviews." *European Journal of Clinical Pharmacology* 69 (2013): 295–307.

Pougens, M.-J.-F.-Alexandre. *L'art de conserver la santé, de vivre long-temps et heureusement*. Montpellier: Picot, 1825.

Powell, Sharla L., T. Gödecke, D. Nikolic, et al. "*In vitro* Serotonergic Activity of Black Cohosh and Identification of N_ω-Methylserotonin as a Potential Active Constituent."

Journal of Agricultural and Food Chemistry 56 (2008): 11718–11726.

Prance, Ghillean T., Derek J. Chadwick, and Joan Marsh, eds. *Ethnobotany and the Search for New Drugs.* Ciba Foundation Symposium 185. Chichester: Wiley, 1994.

Prior, R. C. A. *On the Popular Names of British Plants.* 2nd ed. London: Williams and Norgate, 1870.

Qin, Yu, Kai Niu, Yuan Zeng, et al. "Isoflavones for Hypercholesterolaemia in Adults." *Cochrane Database of Systematic Reviews* (2013): CD009518.

Rahman, Khalid, and Gordon M. Lowe. "Garlic and Cardiovascular Disease: A Critical Review." *Journal of Nutrition* 136 (2006): 736S–740S.

Rajakumar, N., and M. B. Shivanna. "Ethnomedicinal Application of Plants in the Eastern Region of Shimoga District, Karnataka, India." *Journal of Ethnopharmacology* 126 (2009): 64–73.

Rambaldi, A., B. P. Jacobs, and C. Gluud. "Milk Thistle for Alcoholic and/or Hepatitis B or C Virus Liver Diseases." *Cochrane Database of Systematic Reviews* (2007): CD003620.

Rapport, David, and Luisa Maffi. "The Dual Erosion of Biological and Cultural Diversity: Implications for the Health of Ecocultural Systems." In *Nature and Culture: Rebuilding Lost Connections*, edited by Sarah Pilgrim and Jules Pretty, 103–119. London: Earthscan, 2010.

Raskin, Jonah. *American Scream: Allen Ginsberg's "Howl" and the Making of the Beat Generation.* Berkeley: University of California Press, 2004.

Rätsch, Christian. *The Encyclopedia of Psychoactive Plants: Ethnopharmacology and Its Applications.* Rochester, Vt.: Park Street Press, 2005.

Ravindran, P. N., and K. Nirmal Babu. "Introduction." In *Ginger: The Genus Zingiber*, edited by P. N. Ravindran and K. Nirmal Babu, 1–14. Boca Raton, Fla.: CRC Press, 2005.

Ravishankar, B., and V. J. Shukla. "Indian Systems of Medicine: A Brief Profile." *African Journal of Traditional Complementary and Alternative Medicine* 4, no. 3 (2007): 319–337.

Rawcliffe, Carole. *Medicine and Society in Later Medieval England.* Stroud: Sutton, 1995.

Read, Selina, and Jill Eckert. "Opioids and the Law." In *Comprehensive Treatment of Chronic Pain by Medical, Interventional, and Integrative Approaches*, edited by Timothy R. Deer, Michael S. Leon, Asokumar Buvanendran, et al., 135–144. New York: Springer, 2013.

Reame, Nancy E., Jane L. Lukacs, V. Padmanabhan, et al. "Black Cohosh Has Central Opioid Activity in Postmenopausal Women: Evidence from Naloxone Blockade and PET Neuroimaging Studies." *Menopause* 15 (2008): 832–840.

Reddy, K. R., S. H. Belle, M. W. Fried, et al. "Rationale, Challenges, and Participants in a Phase II Trial of a Botanical Product for Chronic Hepatitis C." *Clinical Trials* 9 (2012): 102–112.

Reeds, Karen. "Saint John's Wort (*Hypericum perforatum* L.) in the Age of Paracelsus and the Great Herbals: Assessing the Historical Claims for a Traditional Remedy." In *Herbs and Healers from the Ancient Mediterranean Through the Medieval West: Essays in Honor of John M. Riddle*, edited by Anne Van Arsdall and Timothy Graham, 265–305. Farnham: Ashgate, 2012.

Reilly, Robert F., and Edwin K. Jackson. "Regulation of Renal Function and Vascular Volume." In *Goodman and Gilman's The Pharmacological Basis of Therapeutics*, 12th ed., edited by Laurence L. Brunton, Bruce A. Chabner, and Björn C. Knollmann, 671–720. New York: McGraw-Hill, 2011.

Remadevi, R., E. Surendran, and Takeatsu Kimura. "Turmeric in Traditional Medicine." In *Turmeric: The Genus Curcuma*, edited by P. N. Ravindran, K. Nirmal Babu, and K. Sivaraman, 409–436. Boca Raton, Fla.: CRC Press, 2007.

Reveal, James L. "What's in a Name: Identifying Plants in Pre-Linnaean Botanical Literature." In *Prospecting for Drugs in Ancient and Medieval European Texts: A Scientific Approach*, edited by Bart K. Holland, 57–90. Amsterdam: Harwood, 1996.

Reynolds, T., and A. C. Dweck. "*Aloe vera* Leaf Gel: A Review Update." *Journal of Ethnopharmacology* 68 (1999): 3–37.

Reynolds, Tom. "Aloe Chemistry." In *Aloes: The Genus Aloe*, edited by Tom Reynolds, 39–74. Boca Raton, Fla.: CRC Press, 2004.

Rhyu, Mee-Ra, Jian Lu, Donna E. Webster, et al. "Black Cohosh (*Actaea racemosa, Cimicifuga racemosa*) Behaves as a Mixed Competitive Ligand and Partial Agonist at the Human Mu Opiate Receptor." *Journal of Agricultural and Food Chemistry* 54 (2006): 9852–9857.

Riddle, John. *Dioscorides on Pharmacy and Medicine.* Austin: University of Texas Press, 1985.

——. *Eve's Herbs: A History of Contraception and Abortion in the West.* Cambridge, Mass.: Harvard University Press, 1997.

——. *Goddesses, Elixirs, and Witches: Plants and Sexuality Throughout Human History.* New York: Palgrave Macmillan, 2010.

——. "Historical Data as an Aid in Pharmaceutical Prospecting and Drug Safety Determination." *Journal of Alternative of Complementary Medicine* 5 (1999): 195–201.

——. "The Medicines of Greco-Roman Antiquity as a Source of Medicine for Today." In *Prospecting for Drugs in Ancient and Medieval European Texts: A Scientific Approach*, edited by Bart K. Holland, 7–18. Amsterdam: Harwood, 1996.

——. "Oral Contraceptives and Early-Term Abortifacients During Classical Antiquity and the Middle Ages." *Past & Present* 132 (1991): 3–32.

Riddle, John M., and J. Worth Estes. "Oral Contraceptives in Ancient and Medieval Times." *American Scientist* 80, no. 3 (1992): 226–233.

Ried, K., O. R. Frank, and N. P. Stocks. "Aged Garlic Extract Reduces Blood Pressure in Hypertensives: A Dose-Response Trial." *European Journal of Clinical Nutrition* 67 (2013): 64–70.

Ried, K., O. R. Frank, N. P. Stocks, et al. "Effect of Garlic on Blood Pressure: A Systematic Review and Meta-Analysis." *BMC Cardiovascular Disorders* 8 (2008): 13.

Ried, K., T. R. Sullivan, P. Fakler, et al. "Effect of Cocoa on Blood Pressure." *Cochrane Database of Systematic Reviews* (2012): CD008893.

Ried, K., C. Toben, and P. Fakler. "Effect of Garlic on Serum Lipids: An Updated Meta-Analysis." *Nutrition Reviews* 71 (2013): 282–299.

Rietjens, Ivonne M. C. M., A. M. Sotoca, J. Vervoort, et al. "Mechanisms Underlying the Dualistic Mode of Action of Major Soy Isoflavones in Relation to Cell Proliferation and Cancer Risks." *Molecular Nutrition and Food Research* 57 (2013): 100–113.

Rigano, D., C. Formisano, A. Basile, et al. "Antibacterial Activity of Flavonoids and Phenylpropanoids from *Marrubium globosum* ssp. *libanoticum*." *Phytotherapy Research* 21 (2007): 395–397.

Riksen, Niels P., Paul Smits, and Gerard A. Rongen. "The Cardiovascular Effects of Methylxanthines." In *Methylxanthines: Handbook of Experimental Pharmacology*, edited by Bertil B. Fredholm, 413–437. Berlin: Springer, 2011.

Ritchie, M. R., J. Gertsch, P. Klein, et al. "Effects of Echinaforce® Treatment on *ex vivo*-Stimulated Blood Cells." *Phytomedicine* 18 (2011): 826–831.

Rivlin, Richard S. "Historical Perspective on the Use of Garlic." *Journal of Nutrition* 131 (2001): 951S–954S.

Roehrborn, Claus G. "Benign Prostatic Hyperplasia: An Overview." *Review of Urology* 7 (2005): S3–S14.

Rossi, Ruggero, Silvia Porta, and Brenno Canovi. "Overview on Cranberry and Urinary Tract Infections in Females." *Journal of Clinical Gastroenterology* 44 (2010) S61–S62.

Rucker, Robert. "Nutritional Properties of Cocoa." In *Chocolate: History, Culture, and Heritage*, edited by Louis Evan Grivetti and Howard-Yana Shapiro, 943–946. Hoboken, N.J.: Wiley, 2009.

Ruddock, Patrick S., Mingmin Liao, Brian C. Foster, et al. "Garlic Natural Health Products Exhibit Variable Constituent Levels and Antimicrobial Activity Against *Neisseria gonorrhoeae*, *Staphylococcus aureus*, and *Enterococcus faecalis*." *Phytotherapy Research* 19 (2005): 327–334.

Ruhlen, Rachel L., Grace Y. Sun, and Edward R. Sauter. "Black Cohosh: Insights into Its Mechanism(s) of Action." *Integrative Medicine Insights* 3 (2008): 21–32.

Russo, Ethan B. "Cannabis Treatments in Obstetrics and Gynecology: A Historical Review." *Journal of Cannabis Therapeutics* 2 (2002): 5–35.

——. "Hemp for Headache: An In-Depth Historical and Scientific Review of Cannabis in Migraine Treatment." *Journal of Cannabis Therapeutics* 1 (2001): 21–92.

——. "History of Cannabis and Its Preparations in Saga, Science, and Sobriquet." *Chemistry and Biodiversity* 4 (2007): 1614–1648.

——. "Taming THC: Potential Cannabis Synergy and Phytocannabinoid-Terpenoid Entourage Effects." *British Journal of Pharmacology* 163 (2011): 1344–1364.

Saddiqe, Zeb, Ismat Naeem, and Alya Maimoona. "A Review of the Antibacterial Activity of *Hypericum perforatum* L." *Journal of Ethnopharmacology* 131 (2010): 511–521.

Sahagún, Bernardino de. *Florentine Codex: General History of the Things of New Spain*. Book 8, *Kings and Lords*. Translated by Charles E. Dibble and Arthur J. O. Anderson. Santa Fe, N.Mex.: School of American Research, 1954.

——. *Florentine Codex: General History of the Things of New Spain*. Book 11, *Earthly Things*. Translated by Charles E. Dibble and Arthur J. O. Anderson. Santa Fe, N.Mex.: School of American Research, 1963.

Sahpaz, S., N. Garbacki, M. Tits, et al. "Isolation and Pharmacological Activity of Phenylpropanoid Esters from *Marrubium vulgare*." *Journal of Ethnopharmacology* 79 (2002): 389–392.

Sala, Angelo. *Anatome Essentiarum Vegetabilium*. 1630. Rostock: Johan Hallervord, 1635.

Sampson, Kevin J., and Robert S. Kass. "Anti-Arrhythmic Drugs." In *Goodman and Gilman's The Pharmacological Basis of Therapeutics*, 12th ed., edited by Laurence L. Brunton, Bruce A. Chabner, and Björn C. Knollmann, 815–848. New York: McGraw-Hill, 2011.

Saper, R. B., S. N. Kales, J. Paquin, et al. "Heavy Metal Content of Ayurvedic Herbal Medicine Products." *Journal of the American Medical Association* 292 (2004): 2868–2873.

Sarma, Dandapantula, Marilyn L. Barrett, Mary L. Chavez, et al. "Safety of Green Tea Extracts." *Drug Safety* 31 (2008): 469–484.

Sarma, H., S. Deka, H. Deka, et al. "Accumulation of Heavy Metals in Selected Medicinal Plants." *Reviews of Environmental Contamination and Toxicology* 214 (2011): 63–86.

Sarris, J., A. Panossian, I. Schweitzer, et al. "Herbal Medicine for Depression, Anxiety, and Insomnia: A Review of Psychopharmacology and Clinical Evidence." *European Neuropsychopharmacology* (2011): 841–860.

Sarris, J., C. Stough, C. A. Bousman, et al. "Kava in the Treatment of Generalized Anxiety Disorder." *Journal of Clinical Psychopharmacology* 33 (2013): 643–648.

Sarris, Jerome, and Gerard J. Byrne. "A Systematic Review of Insomnia and Complementary Medicine." *Sleep Medicine Reviews* 15 (2011): 99–106.

Saslis-Lagoudakis, C. H., V. Savolainen, E. M. Williamson, et al. "Phylogenies Reveal Predictive Power of Traditional Medicine in Bioprospecting." *Proceedings of the National Academy of Sciences USA* 109 (2012): 15835–15840.

Sawynok, Jana. "Methylxanthines and Pain." In *Methylxanthines: Handbook of Experimental Pharmacology*, edited by Bertil B. Fredholm, 311–329. Berlin: Springer, 2011.

Scarborough, John. "Theophrastus on Herbals and Herbal Remedies." *Journal of the History of Biology* 11 (1978): 353–385.

Schellenberg, R. "Treatment for the Premenstrual Syndrome with *Agnus castus* Fruit Extract: Prospective, Randomised, Placebo-Controlled Study." *British Medical Journal* 322 (2001): 134–137.

Schmidt, Mathias. "German Court Ruling Reverses Kava Ban; German Regulatory Authority Appeals Decision," *HerbalEGram* 11, no. 7 (2014).

Schmiedlin-Ren, Phyllissa, David J. Edwards, Michael E. Fitzsimmons, et al. "Mechanisms of Enhanced Oral Availability of CYP3A4 Substrates by Grapefruit Constituents." *Drug Metabolism and Disposition* 25 (1997): 1228–1233.

Schneider, Lon S. "*Ginkgo biloba* Extract and Preventing Alzheimer's Disease." *Journal of the American Medical Association* 300 (2008): 2306.

Schulman, Joshua M., and Gary R. Strichartz. "Local Anesthetic Pharmacology." In *Principles of Pharmacology: The Pathophysiologic Basis of Drug Therapy*, 3rd ed., edited by David E. Golan, Armen H. Tashjian, Ehrin J. Armstrong, et al., 147–162. Philadelphia: Lippincott Williams & Wilkins, 2012.

Schultes, Richard Evans. "Amazonian Ethnobotany and the Search for New Drugs." In *Ethnobotany and the Search for New Drugs*, edited by Ghillean T. Prance, Derek J. Chadwick, and Joan Marsh, 106–112. Ciba Foundation Symposium 185. Chichester: Wiley, 1994.

——. "The Importance of Ethnobotany in Environmental Conservation." *American Journal of Economics and Sociology* 53, no. 2 (1994): 202–206.

Schultes, Richard Evans, Albert Hofmann, and Christian Rätsch. *Plants of the Gods: Their Sacred, Healing, and Hallucinogenic Powers*. Rochester, Vt.: Healing Arts, 2001.

Schultes, Richard Evans, William M. Klein, Timothy Plowman, et al. "Cannabis: An Example of Taxonomic Neglect." In *Cannabis and Culture*, edited by Vera Rubin, 21–38. The Hague: Mouton, 1975.

Schütz, Katrin, Reinhold Carle, and Andreas Schieber. "*Taraxacum*: A Review on Its Phytochemical and Pharmacological Profile." *Journal of Ethnopharmacology* 107 (2006): 313–323.

Schwartz, Bryan G., Shereif Rezkalla, and Robert A. Kloner. "Cardiovascular Effects of Cocaine." *Circulation* 122 (2010): 2558–2569.

Scott, Peter. *Physiology and Behaviour of Plants*. Chichester: Wiley, 2008.

Sertürner, F. W. A. "Morphine, a New Salt-forming Base, and Meconic Acid, as the Chief Constituents of Opium." In *Foundations of Anesthesiology*, edited by A. Faulconer and T. E. Keys, 2:1078–1084. Park Ridge, Ill.: Wood Library-Museum of Anesthesiology, 1993.

——. "Über das Morphium, eine neuesalzfahige Grundlage und die Makosaure, als Hauptbestandtheile des Opiums." *Annalen der Physik* 55 (1817): 56–89.

Shah, S. A., S. Sander, C. M. White, et al. "Evaluation of *Echinacea* for the Prevention and Treatment of the Common Cold: A Meta-Analysis." *Lancet Infectious Disease* 7 (2007): 473–480.

Shamloul, Rany. "Natural Aphrodisiacs." *Journal of Sexual Medicine* 7 (2010): 39–49.

Sharkey, Keith A., and John L. Wallace. "Treatment of Disorders of Bowel Motility and Water Flux; Anti-Emetics; Agents Used in Biliary and Pancreatic Disease." In *Goodman and Gilman's The Pharmacological Basis of Therapeutics*, 12th ed., edited by Laurence L. Brunton, Bruce A. Chabner, and Björn C. Knollmann, 1323–1350. New York: McGraw-Hill, 2011.

Sharma, Hari M. "Contemporary Ayurveda." In *Fundamentals of Complementary and Alternative Medicine*, 4th ed., edited by Marc S. Micozzi, 495–508. St. Louis: Saunders Elsevier, 2011.

Sharma, M., R. Schoop, and J. B. Hudson. "*Echinacea* as an Anti-Inflammatory Agent: The Influence of Physiologically Relevant Parameters." *Phytotherapy Research* 23 (2009): 863–867.

Sharma, Ricky A., William P. Steward, and Andreas J. Gescher. "Pharmacokinetics and Pharmacodynamics of Curcumin." In *The Molecular Targets and Therapeutic Uses of Curcumin in Health and Disease*, edited by Bharat B. Aggarwal, Young-Joon Surh, and Shishir Shishodia, 453–470. New York: Springer, 2007.

Sharma, S. M., M. Anderson, S. R. Schoop, et al. "Bactericidal and Anti-Inflammatory Properties of a Standardized *Echinacea* Extract (Echinaforce): Dual Actions Against Respiratory Bacteria." *Phytomedicine* 17 (2010): 563–568.

Shen, P., B. L. Guo, Y. Gong, et al. "Taxonomic, Genetic, Chemical, and Estrogenic Characteristics of *Epimedium* Species." *Phytochemistry* 68 (2007): 1448–1458.

Shergis, J. L., A. L. Zhang, W. Zhou, et al. "*Panax ginseng* in Randomised Controlled Trials: A Systematic Review." *Phytotherapy Research* 27 (2013): 949–965.

Shishodia, S., K. B. Harikumar, S. Dass, et al. "The Guggul for Chronic Diseases: Ancient Medicine, Modern Targets." *Anticancer Research* 28 (2008): 3647–3664.

Shitandi, Ankalo A., Francis Muigai Ngure, and Symon M. Mahungu. "Tea Processing and Its Impact on Catechins, Theaflavin, and Thearubigin Formation." In *Tea in Health and Disease Prevention*, edited by Victor R. Preedy, 193–206. London: Elsevier, 2013.

Shonle, Ruth. "Peyote, the Giver of Visions." *American Anthropologist* 27, no. 1 (1925): 53–75.

Shorter, Edward. "Primary Care." In *The Cambridge History of Medicine*, edited by Roy Porter, 103–135. Cambridge: Cambridge University Press, 2006.

Simmler, Charlotte, Shao-Nong Chen, Jeff Anderson, et al. "Botanical Integrity: The Importance of the Integration of Chemical, Biological, and Botanical Analyses, and the Role of DNA Barcoding." *HerbalEGram* 12 (2015).

Simmonds, P. "Notes on Some Saps and Secretions Used in Pharmacy." *American Journal of Pharmacy* 67, no. 3 (1895): 7–15.

——. "Notes on Some Saps and Secretions Used in Pharmacy." *American Journal of Pharmacy* 67, no. 4 (1895): 7–11

Simoons, Frederick J. *Food in China: A Cultural and Historical Inquiry*. Boca Raton, Fla.: CRC Press, 1991.

Simpson, Beryl, and Molly Ogorzaly. *Economic Botany: Plants in Our World*. 3rd ed. New York: McGraw-Hill, 2001.

Singh, Nirbhay N., Subhashni D. Singh, and Yadhu N. Singh. "Kava: Clinical Studies and Therapeutic Implications." In *Kava: From Ethnology to Pharmacology*, edited by Yadhu N. Singh, 140–164. Boca Raton, Fla.: CRC Press, 2004.

Singh, Seema. "From Exotic Spice to Modern Drug?" *Cell* 130 (2007): 765–768.

Singh, Yadhu N. "Kava: Production, Marketing, and Quality Assurance." In *Kava: From Ethnology to Pharmacology*, edited by Yadhu N. Singh, 29–49. Boca Raton, Fla.: CRC Press, 2004.

——. "Pharmacology and Toxicology of Kava and Kavalactones." In *Kava: From Ethnology to Pharmacology*, edited by Yadhu N. Singh, 104–139. Boca Raton, Fla.: CRC Press, 2004.

Sivin, Nathan. *Traditional Medicine in Contemporary China. A Partial Translation of Revised Outline of Chinese Medicine (1972) with an Introductory Study on Change in Present-Day and Early Medicine*. Science, Medicine, and Technology in East Asia 2. Ann Arbor: Center for Chinese Studies, University of Michigan, 1987.

Slotkin, J. S. "Peyotism, 1521–1891." *American Anthropologist* 57, no. 2 (1955): 202–230.

Slotkin, J. Sydney, and David P. McAllester. "Menomini Peyotism, a Study of Individual Variation in a Primary Group with a Homogeneous Culture." *Transactions of the American Philosophical Society* 42, no. 4 (1952): 565–700.

Smit, Hendrik Jan. "Theobromine and the Pharmacology of Cocoa." In *Methylxanthines: Handbook of Experimental Pharmacology*, edited by Bertil B. Fredholm, 201–234. Berlin: Springer, 2011.

Smith, Tyler. "Cochrane Collaboration Revises 2008 Conclusions on Cranberry for UTI Prevention: Experts, Researchers Clarify Results from Recent Meta-Analysis." *HerbalGram* 97 (2013): 28–31.

——. "Ginkgo and Alzheimer's Disease Prevention: Researchers, Herbal Experts Interpret Results of the Five-Year GuidAge Clinical Trial." *HerbalGram* 97 (2013): 32–37.

Smith, Woodruff. "From Coffee-House to Parlour: The Consumption of Coffee, Tea, and Sugar in Northwestern England in the Seventeenth and Eighteenth Centuries." In *Consuming Habits: Drugs in History and Anthropology*, edited by Jordan Goodman, Paul E. Lovejoy, and Andrew Sherratt, 142–157. London: Routledge, 1995.

Smolinske, Susan C. "Herbal Product Contamination and Toxicity." *Journal of Pharmacy Practice* 18 (2005): 188–208.

Snitz, Beth E., Ellen S. O'Meara, Michelle C. Carlson, et al. "*Ginkgo biloba* for Preventing Cognitive Decline in Older Adults: A Randomized Trial." *Journal of the American Medical Association* 302 (2009): 2663–2670.

Snyder, Rodney, Bradley Foliart Olsen, and Laura Pallas Brindle. "From Stone Metates to Steel Mills: The Evolution of Chocolate Manufacturing." In *Chocolate: History, Culture, and Heritage*, edited by Louis Evan Grivetti and Howard-Yana Shapiro, 611–623. Hoboken, N.J.: Wiley, 2009.

Soejarto, D. D., H. H. S. Fong, G. T. Tan, et al. "Ethnobotany/Ethnopharmacology and Mass Bioprospecting: Issues on Intellectual Property and Benefit Sharing." *Journal of Ethnopharmacology* 100 (2005): 15–22.

Southall, Ashley, and Jack Healey. "U.S. Won't Sue to Reverse States' Legalization of Marijuana." *New York Times*, August 29, 2013.

Spence, Jonathan D. *The Search for Modern China*. 3rd ed. New York: Norton, 2013.

Spillane, Joseph F. *Cocaine: From Medical Marvel to Modern Menace in the United States, 1884–1920*. Baltimore: Johns Hopkins University Press, 2000.

Sproll, C., R. C. Perz, R. Buschmann, et al. "Guidelines for Reduction of Morphine in Poppy Seed Intended for Food Purposes." *European Food Research Technology* 226 (2007): 307–310.

Stabler, S. N., A. M. Tejani, F. Huynh, et al. "Garlic for the Prevention of Cardiovascular Morbidity and Mortality in Hypertensive Patients." *Cochrane Database of Systematic Reviews* (2012): CD007653.

Standaert, David G., and Ryan R. Walsh. "Pharmacology of Dopaminergic Neurotransmission." In *Principles of Pharmacology: The Pathophysiologic Basis of Drug Therapy*, 3rd ed., edited by David E. Golan, Armen H. Tashjian, Ehrin J. Armstrong, et al., 186–206. Philadelphia: Lippincott Williams & Wilkins, 2012.

Stannard, Jerry. "The Herbal as a Medical Document." *Bulletin of the History of Medicine* 43 (1969): 212–220.

——. "Hippocratic Pharmacology." *Bulletin of the History of Medicine* 35 (1961): 497–518.

——. "The Theoretical Bases of Medieval Herbalism." *Medical Heritage* 1 (1985): 186–198.

Steppuhn, Anke, Klaus Gase, Bernd Krock, et al. "Nicotine's Defensive Function in Nature." *PLoS Biology* 2 (2004): e217.

Stewart, Grace. "A History of the Medicinal Use of Tobacco: 1492–1860." *Medical History* 11 (1967): 228–268.

Stewart, K. M. "The African Cherry (*Prunus africana*): Can Lessons Be Learned from an Overexploited Medicinal Tree?" *Journal of Ethnopharmacology* 89 (2003): 3–13.

——. "The African Cherry (*Prunus africana*): From Hoe-Handles to the International Herb Market." *Economic Botany* 57 (2003): 559–569.

Stoicov, Calin, and JeanMarie Houghton. "Green Tea and Protection Against *Helicobacter* Infection." In *Tea in Health and Disease Prevention*, edited by Victor R. Preedy, 593–602. London: Elsevier, 2013.

Stuart, David. "The Language of Chocolate: References to Cacao on Classic Maya Drinking Vessels." In *Chocolate in Mesoamerica: A Cultural History of Cacao*, edited by Cameron L. McNeil, 184–201. Gainesville: University Press of Florida, 2006.

Sumner, Judith. *The Natural History of Medicinal Plants*. Portland, Ore.: Timber Press, 2000.

Surh, Young-Joon, and Kyung-Soo Chun. "Cancer Chemopreventive Effects of Curcumin." In *The Molecular Targets and Therapeutic Uses of Curcumin in Health and Disease*, edited by Bharat B. Aggarwal, Young-Joon Surh, and Shishir Shishodia, 149–172. New York: Springer, 2007.

Swann, John. "The Evolution of the American Pharmaceutical Industry." *Pharmacy in History* 37 (1995): 76–86.

Swift, Aisling. "Palmetto Berries: Industry Has Health Benefits." *Naples (Fla.) Daily News*, September 28, 2013.

Symons, Van Jay. *Ch'ing Ginseng Management: Ch'ing Monopolies in Microcosm.* Tempe: Center for Asian Research, Arizona State University, 1981.

Tacklind, J., R. MacDonald, I. Rutks, et al. "*Serenoa repens* for Benign Prostatic Hyperplasia." *Cochrane Database of Systematic Reviews* (2012): CD001423.

Taku, K., M. K. Melby, N. Nishi, et al. "Soy Isoflavones for Osteoporosis: An Evidence-Based Approach." *Maturitas* 70 (2011): 333–338.

Tam, S. William, Manuel Worcel, and Michael Wyllie. "Yohimbine: A Clinical Review." *Pharmacology and Therapeutics* 91 (2001): 215–243.

Tamayo, Carmen, and Suzanne Diamond. "Review of Clinical Trials Evaluating Safety and Efficacy of Milk Thistle (*Silybum marianum* [L.] Gaertn.)" *Integrative Cancer Therapies* 6 (2007): 146–157.

Tanner, George W., J. Jeffrey Mullahey, and David Maehr. "Saw Palmetto: An Ecologically and Economically Important Native Palm." University of Florida Extension Publication WEC-109. June 1999. http://www.plantapalm.com/vpe/misc/saw-palmetto.pdf.

Taylor, David A. "Botanical Supplements: Weeding out the Health Risks." *Environmental Health Perspectives* 112 (2004): A750–A753.

——. *Ginseng, the Divine Root.* Chapel Hill, N.C.: Algonquin, 2006.

Temkin, Owsei. *Galenism: Rise and Decline of a Medical Philosophy.* Ithaca, N.Y.: Cornell University Press, 1973.

Ten Berge, Jos. "The *Belle Epoque* of Opium." In *Smoke: A Global History of Smoking*, edited by Sander L. Gilman and Zhou Xun, 108–117. London: Reaktion, 2004.

Terry, R., P. Posadzki, L. K. Watson, et al. "The Use of Ginger (*Zingiber officinale*) for the Treatment of Pain: A Systematic Review of Clinical Trials." *Pain Medicine* 12 (2011): 1808–1818.

Teschke, R., J. Sarris, X. Glass, et al. "Kava, the Anxiolytic Herb: Back to Basics to Prevent Liver Injury?" *British Journal of Clinical Pharmacology* 71 (2011): 445–448.

Tétényi, Péter. "Opium Poppy (*Papaver somniferum*): Botany and Horticulture." *Horticultural Reviews* 19 (1997): 373–408.

Than, Tammy P., and Jimmy Bartlett. "Local Anesthetics." In *Clinical Ocular Pharmacology*, 5th ed., edited by Jimmy Bartlett and Siret Jaanus, 85–96. St. Louis Butterworth-Heinemann, 2008.

Thangapazham, Rajesh L., Anuj Sharma, and Radha K. Maheshwari. "Beneficial Role of Curcumin in Skin Diseases." In *The Molecular Targets and Therapeutic Uses of Curcumin in Health and Disease*, edited by Bharat B. Aggarwal, Young-Joon Surh, and Shishir Shishodia, 343–358. New York: Springer, 2007.

Thomas, Valerie. "Do Modern-Day Medical Herbalists Have Anything to Learn from Anglo-Saxon Medical Writings?" *Journal of Herbal Medicine* 1 (2011): 42–52.

Thompson, Reginald Campbell. *A Dictionary of Assyrian Botany.* London: British Academy, 1949.

Thomson, Samuel. *New Guide to Health or Botanic Family Physician.* 2nd ed. Boston: House, 1825.

Tilley, Stephen L. "Methylxanthines in Asthma." In *Methylxanthines: Handbook of Experimental Pharmacology*, edited by Bertil B. Fredholm, 439–456. Berlin: Springer, 2011.

Timbrook, Jan. "Ethnobotany of Chumash Indians, California, Based on Collections by John P. Harrington." *Economic Botany* 44 (1990): 236–253.

Timmer, A., J. Günther, E. Judith, et al. "*Pelargonium sidoides* Extract for Treating Acute Respiratory Tract Infections." *Cochrane Database of Systematic Reviews* (2013): CD006323.

Tobyn, Graeme, Alison Denham, and Margaret Whiteleg. *The Western Herbal Tradition: Two Thousand Years of Medicinal Plant Knowledge.* Edinburgh: Churchill Livingstone/Elsevier, 2009.

Tomasi, Lucia Tongiorgio. "The Renaissance Herbal." In *The Renaissance Herbal*, 9–19. Bronx: New York Botanical Garden, 2013.

Tominaga, Makoto, and Michael J. Caterina. "Thermosensation and Pain." *Journal of Neurobiology* 61 (2004): 3–12.

Tortora, Gerard J., and Bryan H. Derrickson. *Principles of Anatomy and Physiology.* 12th ed. Hoboken, N.J.: Wiley, 2009.

Toyang, Ngeh J., and Rob Verpoorte. "A Review of the Medicinal Potentials of Plants of the Genus *Vernonia* (Asteraceae)." *Journal of Ethnopharmacology* 146 (2013): 681–723.

Tradtrantip, Lukmanee, Wan Namkung, and A. S. Verkman. "Crofelemer, an Antisecretory Antidiarrheal Proanthocyanidin Oligomer Extracted from *Croton lechleri*, Targets Two Distinct Intestinal Chloride Channels." *Molecular Pharmacology* 77 (2010): 69–78.

Trescot, Andrea. "Clinical Use of Opioids." In *Comprehensive Treatment of Chronic Pain by Medical, Interventional, and Integrative Approaches*, by Timothy R. Deer, Michael S. Leong, Asokumar Buvanendran, et al., 99–110. New York: Springer, 2013.

Trescot, Andrea M., Sukdeb Datta, Marion Lee, et al. "Opioid Pharmacology." *Pain Physician* 11 (2008): S133–S153.

Tsien, Tsuen-Hsuin. "Raw Materials for Papermaking in China." *Journal of the American Oriental Society* 93, no. 4 (1973): 510–519.

Tunaru, Sorin, Till F. Althoff, Rolf M. Nüsing, et al. "Castor Oil Induces Laxation and Uterus Contraction via Ricinoleic Acid Activating Prostaglandin EP_3 Receptors." *Proceedings of the National Academy of Sciences USA* 109 (2012): 9179–9184.

Turner, Ronald B., Rudolf Bauer, Karin Woelkart, et al. "An Evaluation of *Echinacea angustifolia* in Experimental Rhinovirus Infections." *New England Journal of Medicine* 353 (2005): 341–348.

Tzellos, T. G., C. Sardeli, A. Lallas, et al. "Efficacy, Safety, and Tolerability of Green Tea Catechins in the Treatment of External Anogenital Warts: A Systematic Review and Metaanalysis." *Journal of the European Academy of Dermatology and Venereology* 25 (2011): 345–353.

Ukers, William H. *All About Coffee*. New York: Tea and Coffee Trade Journal Company, 1922.

——. *All About Tea*. New York: Tea and Coffee Trade Journal Company, 1935.

Ulbricht, C., F. Basch, H. Boon, et al. "Safety Review of Kava (*Piper methysticum*) by the Natural Standard Research Collaboration." *Expert Opinion on Drug Safety* 4 (2005): 779–794.

United States Department of Agriculture. "USDA National Nutrient Database for Standard Reference, Release 26." 2013.

United States Department of Health and Human Services. *How Tobacco Smoke Causes Disease: The Biology and Behavioral Basis for Smoking-Attributable Disease. A Report of the Surgeon General*. Atlanta: Department of Health and Human Services, Centers for Disease Control and Prevention, National Center for Chronic Disease Prevention and Health Promotion, Office on Smoking and Health, 2010.

Unschuld, Paul U. *Huang Di Nei Jing Su Wen: Nature, Knowledge, Imagery in an Ancient Chinese Medical Text*. Berkeley: University of California Press, 2003.

——. *Medicine in China: A History of Pharmaceutics*. Berkeley: University of California Press, 1986.

Vakil, Rustom Jal. "*Rauwolfia serpentina* in the Treatment of High Blood Pressure: A Review of the Literature." *Circulation* 12 (1955): 220–229.

Van Andel, T., J. Behari-Ramdas, R. Havinga, et al. "The Medicinal Plant Trade in Suriname." *Ethnobotany Research and Applications* 5 (2007): 351–372.

Van Arsdall, Anne. "The Medicines of Medieval and Renaissance Europe as a Source of Medicines for Today." In *Prospecting for Drugs in Ancient and Medieval European Texts: A Scientific Approach*, edited by Bart K. Holland, 19–37. Amsterdam: Harwood, 1996.

——. *Medieval Herbal Remedies: The* Old English Herbarium *and Anglo-Saxon Medicine*. New York: Routledge, 2002.

van Die, M. Diana, H. G. Burger, H. Teede, et al. "*Vitex agnus-castus* Extracts for Female Reproductive Disorders: A Systematic Review of Clinical Trials." *Planta Medica* 79 (2013): 562–575.

Van Wyk, Ben-Erik, and Michael Wink. *Medicinal Plants of the World*. Portland, Ore.: Timber Press, 2004.

Vega, Fernando. "The Rise of Coffee." *American Scientist* 96 (2008): 138–145.

Vetter, J. "Poison hemlock (*Conium maculatum* L.)." *Food and Chemical Toxicology* 42 (2004): 1373–1382.

Vinetz, Joseph M., Jérôme Clain, Viengngeun Bounkeua, et al. "Chemotherapy of Malaria." In *Goodman and Gilman's The Pharmacological Basis of Therapeutics*, 12th ed., edited by Laurence L. Brunton, Bruce A. Chabner, and Björn C. Knollmann, 1383–1418. New York: McGraw-Hill, 2011.

von Gernet, Alexander. "Nicotian Dreams: The Prehistory and Early History of Tobacco in Eastern North America." In *Consuming Habits: Drugs in History and Anthropology*, edited by Jordan Goodman, Paul E. Lovejoy, and Andrew Sherratt, 67–87. London: Routledge, 1995.

——. "North American Indigenous *Nicotiana* Use and Tobacco Shamanism: The Early Documentary Record, 1520–1660." In *Tobacco Use by Native North Americans: Sacred Smoke and Silent Killer*, edited by Joseph Winter, 59–80. Norman: University of Oklahoma Press, 2000.

von Tschudi, J. J. *Travels in Peru During the Years 1838–1842*. Translated by Thomasina Ross. London: Bogue, 1847.

Voss, Richard W. "Native American Healing." In *Fundamentals of Complementary and Alternative Medicine*, 4th ed., edited by Marc S. Micozzi, 531–550. St. Louis: Saunders Elsevier, 2011.

Vyas, A., P. Dandawate, S. Padhye, et al. "Perspectives on New Synthetic Curcumin Analogs and Their Potential Anticancer Properties." *Current Pharmaceutical Design* 19 (2013): 2047–2069.

Wachira, F. N., S. Kamunya, S. Karori, et al. "The Tea Plants: Botanical Aspects." In *Tea in Health and Disease Prevention*, edited by Victor R. Preedy, 3–18. London: Elsevier, 2013.

Wachira, F. N., S. Karori, L. C. Kerio, et al. "Cultivar Type and Antioxidant Potency of Tea Product." In *Tea in Health and Disease Prevention*, edited by Victor R. Preedy, 91–102. London: Elsevier, 2013.

Walker, Karen M., and Wendy L. Applequist. "Adulteration of Selected Unprocessed Botanicals in the U.S. Retail Herbal Trade." *Economic Botany* 66 (2012): 321–327.

Walton, Robert P. *Marihuana: America's New Drug Problem*. Philadelphia: Lippincott, 1938.

Wan, Xiaochun, Zhengzhu Zhang, and Daxiang Li. "Chemistry and Biological Properties of Theanine." In *Tea and Tea Products: Chemistry and Health-Promoting Properties*, edited by Chi-Tang Ho, Jen-Kun Lin, and Fereidoon Shahidi, 255–274. Boca Raton, Fla.: CRC Press, 2009.

Wang, X., H. Zhang, L. Chen, et al. "Liquorice, a Unique 'Guide Drug' of Traditional Chinese Medicine: A Review of Its Role in Drug Interactions." *Journal of Ethnopharmacology* 150 (2013): 781–790.

Ward, Jeanine, Kishan Kapadia, Eric Brush, et al. "Amatoxin Poisoning: Case Reports and Review of Current Therapies." *Journal of Emergency Medicine* 44 (2013): 116–121.

Watkins, Frances, Barbara Pendry, Olivia Corcoran, et al. "Anglo-Saxon Pharmacopoeia Revisited: A Potential Treasure in Drug Discovery." *Drug Discovery Today* 16 (2011): 1069–1075.

Webster, Charles. *Paracelsus: Medicine, Magic, and Mission at the End of Time*. New Haven, Conn.: Yale University Press, 2008.

Webster, D. E., J. Lu, S.-N. Chen, et al. "Activation of the -opiate Receptor by *Vitex agnus-castus* Methanol Extracts: Implication for Its Use in PMS." *Journal of Ethnopharmacology* 106 (2006): 216–221.

Webster, Herbert T. *Dynamical Therapeutics*. 2nd ed. San Francisco: Webster Medical, 1898.

Weekley, Ernest. *An Etymological Dictionary of Modern English*. London: Murray, 1921.

Wei, P., M. Liu, Y. Chen, et al. "Systematic Review of Soy Isoflavone Supplements on Osteoporosis in Women." *Asian Pacific Journal of Tropical Medicine* (2012): 243–248.

Weid, M., J. Ziegler, and T. M. Kutchan. "The Roles of Latex and the Vascular Bundle in Morphine Biosynthesis in the Opium Poppy, *Papaver somniferum*." *Proceedings of the National Academy of Sciences USA* 101 (2004): 13957–13962.

Weinberg, Bennett Alan, and Bonnie K. Bealer. *The World of Caffeine: The Science and Culture of the World's Most Popular Drug*. New York: Routledge, 2001.

Weiss, J., N. Ainsworth, and I. Faithfull. *Horehound*, Marrubium vulgare. Publication of the Cooperative Research Centre for Weed Management Systems, University of Adelaide, Australia. September 2000.

Weisser, H., S. Tunn, B. Behnke, et al. "Effects of the *Sabal serrulata* Extract IDS 89 and Its Subfractions on 5-Reductase Activity in Human Benign Prostatic Hyperplasia." *Prostate* 28 (1996): 300–306.

Wells, Pete. "A Liquor of Legend Makes a Comeback." *New York Times*, December 5, 2007.

Westfall, Thomas C., and David P. Westfall. "Adrenergic Agonists and Antagonists." In *Goodman and Gilman's The Pharmacological Basis of Therapeutics*, 12th ed., edited by Laurence L. Brunton, Bruce A. Chabner, and Björn C. Knollmann, 277–334. New York: McGraw-Hill, 2011.

——. "Neurotransmission: The Autonomic and Somatic Motor Nervous Systems." In *Goodman and Gilman's The Pharmacological Basis of Therapeutics*, 12th ed., edited by Laurence L. Brunton, Bruce A. Chabner, and Björn C. Knollmann, 171–218. New York: McGraw-Hill, 2011.

White, Brett. "Ginger: An Overview." *American Family Physician* 75, no. 11 (2007): 1689–1691.

White, Nicholas J. "Artemisinin Resistance: The Clock Is Ticking." *Lancet* 376 (2010): 2051–2052.

——. "*Qinghaosu* (Artemisinin): The Price of Success." *Science* 320 (2008): 330–334.

Wilbert, Johannes. *Tobacco and Shamanism in South America*. New Haven, Conn.: Yale University Press, 1987.

Wilcox, James R. "World Distribution and Trade of Soybean." In *Soybeans: Improvement, Production, and Uses*, 3rd ed., edited by H. Roger Boerma and James E. Specht, 3–14. Madison, Wis.: American Society of Agronomy, Crop Science Society of America, Soil Science Society, 2004.

Wilken, D., and L. Hannah. Marrubium vulgare (Lamiaceae) *Common Horehound, White Horehound*. Publication of the Santa Barbara Botanic Garden, for Channel Islands National Park. December 15, 1998.

Williams, Edward Huntington. "Negro Cocaine 'Fiends' Are a New Southern Menace." *New York Times*, February 8, 1914.

Williams, Roger. *A Key into the Language of America*. London: Gregory Dexter, 1643.

Williamson, E. M. "Synergy and Other Interactions in Phytomedicines." *Phytomedicine* 8 (2001): 401–409.

Wilson, Philip K., and W. Jeffrey Hurst. *Chocolate as Medicine: A Quest over the Centuries*. London: Royal Society of Chemistry, 2012.

Wilt, Timothy, and Areef Ishani. "*Pygeum africanum* for Benign Prostatic Hyperplasia." *Cochrane Database of Systematic Reviews* (2002): CD001423.

Wink, Michael, and Ben-Erik van Wyk. *Mind-Altering and Poisonous Plants of the World*. Portland, Ore.: Timber Press, 2008.

Winter, Joseph. "Introduction to the North American Tobacco Species." In *Tobacco Use by Native North Americans: Sacred Smoke and Silent Killer*, edited by Joseph Winter, 3–8. Norman: University of Oklahoma Press, 2000.

——. "Traditional Uses of Tobacco by Native Americans." In *Tobacco Use by Native North Americans: Sacred Smoke and Silent Killer*, edited by Joseph Winter, 9–58. Norman: University of Oklahoma Press, 2000.

Withering, William. *An Account of the Foxglove and Some of Its Medical Uses*. Birmingham, 1785.

Wong, R. Bin. "Opium and Modern Chinese State-Making." In *Opium Regimes: China, Britain, and Japan, 1839–1952*, edited by Timothy Brook and Bob Tadashi Wakabayashi, 189–211. Berkeley: University of California Press, 2000.

Wood, George W., and Franklin Bache. *The Dispensatory of the United States of America*. 8th ed. Philadelphia: Grigg, Elliot, 1849.

Wood, H. C., Joseph P. Remington, and Samuel P. Sadtler. *The Dispensatory of the United States of America*. 16th ed. Philadelphia: Lippincott, 1892.

——. *The Dispensatory of the United States of America*. 19th ed. Philadelphia: Lippincott, 1907.

World Health Organization. *WHO Monographs on Selected Medicinal Plants*. 2 vols. Geneva: World Health Organization, 1999, 2004.

World Health Organization, International Agency for Research on Cancer. *Some Traditional Herbal Medicines, Some Mycotoxins, Naphthalene, and Styrene*. IARC Monographs on the Evaluation of Carcinogenic Risks to Humans 82. Lyon: IARC, 2002.

Worling, Peter M. "Pharmacy in the Early Modern World, 1617 to 1841 A.D." In *Making Medicines: A Brief History of Pharmacy and Pharmaceuticals*, edited by Stuart Anderson, 57–76. London: Pharmaceutical Press, 2005.

Wright, C. I., L. Van-Buren, C. I. Kroner, et al. "Herbal Medicines as Diuretics: A Review of the Scientific Evidence." *Journal of Ethnopharmacology* 114 (2007): 1–31.

Wright, C. W., P. A. Linley, R. Brun, et al. "Ancient Chinese Methods Are Remarkably Effective for the Preparation of Artemisinin-Rich Extracts of *Qing Hao* with Potent Antimalarial Activity." *Molecules* 15 (2010): 804–812.

Wright, G. A., D. D. Baker, M. J. Palmer, et al. "Caffeine in Floral Nectar Enhances a Pollinator's Memory of Reward." *Science* 339 (2013): 1202–1204.

Wright, Hamilton. *Report on the International Opium Commission and on the Opium Problem as Seen Within the United States and Its Possessions.* Senate Document No. 377, 61st Cong., 2nd sess. Washington, D.C.: Government Printing Office, 1910.

Wuttke, W., H. Jarry, V. Christoffel, et al. "Chaste Tree (*Vitex agnus-castus*): Pharmacology and Clinical Indications." *Phytomedicine* 10 (2003): 348–357.

Wuttke, W., H. Jarry, J. Haunschild, et al. "The Nonestrogenic Alternative for the Treatment of Climacteric Complaints: Black Cohosh (*Cimicifuga* or *Actaea racemosa*)." *Journal of Steroid Biochemistry and Molecular Biology* 139 (2014): 302–310.

Xue, Jian, Lili Hao, and Fei Peng. "Residues of Eighteen Organochlorine Pesticides in Thirty Traditional Chinese Medicines." *Chemosphere* 71 (2008): 1051–1055.

Yaksh, Tony L., and Mark S. Wallace. "Opioids, Analgesia, and Pain Management." In *Goodman and Gilman's The Pharmacological Basis of Therapeutics*, 12th ed., edited by Laurence L. Brunton, Bruce A. Chabner, and Björn C. Knollmann, 481–526. New York: McGraw-Hill, 2011.

Yale, S. H., and K. Liu. "*Echinacea purpurea* Therapy for the Treatment of the Common Cold: A Randomized, Double-Blind, Placebo-Controlled Clinical Trial." *Archives of Internal Medicine* 164 (2004): 1237–1241.

Yap, S. P., P. Shen, J. Li, et al. "Molecular and Pharmacodynamic Properties of Estrogenic Extracts from the Traditional Chinese Medicinal Herb *Epimedium*." *Journal of Ethnopharmacology* 113 (2007): 218–224.

Yarnell, Eric. "Botanical Medicines for the Urinary Tract." *World Journal of Urology* 20 (2002): 285–293.

Young, Allen M. *The Chocolate Tree: A Natural History of Chocolate.* Gainesville: University Press of Florida, 2007.

Young, George. *A Treatise on Opium Founded upon Practical Observations.* London: Millar, 1753.

Young, James Harvey. *The Toadstool Millionaires: A Social History of Patent Medicines in America Before Federal Regulation.* Princeton, N.J.: Princeton University Press, 1961.

Zent, Stanford, and Egleé L. Zent. "On Biocultural Diversity from a Venezuelan Perspective: Tracing the Interrelationships Among Biodiversity, Culture Change, and Legal Reforms." In *Biodiversity and the Law: Intellectual Property, Biotechnology, and Traditional Knowledge*, edited by Charles R. McManis, 91–114. London: Earthscan, 2007.

Zhang, Aihua, H. Sun, Y. Yuan, et al. "An *in vivo* Analysis of the Therapeutic and Synergistic Properties of Chinese Medicinal Formula Yin-Chen-Hao-Tang Based on Its Active Constituents." *Fitoterapia* 82 (2011): 1160–1168.

Zhang, Jinghong. *Puer Tea: Ancient Caravans and Urban Chic.* Seattle: University of Washington Press, 2014.

Zhang, Junhua, Barbara Wider, Hongcai Shang, et al. "Quality of Herbal Medicines: Challenges and Solutions." *Complementary Therapies in Medicine* 20 (2012): 100–106.

Zhang, Liang, Zheng-zhu Zhang, Ya-ning Lu, et al. "L-Theanine from Green Tea: Transport and Effects on Health." In *Tea in Health and Disease Prevention*, edited by Victor R. Preedy, 425–436. London: Elsevier, 2013.

Zheng, Yangwen. *The Social Life of Opium in China.* New York: Cambridge University Press, 2005.

Zhou, H., C. S. Beevers, and S. Huang. "Targets of Curcumin." *Current Drug Targets* 12 (2011): 332–347.

Zhou, S.-F. "Structure, Function, and Regulation of P-glycoprotein and Its Clinical Relevance in Drug Disposition." *Xenobiotica* 38 (2008): 802–832.

Zhu, Feng, Chu Qin, Lin Tao, et al. "Clustered Patterns of Species Origins of Nature-Derived Drugs and Clues for Future Bioprospecting." *Proceedings of the National Academy of Sciences USA* 108 (2011): 12943–12948.

Zohary, Daniel, Maria Hopf, and Ehud Weiss. *Domestication of Plants in the Old World: The Origin and Spread of Domesticated Plants in Southwest Asia, Europe, and the Mediterranean Basin.* 4th ed. New York: Oxford University Press, 2012.

Zysk, Kenneth G. "Traditional Medicine of India: Ayurveda and Siddha." In *Fundamentals of Complementary and Alternative Medicine*, 4th ed., edited by Marc S. Micozzi, 455–467. St. Louis: Saunders Elsevier, 2011.

Index